Lecture Notes in Economics and Mathematical Systems 688

More information about this series at http://www.springer.com/series/300

Jiuping Xu • Zongmin Li • Zhimiao Tao

Random-Like Bi-level Decision Making

Springer

Jiuping Xu
Business School
Sichuan University
Chengdu, Sichuan, China

Zongmin Li
Business School
Sichuan University
Chengdu, Sichuan, China

Zhimiao Tao
Business School
Sichuan University
Chengdu, Sichuan, China

ISSN 0075-8442 ISSN 2196-9957 (electronic)
Lecture Notes in Economics and Mathematical Systems
ISBN 978-981-10-1767-4 ISBN 978-981-10-1768-1 (eBook)
DOI 10.1007/978-981-10-1768-1

Library of Congress Control Number: 2016949403

Printed on acid-free paper

This Springer imprint is published by Springer Nature
The registered company is Springer Science+Business Media Singapore Pte Ltd.

Preface

Thanks to Stackelberg's pioneering contributions, conflicts in decision-making problems have received significant attention. In the real world, many practical decision-making problems involve hierarchical decision structures, the complexity of which has motivated the development of mathematical programming models to exploit the structural features. To represent the conflict in hierarchical organizations, multilevel mathematical programming has often been used with the concept of Stackelberg equilibrium. The most frequently discussed multilevel model is the bi-level model, and a n-level model can be built from $n-1$ bi-level models. In a bi-level model, the two decision makers are called the leader and follower, respectively.

When formulating the mathematical models, it is assumed that both decision makers make rational judgments. A bi-level mathematical programming problem can be interpreted as a static Stackelberg game, which assumes there is no communication between the decision makers, or they do not have a binding agreement even if there is such communication.

In addition to the hierarchical structure, in real-world problems, the evaluation diversity has become more important, as it is natural to assume that decision makers seek to attain several simultaneous goals, i.e., they have multiple objectives and alternatives, so that they must consider trade-offs between these objectives. Assisting decision makers in dealing with multiple objective decision-making problems has been the subject of intensive studies since the 1970s, but many theoretical concepts had been defined much earlier. The concept of the Pareto optimal solution has played an important role in the development of multiple objective decision making. There are several basic solution approaches, such as the weighted-sum approach, the utility function approach, the compromise approach, and the lexicographic ordering approach. Based on the basic solution approaches, many researchers have developed some comprehensive approaches to multiple objective decision-making problems.

Further, people often need to make decisions under uncertainties. This monograph focuses on so-called random-like uncertainty including random phenomenon, random-overlapped random (Ra-Ra) phenomenon, and fuzzy-overlapped random

(Ra-Fu) phenomenon. It is generally believed that the study of probability theory commenced in 1654 by Pascal and Fermat, when they succeeded in deriving the exact probabilities for certain gambling problems. From then on, people paid close attention to random phenomenon, and probability theory began to be widely applied to many social and technological problems, such as vital statistics, premium theory, astro observation, the theory of errors, and quality control. From the 17th to the 19th century, many distinguished scholars, such as Bernoulli, De-Moivre, Laplace, Gauss, Poisson, Tchebychev, and Markov, contributed to the development of probability theory, which, on reflection, was fascinating. However, as probability theory was applied to increasingly more real-life problems across many fields, the basic definition offered by Laplace proved to be limiting, as it was unable to effectively deal with usual random events. Great progress was achieved in 1931 when von Mises initialized the concept of the sample space, which filled the gap between probability theory and measure theory. However, a strict theoretical principle did not still exist until 1933, when it was proposed by the outstanding mathematician Kolmogorov from the former Soviet Union. At this time, he published his famous book *The Basic Concept of Probability Theory*, in which he proposed the axiomatization structure, which was considered as a milestone and was the foundation for the consequent development of probability theory. In 1965, Zadeh initialized fuzzy set theory. Generally, fuzzy events are regarded as subjective uncertainty, for example, "cold" vs. "hot," "short" vs. "high," and "bad" vs. "good," which are not marked by a crisp number. Many scholars believed that traditional single-fold uncertain variables (random variable and fuzzy variable) faced difficulties when seeking to clearly describe complex and changing realistic problems. In 1978, Kwakernaak combined randomness with fuzziness, formulated fuzzy random variable concept, and introduced the basic definitions and properties. This viewpoint, which combined two different uncertain variables to describe complex events, was commended by many scholars and encouraged the move toward the study of uncertain events. Since then, the concepts of the Ra-Ra variable and the Ra-Fu variable have been proposed by scholars.

When considering the hierarchical structure, the multiple objectives, and the random-like uncertainties, we can employ a general model as follows:

$$\begin{cases} \min\limits_{x\in R^{n_1}} [F_1(\boldsymbol{x},\boldsymbol{y},\boldsymbol{\xi}), F_2(\boldsymbol{x},\boldsymbol{y},\boldsymbol{\xi}), \cdots, F_m(\boldsymbol{x},\boldsymbol{y},\boldsymbol{\xi})] \\ \text{s.t.} \begin{cases} G_i(\boldsymbol{x},\boldsymbol{y},\boldsymbol{\xi}) \le 0, i = 1,2,\cdots,q \\ \text{where } \boldsymbol{y} \text{ solves:} \\ \begin{cases} \min\limits_{y\in R^{n_2}} [f_1(\boldsymbol{x},\boldsymbol{y},\boldsymbol{\xi}), f_2(\boldsymbol{x},\boldsymbol{y},\boldsymbol{\xi}), \cdots, f_m(\boldsymbol{x},\boldsymbol{y},\boldsymbol{\xi})] \\ \text{s.t. } g_j(\boldsymbol{x},\boldsymbol{y},\boldsymbol{\xi}) \le 0, j = 1,2,\cdots,p, \end{cases} \end{cases} \end{cases} \tag{1}$$

where $\boldsymbol{x} \in R^{n_1}$ and $\boldsymbol{y} \in R^{n_2}$ are decision vectors for the lower-level decision maker and the upper-level decision maker, respectively; $F(\boldsymbol{x},\boldsymbol{y},\boldsymbol{\xi})$ and $f(\boldsymbol{x},\boldsymbol{y},\boldsymbol{\xi})$ are objective functions of the upper-level and the lower-level model; $G_i(\boldsymbol{x},\boldsymbol{y},\boldsymbol{\xi})$ and

$g_j(\boldsymbol{x}, \boldsymbol{y}, \boldsymbol{\xi})$ are the constraints of the upper-level and the lower-level model; $\boldsymbol{\xi}$ is a random-like (random, birandom, or random fuzzy) vector.

This monograph takes real-life problems as the background and develops a general analytical framework for random-like bi-level decision-making problems, including basic theories, models, algorithms, and practical applications.

This monograph contains five chapters. Chapter 1 provides some basic concepts for the random sets theory and bi-level programming that is the prerequisite for understanding the concepts and examples in this book. Chapter 2 deals with bi-level multiple objective decision-making problems with random phenomena and the application of the model to a regional water resources allocation problem. To obtain the deterministic models mathematically, three model categories of models are proposed, i.e., random EEEE, random CECC, and random DEDC, for which the first two letters represent the method for dealing with the upper-level model and the latter two letters stand for the method for dealing with the lower-level model. For example, CECC denotes an expected value operator that is used to deal with the randomness in the constraints of upper-level programming and a chance constraint operator that is used to deal with the randomness in the upper-level objectives and the lower-level programming. For each model, we develop an equivalent model in special cases and propose a technique for random simulation-based genetic algorithms for general cases. A real application to an irrigation district on the Gan-Fu plain is given in the last section of the chapter. Chapter 3 focuses on a bi-level multiple objective decision-making problem with Ra-Ra phenomenon and its application to a transport flow distribution problem. Ra-Ra EEDE model, Ra-Ra ECEC model, and Ra-Ra DCCC model are proposed to handle the original Ra-Ra model. In some special cases, we obtain the equivalent models and solve them using traditional bi-level programming methods. For general cases, Ra-Ra simulation-based particle swarm optimization algorithms are proposed. The SBY Hydropower Project is taken as an example in the last part of the chapter to illustrate the effectiveness of the proposed models and algorithms. Chapter 4 introduces a bi-level multiple objective decision-making problem with Ra-Fu phenomenon and its application to a construction site security planning problem. In this chapter, three categories of models, Ra-Fu EECC, Ra-Fu CCDD, and Ra-Fu DDEE, are proposed to deal with the original Ra-Fu uncertainty. For each category, we develop an equivalent model in some special cases and design a Ra-Fu simulation-based plant growth simulation algorithm to solve the model. In the last section, the proposed models and algorithms are applied to a security planning problem at the LT hydropower construction project. In the conclusion, Chap. 5 summarizes the methodologies adopted in this monograph from an equilibria viewpoint, including equilibrium motivation, equilibrium in real-world problems, model system in equilibrium, equilibrium algorithms, and perspectives for the models, theories, algorithms, and applications.

Additional materials including the MATLAB® codes for some numerical examples are listed in the Appendix.

This monograph is supported by the National Natural Science Foundation of China (grant no. 71401114), the "985" Program of Sichuan University "Innovative Research Base for Economic Development and Management," the Fundamental

Research Funds for the Central Universities (grant no. skqy201524), and the Research Foundation of the Ministry of Education for the Doctoral Program of Higher Education of China (grant no. 20130181110063). A number of researchers from the Uncertainty Decision-Making Laboratory of Sichuan University contributed valuable insights and information, particularly Yan Tu, Yanfang Ma, Xiaoling Song, Cuiying Feng, Ziqiang Zeng, and Xin Yang. Finally, the authors express their deep gratitude to the editorial staffs of Springer for the wonderful cooperation and helpful comments.

Chengdu, Sichuan
April, 2016

Jiuping Xu
Zongmin Li
Zhimiao Tao

Contents

Chapter 1
Foundations of Random-Like Bi-Level Decision Making

Abstract The investigation of bi-level (decision-making) programming problems has been strongly motivated by real world applications (Dempe, Foundations of bilevel programming. Kluwer Academic, Dordrecht, 2002). The complex nature of decision-making requires practitioners to make decisions based on a wide variety of cost considerations, benefit analyses and purely technical considerations. As we have entered the era of Big Data, which is the next frontier for innovation, competition and productivity, a new scientific revolution is about to begin. Fortunately, we will be in a position to witness the coming technological leapfrog. Many fields and sectors, ranging from economic and business activities to public administration, from national security to scientific research, are involved in Big Data problems. Because of the excessive data, inherent laws cannot be identified or the inherent data laws are changing so dynamically with "uncertainty" an outward manifestation. Therefore, practical situations are usually too complicated to be described using determinate variables and different random phenomena need to be considered for different practical problems. The development of analytical approaches, such as mathematical models with random-like uncertainty and algorithms able to solve the bi-level decision making problems are still largely unexplored. This book is an attempt to elucidate random-like bi-level decision making to all these aspects. In this chapter, we present some elements of random sets theory, including some fundamental concepts, the definitions of random variables, fuzzy variables, random-overlapped random (Ra-Ra) variables and fuzzy-overlapped random (Ra-Fu) variables. Subsequently, the elements of bi-level programming are introduced. This part is the prerequisite for reading this book. Since these results are well-known, we only introduce the results and readers can refer to correlative research such as Billingsley (Probability and measure, John Wiley & Sons, New York, 1965), Cohn (Measure theory, Birkhäuser, Boston, 1980), Halmos (Measure theory, Van Nostrad Reinhold Company, New York, 1950), and Krickeberg (Probability theory, Addison-Wesley, Reading, 1965).

Keywords Random Sets Theory • Random variables • Fuzzy variables • Ra-Ra variables • Ra-Fu variables • Bi-level Programming • Literature Review

J. Xu et al., *Random-Like Bi-level Decision Making*, Lecture Notes in Economics and Mathematical Systems 688, DOI 10.1007/978-981-10-1768-1_1

1.1 Random Sets Theory

In this part, we introduce the elements of random sets theory, including some fundamental concepts, the definitions of random variables, fuzzy variables, Ra-Ra variables and Ra-Fu variables.

1.1.1 General Concepts

The concepts of the measure space, outer measures, probability space, product probability space, conditional probability and independence are introduced.

Measure Space

The concept of measurable sets and measure space are introduced and some properties exhibited.

Definition 1.1 (Halmos [59]). Let Ω be a nonempty set, and $\mathscr{A}$ a σ-algebra over Ω. Then $(\Omega, \mathscr{A})$ is called a measurable space, and the sets in $\mathscr{A}$ are called measurable sets.

Definition 1.2 (Cohn [33]). Let Ω be a nonempty set, and $\mathscr{A}$ a σ-algebra over Ω. A function π whose domain is the σ-algebra $\mathscr{A}$ and whose values belong to the extended half-line $[0, +\infty)$ is said to be countably additive if it satisfies

$$\pi\left(\bigcup_{i=1}^{\infty} A_i\right) = \sum_{i=1}^{\infty} \pi(A_i) \tag{1.1}$$

for each infinite sequence $\{A_i\}$ of disjointed sets that belong to $\mathscr{A}$.

Definition 1.3 (Cohn [33]). Let $(\Omega, \mathscr{A})$ be a measurable space. A function π is a measure if it satisfies

(a) $\pi(\Phi) = 0$;
(b) π is countably additive.

Another related concept is of interest. Assume that $\mathscr{A}$ is algebra (not necessarily a σ-algebra) on the set Ω. A function π, whose domain is $\mathscr{A}$ and whose values belong to $[0, +\infty)$, is finitely additive if it satisfies

$$\pi\left(\bigcup_{i=1}^{n} A_i\right) = \sum_{i=1}^{n} \pi(A_i) \tag{1.2}$$

for each finite sequence $A_1, A_2, \cdots, A_n$ of disjointed sets that belong to $\mathscr{A}$. A finitely additive measure on algebra $\mathscr{A}$ is a function $\pi : \mathscr{A} \to [0, \infty)$ that satisfies $\pi(\Phi) = 0$ and is finitely additive.

Finitely additivity appears to be a more natural property than countable additivity. However, countable additive measures are sufficient for almost all applications, and can support much more powerful theories of integration than finite additive measures. It should be emphasized that a measure is always a countably additive measure. The expression "finitely additive measure" will always be written out in full.

Definition 1.4 (Cohn [33]). Let Ω be a nonempty set, and $\mathscr{A}$ a σ-algebra over Ω. If π is a measure on $\mathscr{A}$, then the triple $(\Omega, \mathscr{A}, \pi)$ is called a measure space.

Usually, if $(\Omega, \mathscr{A}, \pi)$ is a measure space, π is said to be a measure on the measurable space $(\Omega, \mathscr{A})$, or if the σ-algebra $\mathscr{A}$ is clear from context, a measure on Ω. Let's turn to some examples.

Example 1.1. Let $(\Omega, \mathscr{A}, \pi)$ be a measure space, where π is defined as a function $\pi : \mathscr{A} \to [0, +\infty)$ by letting $\pi(A)$ be n if A is a finite set with n elements, and letting $\pi(A)$ be $+\infty$ if A is an infinite set. Then π is a measure, and is often called a *counting measure* on $(\Omega, \mathscr{A})$.

Example 1.2. Let Ω be a nonempty set, $\mathscr{A}$ be a σ-algebra on Ω and x be a member of Ω. Define a function $\delta_x : \mathscr{A} \to [0, +\infty)$ as follows

$$\delta_x = \begin{cases} 1, \text{ if } x \in A, \\ 0, \text{ if } x \notin A. \end{cases}$$

Then δ_x is a measure, and is called a *point mass* concentrated at x.

Example 1.3. Consider the set $\mathbf{R}$ of all real numbers, and the σ-algebra $\mathscr{B}(\mathbf{R})$ for the Borel subsets of $\mathbf{R}$. In the following part we construct a measure on $\mathscr{B}(\mathbf{R})$ that assigns its length to each subinterval of $\mathbf{R}$. This measure is known as the *Lebesgue measure*.

Next, an important result is listed here with proof and interested readers can also consult the original book related to measure theory such as Halmos [59, 60].

Theorem 1.1 (Halmos [60]). *There is a unique measure π on the Borel algebra of* $\mathbf{R}$ *such that* $\pi\{(a, b]\} = b - a$ *for any interval* $(a, b]$ *of* $\mathbf{R}$.

Example 1.4. Let Ω be the set of all positive integers, and let $\mathscr{A}$ be the collection of all subsets A of Ω such that either A or A^c is finite. Then $\mathscr{A}$ is an algebra, but not a σ-algebra (see Example 1.6). Define a function $\pi : \mathscr{A} \to [0, +\infty)$ by

$$\pi(A) = \begin{cases} 1, \text{ if } A \text{ is infinite}, \\ 0, \text{ if } A \text{ is finite}. \end{cases}$$

It is easy to check that π is a finitely additive measure; however, it is impossible to extend π to a countably additive measure on the σ-algebra generated by $\mathscr{A}$ (if $A_k = k$) for each k, then $\pi(\bigcup_{k=1}^{\infty}) = \pi(\Omega) = 1$, while $\sum_{i=1}^{\infty} \pi(A_k) = 0$.

Example 1.5. Let Ω be an arbitrary set, and let $\mathscr{A}$ be an arbitrary σ-algebra on Ω. Define a function $\pi : \mathscr{A} \to [0, +\infty)$ by

$$\pi(A) = \begin{cases} +\infty, \text{ if } A \neq \Phi, \\ 0, \quad \text{ if } A = \Phi. \end{cases}$$

Then π is a measure.

Example 1.6. Let Ω be a set that has at least two members, and let $\mathscr{A}$ be an σ-algebra consisting of all subsets of Ω. Define a function $\pi : \mathscr{A} \to [0, +\infty)$ by

$$\pi(A) = \begin{cases} 1, \text{ if } A \neq \Phi, \\ 0, \text{ if } A = \Phi. \end{cases}$$

Then π is not a measure, nor even a finitely additive measure, for if A_1 and A_2 are disjoint nonempty subsets of Ω, then $\pi(A_1 \cup A_2) = 1$, while $\pi(A_1) + \pi_2 = 2$.

The monotone class theorem, the Carathéodory extension theorem, and the approximation theorem are listed here without proof. An interested reader may consult books related to measure theory such as Berberian [14], Halmos [59, 60], and Jacobs [67].

Theorem 1.2 (Monotone Class Theorem (Srivastava [101])). *Assume that $\mathscr{A}_0$ is an algebra over Ω, and $\mathscr{C}$ is a monotone class of subsets of Ω (if $\mathscr{A}_i \in \mathscr{C}$ and $A_i \uparrow A$ or $A_i \downarrow A$, then $A \in \mathscr{C}$). If $\mathscr{A}_0 \subset \mathscr{C}$ and $\sigma(\mathscr{A}_0)$ is the smallest σ-algebra containing $\mathscr{A}_0$, then $\sigma(\mathscr{A}_0) \subset \mathscr{C}$.*

Theorem 1.3 (Approximation Theorem (Jacobs [67])). *Let $(\Omega, \mathscr{A}, \pi)$ be a measure space, and let $\mathscr{A}_0$ be an algebra over Ω such that $\mathscr{A}$ is the smallest σ-algebra containing $\mathscr{A}_0$. If π is σ-finite and $A \in \mathscr{A}$ has finite measure, then for any given $\varepsilon > 0$, there exists a set $A_0 \in \mathscr{A}_0$ such that $\pi\{A/A_0\} < \varepsilon$.*

Theorem 1.4. *Let $(\Omega, \mathscr{A}, \pi)$ be a measure space, and let A and B be subsets of Ω that belong to $\mathscr{A}$ such that $A \subset B$. Then $\pi(A) \leq \pi(B)$. If $\pi(A) < +\infty$, then $\pi(B) = \pi(B - A) + \pi(A)$.*

Proof. Since $B = A \bigcup B - A$ satisfying that A and $B - A$ are disjoint. Since $A \subset B$, it means $\pi(B - A) \geq 0$. Thus it follows from the additivity of π that

$$\pi(B) = \pi(B - A) + \pi(A). \tag{1.3}$$

Let π be a measure on a measurable space $(\Omega, \mathscr{A})$. If $\pi(\Omega) < +\infty$, π is said to be a *finite* measure. Further it is said to be a σ-*finite* measure if Ω is the union of a sequence $A_1, A_2, \cdots$ of sets that belong to $\mathscr{A}$ and $\pi(A_i) < +\infty$ for each i. More generally, a set in $\mathscr{A}$ is *σ-finite* under π if it is the union of a sequence of sets that belong to $\mathscr{A}$ and have a finite measure under π. The measure space $(\Omega, \mathscr{A}, \pi)$ is also called *finite* or *σ-finite* if π is finite or σ-finite. Most of the constructions and basic properties that we consider here are valid for all measures. For a few important theorems, however, we need to assume that the measures involved are finite or σ-finite.

Theorem 1.5 (Halmos [60]). *Let $(\Omega, \mathscr{A}, \pi)$ be a measure space. If $\{A_k\}$ Cohn [33] is an arbitrary sequence of sets that belong to $\mathscr{A}$, then*

$$\pi\left(\bigcup_{k=1}^{\infty} A_k\right) \le \sum_{k=1}^{\infty} \pi(A_k). \tag{1.4}$$

Theorem 1.6 (Cohn [33]). *Let $(\Omega, \mathscr{A}, \pi)$ be a measure space.*

(a) If $\{A_k\}$ is an increasing sequence of sets that belong to $\mathscr{A}$, then $\pi(\bigcup_k A_k) = \lim_k \pi(A_k)$.
(b) If $\{A_k\}$ is an decreasing sequence of sets that belong to $\mathscr{A}$, and if $\pi(A_n) < +\infty$ holds for some n, then $\pi(\cap_k A_k) = \lim_k \pi(A_k)$.

Theorem 1.7 (Cohn [33]). *Let $(\Omega, \mathscr{A})$ be a measurable space, and let π be a finitely additive measure on $(\Omega, \mathscr{A})$. Then π is a measure if either*

(a) $\lim_k \pi(A_k) = \pi(\bigcup_k A_k)$ holds for each increasing sequence $\{A_k\}$ of sets that belong to $\mathscr{A}$.
(b) $\lim_k \pi(A_k) = 0$ holds for each decreasing sequence $\{A_k\}$ of sets that belong to $\mathscr{A}$ and satisfy $\cap_k A_k = \Phi$.

The preceding theorem will give the continuity of the measure π.

Theorem 1.8 (Berberian [14]). *Let $(\Omega, \mathscr{A}, \pi)$ be a measure space, and $A_1, A_2, \cdots, \in \mathscr{A}$.*

(a) If $\{A_i\}$ is an increasing sequence, then

$$\lim_{i\to\infty} \pi\{A_i\} = \pi\left\{\lim_{i\to\infty} A_i\right\}. \tag{1.5}$$

(b) If $\{A_i\}$ is a decreasing sequence, and $\pi\{A_1\}$ is finite, then

$$\lim_{i\to\infty} \pi\{A_i\} = \pi\left\{\lim_{i\to\infty} A_i\right\}. \tag{1.6}$$

Example 1.7. If $\pi\{A_i\}$ are not finite for any i, then the part (b) of Theorem 1.8 does not hold. For example, let $A_i = [i, +\infty)$ for $i = 1, 2, \cdots$, and let π be the length of intervals. Then $A_i \downarrow \Phi$ as $i \to \infty$. However, $\pi\{A_i\} \equiv +\infty \ne 0 = \pi\{\Phi\}$.

Let $\Omega_1, \Omega_2, \cdots, \Omega_n$ be any sets (not necessarily subsets of the same space). The product $\Omega = \Omega_1 \times \Omega_2 \times \cdots \times \Omega_n$ is the set of all ordered n-tuples of the form $(x_1, x_2, \cdots, x_n)$, where $\xi_i \in \Omega_i$ for $i = 1, 2, \cdots, n$.

Definition 1.5 (Evans and Gariepy [47]). Let $\mathscr{A}_i$ be σ-algebras over $\Omega_i, i = 1, 2, \cdots, n$, respectively. Write $\Omega = \Omega_1 \times \Omega_2 \times \cdots \times \Omega_n$. A measurable rectangle in Ω is a set $A = A_1 \times A_2 \times \cdots \times A_n$, where $A_i \in \mathscr{A}_i$ for $i = 1, 2, \cdots, n$. The smallest σ-algebra containing all measurable rectangles of Ω is called the product σ-algebra, denoted by $\mathscr{A} = \mathscr{A}_1 \times \mathscr{A}_2 \times \cdots \times \mathscr{A}_n$.

Note that the product σ-algebra $\mathscr{A}$ is the smallest σ-algebra containing measurable rectangles, rather than the product of $\mathscr{A}_1, \mathscr{A}_2, \cdots, \mathscr{A}_n$.

Remark 1.1. Let $(\Omega_i, \mathscr{A}_i, \pi_i), i = 1, 2, \cdots, n$ be measure spaces. Assume that $\pi_i, i = 1, 2, \cdots, n$ are σ-finite, $\Omega = \Omega_1 \times \Omega_2 \times \cdots \times \Omega_n$, $\mathscr{A} = \mathscr{A}_1 \times \mathscr{A}_2 \times \cdots \times \mathscr{A}_n$. Then there is a unique measure π on $\mathscr{A}$ such that

$$\pi\{A_1 \times A_2 \times \cdots \times A_n\} = \pi_1\{A_1\} \times \pi_2\{A_2\} \times \cdots \times \pi_n\{A_n\} \tag{1.7}$$

for every measurable rectangle $A_1 \times A_2 \times \cdots \times A_n$. The measure π is called the product of $\pi_1, \pi_2, \cdots, \pi_n$, denoted by $\pi = \pi_1 \times \pi_2 \times \cdots \times \pi_n$. The triplet $(\Omega, \mathscr{A}, \pi)$ is called the product measure space.

If the sequence $(\Omega_i, \mathscr{A}_i, \pi_i), i = 1, 2, \cdots,$ is an finite measure spaces such that $\pi_i(\Omega_i) = 1$ for $i = 1, 2, \cdots$. The product $\Omega = \Omega_1 \times \Omega_2 \times \cdots$ is defined as the set of all ordered tuples of the form $(x_1, x_2, \cdots)$, where $x_i \in \Omega_i$ for $i = 1, 2, \cdots$. For this case, we define a *measurable rectangle* as a set of the form $A_1 \times A_2 \times \cdots$, where $A_i \in \mathscr{A}_i$, where $A_i \in \mathscr{A}_i$ for all i and $A_i = \Omega_i$ for all but finitely many i. The smallest σ-algebra containing all measurable rectangles of Ω is called the product σ-algebra, denoted by $\mathscr{A}_1 \times \mathscr{A}_2 \times \cdots$.

Remark 1.2. Assume that $(\Omega_i, \mathscr{A}_i, \pi_i)$ are measure spaces such that $\pi_i\{\Omega_i\} = 1$ for $i = 1, 2, \cdots$. Let $\Omega = \Omega_1 \times \Omega_2 \times \cdots$ and $\mathscr{A}_1 \times \mathscr{A}_2 \times \cdots$. Then there is a unique measure π on $\mathscr{A}$ such that

$$\pi\{A_1 \times \cdots \times A_n \times \Omega_{n+1} \times \Omega_{n+2} \times \cdots\} = \pi_1\{A_1\} \times \pi_2\{A_2\} \times \cdots \tag{1.8}$$

for any measurable rectangle $A_1 \times \cdots \times A_n \times \Omega_{n+1} \times \Omega_{n+2} \times \cdots$ and all $n = 1, 2, \cdots$. The measure π is called the infinite product, denoted by $\pi = \pi_1 \times \pi_2 \times$. The triplet $(\Omega, \mathscr{A}, \pi)$ is called the infinite product measure space.

Outer Measures

In what follows, we develop one of the standard techniques for constructing measures, then we use it to construct Lebesgue measure on $\mathbf{R}^n$.

Definition 1.6 (Cohn [33]). Let Ω be a set, and let $\mathscr{P}(\Omega)$ be the collection of all subsets of Ω. An outer measure on Ω is a function $\pi^* : \mathscr{P}(\Omega) \to [0, +\infty)$ such that

(a) $\pi^*(\Phi) = 0$,
(b) if $A \subset B \subset \Omega$, then $\pi^*(A) \le \pi^*(A)$, and
(c) if $\{A_n\}$ is an infinite sequence of subsets of Ω, then

$$\pi^*(\cup_n A_n) \le \sum_n \pi^*(A_n).$$

Thus an outer measure on Ω is a *monotone* and *countably subadditive* function from $\mathscr{P}(\Omega)$ to $[0, +\infty)$ whose value at Φ is 0.

Note that a measure can fail to be an outer measure, in fact, a measure on Ω is an outer measure if and only if its domain is $\mathscr{P}(\Omega)$ (see Theorems 1.4 and 1.5). On the other hand, an outer measure generally fails to be countably additive, and so fails to be a measure.

In Theorem 1.9 we shall prove that for each outer measure π^* on Ω there is a relatively natural σ-algebra Π_{π^*} on Ω for which the restriction of π^* to Π_{π^*} is countably additive, and hence a measure. Many important measures can be derived from outer measures in this way.

Here are some examples.

Example 1.8. Let Ω be an arbitrary set, and define π^* on $\mathscr{P}(\Omega)$ by $\pi^*(A) = 0$ if $A = \Phi$, and $\pi^*(A) = 1$ otherwise. Then π^* is an outer measure.

Example 1.9. Let Ω be an arbitrary set, and define π^* on $\mathscr{P}(\Omega)$ by $\pi^*(A) = 0$ if A is countable, and $\pi^*(A) = 1$ if A is uncountable. Then π^* is an outer measure.

Example 1.10. Let Ω be an infinite set, and define π^* on $\mathscr{P}(\Omega)$ by $\pi^*(A) = 0$ if A is finite, and $\pi^*(A) = 1$ if A is infinite. Then π^* fails to be countably subadditive, and so is not an outer measure.

Example 1.11. Lebesgue outer measure on $\mathbf{R}$, which we shall denote by λ^*, is defined as follows. For each subset A of $\mathbf{R}$ let ϕ_A be the set of all infinite sequences $\{(a_i, b_i)\}$ of bounded open intervals such that $A \subset \cup_{i=1}^{\infty}(a_i, b_i)$. Then $\lambda^* : \mathscr{P}(\mathbf{R}) \to [0, +\infty)$ is defined by letting $\lambda^*(A)$ be the infimum of the set

$$\left\{ \sum_i (b_i - a_i) | \{(a_i, b_i)\} \in \phi_A \right\}.$$

Note that this set of sums is non-empty, and that the infimum of the set consisting of $+\infty$ alone is $+\infty$.

Lemma 1.1 (Cohn [33]). *Lebesgue outer measure on* $\mathbf{R}$ *is an outer measure, and it assigns to each subinterval of* $\mathbf{R}$ *its length.*

Example 1.12. Let us turn to Lebesgue outer measure on $\mathbf{R}^n$, which we shall denote by λ^* or, if necessary in order to avoid ambiguity, by λ_n^*. An *n-dimensional interval* is a subset of $\mathbf{R}^n$ of the form $I_1 \times \cdots \times I_n$, where $I_1, \cdots, I_n$ are subintervals of $\mathbf{R}$ and $I_1 \times \cdots \times I_n$ is given by

$$I_1 \times \cdots \times I_n = \{(x_1, \cdots, x_n) | x_i \in I_i, i = 1, 2, \cdots, n\}.$$

Note that the intervals $I_1, \cdots, I_n$, and hence the n-dimensional interval $I_1 \times \cdots \times I_n$, can be open, closed, or neither open nor closed. The *volume* of the lengths of the intervals $I_1, \cdots, I_n$, and will be denoted by $vol(I_1 \times \cdots \times I_n)$. For each subset A of $\mathbf{R}^n$ let ϕ_A be the set of all sequences $\{R_i\}$ of bounded and open n-dimensional intervals for which $A \subset \cup_{i=1}^{\infty} R_i$. Then $\lambda^*(A)$, the outer measure of A, is the infimum of the set

$$\left\{ \sum_{i=1}^{\infty} vol(R_i) | \{R_i\} \in \phi_A \right\}.$$

We note the following analogue of Lemma 1.1.

Lemma 1.2 (Cohn [33]). *Lebesgue outer measure on* $\mathbf{R}^n$ *is an outer measure, and it assigns to each n-dimensional interval its volume.*

The proof can be found in [117].

Let Ω be a set, and let π^* be an outer measure on Ω. A subset B of Ω is π^*-*measure* (or *measurable with respect to* π^*) if

$$\pi^*(A) = \pi^*(A \cap B) + \pi^*(A \cap B^c)$$

holds for each subset A of Ω. Thus a π^*-measurable subset of Ω is one that divides each subset of Ω in such a way that the sizes (as measured by π^*) of the pieces add properly.

Definition 1.7 (Cohn [33]). A *Lebesgue measurable* subset of $\mathbf{R}$ or $\mathbf{R}^n$ is of course one that is measurable with respect to Lebesgue outer measure.

Note that the subadditivity of the outer measure π^* implies that

$$\pi^*(A) \leq \pi^*(A \cap B) + \pi^*(A \cap B^c)$$

holds for all subsets A and B of Ω. Thus to check that a subset B of Ω is π^*-measurable we need only check that

$$\pi^*(A) \geq \pi^*(A \cap B) + \pi^*(A \cap B^c) \tag{1.9}$$

holds for each subset A of Ω. Note also that inequality (1.9) certainly holds if $\pi^*(A) = +\infty$. Thus the π^*-measurability of B can be verified by checking that (1.9) holds for each A that satisfies $\pi^*(A) < +\infty$.

Lemma 1.3 (Cohn [33]). *Let Ω be a set, and let π^* be an outer measure on Ω. Then each subset B of Ω that satisfies $\pi^*(B) = 0$ or that satisfies $\pi^*(B^c) = 0$ is π^*-measurable.*

The proof can be found in [117].

It follows that the sets Φ and Ω are measurable for every outer measure on Ω.

The following theorem is the fundamental fact about outer measures. It will be the key to many of out constructions of measures, then we listed the theorem and the proving process as follows. Interested readers can consult books related to the measure.

Theorem 1.9 (Cohn [33]). *Let Ω be a set, and let π^* be an outer measure on Ω, and let Π_{π^*} be the collection of all π^*-measurable subsets of Ω. Then*

(a) Π_{π^*} is a σ-algebra, and
(b) the restriction of π^* to Π_{π^*} is a measure on Π_{π^*}.

Next, we first turn to some applications of Theorem 1.9, and then begin with Lebesgue measure and deduce some properties of it on $\mathbf{R}^n$.

Remark 1.3. Every Borel subset of $\mathbf{R}$ is Lebesgue measurable.

Remark 1.4. Every Borel subset of $\mathbf{R}^n$ is Lebesgue measurable.

Definition 1.8 (Cohn [33]). The restriction of Lebesgue outer measure on $\mathbf{R}$ (or on $\mathbf{R}^n$) to the collection Π_{π^*} of Lebesgue measurable subsets of $\mathbf{R}$ (or of $\mathbf{R}^n$) is called Lebesgue measure, and will be denoted by λ or by λ_n.

Lemma 1.4 (Cohn [33]). *Let A be a Lebesgue measurable subset of $\mathbf{R}^n$. Then*

(a) $\lambda(A) = \inf\{\lambda(U) | U$ is open and $A \subset U\}$, and
(b) $\lambda(A) = \sup\{\lambda(K) | K$ is compact and $K \subset U\}$.

The following lemma will be needed for the proof of Theorem 1.4. In this lemma we shall be dealing with a certain collection of half-open cubes, namely with those that have the form

$$\{(x_1, \cdots, x_n) | j_i 2^{-k} \le x_i < (j_i + 1)2^{-k}, i = 1, 2, \cdots, n\} \tag{1.10}$$

for some integers $j_1, \cdots, j_n$ and some positive integer k.

Lemma 1.5 (Cohn [33]). *Each open subset of $\mathbf{R}^n$ is the union of a countable disjoint collection of half-open cubes, each of which is of the form given in expression (1.10).*

The following theorem is very important for us to know about the Lebesgue measure.

Theorem 1.10 (Cohn [33]). *Lebesgue measure is the only measure on $(\mathbf{R}^n, \mathscr{B}(\mathbf{R}^n))$ that assigns to each n-dimensional interval, or even to each half-open cube of the form given in expression (1.10), its volume.*

For each element $\boldsymbol{x}$ and subset A of $\mathbf{R}^n$ we shall denote by $A+\boldsymbol{x}$ the subset of $\mathbf{R}^n$ defined by

$$A+\boldsymbol{x} = \{y \in \mathbf{R}^n | y = \boldsymbol{a}+\boldsymbol{x} \text{ for some } \boldsymbol{a} \text{ in } A\},$$

the set $A+\boldsymbol{x}$ is called the translate of A by $\boldsymbol{x}$. We turn to the invariance of Lebesgue measure under such translations.

Probability Space

The fundamental concepts of probability theory have rooted in measure theory. Like any branch of mathematics, probability theory has its own terminology and its own tools. The probability theory has been widely pushed forward by many scholars [21, 74, 77, 83, 94]. In this section, we will introduce some of this terminology and study some basic concepts of probability theory.

Let Ω be a nonempty set, and $\mathscr{A}$ a σ-algebra over Ω. If Ω is countable, usually $\mathscr{A}$ is the power set of Ω. If Ω is uncountable, for example $\Omega = [0, 1]$, usually $\mathscr{A}$ is the Borel algebra of Ω. Each element in $\mathscr{A}$ is called an event. In order to present an axiomatic definition of probability, it is necessary to assign to each event A a number $Pr\{A\}$ which indicates the probability that A will occur. In order to ensure that the number $Pr\{A\}$ has certain mathematical properties which we intuitively expect a probability to have, the following three axioms must be satisfied [21]:

Axiom 1. (*Normality*) $Pr\{\Omega\} = 1$.
Axiom 2. (*Nonnegativity*) $Pr\{A\} \geq 0$ for any $A \in \mathscr{A}$.
Axiom 3. (*Countable Additivity*) For every countable sequence of mutually disjoint events $\{A_i\}$, we have

$$Pr\left\{\bigcup_{i=1}^{\infty} A_i\right\} = \sum_{i=1}^{\infty} Pr\{A_i\}. \tag{1.11}$$

Definition 1.9 (Billingsley [21]). The set function Pr is called a probability measure if it satisfies the three axioms. And the triple $(\Omega, \mathscr{A}, Pr)$ is called a probability space.

We note that, if $A_n \in \mathscr{A}, n = 1, 2, \cdots$, then A_n^c, $\bigcup_{n=1}^{\infty} A_n$, $\bigcap_{n=1}^{\infty} A_n$, $\liminf_{n\to+\infty} A_n$, $\limsup_{n\to+\infty} A_n$ and $\lim_{n\to+\infty} A_n$ (if exists) are events. Also, the probability measure Pr is defined on $\mathscr{A}$, and for all events A, A_n

$$Pr\{A\} \geq 0,\ Pr\left\{\bigcup_{i=1}^{\infty} A_i\right\} = \sum_{n=1}^{\infty} Pr\{A_n\} (A_n \text{ 's disjoint}),\ Pr\{\Omega\} = 1.$$

It follows that

$$Pr\{\Phi\} = 0,\ Pr\{A\} \leq Pr\{B\} \text{ for } A \subset B,\ Pr\left\{\bigcup_{n=1}^{\infty} A_n\right\} \leq \sum_{n=1}^{\infty} Pr(A_n).$$

Moreover,

$$Pr\left(\liminf_{n\to\infty} A_n\right) \le \liminf_{n\to\infty} Pr(A_n) \le \limsup_{n\to\infty} Pr(A_n) \le Pr\left(\limsup_{n\to\infty} A_n\right),$$

and if $\lim_{n\to\infty} A_n$ exists, then

$$Pr\left(\lim_{n\to\infty} A_n\right) = \lim_{n\to\infty} Pr(A_n). \tag{1.12}$$

The last result is known as the continuity property of probability measures.

Example 1.13. Let $\Omega = \{\omega_j : j \ge 1\}$, and let $\mathscr{A}$ be the σ-algebra of all subsets of Ω. Let $\{p_j : j \ge 1\}$ be any sequence of nonnegative real numbers satisfying $\sum_{j=1}^{\infty} p_j = 1$. Define Pr on $\mathscr{A}$ by

$$Pr\{E\} = \sum_{\omega \in E} p_j, \quad E \in \mathscr{A}. \tag{1.13}$$

Then Pr defines a probability measure on $(\Omega, \mathscr{A})$, and $(\Omega, \mathscr{A}, Pr)$ is a probability space.

Example 1.14. Let $\Omega = (0, 1]$ and $\mathscr{A} = \mathscr{B}$ be the σ-algebra of Borel sets on Ω. Let λ be the Lebesgue measure on $\mathscr{B}$. Then $(\Omega, \mathscr{A}, \lambda)$ is a probability space.

Lemma 1.6 (Krickeberg [77]). *Let Ω be a nonempty set, $\mathscr{A}$ be a σ-algebra over Ω, and Pr a probability measure. Then we have*

(a) $Pr\{\Phi\} = 0$;
(b) Pr is self-dual, i.e., $Pr\{A\} + Pr\{A^c\} = 1$ for any $A \in \mathscr{A}$;
(c) Pr is increasing, i.e., $Pr\{A\} \le Pr\{B\}$ whenever $A \subset B$;
(d) $0 \le Pr\{A\} \le 1$ for any $A \in \mathscr{A}$.

The usual starting point in the construction of probability measure is that probabilities are assigned to a restricted class of sets. The probability extension theorem gives a method to construct probability measure.

Theorem 1.11 (Probability Extension Theorem (Blackwell [22])). *Let Ω be a nonempty set, $\mathscr{A}_0$ be an algebra over Ω, and Pr a measure on $\mathscr{A}_0$ such that $Pr\{\Omega\} = 1$. Then Pr has a unique extension to a probability measure on the smallest σ-algebra $\mathscr{A}$ containing $\mathscr{A}_0$.*

Product Probability Space

Let $(\Omega_i, \mathscr{A}_i, Pr_i), i = 1, 2, \cdots, n$ be probability spaces, and $\Omega = \Omega_1 \times \Omega_2 \times \cdots \times \Omega_n$, $\mathscr{A} = \mathscr{A}_1 \times \mathscr{A}_2 \times \cdots \times \mathscr{A}_n$, $Pr = Pr_1 \times Pr_2 \times \cdots Pr_n$. Note that the probability measures

$Pr_i, i = 1, 2, \cdots, n$ are finite. It follows from the product measure theorem that there is a unique measure Pr on $\mathscr{A}$ such that

$$Pr\{A_1 \times A_2 \times \cdots \times A_n\} = Pr_1\{A_1\} \times Pr_2\{A_2\} \times \cdots \times Pr_n\{A_n\}, \tag{1.14}$$

for any $A_i \in \mathscr{A}_i, i = 1, 2, \cdots, n$. This conclusion is called the product probability theorem. The measure Pr is also a probability measure since

$$Pr\{\Omega\} = Pr_1\{\Omega_1\} \times Pr_2\{\Omega_2\} \times \cdots \times Pr_n\{\Omega_n\} = 1. \tag{1.15}$$

Such a probability measure is called the product probability measure, denoted by $Pr = Pr_1 \times Pr_2 \times \cdots \times Pr_n$.

Definition 1.10 (Krickeberg [77]). Let $(\Omega_i, \mathscr{A}_i, Pr_i), i = 1, 2, \cdots, n$ be probability spaces, and $\Omega = \Omega_1 \times \Omega_2 \times \cdots \times \Omega_n$, $\mathscr{A} = \mathscr{A}_1 \times \mathscr{A}_2 \times \cdots \times \mathscr{A}_n$, $Pr = Pr_1 \times Pr_2 \times \cdots \times Pr_n$. Then the triplet $(\Omega, \mathscr{A}, Pr)$ is called the product probability space.

If $(\Omega_i, \mathscr{A}_i, Pr_i), i = 1, 2, \cdots$, be an arbitrary sequence of probability spaces, and

$$\Omega = \Omega_1 \times \Omega_2 \times \cdots, \quad \mathscr{A} = \mathscr{A}_1 \times \mathscr{A}_2 \times \cdots. \tag{1.16}$$

It follows from the infinite product measure theorem that there is a unique probability measure Pr on $\mathscr{A}$ such that

$$Pr\{A_1 \times \cdots \times A_n \times \Omega_{n+1} \times \Omega_{n+2} \times \cdots\} = Pr_1\{A_1\} \times \cdots \times Pr_n\{A_n\}, \tag{1.17}$$

for any measurable rectangle $A_1 \times \cdots A_n \times \Omega_{n+1} \times \Omega_{n+2} \times \cdots$ and all $n = 1, 2, \cdots$ The probability measure Pr is called the infinite product of $Pr_i, i = 1, 2, \cdots$ and is denoted by

$$Pr = Pr_1 \times Pr_2 \times \cdots \tag{1.18}$$

Definition 1.11 (Krickeberg [77]). Let $(\Omega_i, \mathscr{A}_i, Pr_i), i = 1, 2, \cdots$ be probability spaces, and $\Omega = \Omega_1 \times \Omega_2 \times \cdots$, $\mathscr{A} = \mathscr{A}_1 \times \mathscr{A}_2 \times \cdots$, $Pr = Pr_1 \times Pr_2 \times \cdots$. Then the triplet $(\Omega, \mathscr{A}, Pr)$ is called the infinite product probability space.

Conditional Probability and Independence

Consider an experiment that consists of flipping a coin twice, noting each time whether the result is heads or tails. The sample space of this experiment can be taken to be the following set of four outcomes:

$$\Omega = \{(H, H), (H, T), (T, H), (T, T)\},$$

where (H, T) means, for example, that the first flip lands heads and the second tails. Suppose now that each of the four possible outcomes is equally likely to occur and thus has probability 1/4. Suppose further that we observe that the first flip lands on heads. Then, given this information, what is the probability that both flips land on heads? To calculate this probability we reason as follows: Given that the initial flip lands heads, there can be at most two possible outcomes for our experiment, namely, (H, H) or (H, T). In addition, as each of these outcomes originally had the same occurrence probability, they should still have equal probabilities. That is, given that the first flip lands heads, the (conditional) probability of each of the outcomes (H, H) and (H, T) is 1/2, whereas the (conditional) probability of the other two outcomes is 0. Hence the desired probability is 1/2.

If we let A and B denote, respectively, an event in which both flips land on heads and an event in which the first flip lands on heads, then the probability obtained above is called the conditional probability of A given that B has occurred and is denoted by

$$Pr(A|B).$$

A general formula for $Pr(A|B)$ that is valid for all experiments and events A and B can be obtained in the same manner as previously given. Namely, if event B occurs, then in order for A to occur, it is necessary that the actual occurrence be a point in both A and B, that is, it must be in AB. Now since we know that B has occurred, it follows that B becomes our new sample space and hence the probability that the event AB occurs equals the probability of AB relative to the probability of B. That is,

$$Pr(A|B) = \frac{Pr(AB)}{Pr(B)}.$$

As indicated in the coin flip example, $Pr(A|B)$, the conditional probability of A, given that B occurred, is not generally equal to $Pr(A)$, the unconditional probability of A. In other words, knowing that B has occurred generally changes the probability that A occurs (what if they were mutually exclusive?). In a special case where $Pr(A|B)$ is equal to $Pr(A)$, we say that A and B are independent. Since $Pr(A|B) = Pr(AB)/Pr(B)$, we see that A is independent of B if

$$Pr(AB) = Pr(A)Pr(B).$$

Since this relation is symmetric in A and B, it follows that whenever A is independent of B, B is independent of A.

Probability Distribution

The probability distribution of a random variable is given as follows.

Definition 1.12 (Durrett [43]). For every $x \in \mathbf{R}$ such that

$$F(x) = Pr\{-\infty \leq \xi \leq x\} = Pr\{\omega \in \Omega | \xi(\omega) \leq x\}.$$

We call $F(x)$ the distribution function that the random variable ξ takes if a value less than or equal to x.

In the following we write $\{\xi \leq x\}$ for the event $\{\omega \in \Omega | \xi(\omega) \leq x\}$. First, recall the following elementary property of a distribution function. In order to easily understand these, we have listed all the proving processes and interested readers can consult related books [21, 77].

Theorem 1.12 (Krickeberg [77]). *The distribution function F for the random variable ξ is a nondecreasing, right-continuous function on* **R** *which satisfies*

$$F(-\infty) = \lim_{x \to -\infty} F(x) = 0$$

and

$$F(+\infty) = \lim_{x \to +\infty} F(x) = 1.$$

Theorem 1.13 (Krickeberg [77]). *A distribution function F is continuous at $x \in$* **R** *if and only if $Pr\{\omega \in \Omega | \xi(\omega) = x\} = 0$.*

Remark 1.5. Let ξ be a random variable, and let g be a Borel-measurable function defined on **R**. Then $g(\xi)$ is also a random variable, the distribution for which is determined by that of ξ.

We now show that a function F on **R** with the properties stated in Theorem 1.12 uniquely determines a probability measure Pr_F on $\mathscr{B}$.

Theorem 1.14 (Krickeberg [77]). *Let F be a nondecreasing, right-continuous function defined on* **R** *and satisfying*

$$F(-\infty) = 0 \text{ and } F(+\infty) = 0.$$

Then there exists a probability measure $Pr = Pr_F$ on $\mathscr{B}$ uniquely determined by the relation

$$Pr_F(-\infty, x] = F(x) \quad \text{for every } x \in \mathbf{R}.$$

Remark 1.6. Let F be a bounded nondecreasing, right-continuous function defined on **R** satisfying $F(-\infty) = 0$. Then, it is the clear from the proof for Theorem 1.14 that there exists a finite measure $\mu = \mu_F$ on $\mathscr{A}$ uniquely determined by $\mu_F(-\infty, x] = F(x), x \in \mathbf{R}$.

Remark 1.7. Let F on **R** satisfy the conditions of Theorem 1.14. Then there exists a random variable ξ on some probability space such that F is the distribution function of ξ. Consider the probability space $(\mathbf{R}, \mathscr{A}, Pr)$, where Pr is the probability measure as constructed in Theorem 1.14. Let $\xi(\omega) = \omega$, for $\omega \in \mathbf{R}$. It is easy to see that F is the distribution function for the random variable ξ.

Let F be a distribution function, and let $x \in \mathbf{R}$ be a discontinuity point of F. Then $p(x) = F(x) - F(x-0)$ is called the *jump* of F at x. A point is said to be a *point of increase* of F if, for every $\varepsilon > 0, F(x+\varepsilon) - F(x-\varepsilon) > 0$.

1.1.2 Random Variable

Considering the above foundation, a random variable on the probability space is defined as follows.

Definition 1.13 (Feller [52]). Let $(\Omega, \mathscr{A}, Pr)$ be a probability space. A real-valued function ξ defined on Ω is said to be a random variable if

$$\xi^{-1}(B) = \{\omega \in \Omega : \xi(\omega) \in B\} \in \mathscr{A}, \text{ for all } B \in \mathscr{B}, \tag{1.19}$$

where $\mathscr{B}$ is the σ-algebra of Borel sets in $\mathbf{R} = (-\infty, +\infty)$, that is, a random variable ξ is a measurable transformation of $(\Omega, \mathscr{A}, Pr)$ into $(\mathbf{R}, \mathscr{B})$.

It is necessary that $\xi^{-1}(I) \in \mathscr{A}$ for all intervals I in $\mathbf{R}$, or for all semiclosed intervals $I = (a, b]$, or for all intervals $I = (-\infty, b]$, and so on. We also note that the random variable ξ defined on $(\mathbf{R}, \mathscr{B})$ includes a measure Pr_ξ on $\mathscr{B}$ defined by the relation

$$Pr_\xi(B) = Pr\{\xi^{-1}(B)\}, \quad B \in \mathscr{B}. \tag{1.20}$$

Clearly Pr_ξ is a probability measure on $\mathscr{B}$ and is called the *probability distribution* or, simply the *distribution* of ξ.

Definition 1.14 (Krickeberg [77]). A random variable ξ is said to be

(a) nonnegative if $Pr\{\xi < 0\} = 0$;
(b) positive if $Pr\{\xi \le 0\} = 0$;
(c) simple if there exists a finite sequence $\{x_1, x_2, \cdots, x_m\}$ such that

$$Pr\{\xi \neq x_1, \xi \neq x_2, \cdots, \xi \neq x_m\} = 0; \tag{1.21}$$

(d) discrete if there exists a countable sequence $\{x_1, x_2, \cdots\}$ such that

$$Pr\{\xi \neq x_1, \xi \neq x_2, \cdots\} = 0. \tag{1.22}$$

Figure 1.1 illustrates a simple random variable.

Definition 1.15 (Krickeberg [77]). Let ξ_1 and ξ_2 be random variables defined on the probability space $(\Omega, \mathscr{A}, Pr)$. Then $\xi_1 = \xi_2$ if $\xi_1(\omega) = \xi_2(\omega)$ for almost all $\omega \in \Omega$.

Discrete Random Variable

In this part, we consider some certain types of discrete random variables.

Fig. 1.1 A random variables

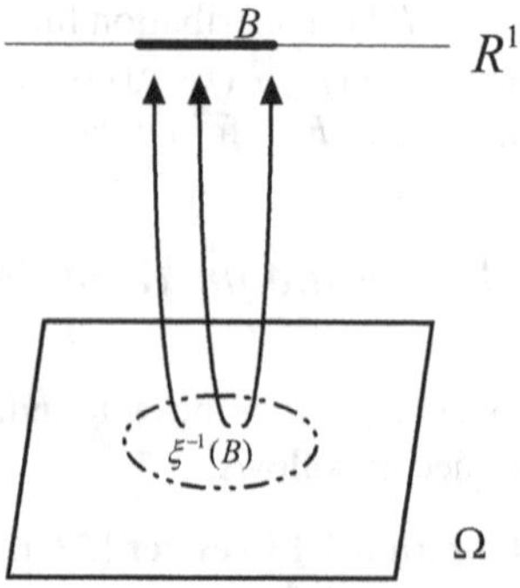

Definition 1.16. Let ξ be a random variable on the probability space $(\Omega, \mathscr{A}, Pr)$. If $\Omega = \{\omega_1, \omega_2, \cdots\}$ is a set combined with finite or infinite discrete elements, where $Pr\{\omega = \omega_i\} = p_i$ and $\sum_{i=1}^{\infty} p_i = 1$, then ξ is called the discrete random variable.

From the above definition, we know that a discrete random variable ξ is a mapping from the discrete probability space Ω to the real space **R**.

Example 1.15. Let $\Omega = \{1, 2, 3, 4\}$ be the probability space and $Pr\{\omega_i = i\} = 0.25, i = 1, \cdots, 4$. If $\xi(\omega_i) = 1/\omega_i$, then ξ is a discrete random variable.

Intuitively, we want to know the distribution of the discrete random variable. The following equation is usually used to describe this distribution,

$$\begin{pmatrix} \xi(\omega_1) & \xi(\omega_2) & \cdots & \xi(\omega_n) \\ p_1 & p_2 & \cdots & p_n \end{pmatrix}$$

In the following part, three special discrete random variables are introduced.

As we know, from the coin tossing trial, the probabilities for the front and the back are both 0.5. Then, we denote ω_1 = 'Front' and ω_1 = 'Back' and let ξ be a mapping from $\{\omega_1, \omega_2\}$ to $\{0, 1\}$ satisfying $\xi(\omega_1) = 1$ and $\xi(\omega_2) = 0$.

Definition 1.17 (Binomial random variable). Suppose that there are n independent trials, each of which results in a "success" with a probability p, to be performed. If ξ represents the number of successes that occur in the n trials, then ξ is said to be a binomial random variable with the parameters (n, p). Its probability mass function is given by

$$Pr\{\xi = i\} = \binom{n}{i} p^i (1-p)^{n-i}, \ i = 0, 1, \cdots, n, \tag{1.23}$$

where

$$\binom{n}{i} = \frac{n!}{i!(n-i)!}$$

is the binomial coefficient, equal to the number of different subsets of i elements that can be chosen from a set of n elements. Obviously, in this example, Ω has n elements combined with the natural number $i = 1, 2, \cdots, n$.

Since we assume that all trials are independent with each other, then the probability of any particular sequence of outcomes results in i successes and $n - i$ failures. Furthermore, it can be seen that Eq. (1.23) is valid since there are $(n, i)^T$ different sequences of the n outcomes that result in i successes and $n - i$ failures, which can be seen by noting that there are $(n, i)^T$ different choices of the i trials that result in successes.

Definition 1.18. A binomial random variable $(1, p)$ is called a Bernoulli random variable.

Since a binomial (n, p) random variable ξ represents the number of successes in n independent trials, each of which results in a success with probability p, we can represent it as follows

$$\xi = \sum_{i=1}^{n} \xi_i, \tag{1.24}$$

where

$$\xi_i = \begin{cases} 1, & \text{if the } i\text{th trial is a success,} \\ 0, & \text{otherwise.} \end{cases}$$

The following recursive formula expressing p_{i+1} in terms of p_i is useful when computing the binomial probabilities:

$$\begin{aligned} p_{i+1} &= \frac{n!}{(n-i-1)!(i+1)!} p^{i+1}(1-p)^{n-i-1} = \frac{n!(n-i)}{(n-i)!i!(i+1)} p^i(1-p)^{n-i} \frac{p}{1-p} \\ &= \frac{n-i}{i+1} \cdot \frac{p}{1-p} p_i. \end{aligned} \tag{1.25}$$

Definition 1.19 (Poisson random variable). A random variable ξ that takes on one of the values $0, 1, 2, \cdots$ is said to be a Poisson random variable with parameter λ, $\lambda > 0$, if its probability mass function is given by

$$p_i = Pr\{\xi = i\} = e^{-\lambda} \frac{\lambda^i}{i!}, \quad i = 1, 2, \cdots \tag{1.26}$$

The symbol e, defined by $e = \lim_{n\to\infty} (1 + 1/n)^n$, is a famous constant in mathematics that is roughly equal to 2.7183.

Poisson random variables have a wide range of applications. One reason for this is that such random variables may be used to approximate the distribution of the number of successes in a large number of trials (which are either independent or at most "weakly dependent") when each trial has a small probability of being a success. To see why this is so, suppose that ξ is a binomial random variable with parameters (n, p), and so represents the number of successes in n independent trials when each trial is a success with probability p, and let $\lambda = np$. Then

$$\begin{aligned} Pr\{\xi = i\} &= \frac{n!}{(n-i)! \cdot i!} p^i (1-p)^{n-i} = \frac{n!}{(n-i)! \cdot i!} \left(\frac{\lambda}{n}\right)^i \left(1 - \frac{\lambda}{n}\right)^{n-i} \\ &= \frac{n(n-1)\cdots(n-i+1)}{n^i} \cdot \frac{\lambda^i}{i!} \cdot \frac{(1-\lambda/n)^n}{(1-\lambda/n)^i}. \end{aligned} \tag{1.27}$$

Now for n large and p small,

$$\lim_{n\to\infty} \left(1 - \frac{\lambda}{n}\right)^n \to e^{-\lambda},$$

$$\lim_{n\to\infty} \frac{n(n-1)\cdots(n-i+1)}{n^i} \to 1,$$

$$\lim_{n\to\infty} \left(1 - \frac{\lambda}{n}\right)^i \to 1.$$

Hence, for n large and p small,

$$Pr\{\xi = i\} \approx e^{-\lambda} \frac{\lambda^i}{i!}.$$

To compute the Poisson probabilities we make use of the following recursive formula:

$$\frac{p_{i+1}}{p_i} = \frac{\dfrac{e^{-\lambda}\lambda^{(i+1)}}{(i+1)!}}{\dfrac{e^{-\lambda}\lambda^i}{i!}} = \frac{\lambda}{i+1},$$

or equivalently,

$$p_{i+1} = \frac{\lambda}{i+1} p_i, \quad i \geq 0.$$

Consider independent trials, each of which is a success with probability p. If ξ represents the number of the first trial that is a success, then

$$Pr\{\xi = n\} = p(1-p)^{n-1}, \tag{1.28}$$

which is easily obtained by noting that in order for the first success to occur on the nth trial, the first $n-1$ must all be failures and the nth a success. Equation (1.28) is said to be a geometric random variable with parameter p.

If we let ξ denote the number of trials needed to amass a total of r successes when each trial is independently a success with probability p, then ξ is said to be a negative binomial, sometimes called a Pascal random variable with parameters p and r. The probability mass function of such a random variable is given as follows,

$$Pr\{\xi = n\} = \binom{n-1}{r-1} p^r (1-p)^{n-r}, \quad n \geq r. \tag{1.29}$$

To see why Eq. (1.29) is valid note that in order for it to take exactly n trials to amass r successes, the first $n-1$ trials must result in exactly $r-1$ successes, and the probability of this is $\binom{n-1}{r-1} p^r(1-p)^{n-r}$, and then the nth trial must be a success, and the probability of this is p.

Consider an urn containing $N+M$ balls, of which N are light colored and M are dark colored. If a sample of size n is randomly chosen (in the sense that each of the $\binom{N+M}{n}$ subsets of size n is equally likely to be chosen) then ξ, the number of light colored balls selected, has probability mass function,

$$Pr\{\xi = i\} = \frac{\binom{N}{i}\binom{M}{n-i}}{\binom{N+M}{n}}.$$

A random variable ξ whose probability mass function is given by the preceding equation is called a hypergeometric random variable.

Continuous Random Variable

In this part, we consider certain types of continuous random variables.

Definition 1.20 (Uniform random variable). A random variable ξ is said to be uniformly distributed over the interval (a, b), $a < b$, if its probability density function is given by

$$f(x) = \begin{cases} \dfrac{1}{b-a}, & \text{if } a < x < b, \\ 0, & \text{otherwise.} \end{cases}$$

In other words, ξ is uniformly distributed over (a, b) if it puts all its mass on that interval and it is equally likely to be "near" any point on that interval.

The distribution function of ξ is given, for $a < x < b$, by

$$F(x) = Pr\{\xi \le x\} = \int_a^x (b-a)^{-1} dx = \frac{x-a}{b-a}.$$

Definition 1.21 (Normal distributed random variable). A random variable ξ is said to be normally distributed with mean μ and variance σ^2 if its probability density function is given by

$$f(x) = \frac{1}{\sqrt{2\pi}\sigma} e^{-(x-\mu)^2/2\sigma^2}, \quad -\infty < x < \infty.$$

The normal density is a bell-shaped curve that is symmetric about μ.

An important fact about normal random variables is that if ξ is normal with mean μ and variance σ^2, then for any constants a and b, $a\xi + b$ is normally distributed with mean $a\mu + b$ and variance $a^2\sigma^2$. It follows from this that if ξ is normal with mean μ and variance σ^2, then

$$\zeta = \frac{\xi - \mu}{\sigma}$$

is normal with mean 0 and variance 1. Such a random variable ζ is said to have a standard (or unit) normal distribution. Let Φ denote the distribution function of a standard normal random variable, that is

$$\Phi(x) = \frac{1}{\sqrt{2\pi}} \int_{-\infty}^x e^{-x^2/2} dx, \quad -\infty < x < \infty.$$

The result that $\zeta = (\xi - \mu)/\sigma$ has a standard normal distribution when ξ is normal with mean μ and variance σ^2 is quite useful because it allows us to evaluate all probabilities concerning ξ in terms of Φ. For example, the distribution function of ξ can be expressed as

$$\begin{aligned}
F(x) &= Pr\{\xi \le x\} \\
&= Pr\left\{\frac{\xi - \mu}{\sigma} \le \frac{x - \mu}{\sigma}\right\} \\
&= Pr\left\{\zeta \le \frac{x - \mu}{\sigma}\right\} \\
&= \Phi\left(\frac{x - \mu}{\sigma}\right).
\end{aligned}$$

The value of $\Phi(x)$ can be determined either by looking it up in a table or by writing a computer program to approximate it. For a in the interval $(0, 1)$, let ζ_a be such that

$$Pr\{\zeta > \zeta_a\} = 1 - \Phi(\zeta_a) = a.$$

That is, a standard normal will exceed ζ_a with probability a. The value of ζ_a can be obtained from a table of the values of Φ. For example, since $\Phi(1.64) = 0.95$, $\Phi(1.96) = 0.975$, $\Phi(2.33) = 0.99$, we see that $\zeta_{0.05} = 1.64$, $\zeta_{0.025} = 1.96$, $\zeta_{0.01} = 2.33$.

The wide applicability of normal random variables results from one of the most important theorems of probability theory - the central limit theorem, which asserts that the sum of a large number of independent random variables has approximately a normal distribution.

Definition 1.22 (Exponential random variable). A continuous random variable having probability density function,

$$f(x) = \begin{cases} \lambda e^{-\lambda x}, & 0 \le x < \infty, \\ 0, & \text{otherwise}, \end{cases}$$

for some $\lambda > 0$ is said to be an exponential random variable with parameter λ.

Its cumulative distribution is given by

$$F(x) = \int_0^x \lambda e^{-\lambda x} dx = 1 - e^{-\lambda x}, \ 0 < x < \infty.$$

The key property of exponential random variables is that they possess the "memoryless property", where we say that the nonnegative random variable ξ is memoryless if

$$Pr\{\xi > s + t | \xi > s\} = Pr\{\xi > t\}, \text{ for all } s, t \ge 0. \tag{1.30}$$

To understand why the above is called the memoryless property, imagine that ξ represents the lifetime of some unit, and consider the probability that a unit of age s will survive an additional time t. Since this will occur if the lifetime of the unit exceeds $t + s$ given that it is still alive at time s, we see that

$$Pr\{\text{additional life of an item of age } s \text{ exceeds } t\} = Pr\{\xi > s + t | \xi > s\}.$$

Thus, Eq. (1.30) is a statement of fact that the distribution of the remaining life of an item of age s does not depend on s. That is, it is not necessary to remember the age of the unit to know its distribution of remaining life. Equation (1.30) is equivalent to

$$Pr\{\xi > s + t\} = Pr\{\xi > s\} Pr\{\xi > t\}.$$

As the above equation is satisfied whenever ξ is an exponential random variable - since, in this case, $Pr\{\xi > x\} = e^{-\lambda x}$ - we see that exponential random variables are memoryless (and indeed it is not difficult to show that they are the only memoryless random variables).

Another useful property of exponential random variables is that they remain exponential when multiplied by a positive constant. To see this suppose that ξ is exponential with parameter λ, and let c be a positive number. Then

$$Pr\{c\xi \le x\} = Pr\left\{\xi \le \frac{x}{c}\right\} = 1 - e^{-\lambda x/c},$$

which shows that $c\xi$ is exponential with parameter λ/c.

Let $\xi_1, \xi_2, \cdots, \xi_n$ be independent exponential random variables with respective rates $\lambda_1, \lambda_2, \cdots, \lambda_n$. A useful result is that $\min\{\xi_1, \xi_2, \cdots, \xi_n\}$ is exponential with rate $\sum_i \lambda_i$ and is independent of which one of the ξ_i is the smallest. To verify this, let $M = \min\{\xi_1, \xi_2, \cdots, \xi_n\}$. Then

$$\begin{aligned} Pr\left\{\xi_j = \min_i \xi_i | M > t\right\} &= Pr\{\xi_j - t = \min_i(\xi_i - t) | M > t\} \\ &= Pr\{\xi_j - t = \min_i(\xi_i - t) | \xi_i > t, i = 1, 2, \cdots, n\} \\ &= Pr\{\xi_j = \min_i \xi_i\}. \end{aligned}$$

The final equality follows because, by the lack of memory property of exponential random variables, given that ξ_i exceeds t, the amount by which it exceeds it is exponential with rate λ_i. Consequently, the conditional distribution of $\xi_1 - t, \cdots, \xi_n - t$ given that all the ξ_i exceed t is the same as the unconditional distribution of $\xi_1, \cdots, \xi_n$. Thus, M is independent of which of the ξ_i is the smallest.

The result that the distribution of M is exponential with rate $\sum_i \lambda_i$ follows from

$$Pr\{M > t\} = Pr\{\xi_i > t, i = 1, 2, \cdots, n\} = \prod_{i=1}^{n} Pr\{\xi_i > t\} = e^{-\sum\limits_{i=1}^{n} \lambda_i t}.$$

The probability that ξ_j is the smallest is obtained from

$$\begin{aligned} Pr\{\xi_j = M\} &= \int Pr\{\xi_j = M | \xi_j = t\} \lambda_j e^{-\lambda_j t} dt \\ &= \int Pr\{\xi_j > t, i \ne j | \xi_j = t\} \lambda_j e^{-\lambda_j t} dt \\ &= \int Pr\{\xi_j > t, i \ne j\} \lambda_j e^{-\lambda_j t} dt \\ &= \int \left(\prod_{i \ne j} e^{-\lambda_i t}\right) e^{-\lambda_j t} dt \\ &= \lambda_j \int e^{-\sum\limits_i \lambda_i t} dt \\ &= \frac{\lambda_j}{\sum\limits_i \lambda_i}. \end{aligned}$$

There are also other special random variables that follow other distributions, and readers can refer to the related research.

Random Vector

An idea extending the real space to n-dimensional real space to define a random vector is now considered.

Definition 1.23 (Krickeberg [77]). An n-dimensional random vector is a measurable function from a probability space $(\Omega, \mathscr{A}, Pr)$ to the set of n-dimensional real vectors.

Since a random vector $\boldsymbol{\xi}$ is a function from Ω to $\mathbf{R}^n$, we can write $\boldsymbol{\xi}(\omega) = (\xi_1(\omega), \xi_2(\omega), \cdots, \xi_n(\omega))$ for every $\omega \in \Omega$, where $\xi_1, \xi_2, \cdots, \xi_n$ are functions from Ω to $\mathbf{R}$. Are $\xi_1, \xi_2, \cdots, \xi_n$ random variables in the same sense as that in Definition 1.13? Conversely, we assume that $\xi_1, \xi_2, \cdots, \xi_n$ are random variables. Is $(\xi_1, \xi_2, \cdots, \xi_n)$ a random vector in the same sense as that in Definition 1.23? The answer is in the affirmative. In fact, we have the following theorem that is used as an important measure theory and set theory.

Theorem 1.15 (Krickeberg [77]). *The vector $(\xi_1, \xi_2, \cdots, \xi_n)$ is a random vector if and only if $\xi_1, \xi_2, \cdots, \xi_n$ are random variables.*

1.1.3 Fuzzy Variable

Fuzzy set theory was developed to solve problems in which the descriptions of the activities and observations are imprecise, vague, and uncertain. The term "fuzzy" refers to a situation in which there are no well-defined boundaries for the set of activities or observations to which the descriptions apply. For example, a person seven feet tall can be easily assigned to the "class of tall men". However, it would be difficult to justify the inclusion or exclusion of a 5.7 feet tall person to that class, because the term "tall" does not constitute a well-defined boundary. This notion of fuzziness exists almost everywhere in our daily lives, such as a "class of red flowers," a "class of good shooters," a "class of comfortable speeds for traveling, the "numbers close to 10," etc. These classes of objects are poorly represented using classical set theory. In classical set theory, an object is either in a set or not in a set, so an object cannot partially belong to a set.

To cope with this difficulty, Zadeh [124] proposed fuzzy set theory in 1965. A fuzzy set is a class of objects with a continuum of membership grades. A membership function, which assigns to each object a membership grade, is associated with each fuzzy set. Usually, the membership grades are in $[0, 1]$. When the membership grade for an object in a set is one, this object is absolutely in that set; when the membership grade is zero, the object is absolutely not in that set. Borderline cases are assigned numbers between zero and one. Precise membership grades do not convey any absolute significance as they are context-dependent and can be subjectively assessed.

Let U be the universe, which is a classical set of objects, and the generic elements are denoted by x. The membership in a crisp subset of A is often viewed as the characteristic function π_A from U to $\{0, 1\}$ such that:

$$\pi_A(x) = \begin{cases} 1, \text{ iff } x \in A, \\ 0, \text{ otherwise}, \end{cases} \tag{1.31}$$

where $\{0, 1\}$ is called a valuation set.

If the valuation set is allowed to be the real interval $[0, 1]$, A is called a fuzzy set. $\mu_A(x)$ is the degree of membership of x in fuzzy set A. The closer the value of $\mu_A(x)$ is to 1, the more x belongs to A. Therefore, A is characterized by a set of ordered pairs:

$$A = \{(x, \mu_A(x))|x \in U\}. \tag{1.32}$$

Sometimes, we might only need the objects of a fuzzy set instead of its characteristic function, that is, to transfer the fuzzy set into a crisp set. In order to do so, we need two concepts, support and α-cut.

It is often necessary to consider those elements in a fuzzy set which have non-zero membership grades. These elements are the support for that fuzzy set.

Definition 1.24 (Zadeh [125]). Given a fuzzy set A, its support $S(A)$ is an ordinary crisp subset on U defined as

$$S(A) = \{x|\mu_A(x) > 0 \text{ and } x \in U\}. \tag{1.33}$$

Definition 1.25 (Zadeh [125]). Given a fuzzy set A, its α-cut A_α defined as

$$A_\alpha = \{x|\mu_A(x) \geq \alpha \text{ and } x \in U\}, \tag{1.34}$$

where, α is the confidence level.

It is obvious that the α-cut of a fuzzy set A is an ordinary crisp subset, the elements of which belong to fuzzy set A - at least to the degree of α. That is, for fuzzy set A, its α-cut is defined as (1.34). The α-cut is a more general case of the support for a fuzzy set, when $\alpha = 0$, $A_\alpha = \text{supp}(A)$.

The term 'fuzzy number' is used to handle imprecise numerical quantities, such as "close to 10," "about 60," "several," etc. A general definition for fuzzy numbers was given by Dubois and Prade [39, 40]: any fuzzy subset $M = \{(x, \mu(x))\}$ where x takes its number on the real line R and $\mu_M(x) \in [0, 1]$.

Definition 1.26 (Dubois and Prade [39]). Let A be a fuzzy set, its membership function is $\mu_A : R \to [0, 1]$, if

(a) A is upper semi-continuous, i.e., α-cut A_α is close set, for $0 < \alpha \leq 1$;
(b) A is normal, i.e., $A_1 \neq \emptyset$;
(c) A is convex, i.e., A_α is a convex subset of R, for $0 < \alpha \leq 1$;

(d) The closed convex hull of A $A_0 = cl[co\{x \in R, \mu_A(x) > 0\}]$ is cored;
then A is a fuzzy number.

By Definition 1.26, the α-cut A_α of the fuzzy number A is actually a close interval of the real number field, that is,

$$A_\alpha = \{x \in R | \mu_A(x) \geq \alpha\} = [A^{\mathrm{L}}_\alpha, A^{\mathrm{R}}_\alpha], \quad \alpha \in [0, 1],$$

where A^{L}_α and A^{R}_α are the left and the right extreme points of the close interval.

Example 1.16. Given fuzzy number A with membership function

$$\mu_{\tilde{A}}(x) = \begin{cases} L(\frac{a-x}{l}), & \text{if } a - l \leq x < a, l > 0, \\ 1 & \text{if } x = a, \\ R(\frac{x-a}{r}), & \text{if } a < x \leq a + r, r > 0, \end{cases}$$

and the basis functions $L(x)$, $R(x)$ are continuous un-increasing functions, and $L, R : [0, 1] \to [0, 1]$, $L(0) = R(0) = 1$, $L(1) = R(1) = 0$, then $\tilde{A}$ is LR fuzzy number, denoted by $\tilde{A} = (a, l, r)_{LR}$, where a is the central value of $\tilde{A}$, $l, r > 0$ is the left and the right spread. α-cut A_α of the LR fuzzy number $\widetilde{A}$ is

$$A_\alpha = [A^L_\alpha, A^R_\alpha] = [a - L^{-1}(\alpha)l, a + R^{-1}(\alpha)r], \alpha \in [0, 1].$$

After that, the concept of fuzzy variable was proposed. Let's introduce the basic knowledge about fuzzy variable, which including the measure, the definition and the properties of fuzzy random variables. We give some basic knowledge about fuzzy variable. Since its introduction in 1965 by Zadeh [124], fuzzy set theory has been well developed and applied in a wide variety of real problems. The term fuzzy variable was first introduced by Kaufmann [71], then it appeared in Zadeh [126, 127] and Nahmias [90]. Possibility theory was proposed by Zadeh [126], and developed by many researchers such as Dubois and Prade [39].

In order to provide an axiomatic theory to describe fuzziness, Nahmias [90] suggested a theoretical framework. Let us give the definition of possibility space (also called pattern space by Nahmias).

Definition 1.27 (Dubois and Prade [39]). Let Θ be a nonempty set, and $P(\Theta)$ be the power set of Θ. For each $A \subseteq P(\Theta)$, there is a nonnegative number $Pos\{A\}$, called its possibility, such that

(a) $Pos\{\emptyset\} = 0$;
(b) $Pos\{\Theta\} = 1$;
(c) $Pos\{\bigcup_k A_k\} = \sup_k Pos\{A_k\}$ for any arbitrary collection $\{A_k\}$ in $P(\Theta)$.

The triplet $(\Theta, P(\Theta), Pos)$ is called a possibility space, and the function Pos is referred to as a possibility measure.

It is easy to obtain the the following properties of *Pos* from the axioms above.

Property 1.1. The properties of *Pos* measure:

(a) $0 \leq Pos\{A\} \leq 1, \forall A \in P(\Theta)$;
(b) $Pos\{A\} \leq Pos\{B\}$, if $A \subseteq B$.

Several researchers have defined fuzzy variable in different ways, such as Kaufman [71], Zadah [126, 127] and Nahmias [90]. In this book we use the following definition of fuzzy variable.

Definition 1.28 (Nahmias [90]). A fuzzy variable is defined as a function from the possibility space $(\Theta, P(\Theta), Pos)$ to the real line $\mathbf{R}$.

Definition 1.29 (Dubois and Prade [39]). Let ξ be a fuzzy variable on the possibility space $(\Theta, P(\Theta), Pos)$. Then its membership function $\mu : \mathbf{R} \mapsto [0, 1]$ is derived from the possibility measure *Pos* by

$$\mu(x) = Pos\{\theta \in \Theta | \xi(\theta) = x\}. \tag{1.35}$$

In order to measure the chances of occurrence of fuzzy events, the possibility and necessity of a fuzzy event is given as follows.

Definition 1.30 (Dubois [42]). Let $\tilde{a}_1, \tilde{a}_2, \cdots, \tilde{a}_n$ be fuzzy variables, and $f : \mathbf{R}^n \to \mathbf{R}$ be continuous functions. Then the possibility of the fuzzy event characterized by $f(\tilde{a}_1, \tilde{a}_2, \cdots, \tilde{a}_n) \leq 0$ is

$$\begin{aligned} & Pos\{f(\tilde{a}_1, \tilde{a}_2, \cdots, \tilde{a}_n) \leq 0\} \\ = & \sup_{x_1, x_2, \cdots, x_n \in R} \left\{ \min_{1 \leq i \leq n} \mu_{\tilde{a}_i}(x_i) | f(x_1, x_2, \cdots, x_n) \leq 0 \right\}. \end{aligned} \tag{1.36}$$

The necessity measure of a set A is defined as the impossibility of the opposite set A^c.

Definition 1.31 (Dubois [42]). Let $(\Theta, P(\Theta), Pos)$ be a possibility space, and A be a set in $P(\Theta)$. Then the necessity measure of A is

$$Nec\{A\} = 1 - Pos\{A^c\}.$$

Thus the necessity measure is the dual of possibility measure, that is, $Pos\{A\} + Nec\{A^c\} = 1$ for any $A \in P(\Theta)$.

In order to measure the mean of a fuzzy variable, several researchers defined an expected value for fuzzy variables with different ways, such as Campos and Verdegay [26], Dubois and Prade [41], González [57] and Yager [119, 120]. Readers could refer to them for details. Next, we will present the definition of random-overlapped random (Ra-Ra) variable.

1.1.4 Ra-Ra Variable

The concept of Ra-Ra variable is initialized by Peng and Liu [93] in 2007. It is an useful tool to deal with some uncertain real-life problems. There do exist many scenes of such birandom phenomena in our real world. As a general mathematical description for this kind of stochastic phenomenon with incomplete statistical information, Ra-Ra variables is defined as a mapping with some kind of measurability from a probability space to a collection of random variables. In this section, we firstly recall some basic definitions and properties of Ra-Ra variables and give some deeper deduction.

Roughly speaking, a Ra-Ra variable is a "random variable" taking random variable values. In other words, a Ra-Ra variable is a mapping from a probability space to a collection of random variables. In this section, we will define the concept with a general way.

Definition 1.32. ξ is said to be a Ra-Ra variable with respect to $\bar{\mu}$, if and only if $\forall \omega \in \Omega$, $\xi(\omega)$ is a random variable with the density function $\bar{d}(x, \bar{\mu}(\omega))$, i.e.,

$$\bar{F}(y) \equiv: Pr\{-\infty \le \xi(\theta) \le y\} = \int_{-\infty}^{y} \bar{d}(x, \bar{\mu}(\omega))dx,$$

where $\bar{\mu}$ is a random variable on the probability space (Ω, A, Pr); and the function $\bar{d}(x, \bar{\mu}(\omega))$ satisfies $\bar{d}(x, \bar{\mu}(\omega)) \ge 0$ and $\int_{-\infty}^{+\infty} \bar{d}(x, \bar{\mu}(\omega))dx = 1$ for $\theta \in \Theta$.

In other word, Ra-Ra variable is a random variable with a random parameter. For each given Borel subset B of the real $\mathbf{R}$, the function $Pr\{\xi(\omega) \in B\}$ is a random variable defined on the probability space $(\Omega, \mathscr{A}, Pr)$.

Example 1.17. Let $\Omega = \{\omega_1, \omega_2, \cdots, \omega_n\}$, and $Pr\{\omega_i\} = p_i$, where $\sum_{i=1}^{n} p_i = 1$. Then $(\Omega, \mathscr{A}, Pr)$ is a probability space on which we define a function as

$$\xi(\omega) = \begin{cases} \xi_1, \text{ if } \omega = \omega_1, \\ \xi_2, \text{ if } \omega = \omega_2, \\ \cdots \\ \xi_n, \text{ if } \omega = \omega_n, \end{cases}$$

where ξ_i is a normally distributed random variable. Then the function ξ is a Ra-Ra variable.

Discrete Ra-Ra Variable

Having recalled these basic definitions and properties, we will respectively define two kinds of Ra-Ra variables, i. e. discrete Ra-Ra variables and continuous variables, and then six special Ra-Ra variables are presented and their basic properties are respectively exhibited. Based on the Definition 1.32 of Ra-Ra variables, we know that they can be divided into two kinds.

Definition 1.33 (*m*-dimensional discrete Ra-Ra variable [117]). Let $\Omega = \{\omega_1, \omega_2, \cdots, \omega_n\}$, and $Pr\{\omega_i\} = p_i, i = 1, 2, \cdots, m$ and $\sum_{i=1}^{m} Pr\{\omega_i\} = 1$, then $(\Omega, \mathscr{A}, Pr)$ is a probability space. A m-dimensional discrete Ra-Ra variable is defined as a function on $(\Omega, \mathscr{A}, Pr)$ as follows,

$$\xi(\omega) = \begin{cases} \xi_1, & \text{if } \omega = \omega_1, \\ \xi_2, & \text{if } \omega = \omega_2, \\ \quad\vdots & \\ \xi_n, & \text{if } \omega = \omega_n, \end{cases}$$

where $\xi_i (i = 1, 2, \cdots, m)$ are random variables defined on the probability space $(\Omega_i, \mathscr{A}_i, Pr_i)$, respectively.

Obviously, ξ is a mapping from a probability space to a set combined with random variables. The set can be combined with discrete random variables or continuous random variables or both of them defined on the probability space $(\Omega_i, \mathscr{A}_i, Pr_i)$, respectively. If the probability space $(\Omega, \mathscr{A}, Pr)$ is infinite dimension, then we get the following definition.

Definition 1.34 (Infinite dimensional discrete Ra-Ra variable [117]). Let $\Omega = \{\omega_1, \omega_2, \cdots\}$, and $Pr\{\omega_i\} = p_i$, and $\sum_{i=1}^{\infty} Pr\{\omega_i\} = 1$, then $(\Omega, \mathscr{A}, Pr)$ is an infinite dimensional probability space. An infinite dimensional discrete Ra-Ra variable is defined as a function on $(\Omega, \mathscr{A}, Pr)$ as follows,

$$\xi(\omega) = \begin{cases} \xi_1, & \text{if } \omega = \omega_1, \\ \xi_2, & \text{if } \omega = \omega_2, \\ \quad\vdots & \end{cases}$$

where ξ_i are random variables defined on the probability space $(\Omega_i, \mathscr{A}_i, Pr_i)$, respectively, $i = 1, 2, \cdots$.

In general, the following formula is used to describe the distribution of discrete Ra-Ra variables,

$$\begin{pmatrix} \xi_1, \xi_2, \cdots, \xi_n, \cdots \\ p_1, p_2, \cdots, p_n, \cdots \end{pmatrix} \tag{1.37}$$

then (1.37) is called *distribution sequence* of Ra-Ra variable ξ.

Example 1.18. Let $\Omega = \{\omega_1, \omega_2\}$, and $Pr\{\omega_1\} = 0.6$, $Pr\{\omega_2\} = 1/2$. Then $(\Omega, \mathscr{A}, Pr)$ is a probability space on which we define a function as

$$\xi(\omega) = \begin{cases} \xi_1, \text{ if } \omega = \omega_1, \\ \xi_2, \text{ if } \omega = \omega_2, \end{cases}$$

where ξ_1 is a uniformly distributed random variable on [0, 1], and ξ_2 is a normally distributed random variable. Then the function ξ is a Ra-Ra variable.

Example 1.19. Let $\Omega = \{\omega_1, \omega_2, \cdots\}$, and $Pr\{\omega_i\} = \frac{\lambda^{i-1}}{(i-1)!}e^{-\lambda}$. Then $(\Omega, \mathscr{A}, Pr)$ is a probability space on which we define a function as

$$\xi(\omega) = \xi_i, \quad \text{if } \omega = \omega_i,$$

where $\xi_i (i = 1, 2, \cdots)$ is a binomially distributed random variable defined on the probability space $(\Omega_i, \mathscr{A}_i, Pr_i)$, respectively. Then the function ξ is an infinite Ra-Ra variable.

Next, we will introduce some special Ra-Ra variables and induce their properties. In many problems, our interest is usually gathered on whether the event occurs in one trial. For example, when we sample to examine products, whether the product is good or bad greatly attracts us. When throwing the coin, we usually pay more attention on that its front side or back side is above. However, in some trials, the probability that the event occurs is not clear and we only know some historical data, then Ra-Ra event is used to describe them.

Definition 1.35 (Ra-Ra 0-1 distribution [117]). Let $\Omega = \{0, 1\}$ and $Pr\{\omega = 1\} = \bar{p}, Pr\{\omega = 0\} = 1 - \bar{p}$, where $\bar{p}$ is a random variable defined on the probability space $(\Omega^*, \mathscr{A}^*, Pr^*)$ such that $0 \leq \bar{p}(\omega^*) \leq 1$ for $\omega^* \in \Omega^*$. Then $(\Omega, \mathscr{A}, Pr)$ is a probability space on which we define a Ra-Ra variable subject to 0-1 distribution as follows

$$\xi(\omega) = \begin{cases} \bar{p}, & \text{if } \omega = 1, \\ \bar{q}, & \text{if } \omega = 0, \end{cases} \tag{1.38}$$

where $\bar{q} = 1 - \bar{p}$ is also a random variable. Then ξ is a Ra-Ra variable subject to 0-1 distribution on the probability space $(\Omega, \mathscr{A}, Pr)$, denoted by $\xi \sim B(0, 1)$.

Obviously, Ra-Ra 0-1 distribution is a special discrete Ra-Ra variables, and its *distribution sequence* is as follows,

$$\begin{pmatrix} 0, & 1 \\ \bar{q}, & \bar{p} \end{pmatrix} \tag{1.39}$$

where $\bar{q} = 1 - \bar{p}$ and $\bar{p}$ is a random variable on the probability space $(\Omega^*, \mathscr{A}^*, Pr^*)$. In (1.39), if $\bar{p}$ is a fixed number, then $Pr\{\omega = 1\}(0 < Pr\{\omega = k\} < 1)$ is also a fixed number and ξ degenerates a random variable subject 0-1 distribution from $(\Omega, \mathscr{A}, Pr)$ to $\mathbf{R}$.

As shown in the probability theory, when we do the Bernoulli experiment n times, how much is the probability that the event A occurs k times? Then the binomial distribution of random variable is brought forward. Similarly, in this section we will introduce the binomial distribution of Ra-Ra variable.

In the Bernoulli trial, if the probability $\bar{p}$ that the event A occurs is stochastic, then the event that Bernoulli experiment done n times is subject to Ra-Ra binomial distribution. Now let's give the definition of Ra-Ra binomial distribution.

Definition 1.36 (Ra-Ra binomial distribution [117]). Let $\Omega = \{0, 1, \cdots, n\}$ and $Pr\{\omega = k\} = \binom{n}{k}\bar{p}^k\bar{q}^{n-k}, k = 0, 1, \cdots, n$, where $\bar{p}$ is a random variable defined on the probability space $(\Omega^*, \mathscr{A}^*, Pr^*)$ such that $0 \leq \bar{p}(\omega^*) \leq 1$ for $\omega^* \in \Omega^*$. Then $(\Omega, \mathscr{A}, Pr)$ is a probability space on which we define a Ra-Ra variable subject to 0-1 distribution as follows

$$\xi(\omega) = \binom{n}{k}\bar{p}^k\bar{q}^{n-k}, \text{ if } \omega = k, \tag{1.40}$$

where $\bar{q} = 1 - \bar{p}$ is also a random variable. Then ξ is a Ra-Ra variable subject to binomial distribution on the probability space $(\Omega, \mathscr{A}, Pr)$, denoted by $\xi \sim B(k, n, \bar{p})$.

Obviously, $\bar{p}$ and $\bar{q}$ are two mappings from the probability space $(\Omega^*, \mathscr{A}^*, Pr^*)$ to $\mathbf{R}$. Then for any $\omega^* \in \Omega^*$, $\bar{p}(\omega^*)$ and $\bar{q}(\omega^*)$ become two certain probability value. It follows that

$$\sum_{i=1}^{n}\binom{n}{k}\bar{p}^k(\omega^*)\bar{q}^{n-k}(\omega^*) = (\bar{p}(\omega^*) + \bar{q}(\omega^*))^n = 1,$$

then Definition 1.36 is well defined. In (1.40), if $\bar{p}$ is a fixed number, then $Pr\{\omega = k\}(0 < Pr\{\omega = k\} < 1)$ is also a fixed number and ξ degenerates a random variable subject binomial distribution from $(\Omega, \mathscr{A}, Pr)$ to $\mathbf{R}$.

Example 1.20. In Definition 1.36, assume that $\bar{p}$ has a uniform distribution in $[0.1, 0.5]$, i.e. $\bar{p} \sim \mathscr{U}(0.1, 0.5)$ and $n = 20$, then we can get the numerical result as shown in Table 1.1.

In order to help readers understand easily, we give the intuitive figure of the data of Table 1.1 and Fig. 1.2.

For many real-life problems, we usually encounter many Bernoulli experiments, but n is great and $E[\bar{p}]$ is small. In this case, it is difficult to deal with it by the Ra-Ra binomial distribution. Then Ra-Ra poisson distribution is proposed to deal with some real problems.

Theorem 1.16. *On the probability space $(\Omega, \mathscr{A}, Pr)$ there is a Ra-Ra variable ξ such that $\xi \sim B(n, k, \bar{p})$, where $\bar{p}$ is a random variable defined on the probability space $(\Omega^*, \mathscr{A}^*, Pr^*)$. Assume that $\bar{\lambda}$ is also a random variable on the probability space $(\Omega^*, \mathscr{A}^*, Pr^*)$ such that $\bar{\lambda}(\omega^*) > 0$ for $\omega^* \in \Omega^*$ and satisfies,*

$$\lim_{n\to\infty} n\bar{p}_n \to^d \bar{\lambda}, \tag{1.41}$$

Table 1.1 The numerical result of $\xi \sim B(k, 20, \bar{\lambda})$

	$b(k,20,\bar{p})$				$b(k,20,\bar{p})$		
k	$\bar{p}(\omega_1)=0.1$	$\bar{p}(\omega_2)=0.3$	$\bar{p}(\omega_3)=0.5$	k	$\bar{p}(\omega_1)=0.1$	$\bar{p}(\omega_2)=0.3$	$\bar{p}(\omega_3)=0.5$
0	0.1216	0.0008	–	11	–	0.0120	0.1602
1	0.2702	0.0068	–	12	–	0.0039	0.1201
2	0.2852	0.278	0.0002	13	–	0.0010	0.0739
3	0.1901	0.0716	0.0011	14	–	0.0002	0.0370
4	0.898	0.1304	0.0046	15	–	–	0.0148
5	0.0319	0.1789	0.0148	16	–	–	0.0046
6	0.0089	0.1916	0.0370	17	–	–	0.0011
7	0.0020	0.1643	0.0739	18	–	–	0.0002
8	0.0004	0.1144	0.1201	19	–	–	–
9	0.0001	0.0654	0.1602	20	–	–	–
10	–	0.0308	0.1762				

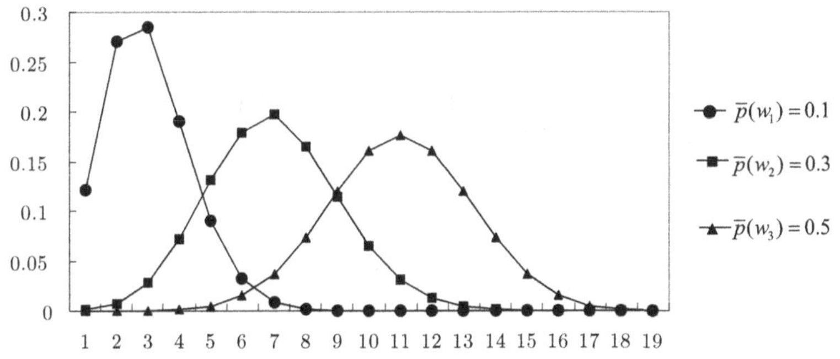

Fig. 1.2 Ra-Ra binomial distribution

then for any nonnegative integer number k,

$$\lim_{n\to\infty} \binom{n}{k} \bar{p}_n^k \bar{q}_n^{n-k} \to^d e^{-\bar{\lambda}} \frac{\bar{\lambda}^k}{k!}. \tag{1.42}$$

Proof. Since $\bar{p}_n$ and $\bar{\lambda}$ are both random variables on $(\Omega^*, \mathscr{A}^*, Pr^*)$, and satisfy that

$$\lim_{n\to\infty} n\bar{p}_n \to^d \bar{\lambda}.$$

This means that $n\bar{p}$ is convergent in distribution, and for any ω^* the following formula holds,

$$\lim_{n\to\infty} n\bar{p}_n(\omega^*) = \bar{\lambda}(\omega^*).$$

Denote $\bar{\lambda}_n = n\bar{p}_n$. It follows that, for any $\omega^* \in \Omega^*$,

$$
\begin{aligned}
B(k,n,\bar{p}_n(\omega^*)) &= \binom{n}{k}\bar{p}_n^k(\omega^*)\bar{q}_n^{n-k}(\omega^*) \\
&= \frac{(n)(n-1)\cdots(n-k+1)}{k!}\left(\frac{\lambda(\omega^*)}{n}\right)^k\left(1-\frac{\lambda(\omega^*)}{n}\right)^{n-k} \\
&= \frac{\lambda^k(\omega^*)}{k!}\left(1-\frac{1}{n}\right)\left(1-\frac{2}{n}\right)\cdots\left(1-\frac{k-1}{n}\right)\left(1-\frac{\lambda(\omega^*)}{n}\right)^{n-k}.
\end{aligned}
$$

For the fixed k, we have

$$
\lim_{n\to\infty}\lambda_n^k(\omega^*) = \lambda^k(\omega^*),\ \lim_{n\to\infty}\left(1-\frac{k-1}{n}\right)\left(1-\frac{\lambda(\omega^*)}{n}\right)^{n-k} = e^{-\lambda(\omega^*)}
$$

and

$$
\lim_{n\to\infty}\left(1-\frac{1}{n}\right)\left(1-\frac{2}{n}\right)\cdots\left(1-\frac{k-1}{n}\right) = 1,
$$

thus,

$$
\lim_{n\to\infty} B(k,n,\bar{p}_n(\omega^*)) = \frac{\lambda^k(\omega^*)}{k!}e^{-\lambda(\omega^*)}. \tag{1.43}
$$

Because (1.43) holds for any $\omega^* \in \Omega^*$, $B(k,n,\bar{p}_n)$ is convergent to $\frac{\lambda^k}{k!}e^{-\lambda}$ in distribution, i.e.,

$$
\lim_{n\to\infty}\binom{n}{k}\bar{p}_n^k\bar{q}_n^{n-k} \to^d e^{-\bar{\lambda}}\frac{\bar{\lambda}^k}{k!}.
$$

The theorem is proved. □

Obviously, $P(k,\bar{\lambda}(\omega^*)) > 0$, $\sum_{k=0}^{\infty} P(k,\bar{\lambda}(\omega^*)) = e^{-\bar{\lambda}(\omega^*)}$, $\sum_{k=0}^{\infty}\frac{\bar{\lambda}^k(\omega^*)}{k!} = 1$.

Definition 1.37 (Ra-Ra poisson distribution [117]). Let $P(k,\bar{\lambda}) = e^{-\bar{\lambda}}\frac{\bar{\lambda}^k}{k!}$, $k = 0,1,2,\cdots$, where $\bar{\lambda}$ is a random variable on $(\Omega^*,\mathscr{A}^*,Pr^*)$. Let $Pr\{\xi = \omega_k\} = P(k,\bar{\lambda})$ where $\omega_k \in \{\omega_0,\omega_1,\omega_2,\cdots\}$. Then ξ is called the Ra-Ra variable subject to poisson distribution on the probability space $(\Omega,\mathscr{A},Pr)$, denoted by $\xi \sim P(k,\bar{\lambda})$.

Example 1.21. In the above definition, let $\bar{\lambda} \sim \mathscr{N}(1,4)$, then $\xi \sim B(k,3.5\bar{\lambda})$ is a Ra-Ra variable with normally distributed parameter.

As we know, the random variable with a poisson distribution is widely applied in many fields, such as how many times an accident occurs in a period, or how many

Table 1.2 The numerical result of $\xi \sim B(k, 4\bar{\lambda})$

$\bar{\lambda} \sim exp(4), \xi \sim P(k, 4\bar{\lambda}), e = 2.718$							
k	$P(k,4)$	$P(k,3)$	$P(k,2)$	$P(k,1)$	$P(k,0.8)$	$P(k,0.5)$	$P(k,0.2)$
0	0.018	0.049	0.135	0.368	0.449	0.607	0.819
1	0.073	0.015	0.271	0.368	0.359	0.304	0.164
2	0.147	0.224	0.271	0.184	0.144	0.076	0.016
3	0.195	0.224	0.180	0.061	0.038	0.013	0.001
4	0.195	0.168	0.090	0.113	0.008	0.002	–
5	0.156	0.101	0.036	0.003	0.001	–	–

radioactive materials produce particles and so on. However, historical data may have uncertainty; thus, the Ra-Ra variable is useful to deal with this data and help assist decision makers. Assume $\lambda \sim exp(4)$, Table 1.2 shows some results for the Ra-Ra variable $\xi \sim B(k, 4\bar{\lambda})$.

Continuous Ra-Ra Variable

There is another kind of Ra-Ra variables other than discrete variables, that is, continuous Ra-Ra variables. They are also mappings from the probability space $(\Omega, \mathscr{A}, Pr)$ to the set combined with random variables on one or many probability spaces, but it is different from the set Ω when ξ is a discrete Ra-Ra variable. Ω is a continuous set combined with all intervals of the form $(-\infty, a]$, $(a, b]$, (b, ∞) and $\mathbf{R}^n$. Then let's give the definition of continuous variables.

Definition 1.38. Let Ω is a continuous set combined with the interval of the form $(-\infty, a]$, $(a, b]$, (b, ∞) and $\mathbf{R}^n$. $Pr\{-\infty < \xi \leq x\} = \bar{F}(x)$, where $\bar{F}(x)$ is a continuous random function with the random density function $\bar{p}(x)$, i.e.,

$$\bar{F}(x) = \int_{-\infty}^{x} \bar{p}(y)dy, \tag{1.44}$$

where $\bar{p}(y)$ is a random variable on another probability space $(\Omega^*, \mathscr{A}^*, Pr^*)$. Then we call ξ a continuous random variable.

By the property of the distribution function, we have that, for any $\omega^* \in \Omega^*$,

$$\bar{p}(x)(\omega^*) \geq 0 \tag{1.45}$$

and

$$\int_{-\infty}^{+\infty} \bar{p}(x)(\omega^*)dx = 1. \tag{1.46}$$

Conversely, for any random function $\bar{p}(x)$ on $(\Omega^*, \mathscr{A}^*, Pr^*)$, they all satisfy (1.45) and (1.46), then the random function $\bar{F}(x)$ defined by (1.44) could be a distribution function when $\omega^* \in \Omega^*$.

Similarly, for any fixed ω, by Eq. (1.12) random function $\bar{F}(x)$ satisfies that for $a, b \in \mathbf{R}$,

$$Pr\{a \le \xi(\omega) \le b\} = \bar{F}^b(\omega^*) - \bar{F}^a(\omega^*) = \int_a^b \bar{p}(\omega^*)(x)dx. \tag{1.47}$$

Similarly with random variables, for any fixed ω, we have

$$Pr\{\xi(\omega) = c\} = 0. \tag{1.48}$$

Example 1.22. Assume that $\bar{a}$ and $\bar{b}$ are two random variables defined on $(\Omega^*, \mathscr{A}^*, Pr^*)$ and for any $\omega^* \in \Omega^*$, $\bar{b}(\omega^*) \ge \bar{a}(\omega^*)$ holds. Define the following density function,

$$\bar{p}(\omega^*)(x) = \begin{cases} \frac{1}{\bar{b}(\omega^*) - \bar{a}(\omega^*)}, & \text{if } \bar{a}(\omega^*) \le x \le \bar{b}(\omega^*), \\ 0, & \text{others}. \end{cases} \tag{1.49}$$

Then ξ is a continuous Ra-Ra variable.

Example 1.23. Let ξ be a Ra-Ra variable defined on the probability space $(\Omega, \mathscr{A}, Pr)$ satisfying $\xi \sim \mathscr{N}(\bar{\mu}, \sigma^2)$, where $\bar{\mu}$ is also a normally distributed random variable on $(\Omega^*, \mathscr{A}^*, Pr^*)$ with the mean μ and variance σ^{*2}. Then ξ is a continuous Ra-Ra variable.

Next, let's discuss some special continuous Ra-Ra variables and their properties. In Examples 1.22 and 1.23, we have simply introduced the uniformly distributed Ra-Ra variables. In this section, we will give its detailed definition.

Definition 1.39 (Ra-Ra uniform distribution [117]). Let ξ be a Ra-Ra variable on the probability space $(\Omega, \mathscr{A}, Pr)$, where $Pr = \bar{F}(x)$ with the following density function,

$$\bar{p}(x) = \begin{cases} \frac{1}{\bar{b} - \bar{a}}, & \text{if } \bar{a}(\omega^*) \le x \le \bar{b}(\omega^*), \\ 0, & \text{others}, \end{cases} \tag{1.50}$$

where $\bar{a}$ and $\bar{b}$ are both random variables on $(\Omega^*, \mathscr{A}^*, Pr^*)$ such that $\bar{b}(\omega^*) > \bar{a}(\omega^*)$ for any $\omega^* \in \Omega^*$. Then ξ is a uniformly distributed Ra-Ra variable, denoted by $\xi \sim \mathscr{U}(\bar{a}, \bar{b})$.

As we know, a Ra-Ra variable is a mapping from a probability space to a set combined with random variables or random functions, so does the uniform Ra-Ra variable. From Definition 1.39, we know that $\xi(\omega) = \bar{F}(x)(\omega \in \Omega)$ is a random function. It holds that

$$Pr\{-\infty < \xi(\omega) < +\infty\} = \int_{-\infty}^{+\infty} \bar{p}(\omega^*)(x)dx$$
$$= \int_{\bar{a}(\omega^*)}^{\bar{b}(\omega^*)} \frac{1}{\bar{b}(\omega^*) - \bar{a}(\omega^*)} dx = 1. \quad (1.51)$$

Example 1.24. Let $\Omega = \Omega^* = \mathbf{R}$, $\bar{a} \sim \mathscr{U}(1,4)$ and $\bar{b} \sim \mathscr{U}(5,6)$. For $\omega^* \in \Omega^*$, we have

$$F^b(\omega^*) = \begin{cases} 0, & \text{if } \omega^* \leq 1, \\ \frac{\omega^*-1}{3}, & \text{if } 1 < \omega^* \leq 4, \\ 1, & \text{if } \omega^* > 4, \end{cases}$$

and

$$F^a(\omega^*) = \begin{cases} 0, & \text{if } \omega^* \leq 5, \\ \omega^* - 5, & \text{if } 5 < \omega^* \leq 6, \\ 1, & \text{if } \omega^* > 6. \end{cases}$$

Obviously, for any $\omega^* \in \Omega^*$, $\bar{b}(\omega^*) \geq \bar{a}(\omega^*)$ holds. Define the density function as Definition 1.39, then ξ is a uniformly distributed Ra-Ra variable.

Example 1.25. Let $\bar{\xi}$ be a random variable on $(\Omega^*, \mathscr{A}^*, Pr^*)$, then $\xi \sim \mathscr{U}(\bar{\xi} + 2, \bar{\xi} + 4)$ is a uniformly distributed Ra-Ra variable.

Similarly, we can also define the normally distributed Ra-Ra variable.

Definition 1.40 (Ra-Ra normal distribution [117]). If ξ is a Ra-Ra variable on $(\Omega, \mathscr{A}, Pr)$, where $Pr = \bar{F}(x)$ with the following density function,

$$\bar{p}(x) = \frac{1}{\sqrt{2\pi}\bar{\sigma}} e^{-\frac{(x-\bar{\mu})^2}{2\bar{\sigma}^2}}, -\infty < x < +\infty, \quad (1.52)$$

where $\bar{\mu}$ or $\bar{\sigma}$ or both of them are random variables on $(\Omega^*, \mathscr{A}^*, Pr^*)$. Then ξ is a normally distributed Ra-Ra variable, denoted by $\xi \sim \mathscr{N}(\bar{\mu}, \bar{\sigma}^2)$.

This definition is well defined and still satisfies the condition of density function and distribution function. Take $\xi \sim \mathscr{N}(\bar{\mu}, \sigma^2)$ as an example. For any $\omega^* \in \Omega^*$, we have

$$\bar{p}(\omega^*)(x) = \frac{1}{\sqrt{2\pi}\sigma} e^{-\frac{(x-\bar{\mu}(\omega^*))^2}{2\sigma^2}} \geq 0; \quad (1.53)$$

and

$$\int_{-\infty}^{+\infty} \bar{p}(\omega^*)(x)dx = \frac{1}{\sqrt{2\pi}\sigma} \int_{-\infty}^{+\infty} e^{-\frac{(x-\bar{\mu}(\omega^*))^2}{2\sigma^2}} dx = 1. \quad (1.54)$$

Its distribution function can be obtained as follows,

$$\bar{F}(x) = \int_{-\infty}^{x} \frac{1}{\sqrt{2\pi}\bar{\sigma}} e^{-\frac{(y-\bar{\mu})^2}{2\bar{\sigma}^2}} dy. \tag{1.55}$$

Obviously, $\bar{F}(x)$ is a random function. If $\bar{\mu}$ and $\bar{\sigma}$ degenerates to be a certain number, ξ degenerates to be a random variable following the normal distribution.

The Ra-Ra exponential distribution can be defined as follows.

Definition 1.41 (Ra-Ra exponential distribution [117]). If ξ is a Ra-Ra variable on $(\Omega, \mathscr{A}, Pr)$, where $Pr = \bar{F}(x)$ with the following density function,

$$\bar{p}(x) = \begin{cases} \bar{\lambda} e^{-\bar{\lambda} x}, & x \geq 0, \\ 0, & x < 0, \end{cases} \tag{1.56}$$

where $\bar{\lambda}$ is a random variable on $(\Omega^*, \mathscr{A}^*, Pr^*)$ and $\bar{\lambda}(\omega^*) > 0$ for any $\omega^* \in \Omega^*$. Then ξ is an exponentially distributed Ra-Ra variable, denoted by $\xi \sim exp(\bar{\lambda})$.

For any $\omega^* \in \Omega^*$, $\bar{\lambda}(\omega^*)$ is a fixed number, then we have

$$\bar{p}(\omega^*)(x) \geq 0$$

and

$$\int_{-\infty}^{+\infty} \bar{p}(\omega^*)(x) dx = \int_{0}^{+\infty} \bar{\lambda}(\omega^*) e^{-\bar{\lambda}(\omega^*)x} = 1.$$

Its distribution function can be got as follows,

$$\bar{F}(x) = \begin{cases} 1 - e^{-\bar{\lambda} x}, & x \geq 0, \\ 0, & x < 0. \end{cases} \tag{1.57}$$

It's obvious that $\bar{F}(x)$ is a random function about the random parameter $\bar{\lambda}$. If $\bar{\lambda}$ degenerates to be a certain number, ξ degenerates to be a random variable.

Example 1.26. Let $\bar{\lambda} \sim \mathscr{U}(2, 4)$ is a random variable on $(\Omega^*, \mathscr{A}^*, Pr^*)$. Then $\xi \sim exp(\bar{\lambda})$ is an exponentially distributed Ra-Ra variable.

Example 1.27. Let $\bar{\lambda}$ is a random variable on $(\Omega^*, \mathscr{A}^*, Pr^*)$ with the following distribution,

$$\bar{\lambda} \sim \begin{pmatrix} 4 & 5 \\ 0.4 & 0.6 \end{pmatrix} \tag{1.58}$$

Table 1.3 The numerical result of $\xi \sim exp(\bar{\lambda})$

$\xi \sim exp(\bar{\lambda})$					
x	$\bar{\lambda} = 4$	$\bar{\lambda} = 5$	x	$\bar{\lambda} = 4$	$\bar{\lambda} = 5$
0	0	0	0.8	0.9592	0.9817
0.1	0.3297	0.3934	1.0	0.9817	0.9933
0.2	0.5506	0.6321	1.5	0.9975	0.9994
0.3	0.6988	0.7768	2.0	0.9997	0.9999
0.4	0.7981	0.8646	2.5	0.9999	0.9999
0.5	0.8646	0.9179	3.0	0.9999	0.9999
0.6	0.9093	0.9502	3.5	0.9999	0.9999
0.7	0.9392	0.9698	4.0	0.9999	0.9999

Then $\xi \sim exp(\bar{\lambda})$ is an exponentially distributed Ra-Ra variable. Its numerical result is listed in Table 1.3.

Because of the existence of the random parameters, it is usually difficult to find a precise decision for a complicated real-life problem. Hence, an efficient tool is needed to convert a random parameter into a crisp parameter. There are three available operators that can handle the objectives and constraints in the upper and lower level of a bi-level decision making problems.

1.1.5 Ra-Fu Variable

Since the fuzzy set was initialized by Zadeh [124], it has been applied to many fields. Later, scholars proposed the concept of two-fold uncertain variables combined with a fuzzy variable and a random variable. Two definitions are available which approach the problem from different viewpoints. The first comes from Kwakernaak [80], who coined the term "fuzzy random variable", and who regarded a fuzzy random variable as a random variable with a value that is not real, but fuzzy. Because Kruse and Meyer [78] worked on an expanded version of a similar model, they are often mentioned along with Kwakernaak. The second viewpoint comes from Puri and Ralescu [95] who regarded fuzzy random variables as random fuzzy sets. Once again, because of Klement et al. [72], and other collaborations, these three authors are often jointly credited with the second definition. In this chapter, we mainly take Kwakernaak's view of the random fuzzy variable or fuzzy-overlapped random variable (Abbr. Ra-Fu variable), which was renamed by some scholars [24, 86, 91, 115] to avoid confusion. This idea has been widely extended to many fields. Xu and Liu [115] discussed a class of supply chain network optimal problems with random fuzzy shipping cost and customer-demand, and proposed a random fuzzy multi-objective mixed-integer non-linear programming model to determine the optimal strategy. Xu and He [114] provided an auxiliary programming model for the random fuzzy programming, converted it to a deterministic mixed 0-1 integer programming model, for which the solutions were proved to exist, and showed its efficiency through a supply chain problem application. Zhou and Xu [130] discussed a class of integrated logistics network models under a random fuzzy environment and applied these to a Chinese beer company. Ferrero et al. [51] studied conditional random-fuzzy variables representing measurement results. Xu et al. [118] developed a new model for a 72-h post-earthquake emergency logistics location-routing problem under a random fuzzy environment.

Consider the following question. A client's retirement savings plan has approximately 100 participants. Several of their portfolios have excessive risk. What is the probability that a participant chosen at random is maintaining a portfolio that has too much risk? If the descriptive variables, namely "approximately", "several" and "too much", were crisp numbers, the answer to the question would be a numerical probability. However, since these terms are fuzzy, rather than crisp values, the solution, like the data upon which it is based, is a fuzzy number. Situations of this sort, which involve a function from a possibility space to a set of random variables, give rise to the notion of a Ra-Fu variable. In this section, we mainly referred to these literatures [24, 72, 78, 80, 81, 95].

Definition 1.42. ξ is said to be a Ra-Fu variable with respect to $\tilde{\rho}$, if and only if $\forall\theta \in \Theta$, $\xi(\theta)$ is a random variable with the density function $\bar{d}(x, \tilde{\rho}(\theta))$, i.e.,

$$\bar{F}(y) \equiv: Pr\{-\infty \leq \xi(\theta) \leq y\} = \int_{-\infty}^{y} \bar{d}(x, \tilde{\rho}(\theta))dx,$$

where $\tilde{\rho}$ is a fuzzy variable on the possibility space $(\Theta, P(\Theta), Pos)$; and the function $\bar{d}(x, \tilde{\rho}(\theta))$ satisfies $\bar{d}(x, \tilde{\rho}(\theta)) \geq 0$ and $\int_{-\infty}^{+\infty} \bar{d}(x, \tilde{\rho}(\theta))dx = 1$ for $\theta \in \Theta$.

In other word, a Ra-Fu variable is a random variable with a fuzzy parameter. For example, if $\xi \sim \mathcal{N}(\mu, \sigma^2)$ is a normally distributed random variable, where μ is a fuzzy variable, then ξ is a Ra-Fu variable as shown in Fig. 1.3. Hence, it is available to take independence, distribution, expected value, and variance into account.

Example 1.28. Let $\eta_1, \eta_2, \cdots, \eta_m$ be random variables and $u_1, u_2, \cdots, u_m$ be real numbers in [0,1]. Then

$$\xi = \begin{cases} \eta_1, & \text{with possibility } u_1, \\ \eta_2, & \text{with possibility } u_2, \\ \cdots \\ \eta_m, & \text{with possibility } u_m, \end{cases}$$

is clearly a Ra-Fu variable. Is it a function from a possibility space $(\Theta, \mathscr{P}(\Theta), Pos)$ to a collection of random variables $\mathscr{R}$? Yes. For example, we define $\Theta = \{1, 2, \cdots, m\}$, $Pos\{i\} = u_i, i = 1, 2, \cdots, m$, $\mathscr{R} = \{\eta_1, \eta_2, \cdots, \eta_m\}$, and the function is $\xi(i) = \eta_i, i = 1, 2, \cdots, m$.

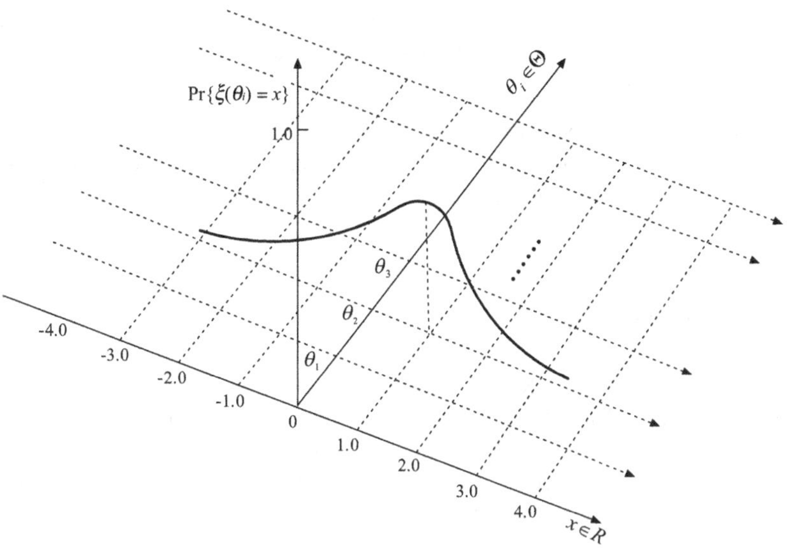

Fig. 1.3 Representation of a Ra-Fu variable

Example 1.29. If η is a random variable, and $\tilde{a}$ is a fuzzy variable defined on the possibility space $(\Theta, \mathscr{P}(\Theta), Pos)$, then $\xi = \eta + \tilde{a}$ is a fuzzy random variable. In fact, ξ is also a Ra-Fu variable, defined by

$$\xi(\theta) = \eta + \tilde{a}(\theta), \forall \theta \in \Theta.$$

Example 1.30. In many statistics problems, the probability distribution is completely known except for the values of one or more parameters. For example, it might be known that the lifetime ξ of a modern engine is an exponentially distributed variable with an unknown mean θ,

$$\phi(x) = \begin{cases} \frac{1}{\theta}e^{-x/\theta}, & \text{if } 0 \le x < \infty, \\ 0, & \text{otherwise.} \end{cases}$$

Usually, there is some relevant information in practice. It is thus possible to specify an interval in which the value of θ is likely to lie, or to give an approximate estimate of the value of θ, but it is typically not possible to determine the value of θ exactly. If the value of θ is provided as a fuzzy variable defined on the possibility space $(\Theta, \mathscr{P}(\Theta), Pos)$, then ξ is a Ra-Fu variable defined as

$$\xi(\theta) \sim exp(\theta), \theta \in \Theta.$$

Example 1.31. Let $\xi \sim \mathcal{N}(\rho, 1)$, where ρ is a fuzzy variable with the membership function $\mu_\rho(x) = [1 - |x - 2|] \vee 0$. Then ξ is a Ra-Fu variable which takes the "normally distributed variable $\mathcal{N}(\rho, 1)$" values.

Remark 1.8. Roughly speaking, if Θ consists of a single element, then the Ra-Fu variable degenerates to a random variable. If $\mathcal{R}$ is a collection of real numbers (rather than random variables), then the Ra-Fu variable degenerates to a fuzzy variable.

Discrete Ra-Fu Variable

From the definition for the Ra-Fu variable, we know that Ra-Fu variables can be divided into discrete Ra-Fu variables, and continuous Ra-Fu variables. There is also a class of special Ra-Fu variables, which are functions from the possibility space $(\Theta, \mathcal{P}(\Theta), Pos)$ to a collection of discrete random variables or a discrete random variable. Next, let's discuss the detail.

Definition 1.43 (Discrete Ra-Fu variable [117]). Let ξ be a Ra-Fu variable on the possibility space $(\Theta, \mathcal{P}(\Theta), Pos)$. If $\xi(\theta)$ is a discrete random variable for any $\theta \in \Theta$, then ξ is said to be a discrete Ra-Fu variable.

Example 1.32. Let ξ be a 0-1 distributed random variable with the probability p of success. Now we assume that p is not exactly known and is to be estimated from a fuzzy space Θ. We substitute a fuzzy number $\tilde{p}$ for p, then ξ is obviously a Ra-Fu variable.

Example 1.33. Let ξ be a Ra-Fu variable on $(\Theta, \mathcal{P}(\Theta), Pos)$, where $\Theta = \{\theta_1, \theta_2\}$, $Pos\{\theta = \theta_1\} = Pos\{\theta = \theta_2\} = 0.5$, and ξ is defined as follows

$$\xi(\theta) = \begin{cases} \eta_1, & \text{with the possibility } 0.5, \\ \eta_2, & \text{with the possibility } 0.5, \end{cases}$$

where η_1 is a binomially distributed random variable and η_2 is a Poisson distributed random variable. Obviously, ξ is a discrete Ra-Fu variable on $(\Theta, \mathcal{P}(\Theta), Pos)$.

In the following part, we discuss some special Ra-Fu variables.

The crisp binomial probability function, usually written $b(n, p)$ where n is the number of independent experiments and p is the probability of a "success" in each experiment, has one parameter p. In these experiments assume that p is not precisely known and needs to be estimated, or obtained from expert opinion. Therefore, the p value is uncertain and we substitute a fuzzy number $\tilde{p}$ for p to get the Ra-Fu binomial distribution.

Definition 1.44 (Ra-Fu binomial distribution). Let ξ be a discrete Ra-Fu variable on $(\Theta, \mathcal{P}(\Theta), Pos)$, then $\xi(\theta)$ is a random variable for $\theta \in \Theta$. Assume that $\xi(\theta)$ has a binomial distribution with the following probability,

$$Pr\{\xi(\theta) = k\} = \binom{n}{k} \tilde{p}^k \tilde{q}^{n-k}, \; k = 0, 1, \cdots, n, \tag{1.59}$$

where $\tilde{p}$ is a fuzzy variable from $(\Theta, \mathscr{P}(\Theta), Pos)$ to (0,1) and $\tilde{q} = 1 - \tilde{p}$. Then ξ is said to be a binomially distributed Ra-Fu variable, denoted by $b(n, \tilde{p})$.

Since $\tilde{p}$ is a fuzzy variable, then $Pr\{\xi(\theta) = k\}$ is a function of the fuzzy parameter $\tilde{p}$. Through the fuzzy arithmetic, we know $\binom{n}{k} \tilde{p}^k \tilde{q}^{n-k}$ is a fuzzy variable defined on the product space. Obviously, $\tilde{p}$ and $\tilde{q}$ are two mappings from the possibility space $(\Theta, \mathscr{P}(\Theta), Pos)$ to $(0, 1)$. Then for any $\theta^* \in \theta^*$, $\tilde{p}(\theta^*)$ and $\tilde{q}(\theta^*)$ become two certain probability values. It follows that

$$\sum_{i=1}^{n} \binom{n}{k} \tilde{p}^k(\theta^*) \tilde{q}^{n-k}(\theta^*) = (\tilde{p}(\theta^*) + \tilde{q}(\theta^*))^n = 1,$$

then Definition 1.44 is well defined. In Eq. (1.59), if $\tilde{p}$ is a fixed number, then $Pr\{\xi(\theta) = k\}(0 < Pr\{\xi(\theta) = k\} < 1)$ is also a fixed number and ξ degenerates to a random variable subject to binomial distribution from $(\Omega, \mathscr{A}, Pr)$ to $\mathbf{R}$.

Example 1.34. Let $\xi \sim b(n, \tilde{p})$ be a Ra-Fu variable, where $\tilde{p}$ is a fuzzy variable on $(\Theta, \mathscr{P}(\Theta), Pos)$ with the following membership function (Fig. 1.4),

$$\mu_{\tilde{p}}(x) = \begin{cases} 0, & x \leq 1, \\ \left(1 + \frac{100}{(x-1)^2}\right)^{-1}, & x > 1, \end{cases} \tag{1.60}$$

where $\Theta = (-\infty, \infty)$. Then we get the distribution for ξ, see Table 1.4.

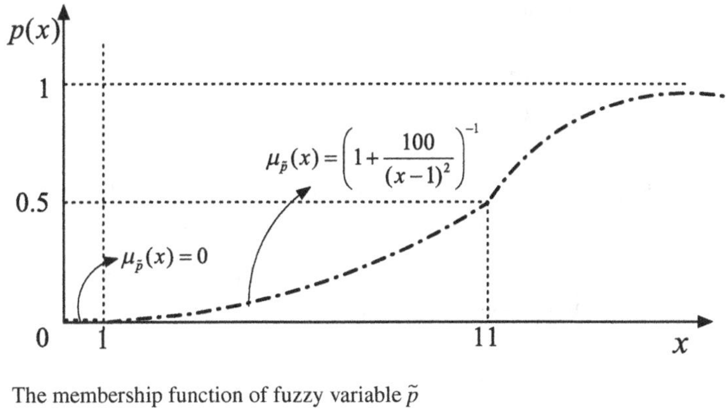

Fig. 1.4 The membership function of fuzzy variable $\tilde{p}$

Table 1.4 The numerical result of $\xi \sim b(20, \tilde{p})$

	$b(20,\tilde{p})$				$b(20,\tilde{p})$		
k	$\theta_1 \leq 1$	$\theta_2 = 11$	$\theta_3 = 70$	k	$\theta_1 \leq 1$	$\theta_2 = 11$	$\theta_3 = 70$
0	–	–		11	–	0.1602	
1	–	–		12	–	0.1201	
2	–	0.0002		13	–	0.0739	
3	–	0.0011		14	–	0.0370	
4	–	0.0046		15	–	0.0148	
5	–	0.0148		16	–	0.0046	
6	–	0.0370		17	–	0.0011	
7	–	0.0739		18	–	0.0002	
8	–	0.1201		19	–	–	–
9	–	0.1602		20	–	–	–
10	–	0.1762					

For a random Poisson, we know that when n is a large number and p is a small number, the random binomial distribution becomes a Poisson distribution, so the Ra-Fu variable is also a Poisson distribution.

Definition 1.45 (Ra-Fu Poisson distribution). Let ξ be a random variable which has a Poisson probability function. If $Pr\{\xi = k\}$ stands for the probability that $\xi = k$, then

$$Pr\{\xi = k\} = \frac{\lambda^k e^{-\lambda}}{k!},$$

for $k = 0, 1, 2, 3, \cdots$, and parameter $\lambda > 0$. Now substitute fuzzy number $\tilde{\lambda}$ for λ to produce the Ra-Fu Poisson probability function, denoted by $\xi \sim P(\tilde{\lambda})$.

From the definition, we know that $Pr\{\xi(\theta) = k\}$ is a fuzzy function of $\tilde{\lambda}$, where $\tilde{\lambda}$ is a fuzzy variable from the possibility space $(\Theta, \mathscr{P}(\Theta), Pos)$ to $(0, +\infty)$. For any $\theta^* \in \Theta$, $\tilde{\lambda}(\theta^*)$ is a real number in $(0, +\infty)$, then

$$\sum_{k=0}^{\infty} \frac{\tilde{\lambda}(\theta^*)^k}{k!} e^{-\tilde{\lambda}(\theta^*)} = e^{\tilde{\lambda}(\theta^*)} \cdot e^{-\tilde{\lambda}(\theta^*)} = 1.$$

So Definition 1.45 is well defined. If $(\tilde{\lambda})$ degenerates to a fixed number, then ξ degenerates to a random variable with a Poisson distribution.

Example 1.35. Let $\xi \sim P(\tilde{\lambda})$ be a Poisson distributed Ra-Fu variable, where $\tilde{\lambda}$ is an L-R fuzzy variable with the following membership function,

$$\mu_{\tilde{\lambda}} = \begin{cases} L\left(\frac{\lambda - x}{\alpha}\right), & x \leq \lambda, \alpha > 0, \\ R\left(\frac{x - \lambda}{\beta}\right), & x > \lambda, \beta > 0, \end{cases}$$

where λ is the "mean" of $\tilde{\lambda}$, and α β are the left and right spread of ξ. If $L(x)$ and $R(x)$ are linear functions, then $\tilde{\lambda}$ is a triangular fuzzy variable.

Continuous Ra-Fu Variable

In additional to discrete Ra-Fu variable, the concept of continuous Ra-Fu variable is defined as follows.

Definition 1.46 (Continuous Ra-Fu variable). Let ξ be a Ra-Fu variable on the possibility space $(\Theta, \mathscr{P}(\Theta), Pos)$. If $\xi(\theta)$ is a continuous random variable for any $\theta \in \Theta$, then ξ is said to be a continuous Ra-Fu variable.

Example 1.36. Let ξ be a Ra-Fu variable on $(\Theta, \mathscr{P}(\Theta), Pos)$ with the following density function,

$$\phi(x) = \begin{cases} \frac{1}{\tilde{\lambda}} e^{-x/\tilde{\lambda}}, & \text{if } 0 \le x < \infty, \\ 0, & \text{otherwise,} \end{cases}$$

where $\tilde{\lambda}$ is a triangular fuzzy variable on $(\Theta, \mathscr{P}(\Theta), Pos)$. Then ξ is a continuous Ra-Fu variable.

Definition 1.47 (Ra-Fu uniform distribution). Let ξ be a Ra-Fu variable on $(\Theta, \mathscr{P}(\Theta), Pos)$ with the following density function,

$$\bar{p}(x) = \begin{cases} \frac{1}{\tilde{b}-\tilde{a}}, & \text{if } \tilde{a} \le x \le \tilde{b}, \\ 0, & \text{otherwise,} \end{cases}$$

where $\tilde{a}$ and $\tilde{b}$ are fuzzy variables on $(\Theta, \mathscr{P}(\Theta), Pos)$ and $\tilde{a} < \tilde{b}$. Then ξ is said to be a uniformly distributed Ra-Fu variable, denoted by $\xi \sim \mathscr{U}(\tilde{a}, \tilde{b})$.

Since it is necessary to guarantee that $\tilde{b} - \tilde{a} \neq 0$, $\tilde{a}$ and $\tilde{b}$ needs to be two fuzzy variables, with one larger than the other. Then it is obvious that $Pos\{\tilde{a} < \tilde{b}\} = 1$ holds. It follows that

$$Pos\{\tilde{a} < \tilde{b}\} = 1 \Leftrightarrow \sup_{x,y \in \mathbf{R}} \{\mu_{\tilde{a}}(x) \wedge \mu_{\tilde{b}}(y) | x \le y\} = 1, \quad (1.61)$$

where $\mu_{\tilde{a}}(x)$ and $\mu_{\tilde{b}}(y)$ are the membership functions for $\tilde{a}$ and $\tilde{b}$. Let's consider the following example.

Example 1.37. Let $\xi \sim \mathscr{U}(\tilde{a}, \tilde{b})$ be a Ra-Fu variable, and $\tilde{a}$ and $\tilde{b}$ be fuzzy variables with the following membership,

$$\mu_{\tilde{a}}(x) = \frac{10 - x}{10},$$

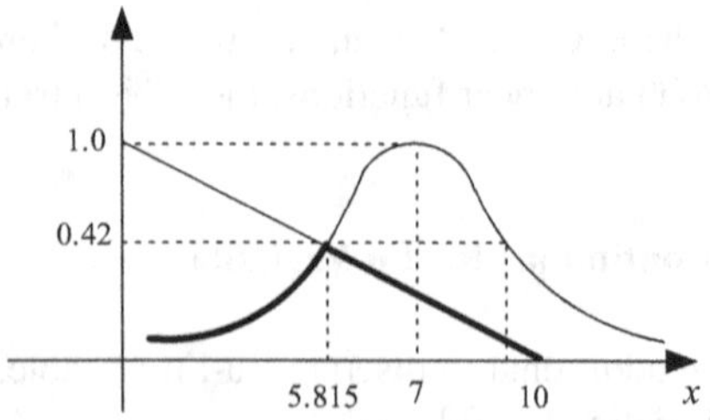

Fig. 1.5 The membership function of $\tilde{a} < \tilde{b}$

$$\mu_{\tilde{b}}(x) = \frac{1}{1+(x-7)^2}.$$

Really,

$$\mu_{\tilde{a}}(x) \wedge \mu_{\tilde{b}}(x) = \begin{cases} \frac{1}{1+(x-7)^2}, & 0 \le x < 5.815, \\ \frac{10-x}{10}, & 5.815 \le x \le 10, \\ 0, x > 10. \end{cases}$$

Then we have $\sup_{x,y \in \mathbf{R}}\{\mu_{\tilde{a}}(x) \wedge \mu_{\tilde{b}}(y) | x \le y\} = 0.42$, see Fig. 1.5. Therefore, ξ cannot be considered a Ra-Fu variable, for there exists the possibility that $\tilde{b} - \tilde{a} = 0$.

Definition 1.48 (Ra-Fu normal distribution). Let $\xi \sim \mathscr{N}(\rho, \sigma^2)$ be a Ra-Fu variable on $(\Theta, \mathscr{P}(\Theta), Pos)$. We then substitute the fuzzy variable $\tilde{\rho}$ for the mean value ρ or the fuzzy variable $\tilde{\sigma}$ for the variance σ, or substitute both. Then ξ is said to be a normally distributed Ra-Fu variable, denoted $\xi \sim \mathscr{N}(\tilde{\rho}, \sigma^2)$ or $\xi \sim \mathscr{N}(\rho, \tilde{\sigma}^2)$.

In fact, there are many cases of Ra-Fu normal distributions in the real world. For example, from statistical data, the amount sold of some seasonal products are subject to a normal distribution; however, the average amount, or the "mean", sold each year, varies from year to year, so is described as a fuzzy number from the historical data, and can be useful for an evaluation of the estimated a mount to be sold in the following year. Next, let's discuss a numerical example.

Example 1.38. Let $\xi \sim \mathscr{N}(\tilde{\rho}, 1)$ be a normally distributed Ra-Fu variable, where $\tilde{\rho}$ is a fuzzy variable with the following membership function,

$$\mu_{\tilde{\rho}}(x) = \begin{cases} x-2, & 2 \le x < 3, \\ -\frac{x}{2} + \frac{5}{2}, & 3 \le x \le 5. \end{cases}$$

From the definition of α-cut set, we know that $\tilde{\rho}(\theta) \in [\alpha + 2, -2\alpha + 5]$, then for any $\theta \in \Theta$, $\xi(\theta) \sim \mathscr{N}(\tilde{\rho}(\theta), 1)$ is a normally distributed random variable, where $\tilde{\rho}(\theta)$ varies between $\alpha + 2$ and $-2\alpha + 5$.

Definition 1.49 (Ra-Fu exponential distribution). Let ξ be an exponentially distributed random variable with the parameter λ. If we substitute a fuzzy variable $\tilde{\lambda}$ for λ, ξ becomes an exponentially distributed Ra-Fu variable, denoted by $\xi \sim exp(\tilde{\lambda})$.

There are many real-life cases which use the Ra-Fu exponential distribution.

Example 1.39. In many statistics problems, the probability distribution is often completely known except for one or more parameters. For example, the lifetime ξ of a modern engine is an exponentially distributed variable with an unknown mean $\tilde{\lambda}$,

$$\phi(x) = \begin{cases} \frac{1}{\tilde{\lambda}} e^{-x/\tilde{\lambda}}, & \text{if } 0 \leq x < \infty. \\ 0, & \text{otherwise.} \end{cases}$$

Usually, there is some relevant information available in practice. It is thus possible to specify an interval in which the value $\tilde{\lambda}$ is likely to lie, or to give an approximate estimate of the value of $\tilde{\lambda}$. However, it is often not possible to exactly determine the value of $\tilde{\lambda}$. If the value of $\tilde{\lambda}$ is provided as a fuzzy variable defined on the possibility space $(\Theta, \mathscr{P}(\Theta), Pos)$, then ξ is a Ra-Fu variable defined as

$$\xi \sim exp(\tilde{\lambda}),$$

where $\tilde{\lambda}$ is a fuzzy variable on $(\Theta, \mathscr{P}(\Theta), Pos)$.

Besides the above two discrete Ra-Fu variables and the three continuous Ra-Fu variables, there are also many other Ra-Fu variables that follow other distributions. While we do not introduce these here, interested readers can define and deduce these using the above definitions.

1.2 Bi-Level Programming

Decentralized planning problems have long been recognized as important decision-making problems [15, 70, 89, 128]. Decision makers often work within a hierarchical administrative structure, each with possible conflicting objectives. Mathematical programming models to solve decentralized planning problems can be traced back to the early development of linear programming. Multi-level programming (MLP) was developed to solve decentralized planning problems. This book concentrates on bi-level programming problems, which are a special case of MLP problems, and have a two-level structure.

The general formulation of a bi-level programming problem is

$$\begin{cases} \min\limits_{x} F(\boldsymbol{x}, \boldsymbol{y}) \\ \text{s.t.} \begin{cases} G(\boldsymbol{x}, \boldsymbol{y}) \leq 0 \\ \min\limits_{y} f(\boldsymbol{x}, \boldsymbol{y}) \\ \text{s.t. } g(\boldsymbol{x}, \boldsymbol{y}) \leq 0, \end{cases} \end{cases} \tag{1.62}$$

where $x \in \mathbb{R}^{n_1}$ and $y \in \mathbb{R}^{n_2}$. The variables of problem (1.62) are divided into two classes, namely the *upper-level variables* $x \in \mathbb{R}^{n_1}$ and the lower-level variables $y \in \mathbb{R}^{n_2}$. Similarly, the functions $F : \mathbb{R}^{n_1} \times \mathbb{R}^{n_2} \to \mathbb{R}$ and $f : \mathbb{R}^{n_1} \times \mathbb{R}^{n_2} \to \mathbb{R}$ are *upper-level and lower-level objective functions* respectively, while the vector-valued functions $G : \mathbb{R}^{n_1} \times \mathbb{R}^{n_2} \to \mathbb{R}$ and $g : \mathbb{R}^{n_1} \times \mathbb{R}^{n_2} \to \mathbb{R}$ are called the *upper-level and lower-level constraints* respectively. Upper-level constraints involve variables from both levels and play a very specific role. Indeed, they must be enforced indirectly, as they do not bind the lower-level decision-maker. In the particular framework of Stackelberg games, the *leader* is the upper-level decision maker who is assumed to anticipate the reactions of the *follower* who is the lower-level decision maker.

A number of applications have been formulated as bi-level programming problems. Based on Colson et al. [34] classification, bi-level programming problems can be classified as follows: (1) Agricultural model and water resource planning. This problem has been used for many practical applications, such as for water supply models [27], multilateral agricultural negotiations [76], multi-reservoir operating policies [58], and water resources allocation [116]; (2) *Government policy.* Bi-level programming applications have been used in such areas as the distribution of government resources [30, 82, 87], and subsidies and penalty strategies [111]; (3) *Economic system.* Bi-level programming applications have been used to study economic systems, such as distribution center location problems [55], oil industry price ceilings decisions [37] and price-based market clearing problems [53]; (4) *Finance model.* Bi-level programming applications have been developed for such financial management areas as bank asset portfolio problems [4, 102]; (5) Warfare, homeland protection or critical infrastructure protection planning [23, 121]; (6) *Transportation.* In many previous papers, highway network systems have been studied using the bi-level programming technique [13, 108].

At the same time, good decision making demands the fulfillment of several competing and yet often conflicting objectives. A multi-objective bi-level optimization problem has upper and lower level multi-objective optimization tasks. The lower level optimization problem is constrained by the upper level optimization problem, such that, a member can be feasible at the upper level only if it is the Pareto-optimal for the lower level optimization problem. A general multi-objective bi-level optimization problem can be described as follows:

$$\begin{cases} \min\limits_{x_u, x_l} \boldsymbol{F}(\boldsymbol{x}) = (F_1(\boldsymbol{x}), \cdots, F_M(\boldsymbol{x})), \\ \text{s.t.} \begin{cases} \boldsymbol{x}_l \in \operatorname{argmin}(\boldsymbol{x}_l)\{\boldsymbol{f}(\boldsymbol{x}) = (f_1(\boldsymbol{x}), \cdots, f_m(\boldsymbol{x})) | \boldsymbol{g}(\boldsymbol{x}) \geq \boldsymbol{0}, \boldsymbol{h}(\boldsymbol{x}) = \boldsymbol{0}\} \\ \boldsymbol{G}(\boldsymbol{x}) \geq \boldsymbol{0}, \boldsymbol{H}(\boldsymbol{x}) = \boldsymbol{0} \\ x_i^{(L)} \leq x_i \leq x_i^{(U)}, i = 1, \cdots, n. \end{cases} \end{cases} \tag{1.63}$$

In the above description of a generic multi-objective bi-level problem, $\boldsymbol{F}(\boldsymbol{x})$ are the upper level objectives and $\boldsymbol{f}(\boldsymbol{x})$ are the lower level objectives. The functions $\boldsymbol{g}(\boldsymbol{x})$ and $\boldsymbol{h}(\boldsymbol{x})$ determine the feasible space for the lower level problem. The decision vector $\boldsymbol{x}$ is formed from two smaller vectors $\boldsymbol{x}_u$ and $\boldsymbol{x}_l$, such that $\boldsymbol{x} = (\boldsymbol{x}_u, \boldsymbol{x}_l)$. At the lower

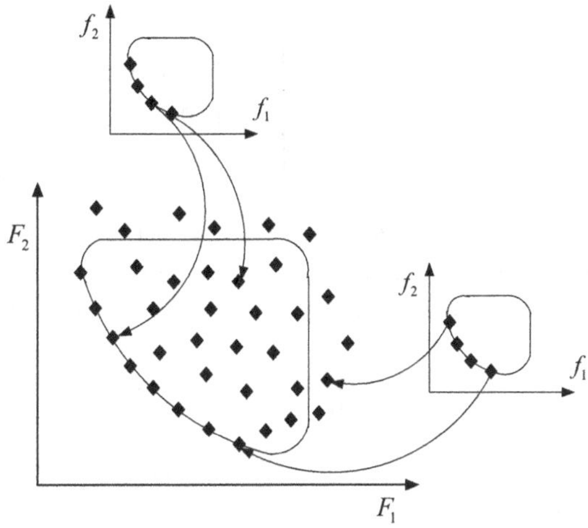

Fig. 1.6 Bi-level minimization problem with two objectives in the upper (F_1, F_2) and lower-level (f_1, f_2)

level, the optimization is performed only in respect to the variables $\boldsymbol{x}_l$; therefore, the solution set for the lower level problem can be represented as a function of $\boldsymbol{x}_u$, or as $\boldsymbol{x}_l^*(\boldsymbol{x}_u)$. This means that the upper level variables $\boldsymbol{x}_u$ act as a parameter for the lower level problem, and therefore, the lower level optimal solutions $\boldsymbol{x}_l^*$ are a function of the upper level vector $\boldsymbol{x}_u$. The functions $\boldsymbol{G}(\boldsymbol{x})$ and $\boldsymbol{H}(\boldsymbol{x})$ along with the Pareto-optimality to the lower level problem determine the feasible space for the upper level optimization problem. Both sets $\boldsymbol{x}_l$ and $\boldsymbol{x}_u$ are decision variables for the upper level problem.

Figure 1.6 shows the feasible regions for a multi-objective bi-level problem. There are two objectives both for the upper and the lower levels. Figure 1.6 shows the possible relationships between the members in the lower level space and their positions in the upper level space.

In what follows, we will review some elements of bi-level programming from a general point of view which include both linear and nonlinear bi-level programming and solving methods.

1.2.1 Linear Bi-Level Programming

The formal formulation of the linear bi-level programming problem has been defined by Candler and Townsley [28] as well as Fortuny-Amat and McCarl [55]. Various versions are presented by some authors [6, 8, 18]. Let a vector of decision variables $(\boldsymbol{x}, \boldsymbol{y}) \in \mathbb{R}$ be partitioned among the two planners. The upper-level decision-maker controls over the vector $\boldsymbol{x} \in \mathbb{R}^{n_1}$ and the lower-level decision-maker has control over the vector $\boldsymbol{y} \in \mathbb{R}^{n_2}$, where $n_1 + n_2 = n$. Furthermore, assuming that $F, f : \mathbb{R}^{n_1} \times \mathbb{R}^{n_2} \to \mathbb{R}^1$ are linear and bounded, the linear bi-level problem takes the form as follows.

$$\begin{cases} \max\limits_{x} F(x,y) = ax + by \\ \text{s.t.} \begin{cases} \text{where } y \text{ solves:} \\ \max\limits_{y} f(x,y) = cx + dy \\ \text{s.t. } \boldsymbol{A}x + \boldsymbol{B}y \le r, \end{cases} \end{cases}$$

where $a, c \in \mathbb{R}^{n_1}, b, d \in \mathbb{R}^{n_2}, r \in \mathbb{R}^m$, $\boldsymbol{A}$ is an $m \times n_1$ matrix, $\boldsymbol{B}$ is an $m \times n_2$ matrix. Denote the problem constraint region by $S = \{(x,y)|\boldsymbol{A}x + \boldsymbol{B}y \le r\}$. For a given x, let $Y(x)$ denote the set of optimal solutions to the lower-level problem,

$$\max_{y \in Q(x)} \tilde{f}(y) = dy \text{ where } Q(x) = \{y|\boldsymbol{B}y \le r - \boldsymbol{A}x\} \tag{1.64}$$

and represent the upper-level decision-maker's solution space, or the set of rational reactions of f over S, as

$$\Psi_f(S) = \{(x,y)|(x,y) \in S, y \in Y(x)\}. \tag{1.65}$$

The hierarchical relationship is here reflected in that the mathematical program related to the users' behaviour is part of the manager's constraints. This is the major feature of bi-level programs: they include two mathematical programs within a single instance, one of these problems being part of the constraints of the other one [34].

The rational reaction set is conceptually similar to the inducible region in the Stackelberg game in Chang and Luh [31], and Ho et al. [65]. Assuming that S and $Q(x)$ are bounded and non-empty. It may also be noted that for a given x, the choice of y is reduced to a linear programming problem [112]. The linear max-min problem studied by Falk [48],

$$\max_{x} \min_{y} \{ax + by : \boldsymbol{A}x + \boldsymbol{B}y \le r\}, \tag{1.66}$$

is a special bi-level linear programming problem in which the lower-level objective function is in direct opposition to the higher objective function, i.e. $f(x,y) = -F(x,y)$.

The definitions of feasibility and optimality for the linear bi-level programming problem are then given by the following:

Definition 1.50 ([104]). A point (x,y) is called feasible if $(x,y) \in \Psi_f(S)$.

Definition 1.51 ([104]). A feasible point (x^*, y^*) is called optimal if $ax^* + by^*$ is unique for all $y^* \in Y(x^*)$, and $ax^* + by^* \ge ax + by$ for all feasible pairs $(x,y) \in \Psi_f(S)$.

Example 1.40. Consider the following bi-level programming problem:

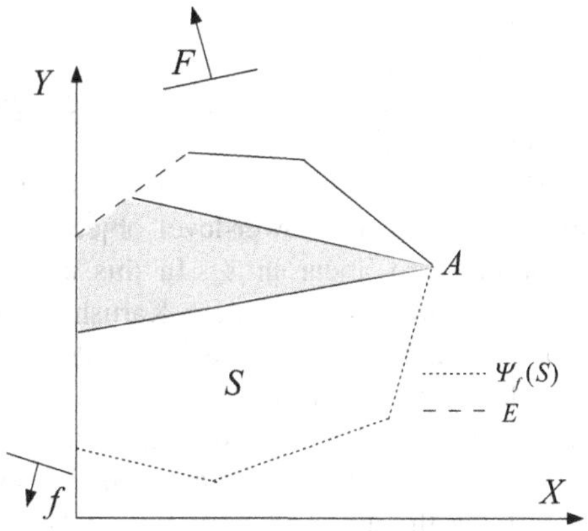

Fig. 1.7 A bi-level programming problem example. Note: $\Psi_f(S)$—the set of rational reactions of over f over S, E—the set of an efficient points to $\max(F, f)$ over S

$$\begin{cases} \max F(x,y) = -2x + 11y \\ \text{s.t.} \begin{cases} \text{where } y \text{ solves} \\ \max f(x,y) = -x - 3y \\ \text{s.t.} \begin{cases} x - 2y \leq 4, \\ 2x - y \leq 24, \\ 3x + 4y \leq 96, \\ x + 7y \leq 126, \\ -4x + 5y \leq 65, \\ x + 4y \geq 8, \\ x \geq 0, \\ y \geq 0. \end{cases} \end{cases} \end{cases}$$

Both the constraint region S and the set of feasible point $\Psi_f(S)$ are depicted in Fig. 1.7. The optimal solution $(x^*, y^*) = (192/11, 120/11)$ is point A in the Fig. 1.7.

1.2.2 Nonlinear Bi-Level Programming

Mathematically, the general nonlinear bi-level programming problem we addressed can be stated as follows:

$$\begin{cases} \min\limits_{x_1,x_2} F(x_1,x_2) \\ \text{s.t.} \begin{cases} G(x_1,x_2) \leq 0 \\ x_2 \in \arg\min\limits_{x_2'} f(x_1,x_2') \\ \text{s.t. } g(x_1,x_2'). \end{cases} \end{cases} \tag{1.67}$$

Several algorithms have been proposed for nonlinear bi-level programming, under various assumptions. Roughly speaking, the methods for solving a general nonlinear case belong to one of three categories: enumeration (branch-and-bound), descent or penalty methods.

The above mentioned branch-and-bound approach can be extended to cover a situation where the lower-level objective is quadratic in x_2, and the lower-level constraints are linear in x_2. In this case, one of these replaces the lower-level problem using the equivalent Karush-Kuhn-Tucker (KKT) system, in which the equalities and inequalities are considered to be linear, with the exception of the complementarity constraint. The complementarity constraint is a difficult type of constraint to deal with, and branch-and-bound methods may be used to defer the introduction of such constraints in the solution process. This is the basic idea underlying the approaches of Edmunds and Bard [44], Al-Khayyal et al. [5] and Thoai et al. [106].

Some authors have proposed descent methods for solving bi-level programs. Assuming that for any x_1 the optimal solution x_2^* for the lower-level problem is unique and defines x_2 as an implicit function of x_1, the bi-level programming problem may be viewed solely in terms of the upper-level variables $x_1 \in \mathbb{R}^{n_1}$. Given a feasible point x_1, an attempt is made to find a direction $d \in \mathbb{R}^{n_1}$ along which the upper-level objective decreases. A new point $x_1 + \alpha d(\alpha > 0)$ is computed so as to ensure a reasonable decrease in F while maintaining the bi-level problem feasibility. However, a major issue here is the availability of a gradient (or a generalized gradient) at a feasible point for the upper-level objective. While Kolstad and Lasdon [75] proposed a method for approximating this term, a slightly different approach was taken by Savard and Gauvin [97], for problems in which there was no upper-level constraints of the type shown in Model (1.67), and for which the lower-level constraints are rewritten in the following way:

$$\begin{aligned} g_i(x_1, x_2) &\leq 0, i \in I, \\ g_j(x_1, x_2) &= 0, j \in J. \end{aligned}$$

They first show that an upper-level descent direction at a given point x_1 is a vector $d \in \mathbb{R}^{n_1}$ such that

$$\nabla_{x_1} F(x_1, x_2^*)d + \nabla_{x_2} F(x_1, x_2^*)w(x_1, d) < 0, \tag{1.68}$$

where $w \in \mathbb{R}^{n_2}$ is a solution to the program

$$\begin{cases} \min\limits_{w} (d^T, w^T)\nabla_{x_1, x_2^*, \lambda}(d, w) \\ \text{s.t.} \begin{cases} \nabla_{x_2} g_i(x_1, x_2^*)w \leq -\nabla_{x_1} g_i(x_1, x_2^*)d, i \in I(x_1), \\ \nabla_{x_2} g_j(x_1, x_2^*)w = -\nabla_{x_1} g_j(x_1, x_2^*)d, j \in J, \\ \nabla_{x_2} f(x_1, x_2^*)w = -\nabla_{x_1} f(x_1, x_2^*)d + \nabla_{x_1} \mathscr{L}(x_1, x_2^*, \lambda)d, \end{cases} \end{cases} \tag{1.69}$$

where $I(x_1) = \{i \in I : g_i(x_1, x_2^*) = 0\}$ and

$$\mathscr{L}(x_1, x_2, \lambda) = f(x_1, x_2) + \sum_{i \in I(x_1) \cup J} \lambda_i g_i(x_1, x_2)$$

is the Lagrangian of the lower-level problem with respect to the active constraints. The steepest descent direction of those defined by (1.68) may coincide with the optimal solution to the linear-quadratic bi-level program

$$\begin{cases} \min\limits_{d} \nabla_{x_1} F(x_1, x_2^*) d + \nabla_{x_2} F(x_1, x_2^*) w(x_1, d) \\ \text{s.t.} \begin{cases} \|d\| \leq 1, \\ w(x_1, d) \text{ solves the quadratic program (1.69),} \end{cases} \end{cases}$$

for which exact algorithms exist (see e.g. Bard and Moore [7]). Falk and Liu [49] presented a bundling method, in which a decrease in the upper-level objective is adapted according to the sub-differential information obtained from the lower-level problem. They called the resulting setup a leader predominate algorithm because of the dominant role taken by the leader in the sequential decision making process. Note that all these descent methods are intended to only find the local optima.

Finally, penalty methods constitute the third category of algorithms. Because of the inherent difficulty in the nonlinear cases, these methods are generally limited to computing stationary points and local minima. An initial step in this direction was achieved by Aiyoshi and Shimizu [2, 3]. However the main drawback to their method was that, for a fixed value of the penalty parameter, it was necessary to compute a solution to the unconstrained lower-level problem at every update of the upper-level variables. Ishizuka and Aiyoshi [66] proposed a double penalty method in which both the objective functions in Model (1.67) are penalized. This method still uses an augmented lower-level objective and the penalty function from Aiyoshi and Shimizu, but replaces the lower-level problem with its stationarity condition, thus transforming the bi-level problem into a single-level problem. For a given value of the penalty parameter, this single-level problem can then be solved using a second penalty function applied to the upper-level objective. Amore recent contribution is proposed in Case's Ph.D. thesis [29] and follows up on ideas of Bi et al. [17]. This approach focuses on a bi-level program in which the lower-level problem is replaced by its KKT conditions. More precisely, Case [29] builds a penalty function $\nu(x_1, x_2, \lambda)$ with respect to the l_1 norm, which can be defined as the sum of the terms associated with each constraint of the KKT reformulation of the original problem. In view of the complex structure of the penalty function, the authors use a trust-region method; i.e., they minimize a simplified version of p, in which the local behavior is expected to model that of p. The trust-region model adopted by [29] corresponds to replace each component function of the penalty with its second-order Taylor expansion around the current iterate.

1.2.3 Complexity and Optimality Conditions

Researchers have proven that in the simplest format for bi-level programming, all the functions are continuous and linear, are still strongly NP-hard [9, 104]. Even if all the functions defining the bi-level programming are continuous and linear, the induced region is a non-convex set [54]. In the presence of upper-level constraints with the form $G(x) \leq 0$, the induced region is a connected set. If we consider the upper-level constraints involve the lower-level variables, in the form $G(x, y) \leq 0$, then the induced region could become a disconnected set [54]. Further, checking local optimality in a continuous linear bi-level programming is an NP-hard problem [110]. It is very easy to construct a linear bi-level programming problem, in which the number of local optima grows as exponentially function of the number of variables [25].

Theorem 1.17. *For any $\varepsilon > 0$, it is NP-hard to find a feasible solution to a linear bi-level programming problem which has no more than ε times the optimal value [36, 104].*

Many research studies have examined the determination of optimality conditions for a bi-level programming. This is a central topic when the lower-level optimization problem acts as a constraint on the upper-level problem. In some conditions, a bi-level programming problem can be transformed to a single-level optimization problem.

If the lower-level problem is convex and the lower-level variables are continuously differentiable, then the bi-level programming has a necessary and sufficient representation in terms of the first-order necessary conditions. The resulting problem becomes a single level optimization problem.

Several approaches replace the lower-level optimization problem with its KKT conditions. This popular approach transforms the bi-level optimization problem into a single-level optimization problem with complementary constraints. KKT conditions are used to identify whether a solution is the optimum for a constrained optimization problem [79]. If a regularity condition is satisfied for the lower-level problem, then the KKT conditions are necessary optimality conditions. Those conditions are also sufficient when the bi-level optimization problem is a convex optimization problem in the y variables for fixed parameters x. This problem can be transformed as:

$$\begin{cases} \min\limits_{x,y,\lambda} F(x,y) \\ s.t. \begin{cases} G(x,y) \leq 0 \\ \nabla_y f(x,y) + \lambda' \nabla_y g(x,y) = 0 \\ g(x,y) \leq 0 \\ \lambda \geq 0, \lambda' g(x,y) = 0. \end{cases} \end{cases}$$

In a constrained optimization problem, two types of optimal solutions are possible; the first lies inside the feasible region and the second lies on the boundary of the feasible region. When the optimum is inside the feasible region and the problem does not contain equality constraints, the gradient of the objective function and the Lagrange multipliers μ are equal to zero. When equality constraints are present in the model, the gradient of the objective function and the Lagrange multipliers for the equality constraints λ can be different to zero. If the optimal solution lies on the boundary, these terms take non-zero values.

1.2.4 Traditional Methods

Bi-level programming problems are difficult to solve. As stated above, even the simplest instance—in which all functions are linear—is strongly NP-hard [61]. More recently, Vicente et al. [109] demonstrated that checking local optimality in a linear bi-level program was also NP-hard. It is therefore no surprise that most algorithmic research to date has focused on the simplest bi-level programming cases; that is, problems that have nice properties such as linear, quadratic or convex objective and/or constraint functions (see Bard [9] for the study of the convex case). In particular, linear bi-level problems have received significant attention over the last two decades. Linear bi-level programs have a property in which a nonempty solution set, contains at least one vertex of the polyhedron.

A wide class of methods has been based on vertex enumeration [18, 28, 107]. Other approaches are based on a replacement of the lower-level problem with its primal-dual optimality conditions, which reduces the original problem to a single-level program involving disjunctive constraints. Other branch-and-bound algorithms were developed by Bard and Falk [6] and Fortuny-Amat and McCarl [55]. By combining branch-and-bound, monotonicity principles and penalization, Hansen et al. [61] developed a code capable of solving medium-size linear bi-level programs. Bialas et al. [19] and Judice and Faustino [68, 69], in contrast, based their solution approach on complementary pivoting theory. Comprehensive reviews related to that particular class of problems may be found in Wen and Hsu [112] and Ben-Ayed [12].

In this section, several traditional methods are listed.

KKT Approach

One of the most popular approaches used to solve the bi-level programming problems is the KKT approach, which transform the original problem to its first level auxiliary problem [122]. In this way, a bi-level programming problem is reduced to a regular mathematical programming problem.

For the sake of simplicity, we call the upper decision maker UDM and the kth lower decision maker LDM_k. The UDM have m_1 objective functions and each LDM have single objective function. The decision mechanism is that the UDM and LDM_k

adopt the leader-follower Stackelberg game [100], It is a non-cooperative Nash equilibrium game among the $\text{LDM}_k(k = 1, 2, \cdots, p)$.

In this part, we show the KKT condition for the bi-level multi-objective programming model, as the multi-objective case exists extensively and the single objective is its simplified form. We place all the constraints together and thus obtain the following crisp multi-objective linear bi-level programming model:

$$\min_x F(x, y) = (a_{11}^T x + b_{11}^{(1)T} y_1 + \cdots + b_{11}^{(p)T} y_p, \cdots, a_{1m_1}^T x + b_{1m_1}^{(1)T} y_1 + \cdots + b_{1m_1}^{(p)T} y_p) \quad (1.70)$$

$$\min_{y_1} f_1(x, y) = a_2^{(1)T} x + b_{21}^{(1)T} y_1 + \cdots + b_{2p}^{(1)T} y_p \quad (1.71)$$

$$\min_{y_2} f_2(x, y) = a_2^{(2)T} x + b_{21}^{(2)T} y_1 + \cdots + b_{2p}^{(2)T} y_p \quad (1.72)$$

$$\vdots$$

$$\min_{y_p} f_2(x, y) = a_2^{(p)T} x + b_{21}^{(p)T} y_1 + \cdots + b_{2p}^{(p)T} y_p \quad (1.73)$$

$$\text{s.t.} c_r^T x + d_{1r}^T y_1 + \cdots + d_{pr}^T y_p \leq e_{2r},\ r = 1, 2, \cdots, R \quad (1.74)$$

$$x, y_k \geq 0,\ k = 1, 2, \cdots, p, \quad (1.75)$$

where x is exclusively controlled by UDM, y_k is exclusively controlled by LDM_k, $x \in R^{n_1}, y_k \in R^{n_{2k}} (k = 1, 2, \cdots, p)$.

Definition 1.52. Assume Ω is the convex constraint set, for any $x(x \in G = \{x|(x, y) \in \Omega\})$ given by the UDM, if the decision-making variable $y_k(y_k \in G_k = \{y_k|(x, y) \in \Omega\})$ at the lower level is the non-inferior solution of LDM_k, then (x, y) is a feasible solution of bi-level multi-objective decision making.

Consider the first division on the second level subject to (1.74) and $y_1 \geq 0$; let ω_r and $\xi_1 \in R^{n_{21}}$ be the KKT multipliers where $\omega_r \geq 0$ and $\xi_1 \geq 0$. Let slack variables S_r^2 be associatedwi th (1.74) and $T_{1h_1}^2$ with $-y_1 \leq 0$. Also $S_r^2 \geq 0$, $T_{1h_1}^2 \geq 0$. Then the Lagrangian function of the first division on second level for maximization objective is formulated as

$$L(\omega, S, y_1, \xi_1, T_1) = a_2^{(1)T} x + b_{21}^{(1)T} y_1 + \cdots + b_{2p}^{(1)T} y_p - \sum_{r=1}^{R} \omega_r (c_r^T x + d_{1r}^T y_1 + \cdots + d_{pr}^T y_p - e_{2r} + S_r^2) - \sum_{h_1=1}^{n_{21}} \xi_{1h_1} (-y_{1h_1} + T_{1h_1}^2), \quad (1.76)$$

where $\omega = \{\omega_1, \omega_1, \cdots, \omega_R\}$, $S = \{S_1, S_2, \cdots, S_R\}$, $y_1 = \{y_{11}, y_{12}, \cdots, y_{1n_{21}}\}$ and $T_1 = \{T_{11}, T_{12}, \cdots, T_{1n_{21}}\}$.

The first-order necessary conditions for the first division optimization problem (1.76) are:

$$\partial L/\partial \omega_r = 0, \quad \partial L/\partial S_R = 0, \quad r = 1, 2, \cdots, R, \tag{1.77}$$

$$\partial L/\partial y_{1h_1} = 0, \quad \partial L/\partial \xi_{1h_1} = 0, \quad \partial L/\partial T_{1h_1} = 0, \quad h_1 = 1, 2, \cdots, n_{21}. \tag{1.78}$$

Thus from Eqs. (1.77) and (1.78) we establish the following, respectively:

$$c_r^T x + d_{1r}^T y_1 + \cdots + d_{pr}^T y_p \leq e_{2r},$$

$$\omega_r(c_r^T x + d_{1r}^T y_1 + \cdots + d_{pr}^T y_p - e_{2r}) = 0,$$

$$-b_{1h_1}^{(1)} + \sum_{r=1}^{R} \omega_r d_{1rh_1} \geq 0, \quad y_{1h_1} \geq 0, \quad h_1 = 1, 2, \cdots n_{21},$$

$$\sum_{h_1=1}^{n_{21}} \left(-b_{1h_1}^{(1)} + \sum_{r=1}^{R} \omega_r d_{1rh_1} \right) y_{1h_1} = 0.$$

Similarly we can obtain the KKT optimality conditions for the other divisions on the second level. Incorporating KKT optimal conditions of 1st division (shown above), 2nd $\sim$ pth division and $x_j \geq 0\,(j = 1, 2, \cdots, n_1)$, the upper level problem is obtained as

$$\min_x F(x, y) = (a_{11}^T x + b_{11}^{(1)T} y_1 + \cdots + b_{11}^{(p)T} y_p, \cdots, a_{1m_1}^T x + b_{1m_1}^{(1)T} y_1 + \cdots + b_{1m_1}^{(p)T} y_p) \tag{1.79}$$

$$\text{s.t.} - b_{kh_k}^{(k)} + \sum_{r=1}^{R} \omega_r d_{krh_k} \geq 0, \; h_k = 1, 2, \cdots, n_{2k}, \; k = 1, 2, \cdots, p, \tag{1.80}$$

$$\sum_{h_k=1}^{n_{2k}} \left(-b_{kh_k}^{(k)} + \sum_{r=1}^{R} \omega_r d_{krh_k} \right) y_{kh_k} = 0, \; k = 1, 2, \cdots, p, \tag{1.81}$$

$$c_r^T x + d_{1r}^T y_1 + \cdots + d_{pr}^T y_p \leq e_{2r}, \tag{1.82}$$

$$\omega_r(c_r^T x + d_{1r}^T y_1 + \cdots + d_{pr}^T y_p - e_{2r}) = 0, \tag{1.83}$$

$$x, y_k, \omega_r \geq 0, \; k = 1, 2, \cdots, p, \; r = 1, 2, \cdots, R. \tag{1.84}$$

Therefore, we have transformed the bi-level multi-objective model into a normal multi-objective programming model. Multi-objective programming solving methods can be used in solving this model.

Brand-and-Bound

When the lower-level problem is convex and regular, it can be replaced by its KKT conditions, yielding the single-level reformulation of problem (1.62) [35]:

$$\begin{cases} \min_{x,y} F(x,y) \\ \text{s.t.} \begin{cases} G(x,y) \leq 0 \\ g(x,y) \leq 0 \\ \lambda_i \geq 0, i = 1, \cdots, m \\ \lambda_i g_i(x,y) = 0, i = 1, \cdots, m \\ \nabla_y \mathscr{L}(x,y,\lambda) = 0, \end{cases} \end{cases} \tag{1.85}$$

where

$$\mathscr{L}(x,y,\lambda) = f(x,y) + \sum_i^m \lambda_i g_i(x,y)$$

is the Lagrangean function associated with the lower-level problem.

Even under suitable convexity assumptions on the functions F, G and the set $X(x \in X)$, the above mathematical program is not easy to solve, due mainly to the non-convexities that occur in the complementarity and Lagrangean constraints. While the Lagrangean constraint is linear in certain important cases (linear or convex quadratic functions), the complementarity constraint is intrinsically combinatorial, and is best addressed by enumeration algorithms, such as branch-and-bound.

In the branch-and-bound scheme, the root node of the tree corresponds to problem (1.85) from which constraint $\lambda_i g_i(x,y) = 0$ is removed. At a generic node of the branch- and-bound tree that does not satisfy the complementarity constraints, separation is performed in the following manner: two children nodes are constructed, one with $\lambda_i = 0$ as an additional constraint, and the other with the constraint $g_i(x,y) = 0$. The optimal values of these problems yield lower bounds valid for the corresponding subtree.

In the absence of upper-level constraints, a rational solution can be computed by solving the lower-level problem resulting from setting x to the partial optimal solution of the relaxed problem. Note that, in contrast with standard branch-and-bound implementations, feasible (i.e., rational) solutions are then generated at every node of the implicit enumeration tree. The upper bound is updated accordingly.

Algorithms based on this idea were proposed by Bard and Falk [10] and Fortuny-Amat and McCarl [55] for solving linear bi-level programming problems. The approach was adapted by Bard and Moore [11] to linear-quadratic problems and by Al-Khayal et al. [5], Bard [9] and Edmunds and Bard [44] to the quadratic case.

Penalty Function

Penalty function methods constitute another important class of algorithms for solving non-linear bi-level programming problems, although they are generally limited to computing stationary points and local minima [35].

An initial step in this direction was achieved by Aiyoshi and Shimizu [3] and Shimizu and Aiyoshi [99]. Their approach consists in replacing the lower-level problem in Eq. (1.62) by the penalized problem

$$\min_{y} p(x, y, r) = f(x, y) + r\phi(g(x, y)), \tag{1.86}$$

where r is a positive scalar, ϕ is a continuous penalty function that satisfies

$$\begin{array}{ll} r\phi(g(x, y)) > 0 & \text{if } y \in \text{int } S(x), \\ r\phi(g(x, y)) \rightarrow +\infty & \text{if } y \rightarrow \text{bd } S(x), \end{array} \tag{1.87}$$

where int $S(x)$ and bd $S(x)$ denote the relative interior and the relative boundary of $S(x) = \{y : g(x, y) \leq 0\}$, respectively. Model (1.62) is then transformed into:

$$\begin{cases} \min\limits_{x,y} F(x, y^*(x, r)) \\ \text{s.t.} \begin{cases} G(x, y^*(x, r)) \leq 0, \\ p(x, y^*(x, r), r) = \min\limits_{y} p(x, y, r). \end{cases} \end{cases} \tag{1.88}$$

Shimizu and Aiyoshi [99] proved that the sequence $\{(x^k, y^*(x^k, r^k))\}$ of optimal solutions to (1.88) converges to the solution of (1.62). The main drawback of this method is that solving (1.88) for a fixed value of r requires the global solution at every update of the upper-level variables. Each subproblem is not significantly easier to solve than the original bi-level program. Ishizuka and Aiyoshi [66] proposed a double penalty method in which both objective functions in Eq. (1.62) are penalized. They still use the augmented lower-level objective (1.86) and the penalty function ϕ characterized by (1.87) but replace the lower-level problem by its stationarity condition $\nabla_y p(x, y, r) = 0$, thus transforming (1.62) into the single-level program

$$\begin{cases} \min\limits_{x,y} F(x, y) \\ \text{s.t.} \begin{cases} G(x, y) \leq 0, \\ \nabla_y p(x, y, r) = 0 \\ g(x, y) \leq 0. \end{cases} \end{cases} \tag{1.89}$$

Note that the last constraint restricts the domain of the function p. For a given r, problem (1.89) is solved using a second penalty function applied to the constraints.

Thereafter, Case [29], follows up on ideas of Bi et al. [17], who themselves extend a technique proposed in Bi et al. [16] for linear bi-level programs. Their approach is based on (1.85), that is, a bi-level program for which the lower-level problem has been replaced by its KKT conditions. Their method involves a penalty function of the form

$$p(x, y, \lambda, \mu) = F(x, y) + \mu v(x, y, \lambda),$$

where μ is a positive penalty parameter and the upper-level objective $F(x, y)$ is augmented by a weighted, nonnegative penalty function associated with the current iterate. More precisely, Case [29] builds a penalty function $v(x, y, \lambda)$ with respect to the ℓ_1 norm, defined as the sum of the terms associated with each constraint of the single-level problem (1.85). The resulting algorithm involves the minimization of the penalty function $p(x, y, \lambda, \mu)$ for a fixed value of μ. In view of the complex structure of the latter function, the authors develop a trust-region method, where the model for p is obtained by replacing each component function of $p(x, y, \lambda, \mu)$ by its second-order Taylor expansion around the current iterate.

Interactive Programming Method

If we can transform the random-like bi-level programming into some solvable forms, namely to eliminate its uncertainty, we can use the interactive programming method to transform it into a single level programming, then we can solve it easily. Therefore, in this section, the interactive programming method proposed in [96] is applied to solve the bi-level programming.

It's natural that we can take the uncertain objective function to evaluate the decision maker's imprecise consideration. For the objective function of every level in Model (1.62), decision maker has fuzzy goals such as "the goal should be more than or equal to a certain value".

Let

$$H_1(x, y) = (u_1^{aT} x + u_1^{bT} y) + \Phi^{-1}(1 - \eta_1)\sqrt{x^T V_1^a x + y^T V_1^b y} + (1 - \zeta_1)(\beta_1^{aT} x + \beta_1^{bT} y),$$

$$H_2(x, y) = (u_2^{aT} x + u_2^{bT} y) + \Phi^{-1}(1 - \eta_2)\sqrt{x^T V_2^a x + y^T V_2^b y} + (1 - \zeta_2)(\beta_2^{aT} x + \beta_2^{bT} y),$$

and

$$S = \left\{ (x, y) \geq 0 \;\middle|\; \begin{array}{l} (u_{1r_1}^c x + u_{1r_1}^d y - u_{1r_1}^e)\Phi^{-1}(\gamma_{1r_1})\sqrt{x^T V_{1r_1}^{cT} x + y^T V_{1r_1}^{dT} y + (\sigma_{1r_1}^e)^2} \\ \quad -(1 - \delta_{1r_1})(\beta_{1r_1}^e + \alpha_{1r_1}^{cT} x + \alpha_{1r_1}^{dT} y) \leq 0,\ r_1 = 1, 2, \cdots, p_1; \\ (u_{2r_2}^c x + u_{2r_2}^d y - u_{2r_2}^e) + \Phi^{-1}(\gamma_{2r_2})\sqrt{x^T V_{2r_2}^{cT} x + y^T V_{2r_2}^{dT} y + (\sigma_{2r_2}^e)^2} \\ \quad -(1 - \delta_{2r_2})(\beta_{2r_2}^e + \alpha_{2r_2}^{cT} x + \alpha_{2r_2}^{dT} y) \leq 0,\ r_2 = 1, 2, \cdots, p_2. \end{array} \right\}.$$

We can denote the maximum and minimum values of each objective functions as follows:

$$H_1^{\max} = \max_{(x,y)\in S} H_1(x,y),\ H_1^{\min} = \min_{(x,y)\in S} H_1(x,y);$$
$$H_2^{\max} = \max_{(x,y)\in S} H_2(x,y),\ H_2^{\min} = \min_{(x,y)\in S} H_2(x,y).$$

The functions $\mu_i(H_i(x,y)), i = 1,2$ vary strictly between $H_i^{\min}$ and $H_i^{\max}$. For the sake of simplicity, we can take the linear function to characterize the goal at each level. They can be defined as follows:

$$\mu_i(H_i(x,y)) = \begin{cases} 1, & H_i(x,y) \geq H_i^{\max} \\ \frac{H_i(x,y)-H_i^{\min}}{H_i^{\max}-H_i^{\min}}, & H_i^{\min} \leq H_i(x,y) < H_i^{\max} \\ 0, & H_i(x,y) < H_i^{\min}. \end{cases} \quad i = 1,2. \tag{1.90}$$

The objective of the upper level can be specified with a minimal satisfactory level $\varepsilon \in [0,1]$ after introducing the membership functions. Then the lower level minimize the objective subjects to the additional condition $\mu_1(H_1(x,y)) \geq \varepsilon$, that is, the lower level should solve the following problem:

$$\begin{cases} \max\ \mu_2(H_2(x,y)) \\ \text{s.t.} \begin{cases} \mu_1(H_1(x,y)) \geq \varepsilon \\ (x,y) \in S. \end{cases} \end{cases} \tag{1.91}$$

In order to obtain the overall satisfactory optimal solution for both upper and lower levels, the upper level have to comprise with the lower level with consideration of his satisfactory level. Thus, a satisfactory degree for both upper and lower levels is defined as

$$\lambda = \min\{\mu_1(H_1(x,y)), \mu_2(H_2(x,y))\} \tag{1.92}$$

and Model (1.91) can be transformed into

$$\begin{cases} \max\ \lambda \\ \text{s.t.} \begin{cases} \mu_1(H_1(x,y)) \geq \lambda \\ \mu_2(H_2(x,y)) \geq \lambda \\ (x,y) \in S. \end{cases} \end{cases} \tag{1.93}$$

By solving Model (1.93), we get the overall satisfactory solution for both upper and lower levels.

Theorem 1.18. *If Model (1.93) has optimal solutions, then solutions of Model (1.93) must be the solutions of Model (1.62).*

Proof. We consider $x = (x_1, x_2, \cdots, x_{n_1})$, $y = (y_1, y_2, \cdots, y_{n_2})$ and λ as decision variables. Let $X^* = (x^*, y^*, \lambda^*)$ be an optimal solution of Model (1.93). Apparently, X^* satisfies all constraints. Assume that X^* is not the optimal solution of

Model (1.62), so in the lower level, there exists a solution $y' = (y'_1, y'_2, \cdots, y'_{n_2})$ such that $H_2(x, y') < H_2(x, y^*)$. Then

$$\mu_2(H_2(x, y')) = \frac{H_2^{\max} - H_2(x, y')}{H_2^{\max} - H_2^{\min}} > \mu_2(H_2(x, y^*)) = \frac{H_2^{\max} - H_2(x, y^*)}{H_2^{\max} - H_2^{\min}}. \tag{1.94}$$

Similarly, in the upper level, x^* satisfies the constraints in the lower level. If x^* is not the optimal solution of Model (1.62), there exists x' such that $H_1(x', y') < H_1(x^*, y^*)$, then

$$\mu_1(H_1(x', y')) = \frac{H_1^{\max} - H_1(x', y')}{H_1^{\max} - H_1^{\min}} > \mu_1(H_1(x^*, y^*)) = \frac{H_1^{\max} - H_1(x^*, y^*)}{H_1^{\max} - H_1^{\min}}. \tag{1.95}$$

According to (1.92), (1.94) and (1.95), we know that the optimal solution λ' for variables x', y' must be subjected to $\lambda' < \lambda^*$. It shows a conflict that λ^* is the optimal solution of Model (1.93).

This completes the proof. □

After proving the above theorem, we know that if Model (1.93) has an optimal solution, the bi-level programming must have optimal solutions. Next, we can apply appropriate algorithm to find the optimal solutions to Model (1.93).

1.2.5 Developments of Algorithms

A simple algorithm using a point-by point approach to solve bi-level optimization problems directly treats the lower level problem as a hard constraint. Every solution $(\boldsymbol{x} = (\boldsymbol{x}_u, \boldsymbol{x}_l))$ must be sent to the lower level problem as an initial point, and an optimization algorithm is then be employed to find the optimal solution $\boldsymbol{x}_l^*$ to the lower level optimization problem. Then, the original solution x from the upper level problem must be repaired as $(\boldsymbol{x}_u, \boldsymbol{x}_l^*)$. The employment of a lower level optimizer within the upper level optimizer for every upper level solution makes the overall search a nested optimization procedure, which may be computationally expensive. Moreover, if this idea is to be extended to multiple conflicting objectives in the lower level, for every upper level solution, multiple Pareto-optimal solutions for the lower level problem need to be found and stored using a suitable multi-objective optimizer.

Another method for handling a lower level optimization problem which has differentiable objectives and constraints is to directly use the explicit KKT conditions from the lower level optimization problem as constraints for the upper level problem [20, 64]. This would then involve Lagrange multipliers for the lower level optimization problem to determine the additional variables for the upper level problem. As KKT points need not always be optimum points, further conditions have to be included to ensure the lower level problem optimality. For multi-objective bi-level problems, corresponding multi-objective KKT formulations need to be used, thereby involving further Lagrange multipliers and optimality conditions as the constraints for the upper level problem. Despite these apparent difficulties, there have been some useful studies, including reviews on bi-level programming [35, 110], nested bi-level linear programming [56], and applications [1, 50, 73], mostly for single-objective bi-level optimization.

Studies by Eichfelder [45, 46] focused on handling multi-objective bi-level problems using classical methods. While the lower level problem employs a numerical optimization technique, the upper level problem is handled using an adaptive exhaustive search method, thereby making the overall procedure computationally expensive for large-scale problems. This method uses a nested optimization strategy to find and store the multiple Pareto-optimal solutions for each of many finite upper level variable vectors.

Another study by Shi and Xia [98] transformed a multi-objective bi-level programming problem into a bi-level ε-constraint approach on both levels by keeping one of the objective functions and converting the remaining objectives to constraints. The ε values for the constraints were supplied by the decision-makers as different levels of 'satisfaction'. The lower-level single-objective constrained optimization problem was rthen eplaced by the equivalent KKT conditions and a variable metric optimization method was used to solve the resulting problem.

From his discussion, it is apparent that greater efforts are needed to develop effective classical methods for multi-objective bi-level optimization, particularly for a more coordinated handling of the upper level and lower level optimization tasks.

Several researchers have proposed evolutionary algorithmic based approaches when seeking to solve single-objective bi-level optimization problems. As early as 1994, Mathieu et al. [88] proposed a genetic algorithm (GA) based approach to solve bi-level linear programming problem with a single objective on each level. The lower level problem was solved using a standard linear programming method, and the upper level was solved using a GA. This early GA study used a nested optimization strategy, which may be computationally expensive when extended to nonlinear or large-scale problems. Yin [123] claimed to have solved non-convex bi-level optimization problems better than existing classical methods using another GA based nested approach, in which the lower level problem was solved using the Frank-Wolfe gradient based linearized optimization method. Oduguwa and Roy [92] suggested a coevolutionary GA approach, in which two different populations are used to handle the variable vectors x_u and x_l independently. A linking procedure is then used to align the the populations. For single-objective bi-level optimization problems, the final outcome is usually a single optimal solution on each level. The proposed co-evolutionary approach is viable when seeking to find a corresponding single solution in $\boldsymbol{x}_u$ and $\boldsymbol{x}_l$ spaces. However, when used to handle multi-objective bi-level programming problems, multiple solutions corresponding to each upper level solution must be found and maintained during the co-evolutionary process. Therefore, it is not clear how this co-evolutionary algorithm can be effectively designed to handle multi-objective bi-level optimization problems. We do not address this issue in this chapter, but recognize that Oduguwa and Roy's study was the first to suggest a coevolutionary procedure for single-objective bi-level optimization problems [92].

Since 2005, there has been a surge in algorithmic development research in various application areas, with most using the nested approach and explicit KKT conditions for the lower level problem [38, 73, 85, 103, 129].

Li et al. [84] proposed particle swarm optimization (PSO) based procedures for both the lower and upper levels, but, instead of using a nested approach, proposed a serial iterative application for the upper and lower levels. This idea is applicable when solving single-objective problems on each level as the sole target is in finding a single optimal solution. As discussed above, when there are multiple conflicting objectives on each level, multiple solutions need to be found and preserved for each upper level solution, so a serial application for the upper and lower level optimization does not make sense in multi-objective bi-level optimization problems. Halter and Mostaghim [62] also used PSO on both levels, but since the lower level problem in their application problem was linear, a specialized linear multi-objective PSO algorithm was used and an overall nested optimization strategy employed on the upper level.

Bi-level programming problems, particularly those with multiple conflicting objectives, have not been paid the kind of attention that has been needed for real world applications. As more studies are performed, the algorithms must be tested and compared.

Meta-heuristics are approximation algorithms which can deal with large-sized problems as they are able to obtain a satisfactory solution in a reasonable time [104, 105]. Due to their complexity, most bi-level optimization problems are dealt with using approaches which involve a model reformulation which masks the bi-level aspect of the problem [54, 113], or involve meta-heuristics. Evolutionary algorithms can be seen to be meta-heuristics which mimick species' evolution. Here, we use several terms related to evolutionary algorithms: an individual is a feasible solution, a population is a set of individuals, and a mutation is the creation of a new individual from an existing individual, while generally retaining some properties. A cross-over is the creation of an individual(s), called offspring, from several other individuals called parents. Generation is the application of the cross-over and mutation operators to a population in order to create a new population. In each generation, the selection step consists of selecting individuals to meet defined goals. Evolutionary algorithms create multiple generations and apply selections until a stopping criterion is met (Fig. 1.8). More details about population-based meta-heuristics can be found in the reference [105].

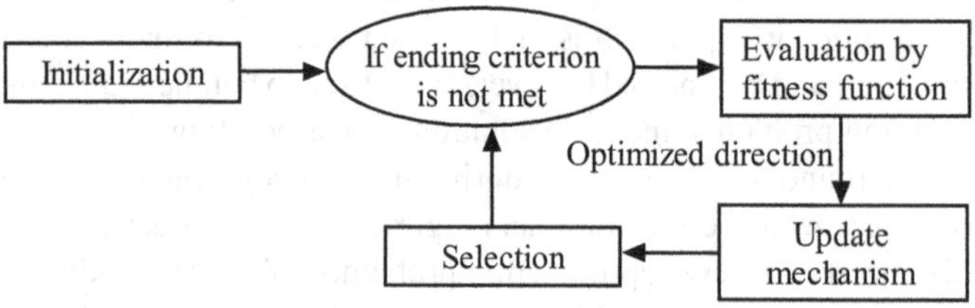

Fig. 1.8 General scheme of an evolutionary algorithm

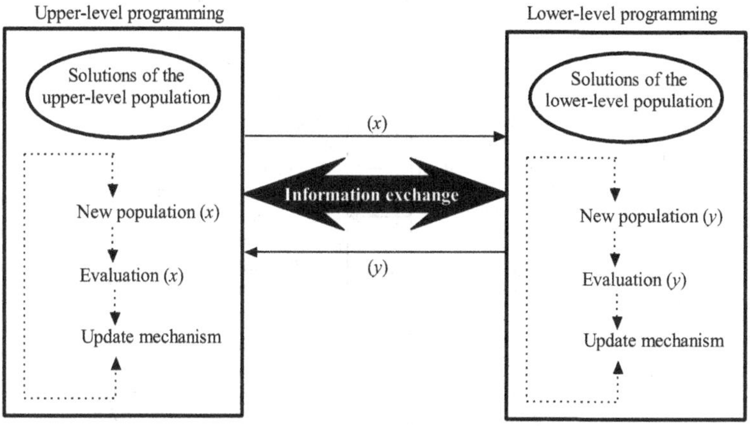

Fig. 1.9 The two metaheuristics evolve in parallel and cooperate via information exchange

In many cases, methodologies based on the nested, multi-objective or reformulation approaches may be practically inefficient. Indeed, most of these traditional approaches were designed for specific bi-level problems or based on specific assumptions (e.g. upper-level or lower-level problem differentials, convex feasible regions, low-level structured problems, upper-level reduced search space). Because of such deficiencies, these approaches cannot be used to solve real-life complex applications (e.g. bi-level problems with non-differentiable objective functions, complex combinatorial bi-level problems). Therefore, some co-evolutionary based meta-heuristics approaches have been developed to solve general bi-level optimization problems as bi-level programming problems without any transformation. In co-evolutionary meta-heuristics, the two levels proceed in parallel. At each level, an optimization strategy is applied. In general, the optimization strategy is a population-based meta-heuristic. Each level attempts to maintain and improve its own population separately (Fig. 1.9), and the two populations evolve in parallel. Different populations evolve to become part of the decision variables, and complete solutions are built using a cooperative exchange of individuals from the populations. In this way, the two levels exchange of information maintains a global view of the bi-level problem.

In designing a co-evolutionary model for any metaheuristic, the same design questions need to be answered [104]:

(1) *The exchange decision criterion (When?)*: the exchange of information between the metaheuristics can be done using either a blind (periodic or probabilistic) or an intelligent adaptive criterion. Periodic exchange occurs in each algorithm after a fixed number of iterations, and is a synchronous type of communication. Probabilistic exchange consists of performing a communication operation with a given probability after each iteration. Adaptive exchanges are guided by search run-time characteristics. For instance, they may depend on an evolution in the quality of the solutions or the search memory. A classical criterion is related to an improvement in the best found local solutions.

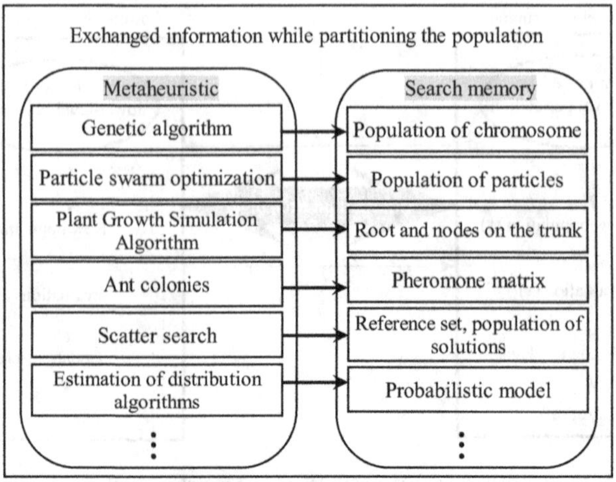

Fig. 1.10 Exchanged information while partitioning the population

(2) *The information exchanged (What?)*: this parameter specifies the information to be exchanged between the metaheuristics. In general, it may be composed of:

- Solutions: this information deals with a selection of the generated and stored solutions during the search. In general, it contains elite solutions that have been found, such as the best solution at the current iteration, local best solutions, the global best solution, or the best diversified solutions. The number of solutions for exchange may be an absolute value or a given percentage of the population. Any selection mechanism can be used to select the solutions, but the most used selection strategy consists of selecting the best solutions for a given criteria (e.g. objective function for the problem, diversity, age) or random solutions.
- Search memory: this information deals with any element of the search memory associated with the involved metaheuristic (Fig. 1.10).

(3) *The integration policy (How?)*: symmetrically with information exchange policies, an integration policy deals with the use of the received information. In general, there is a local copy of the received information, and the local variables are updated using the received variables. For instance, the best found solution for a given solution (x, y) is recombined with the received solution (x, y^*).

1.3 Synthetical Analysis of Literature

To synthetically analyse the research, the Web of Science database (WOS) was chosen as in contains 7 databases with information from thousands of scholarly journals, books, book series, reports, conferences and others, so was able to provide overall impression of random-like bi-level decision making research to date.

There is a long history of uncertain, bi-level decision making research. However, there have been few studied which have investigated random-like bi-level decision making. This is because random bi-level optimization is only applicable to specific problems, in which there are multiple decision makers, conflicts between the different decision makers. In addition, past research has tended to focus on crisp or simple random situations due to mathematic description and modeling simplicity. Therefore, using "complex random" or "random-overlapped random"/"fuzzy-overlapped random" and "bi-level" as the search keywords, we were only able to extract a few recent studies. To enlarge our analysis, we chose the two foci discussed in this book "random/stochastic" and "bi-level/bi-level model/programming/optimization" as the search keywords.

After eliminating repeat articles, proceedings reports, and some papers with minimal focus on our subject, we finally determined an analytical database of 178 papers. After a rigorous review, the document types were sorted into three main categories, book chapters, journal articles and conference papers. Book chapters often summarize predecessors' research contribution. Journal articles describe the characteristics of the complete research process, from manuscript to final publication. Conference papers, which have a quicker review process, are aimed at current topics of interest, or new ideas and new points of view from published research in the field. Using NoteExpress analyses, the final journal article and conference paper database was developed based on quantity and the proportional relationships, as shown in Fig. 1.11. From Fig. 1.11, we can see that there were only 29 book chapters and the ratio of conference proceedings was still significant, indicating that the area is not mature and there are many issues still under discussion.

There were few papers in the database published before 2003, but after 2003, the research quantity rapidly rose.

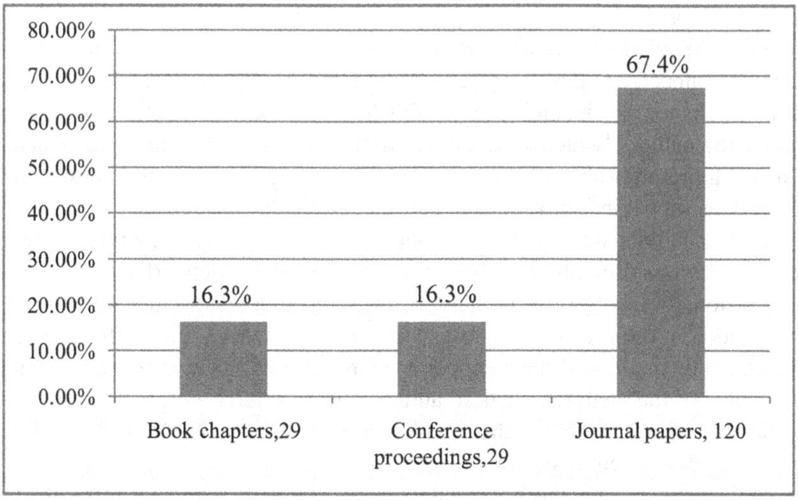

Fig. 1.11 Document type

Table 1.5 Authors articles quantity

Author	Quantity	(%)	Author	Quantity	(%)
Lam, WHK	12	2.58 %	Chen, A	11	2.37 %
Patriksson, M	9	1.94 %	Xu, J P	6	1.29 %
Wong, S C	4	0.86 %	Ye, C M	4	0.86 %
Gao, Z Y	4	0.86 %	Liu, L H	4	0.86 %

A journal's research direction usually determines the foci of the published research as the topics suggested by the journal scope are usually those considered most important to the field at that time. Therefore, the journal scope often focuses on research that can assist in developing a mature theoretical basis in the relevant field. Using NoteExpress, the journals from which the journal articles were extracted numbered in the hundreds. However, regardless of this, about 30 % of the research was found in only 10 journals. The journals which had the largest number of articles were; the *European Journal of Operational Research*, *IEEE Transactions on Power Systems*, *Expert Systems with Applications*, and *Transportation Research Part B-Methodological*. The largest number of documents were focused on operations research and mathematical modeling.

Many scholars have presented multiple research papers on random and bi-level related topics. Using NoteExpress, it was relatively simple to determine the number of articles had been published by each author. More than 11 % of all articles are published only by 8 authors (as shown in Table 1.5), indicating that there are some scholars who have made a prominent contribution to this field. This research was considered worthy of deeper study.

Each of these prominent scholars had a particular focus, but their findings may have contributed significantly to other related fields.

Using a meta-synthesis method, the statistical results from the previous step, in which NoteExpress was used, were put into NodeXL for network analysis. Firstly, the number of all these authors articles was set, and then, each author of each article was connected to develop a network system. The authors were linked to each other by published article, and jointly published articles, to form clusters. The dots represent the authors, which were clustered and color-coded through the Clauset-Newman-Moore algorithm [32] using NodeXL. Using NodeXL, the network was visualized, as shown in Fig. 1.12.

When the authors were sorted by cooperation, a very large, complex network formed, which was difficult to analyze. Some authors in the network may have been only occasionally involved in the field, or their research achievements had not yet been included in the previously chosen database, so their research efforts appears to be relatively small, and their cooperation in the field not widespread. To meet the demands of the analysis, such an author may be filtered out, regardless of the importance of their research results. By filtering the authors using NodeXL, a clearer network was developed as shown in Fig. 1.13.

From the network of authors, several main clusters were found. Of these, the first was clustered by the scholars Chen, Shao, Lam, Wong, Lee, and Kim as the main

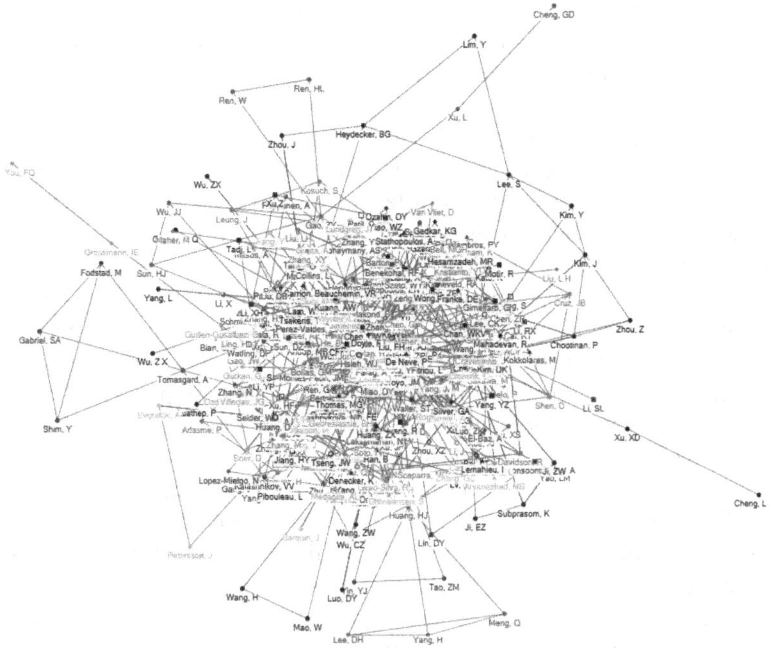

Fig. 1.12 The initial network constructed by the cooperation of all authors in the database

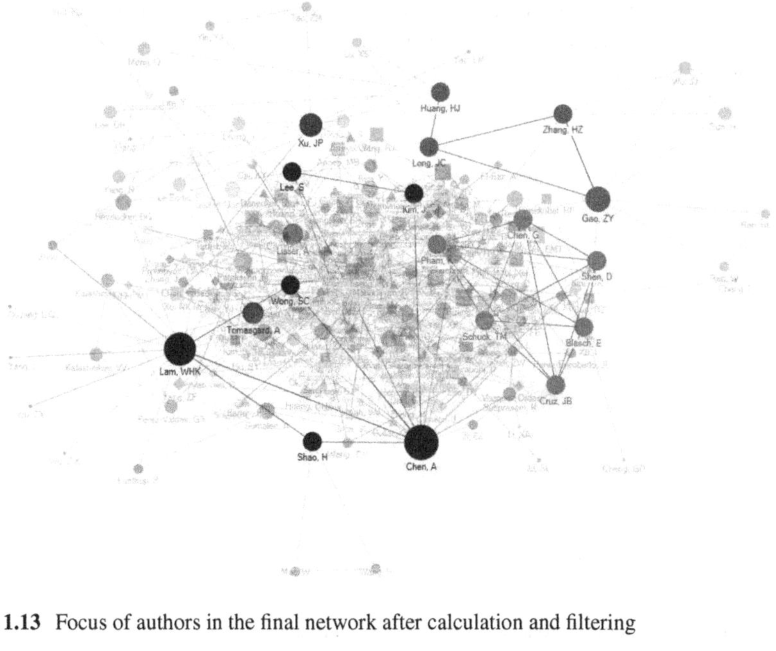

Fig. 1.13 Focus of authors in the final network after calculation and filtering

authors. The second cluster identified Huang, Zhang, Long and Gao. The third was clustered by Chen, Shen, Pham, Schuck, Cruz and Blasch. In addition to these two clusters, some scholars, had conducted more in-depth research in their respective areas, such as Xu, Lisser, Tomasgard and Patriksson.

Keywords in the same article have a close relationship, so because of these relationships, all keywords in the articles formed a large complex network. The importance of different keywords is different, which depends on their frequency of appearance and the number of related words. Keywords which have a higher appearance frequency also have more chance of being connected with other keywords. Some keywords appeared in pairs in different articles, which strengthened the relationship. Some keywords may be low in appearance frequency, but high in degree of centrality, because they serve as important link between some concepts. The identification of the important keywords and the relationships is significant as it allows for an in-depth exploration of this field and points to the research directions. Using NodeXL, the related keywords were input to form an initial network, and then matrix calculated these keywords' degree of centrality. Using the Clauset-Newman-Moore algorithmic [32] clustering analysis, these keywords were automatically clustered with different shapes and colors by the Harel-Koren Fast Multi-scale layout algorithm [63], eventually forming the network as shown in Fig. 1.14.

However, as the network had so many entities, the relationships between them were complex and hard to follow. To clarify the network, a filter was used to fade the less important entities and edges. Through adjustments to the filter degree, a simpler network was developed as shown in Fig. 1.15.

The keywords shown in Fig. 1.15 represent the most important for bi-level and random uncertainty research papers, not only because of their frequency in the documents, but also because of relationships they formed. Using colors and shapes, these keywords were identified, as listed in the Table 1.6.

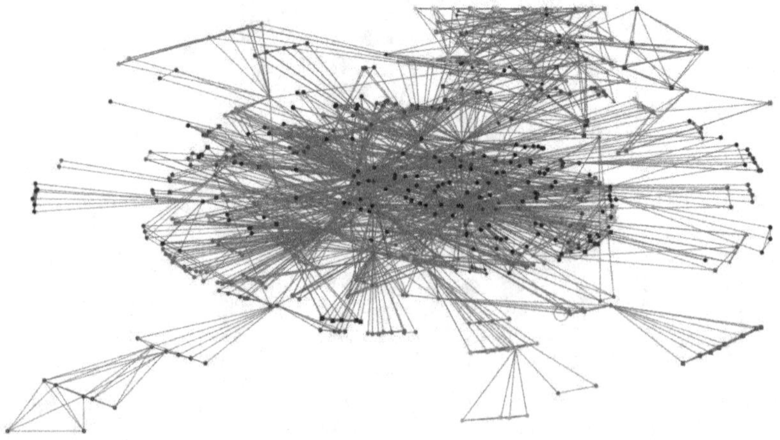

Fig. 1.14 The initial network constructed by the logical relationship between keywords

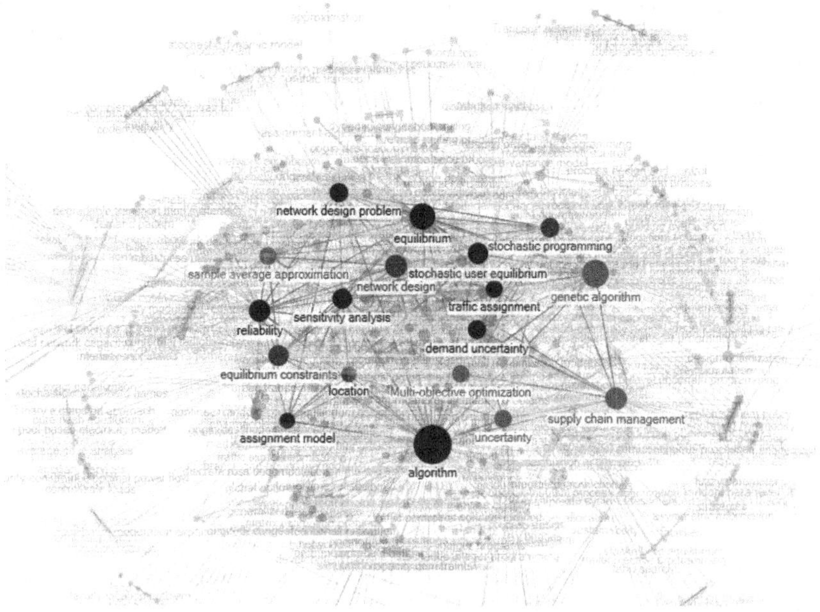

Fig. 1.15 Focuses of keywords in the final network after calculated and filtered

Table 1.6 Keywords and associated keywords quantity

Keywords	Degree	Betweenness Centrality	Eigenvector Centrality	PageRank	Clustering Coefficient
Supply chain management	61	14681.895	0.008	6.790	0.106
Algorithm	195	65478.529	0.029	20.303	0.038
Demand uncertainty	44	6363.183	0.010	4.332	0.183
Reliability	61	18047.826	0.011	5.895	0.133
Genetic algorithm	95	27910.056	0.014	10.818	0.064
Uncertainty	40	9863.936	0.006	4.318	0.155
Network design	65	9963.269	0.015	6.094	0.145
Traffic assignment	38	1427.832	0.010	3.493	0.248
Location	30	3793.865	0.006	3.282	0.205
Sensitivity analysis	50	5211.595	0.010	4.624	0.163
Stochastic programming	48	10016.045	0.009	5.191	0.143
Network design problem	45	7221.254	0.011	4.204	0.207
Equilibrium constraints	54	12656.639	0.008	5.474	0.131
Stochastic user equilibrium	56	4398.029	0.012	5.259	0.151
Equilibrium	86	17770.430	0.017	8.401	0.102
Sample average approximation	40	4339.663	0.007	3.625	0.253
Multi-objective optimization	39	6077.970	0.008	4.002	0.188
Assignment model	30	4326.231	0.007	2.981	0.234

By comparing keyword frequency and other indicators, the research direction of each category was determined. The category with the most keywords was algorithm, which was seen to be a core item with the most important content.

As listed in the Table 1.6, multi-objective optimization is a major focus of bi-level and random uncertainty research, which is also the focus of this book. Algorithm, equilibrium and sensitivity analyses are also major foci for bi-level research. Because bi-level programming problems fall into the class of NP-hard problems, many scholars have attempted to find effective algorithms to solve this type of problems. Equilibrium and sensitivity analyses are also important in parameter analysis and bi-level game analysis. Specific problems, such as supply chain management, reliability, network design, traffic assignment, location problems, and assignment problems were found to be the foci of the most bi-level and random uncertainty problems.

References

1. Abass SA (2005) Bilevel programming approach applied to the flow shop scheduling problem under fuzziness. Comput Manag Sci 4(4):279–293
2. Aiyoshi E, Shimizu K (1981) Hierarchical decentralized systems and its new solution by a barrier method. IEEE Trans Syst Man Cybern 11(6):444–449
3. Aiyoshi E, Shimizu K (1984) A solution method for the static constrained Stackelberg problem via penalty method. IEEE Trans Autom Control 29(12):1111–1114
4. Aliakbarian N, Dehghanian F, Salari M (2015) A bi-level programming model for protection of hierarchical facilities under imminent attacks. Comput Oper Res 64(3):210–224
5. Al-Khayyal FA, Horst R, Pardalos PM (1992) Global optimization of concave functions subject to quadratic constraints: an application in nonlinear bilevel programming. Ann Oper Res 34(1):125–147
6. Bard J, Falk J (1982) An explicit solution to the multi-level programming problem. Comput Oper Res 9(1):77–100
7. Bard J, Moore J (1990) A branch and bound algorithm for the bilevel programming problem. SIAM J Sci Stat Comput 11:281
8. Bard JF (1985) Geometric and algorithmic developments for a hierarchical planning problem. Eur J Oper Res 19(3):372–383
9. Bard J.F (1988) Convex two-level optimization. Math Program 40(1):15–27
10. Bard JF, Falk J (1982) An explicit solution to the multi-level programming problem. Comput Oper Res 9:77–100
11. Bard JF, Moore J (1990) A branch and bound algorithm for the bilevel programming problem. SIAM J Sci Stat Comput 11:281–292
12. Ben-Ayed O (1993) Bilevel linear programming. Comput Oper Res 20(5):485–501
13. Ben-Ayed O, Blair CE, Boyce DE, LeBlanc LJ (1992) Construction of a real-world bilevel linear programming model of the highway network design problem. Ann Oper Res 34(1): 219–254
14. Berberian S (1965) Measure and integration. The Macmillan Co., New York; Collier-Macmillan Ltd., London
15. Beyer I (2016) Information technology-based logistics planning: approaches to developing a coordination mechanism for decentralized planning. Commun Llma. doi:10.1016/j.neucom.2016.01.031

16. Bi Z, Calamai PH, Conn AR (1989) An exact penalty function approach for the linear bilevel programming problem. Tech. rep., Department of Systems Design Engineering, University of Waterloo. Technical Report #167-O-310789
17. Bi Z, Calamai PH, Conn AR (1989) An exact penalty function approach for the nonlinear bilevel programming problem. University of Waterloo. Technical Report No. 167-0-310789
18. Bialas W, Karwan M (1984) Two-level linear programming. Manag Sci 30:1004–1020
19. Bialas W, Karwan M, Shaw J (1980) A parametric complementary pivot approach for two-level linear programming. State University of New York at Buffalo
20. Bianco L, Caramia M, Giordani S (2009) A bilevel flow model for Hazmat transportation network design. Transp Res Part C Emerg Technol 17(2):175–196
21. Billingsley P (1965) Probability and measure. John Wiley & Sons, New York
22. Blackwell D (1956) On a class of probability spaces. In: Proceedings of the third Berkeley symposium on mathematical statistics and probability: contributions to econometrics industrial research and psychometry. University of California Press, Berkeley, p 1
23. Bracken J, Falk JE, Miercort FA (1977) A strategic weapons exchange allocation model. Oper Res 25(6):968–976
24. Buckley J (2006) Fuzzy probability and statistics. Springer, Berlin
25. Calamai P, Vicente L (1993) Generating bilevel and linear-quadratic programming problems. SIAM J Sci Stat Comput 14:770–782
26. Campos L, Verdegay J (1989) Linear programming problems and ranking of fuzzy numbers. Fuzzy Sets Syst 32(1):1–11
27. Candler W, Fortuny-Amat J, McCarl B (1981) The potential role of multilevel programming in agricultural economics. Am J Agric Econ 63(3):521–531
28. Candler W, Townsley R (1982) Linear two-level programming problem. Comput Oper Res 9(1):59–76
29. Case LM (1999) An ℓ_1 penalty function approach to the nonlinear bilevel programming problem. Ph.D. thesis, University of Waterloo, Ontario
30. Cassidy RG, Kirby MJL, Raike WM (1971) Efficient distribution of resources through three levels of government. Manag Sci 17(8):462–473
31. Chang TS, Luh P (1984) Derivation of necessary and sufficient conditions for single-stage stackelberg games via the inducible region concept. IEEE Trans Autom Control 29(1):63–66
32. Clauset A, Newman MEJ, Moore C (2004) Finding community structure in very large networks. Phys Rev E 70(6):066, 111
33. Cohn D (1980) Measure theory. Birkhäuser, Boston
34. Colson B, Marcotte P, Savard G (2005) Bilevel programming: a survey. 4OR: Q J Oper Res 3(2):87–107
35. Colson B, Marcotte P, Savard G (2007) An overview of bilevel optimization. Ann Oper Res 153(1):235–256
36. Deng X (1998) Complexity issues in bilevel linear programming. In: Multilevel optimization: algorithms and applications. Kluwer Academic, Dordrecht, pp 149–164
37. Desilva AH (1978) Sensitivity formulas for nonlinear factorable programming and their application to the solution of an implicitly defined optimization model of US crude oil production. Ph.D. thesis, George Washington University, Washington, DC
38. Dimitriou L, Tsekeris T, Stathopoulos A (2008) Genetic computation of road network design and pricing stackelberg games with multi-class users. In: Giacobini M, Brabazon A, Cagnoni S, Di Caro GA, Drechsler R, Ekárt A, Esparcia-Alcázar AI, Farooq M, Fink A, McCormack J, O'Neill M, Romero J, Rothlauf F, Squillero G, Uyar AŞ, Yang S (eds) EvoWorkshops 2008. LNCS, vol 4974. Springer, Heidelberg, pp 669–678
39. Dubois D, Prade H (1978) Operations on fuzzy numbers. Int J Syst Sci 9(6):613–626
40. DuBois D, Prade H (1980) Fuzzy sets and systems: theory and applications. Academic Press, New York
41. Dubois D, Prade H (1987) Fuzzy numbers: an overview. Anal Fuzzy Inf Math Logics 1:3–39
42. Dubois D, Prade H (1988) Possibility theory: an approach to computerized processing of uncertainty. Plenum Press, New York

43. Durrett R, Durrett R (2010) Probability: theory and examples. Cambridge University Press, Cambridge/New York
44. Edmunds TA, Bard JF (1991) Algorithms for nonlinear bilevel mathematical programs. IEEE Trans Syst Man Cybern 21(1):83–89
45. Eichfelder G (2007) Solving nonlinear multiobjective bilevel optimization problems with coupled upper level constraints. Technical Report Preprint No. 320, Preprint-Series of the Institute of Applied Mathematics, University of Erlangen-Nürnberg
46. Eichfelder G (2008) Multiobjective bilevel optimization. Math Program. doi:10.1007/s10107-008-0259-0
47. Evans L, Gariepy R (1992) Measure theory and fine properties of functions. CRC, Boca Raton
48. Falk JE (1973) A linear max–min problem. Math Program 5(1):169–188
49. Falk JE, Liu J (1995) On bilevel programming, Part I: general nonlinear cases. Math Program 70(1):47–72
50. Fampa M, Barroso LA, Candal D, Simonetti L (2008) Bilevel optimization applied to strategic pricing in competitive electricity markets. Comput Optim Appl 39:121–142
51. Ferrero A, Prioli M, Salicone S (2015) Conditional random-fuzzy variables representing measurement results. IEEE Trans Instrum Meas 64(5):1170–1178
52. Feller W (1971) An introduction to probability theory and its application. Tome II, John Wiley & Sons, New York
53. Fernández-Blanco R, Arroyo J, Alguacil N (2012) A unified bilevel programming framework for price-based market clearing under marginal pricing. IEEE Trans Power Syst 27(11): 517–525
54. Fliege J, Vicente LN (2006) Multicriteria approach to bilevel optimization. J Optim Theory Appl 131(2):209–225
55. Fortuny-Amat J, McCarl B (1981) A representation and economic interpretation of a two-level programming problem. J Oper Res Soc 32:783–792
56. Gaur A, Arora SR (2008) Multi-level multi-attributemulti-objective integer linear programming problem. AMO-Adv Model Optim 10(2):297–322
57. González A (1990) A study of the ranking function approach through mean values. Fuzzy Sets Syst 35(1):29–41
58. Guo XN, Hua TS, Zhang T, Lv YB (2012) Bilevel model for multi-reservoir operating policy in inter-basin water transfer-supply project. J Hydrol 424:252–263
59. Halmos P (1950) Measure theory. Van Nostrad Reinhold Company, New York
60. Halmos P (1974) Naive set theory. Springer, New York
61. Hansen P, Jaumard B, Savard G (1992) New branch-and-bound rules for linear bilevel programming. SIAM J Sci Stat Comput 13:1194
62. Halter W, Mostaghim S (2006) Bilevel optimization of multicomponent chemical systems using particle swarm optimization. In: Proceedings of world congress on computational intelligence (WCCI 2006), pp 1240–1247
63. Harel D, Koren Y (2001) A fast multi-scale method for drawing large graphs//Graph drawing. Springer, Berlin/Heidelberg
64. Herskovits J, Leontiev A, Dias G, Santos G (2000) Contact shape optimization: a bilevel programming approach. Struct Multidisc Optim 20:214–221
65. Ho YC, Luh P, Muralidharan R (1981) Information structure, Stackelberg games, and incentive controllability. IEEE Trans Autom Control 26(2):454–460
66. Ishizuka Y, Aiyoshi E (1992) Double penalty method for bilevel optimization problems. Ann Oper Res 34(1):73–88
67. Jacobs K, Kurzweil J (1978) Measure and integral. Academic Press, New York
68. Judice J, Faustino A (1988) The solution of the linear bilevel programming problem by using the linear complementarity problem. Investigação Operacional 8:77–95
69. Júdice J, Faustino AM (1992) A sequential LCP method for bilevel linear programming. Ann Oper Res 34(1):89–106
70. Katsoulakos NM, Kaliampakos DC (2016) Mountainous areas and decentralized energy planning: insights from Greece. Energy Policy 91:174–188

71. Kaufmann A (1975) Introduction to the theory of fuzzy subsets. Academic Press, New York
72. Klement E, Puri M, Ralescu D (1986) Limit theorems for fuzzy random variables. Proc R Soc Lond Ser A Math Phys Sci 407(1832):171–182
73. Koh A (2007) Solving transportation bi-level programs with differential evolution. In: IEEE congress on evolutionary computation (CEC 2007), pp 2243–2250. IEEE Press
74. Kolmogorov A (1950) Foundations of the theory of probability. Chelsea Publishing Company. New York
75. Kolstad CD, Lasdon LS (1990) Derivative estimation and computational experience with large bilevel mathematical programs. J Optim Theory Appl 65:485–499
76. Kong QF, Yang YN (2011) A political economic analysis on multilateral agricultural negotiation: based on two-level interactive evolutional game model. J Int Trade 6(3): 21–34
77. Krickeberg K (1965) Probability theory. Addison-Wesley, Reading
78. Kruse R, Meyer K (1987) Statistics with vague data. Springer, Dordrecht
79. Kuhn H, Tucker A (1951) Non linear programming. In: Neyman J (ed) 2nd Berkeley Symposium on Mathematical Statistics and Probability, Berkeley, pp 481–492. University of California
80. Kwakernaak H (1978) Fuzzy random variables-I. Definitions and theorems. Information Sciences 15(1):1–29
81. Kwakernaak H (1979) Fuzzy random variables-II. Algorithms and examples for the discrete case. Inf Sci 17(3):253–278
82. Kydland F (1975) Hierarchical decomposition in linear economic models. Manag Sci 21(9):1029–1039
83. Laha R, Rohatgi V (1985) Probability Theory. John Wiley & Sons, New York
84. Li X, Tian P, Min X (2006) A hierarchical particle swarm optimization for solving bilevel programming problems. In: Rutkowski L, Tadeusiewicz R, Zadeh LA, Żurada JM (eds) ICAISC 2006. LNCS (LNAI), vol 4029. Springer, Heidelberg, pp 1169–1178
85. Li H, Wang Y (2007) A genetic algorithm for solving a special class of nonlinear bilevel programming problems. In: Shi Y, van Albada GD, Dongarra J, Sloot PMA (eds) ICCS 2007, Part IV. LNCS, vol 4490. Springer, Heidelberg, pp 1159–1162
86. Liu Y, Liu B (2003) Expected value operator of random fuzzy variable and random fuzzy expected value models. Int J Uncertain Fuzziness Knowl Based Syst 11(2):195–216
87. Lu J, Atamturktur S, Huang Y (2016) Bi-level resource allocation framework for retrofitting bridges in a transportation network. Transp Res Rec J Transp Res Board 2550:31–37
88. Mathieu R, Pittard L, Anandalingam G (1994) Genetic algorithm based approach to bi-level linear programming. Oper Res 28(1):1–21
89. Meijboom BR (1987) Planning in decentralized firms—a contribution to the theory on multilevel decisions. Springer, Berlin/New York
90. Nahmias S (1978) Fuzzy variables. Fuzzy Sets Syst 1(2):97–110
91. Näther W (2000) On random fuzzy variables of second order and their application to linear statistical inference with fuzzy data. Metrika 51(3):201–221
92. Oduguwa V, Roy R (2002) Bi-level optimisation using genetic algorithm. In: IEEE International Conference on Artificial Intelligence Systems (ICAIS 2002), pp 322–327
93. Peng J, Liu B (2007) Birandom variables and birandom programming. Comput Ind Eng 53(3):433–453
94. Pfeiffer P (1990) Probability for applications. Springer
95. Puri M, Ralescu D (1986) Fuzzy random variables. J Math Anal Appl 114(2):409–422
96. Sakawa M, Nishizaki I (2002) Interactive fuzzy programming for decentralized two-level linear programming problems. Fuzzy Set Syst 125(3):301–315
97. Savard G, Gauvin J (1994) The steepest descent direction for the nonlinear bilevel programming problem. Oper Res Lett 15(5):265–272
98. Shi X, Xia H (1997) Interactive bilevel multi-objective decision making. Journal of the Operational Research Society 48(9):943–949
99. Shimizu K, Aiyoshi E (1981) A new computational method for Stackelberg and min-max problems by use of a penalty method. IEEE Trans Autom Control 26:460–466

100. Simaan M, Cruz J.B (1973) On the Stackelberg strategy in nonzero-sum games. J Optim Theory Appl 11(5):533–555
101. Srivastava S (1998) A course on Borel sets. Springer, New York
102. Stoilov T, Stoilova K (2012) Portfolio risk management modelling by bi-level optimization. Handbook on Decision Making, pp 91–110
103. Sun D, Benekohal RF, Waller ST (2006) Bi-level programming formulation and heuristic solution approach for dynamic traffic signal optimization. Comput-Aided Civil Infrastruct Eng 21(5):321–333
104. Talbi E (2014) Metaheuristics for bilevel optimization, Springer, Berlin/New York
105. Talbi E (2009) Metaheuristics: from design to implementation. John Wiley & Sons, Hoboken
106. Thoai NV, Yamamoto Y, Yoshise A (2005) Global optimization method for solving mathematical programs with linear complementarity constraints. J Optim Theory Appl 124(2):467–490
107. Tuy H, Migdalas A, Värbrand P (1993) A global optimization approach for the linear two-level program. J Global Optim 3(1):1–23
108. Vallejo JFC, Sánchez RM (2013) A path based algorithm for solve the hazardous materials transportation bilevel problem. Appl Mech Mater 253–255:1082–1088
109. Vicente L, Savard G, Judice J (1996) Discrete linear bilevel programming problem. J Optim Theory Appl 89(3):597–614
110. Vicente LN, Calamai PH (2004) Bilevel and multilevel programming: a bibliography review. J Global Optim 5(3):291–306
111. Wang HF (2011) Multi-level subsidy and penalty strategy for a green industry sector. In: IEEE ninth international conference on dependable, autonomic and secure computing (DASC 2011), pp 776–783
112. Wen UP, Hsu ST (1991) Linear bi-level programming problems—a review. J Oper Res Soc 42:125–133
113. White DJ, Anandalingam G (1993) A penalty function approach for solving bi-level linear programs. J Global Optim 3(4):397–419
114. Xu J, He Y, Gen M (2009) A class of random fuzzy programming and its application to supply chain design. Comput Ind Eng 56(3):937–950
115. Xu J, Liu Q, Wang R (2008) A class of multi-objective supply chain networks optimal model under random fuzzy environment and its application to the industry of Chinese liquor. Inf Sci 178(8):2022–2043
116. Xu JP, Tu Y, Zeng ZQ (2012) Bi-level optimization of regional water resources allocation problem under fuzzy random environment. J Water Resour Plan Manag 139(3):246–264
117. Xu JP, Yao LM (2011) Random-like multiple objective decision making. Springer, Berlin/Heidelberg
118. Xu JP, Wang Z, Zhang M, et al (2016) A new model for a 72-h post-earthquake emergency logistics location-routing problem under a random fuzzy environment. Transp Lett Int J Transp Res. doi:10.1080/19427867.2015.1126064
119. Yager R (1981) A procedure for ordering fuzzy subsets of the unit interval. Inf Sci 24(2):143–161
120. Yager R (2002) On the evaluation of uncertain courses of action. Fuzzy Optim Decis Making 1(1):13–41
121. Yates J, Casas I (2010) Role of spatial data in the protection of critical infrastructure and homeland defense. Appl Spat Anal Policy 5:1–23
122. Ye J (2006) Constraint qualifications and KKT conditions for bilevel programming problems. Math Opera Res 31(4):811–824
123. Yin Y (2000) Genetic algorithm based approach for bilevel programming models. J Transp Eng 126(2):115–120
124. Zadeh L (1965) Fuzzy sets. Inf control 8(3):338–353
125. Zadeh L (1972) A fuzzy-set-theoretic interpretation of linguistic hedges. Cybern Syst 2(3): 4–34
126. Zadeh L (1978) Fuzzy sets as a basis for a theory of possibility. Fuzzy Sets Syst 1(1):3–28

127. Zadeh L, et al (1975) Calculus of fuzzy restrictions. Fuzzy sets and their applications to cognitive and decision processes. Academic Press, New York, pp 1–39
128. Zannetos ZS (1965) On the theory of divisional structures: aspects of centralization and decentralization of control and decision-making. Manag Sci 12(4):49–68
129. Zhang W, Fan J (2015) Cloud architecture intrusion detection system based on KKT condition and hyper-sphere incremental SVM algorithm. J Comput Appl 35(10):2886–2890
130. Zhou X, Xu J (2009) A class of integrated logistics network model under random fuzzy environment and its application to Chinese beer company. Int J Uncertain Fuzziness Knowl-Based Syst 17(6):807–831

Chapter 2
Bi-Level Decision Making in Random Phenomenon

Abstract Random phenomena are events for which the possible outcomes are known, but the actual outcome in a given situation is not known. However, while individual outcomes are uncertain, a regular distribution can be seen after many repetitions. Random phenomena can be found in many problems, such as random networks (Jae-Hyeok et al., Acs Appl Mater Interfaces 7(3):1560–1567, 2015), stochastic processes (Barone-Adesi, Stochastic processes. Wiley encyclopedia of management. John Wiley & Sons, Hoboken, 2015; Papoulis, Probability, random variables and stochastic processes stochastic processes. McGraw-Hill, New York, 1991), and random noise (Stratonovich and Silverman, Topics in the theory of random noise. Gordon and Breach, New York, 1967). To describe bi-level decision making models with random phenomena, this chapter examines a practical example from a regional water resources allocation problem (RWRAP). The first section introduces the background to the problem, gives the bi-level problem description and discusses the random phenomena in the RWRAP. Then, several bi-level decision making models with random coefficients are developed and the transformation methods and properties are discussed. To solve the bi-level models, several algorithmic designs are outlined and numerical examples are given to illustrate the effectiveness of the algorithms. A real-world RWRAP construction site case is then described to demonstrate the practical effectiveness of the methods.

Keywords Random phenomenon • Random EEEE model • Random CECC model • Random DEDC model • Regional water resources allocation problem

2.1 Regional Water Resources Allocation Problem

As economies develop, the demand for fresh water increases, and, in many regions, the demand is often greater than the supply, restraining economic development and reducing productivity. There are many reasons for these water shortages, such as population growth, industrial pollution and climate change. In this chapter, however, we examine water shortages which are the result of geographically and temporally uneven regional water resources distribution [3]. In these situations, conflicts often arise when the different water users (including the environment) compete for the limited water supply. To achieve sustainable development and a secure society,

J. Xu et al., *Random-Like Bi-level Decision Making*, Lecture Notes in Economics and Mathematical Systems 688, DOI 10.1007/978-981-10-1768-1_2

water allocation policies and methodologies need to be reformed, especially in regions suffering from chronic water shortages. To ensure overall fairness to all users, water must be allocated according to the three key principles of equity, efficiency and sustainability [124], in which equity ensures that water resources within a river basin are shared fairly by all stakeholders; efficiency focuses on the economic use of the water resources with respect to cost minimization and benefit maximization; and sustainability refers to the economic use of water to guarantee that the environment is and will not be not damaged. However, it is difficult to fulfill all three principles when seeking to solve basin level water allocation problems [126]. Effective management and rational allocation are the key to resolving water shortages, so more stringent requirements are needed to ensure optimal regional water resources allocation.

In recent decades, many studies have focused on regional water resources allocation problems (RWRAPs) and, consequently, many models and algorithms have been developed [1, 15, 41, 56, 128]. However, most of these studies have lacked effective incentives to save water, with many resulting in a serious waste of scarce water resources and an aggravation of the supply and demand imbalances. Economic measures, such as the allocation of water rights and the establishment of water markets, have become more popular in ensuring water resources allocation optimization. Water rights refer to the right to use water resources for purposes such as navigation, water withdrawal, and to trade water for personal or financial gain. Water users' original water rights are assigned by a water resources administrator, after which the rights become personal property that can be used or traded in a water market to gain profit. A water market refers to the water rights trading mechanisms that have become important economic tools in achieving the RWRAP optimization, as they can improve water use and allocation efficiency [36]. Since the 2000 Ministerial Meeting of the World Water Forum held at the Hague, water rights theoretical frameworks, policies and regulations, and reform and practices have been actively examined [42, 126]. Water rights management has slowly become a new focus for resources management systems with the incorporation of a mechanism that effectively combines regional authorities macro-control with market regulations. Water trading, however, must be seen as significantly different from other general commodities. Specifically, the water market is only a quasi-market as it cannot be separated from a regional authority's macro-control, whose aim is to optimally allocate the regional water resources to ensure social equity and environmental sustainability. Under river basin water rights management, regional authorities and the sub-area water managers operate relatively independently in a cooperative environment. A RWRAP based on water rights allocated according to market mechanisms, therefore, is considered a game with bi-level decision makers who select their respective strategies through a decision-making consensus process; a process known as a game decision-making problem. Bi-level RWRAP modeling and solutions have become a new operations research area.

In water resources management, precise quantification of the water system performance criteria and the parameter and decision variables is not always possible, nor is it always necessary [83]. In the past few decades, many inexact optimization methods have been developed to deal with the uncertainties inherent in water resources management [57, 61]. While these studies have significantly improved the measurement of the uncertainty, they have not yet been able to fully account for the multiple parameter uncertainties presented as combinations of fuzziness and probability distributions. In practical decision-making processes, water allocation often takes place in a hybrid uncertain environment, the events for which are characterized by both fuzzy uncertainty and randomness; the so-called 'twofold' uncertainty. In these cases, random variables have often been employed. There have been several bi-level optimization methods proposed to deal with a RWRAP under uncertainty. The interval fuzzy bi-level programming (IFBP) approach [85], however, has proved to be increasingly important for efficient and effective regional water resources system planning. While the IFBP has advantages when seeking to address fuzzy information because of its lower data collection and solution generation requirements, the bi-level programming method proposed in this chapter promotes an increased motivation to save water, and alleviates the supply and demand contradictions by introducing the water rights and water market concepts in more complex circumstances which consider the multiple decision makers on the lower level. Specifically, the water demand and stream flow imprecisions are expressed as random variables, which are better able to describe the uncertain environments encountered in a real-life RWRAP. When these actual representations are available, the RWRAP can be realistically solved.

During the last two decades, water resources planning and management research has seen a dramatic increase in the development and application of various types of evolutionary algorithms (EAs) [101, 125] that can be applied to bi-level programming problems such as the RWRAP. The solution methods for bi-level programming problems are generally as follows (as mentioned in the Chap. 1): (1) the vertex enumeration method [7], in which the optimal solution is one of the vertices, or points in a feasible space in the bi-level problem; (2) the Kuhn-Tucker method [118], in which the Kuhn-Tucker conditions are applied to deal with the various levels, thereby converting the bi-level model into the single-level model; (3) the penalty function approach [129], which adds a penalty term to the upper level objective function to satisfy the optimality of the lower level problem; and (4) methods based on meta heuristics, which include genetic algorithms (GAs) [49], simulated annealing (SA) [109], and hybrid tabu-ascent algorithms [44].

Most existing exact techniques can only effectively solve very simple problems, but complex or large-scale problems such as bi-level problems cannot be optimally solved in a reasonable amount of time and tend to have inconsistent performance. Also, because of the need to design meta-heuristics methods for each individual problem, it is often difficult to find a usual or normal pattern [133]. Furthermore, as the methods described above are all designed to find Stackelberg solutions, it is assumed that there is no communication between the decision makers on the two levels, and no binding agreements are made even if such communication exists [117]. However, in most bi-level programming problems with multiple lower level decision makers, the upper and lower level objective functions rely partly on a degree of cross-level interaction or cooperation, even though the information is incomplete and vague. From this then, when the fuzziness of human judgment is considered, both upper and lower level decision makers have fuzzy goals for their objective functions [72, 74]. Since the interactive fuzzy programming technique is a useful tool for dealing with this situation, and has been applied extensively to bi-level and multilevel programming problems [4, 111, 122, 123], it is employed here to convert the bi-level RWRAP programming model into a single-level model. The popularity of evolutionary algorithms (EA) has been due to their potential to solve nonlinear, nonconvex, multimodal, and discrete problems in which deterministic search techniques have difficulties or fail completely. GAs have been the most commonly applied EAs in water resources planning and management literature [101]. At the end of this chapter, an entropy-Boltzmann selection-based GA (EBS-based GA) is developed to solve the generated single-level programming model. Then, results and comparison analyses using a case study are presented to demonstrate the practicality and efficiency of the optimization method.

2.1.1 Random Phenomenon

The need to address uncertainty in water management is widely recognized, as uncertainties exist in a variety of system components, so a linkage to pre-regulated policies formulated by local authorities is desired. As a result, the inherent complexity and stochastic uncertainty existing in real-world water resources decision-making have essentially placed them beyond conventional deterministic optimization methods.

1. Water demand

Previous studies on water resources allocation under market mechanisms have usually considered water demand as deterministic. However, this is not consistent with reality. Since the water use decision making of each sub-area lower level water manager is dependent on rate structures, they have an incentive to reduce water use to reduce the revenues paid to the regional authority in the upper level. However, when seeking to reduce these rates, it is difficult to avoid the influence of the water use reduction on the water users income in each sub-area, so in these cases, water demand decisions tend to fluctuate due to uncertainty, lack of evidence, insufficient information, and the regional water systemą́s dynamic environment. It is therefore often difficult for water managers to provide a crisp description, so these parameters are assumed to be flexible or imprecise. For instance, in this study, because it is very difficult to estimate the exact water demand, it is vaguely defined, with the water managers providing a range of values, in which the most possible value is arandom variable; i.e., viz $\mathcal{N}(\mu, \sigma^2)$. Therefore, it is appropriate to consider water demand as a random variable (see Fig. 2.1).

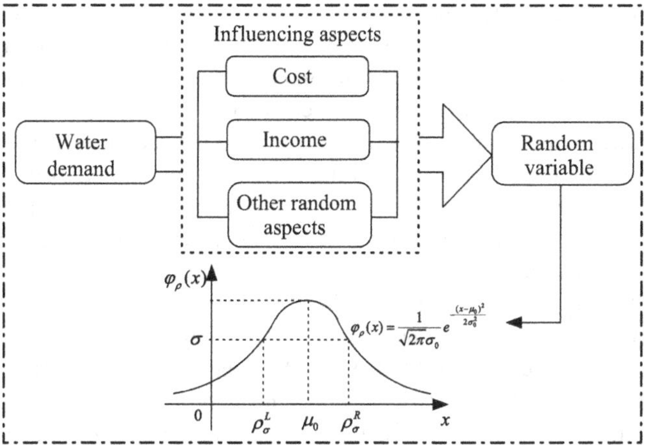

Fig. 2.1 Random description of water demand

2. Stream flow

The regional authority must make a decision about the allocation of the scarce water supplies to competing users. However, making such a decision is somewhat difficult as future flows in the river are uncertain, so only when these uncertain water flows are quantified, can allocations be given. Future water supply availability is uncertain because of stream flow variabilities, which are judged using three different scenarios: low level, medium level, high level; and which are presented as probability distributions [57]. Based on statistical data, the probabilities for the three levels are denoted p_1, p_2, and p_3 respectively. In summary, the stream flow volume ofstream flow (i.e., available water for distribution) is considered a random variable $\tilde{Q}$:

$$\tilde{Q} = \begin{cases} a_1 \text{ with probability } p_1, \\ a_2 \text{ with probability } p_2, \\ a_3 \text{ with probability } p_3. \end{cases}$$

2.1.2 Bi-Level Description

This chapter considers regional water planning design in sub-areas that are all located in the same river basin. Generally, for regional water planning under market mechanisms, because each sub-area is an independent decision maker, water resources allocation volume is not directly determined by the river basin regional authority. Instead, the regional authority allocates the initial water rights to the sub-areas, then, each sub-area water manager makes water trading decisions under a market mechanism based on water use volume and water allocations to promote equitable cooperation in the river basin and to achieve efficient water use. At the same time, the water users pay for water withdrawal. Therefore, the regional authority influences the decision-making of each sub-area water manager by adjusting the initial water rights, while each sub-area water manager strives to meet their individual economic benefit goals by making rational water withdrawal decisions based on the initial decisions made by the regional authority. To ensure the sustainable development of the river basin, basin water resources are primarily used for ecological water, industrial water, municipal water and agricultural water, with ecological water taking priority.

In this situation, the regional authority has the higher authority, so the sub-area water managers are subordinates, even though they are relatively independent decision-makers. Therefore, the RWRAP in this chapter can be abstracted as a bi-level programming problem, with the upper level programming being the main water distribution body from which the regional authority (i.e., the upper level leader) makes decisions about the initial water rights allocations to n sub-areas to maximize total social benefit. The following or lower level decision makers are the sub-area water managers (i.e., the followers), who then decide on the water withdrawal volume required to maximize economic benefits (see Fig. 2.2).

2.2 Bi-Level Decision Making Models with Random Coefficients

In the mathematical model for these practical problems, the random phenomenon is depicted as random coefficients. In this section, we propose several general bi-level models with random coefficients to describe a range of situations.

For the simplest case, the general single-objective bi-level decision making model with random coefficients is formulated as follows,

$$\begin{cases} \max\limits_{x \in R^{n_1}} F(\boldsymbol{x}, \boldsymbol{y}, \boldsymbol{\xi}) \\ \text{s.t.} \begin{cases} G_i(\boldsymbol{x}, \boldsymbol{y}, \boldsymbol{\xi}) \leq 0, i = 1, 2, \cdots, q \\ \text{where } \boldsymbol{y} \text{ solves:} \\ \begin{cases} \max\limits_{y \in R^{n_2}} f(\boldsymbol{x}, \boldsymbol{y}, \boldsymbol{\xi}) \\ \text{s.t. } g_j(\boldsymbol{x}, \boldsymbol{y}, \boldsymbol{\xi}) \leq 0, j = 1, 2, \cdots, p, \end{cases} \end{cases} \end{cases} \tag{2.1}$$

where, $\boldsymbol{x} \in R^{n_1}$ is the decision vector for the upper-level decision maker and $\boldsymbol{y} \in R^{n_2}$ is the decision vector for the lower-level decision maker; $F(\boldsymbol{x}, \boldsymbol{y}, \boldsymbol{\xi})$ is the objective function of the upper-level model and $f(\boldsymbol{x}, \boldsymbol{y}, \boldsymbol{\xi})$ is the objective function of the lower-level model $G_i(\boldsymbol{x}, \boldsymbol{y}, \boldsymbol{\xi})$ is the constraint of the upper-level model and $g_j(\boldsymbol{x}, \boldsymbol{y}, \boldsymbol{\xi})$ is the constraint of the lower-level model; $\boldsymbol{\xi}$ is a random vector.

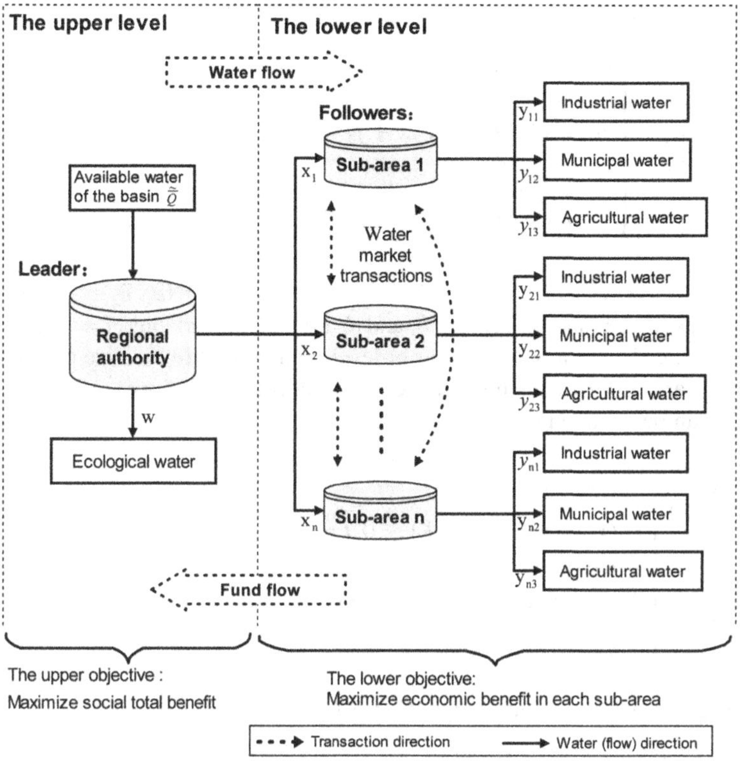

Fig. 2.2 Model structure for RWRAP

If there are more than one objectives for the decision maker, the general form of the bi-level multi-objective decision making (MODM) model with random coefficients are presented as follows.

$$\begin{cases} \min\limits_{x \in R^{n_1}} [F_1(\boldsymbol{x}, \boldsymbol{y}, \boldsymbol{\xi}), F_2(\boldsymbol{x}, \boldsymbol{y}, \boldsymbol{\xi}), \cdots, F_m(\boldsymbol{x}, \boldsymbol{y}, \boldsymbol{\xi})] \\ \text{s.t.} \begin{cases} G_i(\boldsymbol{x}, \boldsymbol{y}, \boldsymbol{\xi}) \leq 0, i = 1, 2, \cdots, q \\ \text{where } \boldsymbol{y} \text{ solves:} \\ \begin{cases} \min\limits_{y \in R^{n_2}} [f_1(\boldsymbol{x}, \boldsymbol{y}, \boldsymbol{\xi}), f_2(\boldsymbol{x}, \boldsymbol{y}, \boldsymbol{\xi}), \cdots, f_m(\boldsymbol{x}, \boldsymbol{y}, \boldsymbol{\xi})] \\ \text{s.t. } g_j(\boldsymbol{x}, \boldsymbol{y}, \boldsymbol{\xi}) \leq 0, j = 1, 2, \cdots, p, \end{cases} \end{cases} \end{cases} \tag{2.2}$$

If all the functions in model (2.2) are linear, the general form of the linear bi-level multi-objective decision making model with random coefficients is rewritten directly as follows,

$$\begin{cases} \min\limits_{x \in R^{n_1}} \left[\tilde{a}_1^T \boldsymbol{x} + \tilde{b}_1^T \boldsymbol{y}, \tilde{a}_2^T \boldsymbol{x} + \tilde{b}_2^T \boldsymbol{y}, \cdots, \tilde{a}_m^T \boldsymbol{x} + \tilde{b}_m^T \boldsymbol{y}\right] \\ \text{s.t.} \begin{cases} \tilde{c}_i^{1T} \boldsymbol{x} + \tilde{c}_i^{2T} \boldsymbol{y} \le \tilde{d}_i, i = 1, 2, \cdots, q \\ \text{where } \boldsymbol{y} \text{ solves:} \\ \min\limits_{y \in R^{n_2}} \left[\tilde{e}_1^T \boldsymbol{x} + \tilde{f}_1^T \boldsymbol{y}, \tilde{e}_2^T \boldsymbol{x} + \tilde{f}_2^T \boldsymbol{y}, \cdots, \tilde{e}_m^T \boldsymbol{x} + \tilde{f}_m^T \boldsymbol{y}\right] \\ \text{s.t.} \begin{cases} \tilde{g}_j^{1T} \boldsymbol{x} + \tilde{g}_j^{2T} \boldsymbol{y} \le \tilde{h}_j, j = 1, 2, \cdots, p \\ \boldsymbol{x} > 0, \boldsymbol{y} > 0. \end{cases} \end{cases} \end{cases} \tag{2.3}$$

where $\tilde{c}_i = (\tilde{c}_{i1}, \tilde{c}_{i2}, \cdots, \tilde{c}_{im})^T$, $\tilde{a}_j = (\tilde{a}_{j1}, \tilde{a}_{j2}, \cdots, \tilde{a}_{jm})^T$, $\tilde{e} = (\tilde{e}_1, \tilde{e}_2, \cdots, \tilde{e}_q)^T$, $\tilde{k} = (\tilde{k}_1, \tilde{k}_2, \cdots, \tilde{k}_p)^T$ are random vectors; $\tilde{b}_i$ and $\tilde{d}_j$ are random variables, $i = 1, 2, \cdots, q$, $j = 1, 2, \cdots, p$. This is a typical linear bi-level random multi-objective problem.

If there are more than one follower, the decentralized forms of (2.1), (2.2) and (2.3) can be obtained as

$$\begin{cases} \max\limits_{x \in R^{n_1}} F(\boldsymbol{x}, \boldsymbol{y}_1, \boldsymbol{y}_2, \cdots, \boldsymbol{y}_m, \boldsymbol{\xi}) \\ \text{s.t.} \begin{cases} G_i(\boldsymbol{x}, \boldsymbol{y}_1, \boldsymbol{y}_2, \cdots, \boldsymbol{y}_m, \boldsymbol{\xi}) \le 0, i = 1, 2, \cdots, q \\ \text{where } \boldsymbol{y}_i \text{ solves:} \\ \begin{cases} \max\limits_{y_i \in R^{n_{i2}}} f_i(\boldsymbol{x}, \boldsymbol{y}_1, \boldsymbol{y}_2, \cdots, \boldsymbol{y}_m, \boldsymbol{\xi}) \\ \text{s.t. } g_j(\boldsymbol{x}, \boldsymbol{y}_1, \boldsymbol{y}_2, \cdots, \boldsymbol{y}_m, \boldsymbol{\xi}) \le 0, j = 1, 2, \cdots, p, \end{cases} \end{cases} \end{cases} \tag{2.4}$$

$$\begin{cases} \min\limits_{x \in R^{n_1}} [F_1(\boldsymbol{x}, \boldsymbol{y}_1, \boldsymbol{y}_2, \cdots, \boldsymbol{y}_m, \boldsymbol{\xi}), F_2(\boldsymbol{x}, \boldsymbol{y}, \boldsymbol{\xi}), \cdots, F_m(\boldsymbol{x}, \boldsymbol{y}, \boldsymbol{\xi})] \\ \text{s.t.} \begin{cases} G_i(\boldsymbol{x}, \boldsymbol{y}_1, \boldsymbol{y}_2, \cdots, \boldsymbol{y}_m, \boldsymbol{\xi}) \le 0, i = 1, 2, \cdots, q \\ \text{where } \boldsymbol{y}_i \text{ solves:} \\ \begin{cases} \min\limits_{y \in R^{n_2}} [f_{i1}(\boldsymbol{x}, \boldsymbol{y}, \boldsymbol{\xi}), f_{i2}(\boldsymbol{x}, \boldsymbol{y}, \boldsymbol{\xi}), \cdots, f_{im}(\boldsymbol{x}, \boldsymbol{y}_1, \boldsymbol{y}_2, \cdots, \boldsymbol{y}_m, \boldsymbol{\xi})] \\ \text{s.t. } g_j(x, \boldsymbol{y}_1, \boldsymbol{y}_2, \cdots, \boldsymbol{y}_m, \xi) \le 0, j = 1, 2, \cdots, p \end{cases} \end{cases} \end{cases} \tag{2.5}$$

and

$$\begin{cases} \min\limits_{x \in R^{n_1}} \left[\tilde{a}_1^T \boldsymbol{x} + \tilde{b}_1^T \boldsymbol{y}, \tilde{a}_2^T \boldsymbol{x} + \tilde{b}_2^T \boldsymbol{y}, \cdots, \tilde{a}_m^T \boldsymbol{x} + \tilde{b}_m^T \boldsymbol{y}\right] \\ \text{s.t.} \begin{cases} \tilde{c}_i^T \boldsymbol{x} \le \tilde{d}_i, i = 1, 2, \cdots, q \\ \text{where } \boldsymbol{y}_i \text{ solves:} \\ \min\limits_{y \in R^{n_{i2}}} \left[\tilde{e}_1^T \boldsymbol{x} + \tilde{f}_1^T \boldsymbol{y}, \tilde{e}_2^T \boldsymbol{x} + \tilde{f}_2^T \boldsymbol{y}, \cdots, \tilde{e}_m^T \boldsymbol{x} + \tilde{f}_m^T \boldsymbol{y}\right] \\ \text{s.t.} \begin{cases} \tilde{g}_j^T \boldsymbol{x} \le \tilde{h}_j, j = 1, 2, \cdots, p \\ \boldsymbol{x} > 0, \boldsymbol{y} > 0, \end{cases} \end{cases} \end{cases} \tag{2.6}$$

where $\boldsymbol{y} = (\boldsymbol{y}_1, \boldsymbol{y}_2, \cdots, \boldsymbol{y}_m)$ is a partitioned matrix.

Table 2.1 36 kinds of bi-level multi-objective models based on random operators

	Lower-level					
Upper-level	EE	EC	DE	DC	CE	CC
EE	EEEE	EEEC	EEDE	EEDC	EECE	EECC
EC	ECEE	ECEC	ECDE	ECDC	ECCE	ECCC
DE	DEEE	DEEC	DEDE	DEDC	DECE	DECC
CE	CEEE	CEEC	CEDE	CEDC	CECE	CECC
CC	CCEE	CCEC	CCDE	CCDC	CCCE	CCCC

All of models (2.1), (2.2), (2.3), (2.4), (2.5) and (2.6) are not well defined mathematically. In fact, For each given decision vector $\boldsymbol{x}, \boldsymbol{y}$, it is meaningless to optimize the objective functions, such as $F(\boldsymbol{x}), f(\boldsymbol{x})$, before we know the value of random vector, $\boldsymbol{\xi}$. Also, we cannot judge whether or not a decision vector $\boldsymbol{x}, \boldsymbol{y}$ are feasible before we know the precise value of the random vector. In order to present a mathematically meaningful random decision making model, we must consider the issues of the treatment of the constraints and objectives in models (2.1), (2.2), (2.3), (2.4), (2.5) and (2.6).

At present, there are two kinds of technique to deal with the constraints in the upper and lower level model, and three techniques to handle the objectives. These techniques are based on expected value operator and chance operator of stochastic events, represented by $f(\boldsymbol{x}, \boldsymbol{y}, \boldsymbol{\xi})$. It follows from the law of permutation and combination, there are 36 kinds of bi-level multi-objective models based on random operators theoretically as shown in Table 2.1. "E" represents Expected Value Operator, "C" represents Chance Constrained Operator, "D" represents Dependent Chance Operator. In this chapter, three kinds of bi-level multi-objective models "EEDE", "ECCE" and "DCCC" are introduced. Those three kinds of models contains all 6 philosophies for a single level programming, namely EE, EC, DE, DC, CE and CC.

This chapter selects a few of these 36 models to illustrates the methodology of bi-level MODM models with random coefficients. Before that, some elements of random operators are provided as follows.

1. Random Expected Value Operator

The *expectation* of a random variable is a central concept in the study of probability. It is the average of all possible values of a random variable, where a value is weighted according to the probability that it will appear. The expectation is sometimes also called *average*. It is also called the *expected value* or the *mean* of the random variable. These terms are all synonymous. The so-called *expected value model* (EVM) means to optimize some expected objective functions subject to some expected constraints, for example, minimizing expected cost, maximizing expected profit, and so on.

Now let us recall the well-known newsboy problem in which a boy operating a news stall has to determine the number x of newspapers to order in advance from the publisher at a cost of c per one newspaper every day. It is known that the selling price is a per one newspaper. However, if the newspapers are not sold at the end of the day, then the newspapers have a small value of b per one newspaper at the recycling center. Assume that the demand for newspapers is denoted by ξ in a day, then the number of newspapers at the end of the day is clearly $x - \xi$ if $x > \xi$ or 0 if $x < \xi$. Thus the profit of the newsboy should be

$$f(x,\xi) = \begin{cases} (a-c)x, & \text{if } x \le \xi, \\ (b-c)x + (a-b)\xi, & \text{if } x > \xi. \end{cases} \tag{2.7}$$

In practice, the demand ξ for newspapers is usually a stochastic variable, so is the profit function $f(x,\xi)$. Since we cannot predict how profitable the decision of ordering x newspapers will actually be, a natural idea is to employ the expected profit, shown as follows,

$$E[f(x,\xi)] = \int_0^x [(b-c)x + (a-b)r]dF(r) + \int_x^{+\infty} (a-c)xdF(r), \tag{2.8}$$

where E denotes the expected value operator and $F(\cdot)$ is the distribution function of demand ξ. The newsboy problem is related to determining the optimal integer number x of newspapers such that the expected profit $E[f(x,\xi)]$ achieves the maximal value, i.e.,

$$\begin{cases} \max E[f(x,\xi)] \\ \text{s.t. } x \ge 0, \text{ integers.} \end{cases} \tag{2.9}$$

This is a typical example of an expected value model.

Then we firstly should give the basic definition of the expected value. For the discrete random variable, we can define its expected value as follows.

Definition 2.1. Let ξ be a discrete random variable on the probability $(\Omega, \mathscr{A}, Pr)$ as follow,

$$\xi(\omega) = \begin{cases} x_1 & \text{if } \omega = \omega_1, \\ x_2 & \text{if } \omega = \omega_2, \\ \cdots & \cdots \end{cases} \tag{2.10}$$

where the probability of $\omega = \omega_i (i = 1, 2, \cdots)$ is p_i. If the series $\sum_{\omega \in \Omega} \xi(\omega_i) Pr\{\omega = \omega_i\}$ is absolutely convergent, then we call it the expected value of ξ, denoted by $E[\xi]$.

For the continuous random variable, its expected value can be defined as follows.

Definition 2.2 (Durrett [34]). Let ξ be a random variable on the probability space $(\Omega, \mathscr{A}, Pr)$. Then the expected value of ξ is defined by

$$E[\xi] = \int_0^{+\infty} Pr\{\xi \ge r\}dr - \int_{-\infty}^0 Pr\{\xi \le r\}dr. \tag{2.11}$$

There is another equivalent definition by the density function.

Definition 2.3 (Durrett [34]). The expected value of a random variable ξ with probability density function $f(x)$ is

$$E[\xi] = \int_{-\infty}^{+\infty} xf(x)dx. \tag{2.12}$$

Expected value, average and mean are the same thing, but median is entirely different. The median is defined below, but only to make the distinction clear. After this, we won't make further use of the median.

Definition 2.4 (Durrett [34]). The median of a random variable ξ is the unique value r in the range of ξ such that $Pr\{\xi < r\} \leq 1/2$ and $Pr\{\xi > r\} < 1/2$.

For example, with an ordinary die, the median thrown value is 4, which not the same as the mean 3.5. The median and the mean can be very far apart. To deeply understand random variables, the variance of a random variable is given as follows.

Definition 2.5 (Durrett [34]). The variance of a random variable ξ is defined by

$$V[\xi] = E[(\xi - E[\xi])^2].$$

The following properties about the expected value and variance of a random variable are very useful to the decision making problems with random parameters [34].

Lemma 2.1. *Let ξ and η be random variables with finite expected values. Then for any numbers a and b, we have*

$$E[a\xi + b\eta] = aE[\xi] + bE[\eta]. \tag{2.13}$$

Lemma 2.2. *For two independent random variables ξ and η, we have*

$$E[\xi\eta] = E[\xi]E[\eta]. \tag{2.14}$$

Lemma 2.3. *For the random variable ξ, we have*

$$V[\xi] = E[\xi^2] - (E[\xi])^2, \tag{2.15}$$

and for $a, b \in \mathbf{R}$, we have

$$V[a\xi + b] = a^2 V[\xi]. \tag{2.16}$$

Let's consider the typical single objective with random parameters,

$$\begin{cases} \max f(\boldsymbol{x}, \boldsymbol{\xi}) \\ \text{s.t.} \begin{cases} g_j(\boldsymbol{x}, \boldsymbol{\xi}) \leq 0, \ j = 1, 2, \cdots, p, \\ \boldsymbol{x} \in X, \end{cases} \end{cases} \tag{2.17}$$

where $f(\boldsymbol{x}, \boldsymbol{\xi})$ and $g_j(\boldsymbol{x}, \boldsymbol{\xi}), j = 1, 2 \cdots, p$ are continuous functions in X and $\boldsymbol{\xi} = (\xi_1, \xi_2, \cdots, \xi_n)$ is a random vector on the probability space $(\Omega, \mathscr{A}, Pr)$. Then it follows from the expected operator that,

$$\begin{cases} \max E[f(\boldsymbol{x}, \boldsymbol{\xi})] \\ \text{s.t.} \begin{cases} E[g_j(\boldsymbol{x}, \boldsymbol{\xi})] \leq 0,\ j = 1, 2, \cdots, p, \\ \boldsymbol{x} \in X. \end{cases} \end{cases} \tag{2.18}$$

After being dealt with by expected value operator, the problem (2.19) has been converted into a certain programming and then decision makers can easily obtain the optimal solution. However, wether the problem (2.20) has optimal solutions is a spot which decision makers pay more attention to, then its convexity is the focus we will discuss in the following part.

Definition 2.6. $\boldsymbol{x}$ is said to be a *feasible solution* of problem (2.20) if and only if $E[g_j(\boldsymbol{x}, \boldsymbol{\xi})] \leq 0 (j = 1, 2, \cdots, p)$. For any feasible solution $\boldsymbol{x}$, if $E[f(\boldsymbol{x}^*, \boldsymbol{\xi})] \geq E[f(\boldsymbol{x}, \boldsymbol{\xi})]$, then $\boldsymbol{x}^*$ is the optimal solution of problem (2.20).

Let's consider the typical bi-level single objective with random parameters,

$$\begin{cases} \max\limits_{\boldsymbol{x}} F(\boldsymbol{x}, \boldsymbol{y}, \boldsymbol{\xi}) \\ \text{s.t.} \begin{cases} G_i(\boldsymbol{x}, \boldsymbol{y}, \boldsymbol{\xi}) \leq 0,\ i = 1, 2, \cdots, q \\ \text{where } y \text{ solves:} \\ \max\limits_{\boldsymbol{y}} f(\boldsymbol{x}, \boldsymbol{y}, \boldsymbol{\xi}) \\ \text{s.t.} \begin{cases} g_j(\boldsymbol{x}, \boldsymbol{y}, \boldsymbol{\xi}) \leq 0,\ j = 1, 2, \cdots, p \\ \boldsymbol{x} \in X, \boldsymbol{y} \in Y, \end{cases} \end{cases} \end{cases} \tag{2.19}$$

where $F(\boldsymbol{x}, \boldsymbol{y}, \boldsymbol{\xi}), f(\boldsymbol{x}, \boldsymbol{y}, \boldsymbol{\xi}),\ G_i(\boldsymbol{x}, \boldsymbol{y}, \boldsymbol{\xi}), i = 1, 2, \cdots, q$ and $g_j(\boldsymbol{x}, \boldsymbol{y}, \boldsymbol{\xi}), j = 1, 2, \cdots, p$ are continuous functions in X, Y and $\boldsymbol{\xi} = (\xi_1, \xi_2, \cdots, \xi_n)$ is a random vector on the probability space $(\Omega, \mathscr{A}, Pr)$. Then it follows from the expected operator that,

$$\begin{cases} \max\limits_{\boldsymbol{x}} E[F(\boldsymbol{x}, \boldsymbol{y}, \boldsymbol{\xi})] \\ \text{s.t.} \begin{cases} E[G_i(\boldsymbol{x}, \boldsymbol{y}, \boldsymbol{\xi})] \leq 0,\ i = 1, 2, \cdots, q \\ \text{where } y \text{ solves:} \\ \max\limits_{\boldsymbol{y}} E[f(\boldsymbol{x}, \boldsymbol{y}, \boldsymbol{\xi})] \\ \text{s.t.} \begin{cases} E[g_j(\boldsymbol{x}, \boldsymbol{y}, \boldsymbol{\xi})] \leq 0,\ j = 1, 2, \cdots, p \\ \boldsymbol{x} \in X, \boldsymbol{y} \in Y. \end{cases} \end{cases} \end{cases} \tag{2.20}$$

After being dealt with by expected value operator, the problem (2.19) has been converted into a certain programming and then decision makers can easily obtain the optimal solution.

Definition 2.7. A point $(\boldsymbol{x},\boldsymbol{y})$ is said to be a *feasible solution* of problem (2.20) if and only if $E[G_i(\boldsymbol{x},\boldsymbol{y},\boldsymbol{\xi})] \leq 0, E[g_j(\boldsymbol{x},\boldsymbol{y},\boldsymbol{\xi})] \leq 0 (i = 1,2,\cdots,q, j = 1,2,\cdots,p)$. A feasible point (x^*,y^*) is called optimal of problem (2.20) if for all feasible pairs (x,y), $E[F(\boldsymbol{x}^*,\boldsymbol{y}^*,\boldsymbol{\xi})] \geq E[F(\boldsymbol{x},\boldsymbol{y},\boldsymbol{\xi})]$.

In many cases, there are usually multiple objectives which decision makers must consider. Thus we have to employ the following expected value model (EVM),

$$\begin{cases} \max\limits_{\boldsymbol{x}}[E[F_1(\boldsymbol{x},\boldsymbol{y},\boldsymbol{\xi})], E[F_2(\boldsymbol{x},\boldsymbol{y},\boldsymbol{\xi})],\cdots,E[F_m(\boldsymbol{x},\boldsymbol{y},\boldsymbol{\xi})]] \\ \text{s.t.} \begin{cases} E[G_i(\boldsymbol{x},\boldsymbol{y},\boldsymbol{\xi})] \leq 0,\ i = 1,2,\cdots,q \\ \text{where } \boldsymbol{y} \text{ solves:} \\ \max\limits_{\boldsymbol{y}}[E[f_1(\boldsymbol{x},\boldsymbol{y},\boldsymbol{\xi})], E[f_2(\boldsymbol{x},\boldsymbol{y},\boldsymbol{\xi})],\cdots,E[f_n(\boldsymbol{x},\boldsymbol{y},\boldsymbol{\xi})]] \\ \text{s.t.} \begin{cases} E[g_j(\boldsymbol{x},\boldsymbol{y},\boldsymbol{\xi})] \leq 0,\ j = 1,2,\cdots,p \\ \boldsymbol{x} \in X, \end{cases} \end{cases} \end{cases} \tag{2.21}$$

where $\boldsymbol{\xi} = (\xi_1,\xi_2,\cdots,\xi_n)$ is a random vector on probability space $(\Omega,\mathscr{A},Pr)$.

Definition 2.8. $(\boldsymbol{x}^*,\boldsymbol{y}^*)$ is the Pareto solution of problem (2.21), if there doesn't exist feasible solutions $(\boldsymbol{x},\boldsymbol{y})$ such that

$$E[F_u(\boldsymbol{x},\boldsymbol{y},\boldsymbol{\xi})] \geq E[F_u(\boldsymbol{x}^*,\boldsymbol{y}^*,\boldsymbol{\xi})], u = (1,2,\cdots,m),$$

and there exists at least one $u (u = 1,2,\cdots,m)$ such that

$$E[F_u(\boldsymbol{x},\boldsymbol{y},\boldsymbol{\xi})] > E[F_u(\boldsymbol{x}^*,\boldsymbol{y}^*,\boldsymbol{\xi})].$$

2. Random Chance-Constrained Operator

In 1959, Charnes and Cooper [21] developed another technique to deal with random programming problems, that is chance-constrained model (CCM). It is a powerful tool to help decision makers to make the decision in the stochastic decision systems with assumption that the random constrains will hold at least α level value, where α is referred to as the confidence level provided as an appropriate safety margin by the decision maker.

In practice, the goal of decision makers is to maximize the objective value on the condition of probability α, where α is predetermined confidence level. Next, we will introduce the concept of the chance measure of random variables. The chance measure of a random event is considered as the probability of the event $f(\boldsymbol{x},\boldsymbol{\xi}) \geq \bar{f}$. Then the chance constraint is considered as $Pr\{f(\boldsymbol{x},\boldsymbol{\xi}) \geq \bar{f}\} \geq \alpha$, where α is the predetermined confidence level, and $\bar{f}$ is called the *critical value*. A natural idea is to provide a confidence level α at which it is desired that the random constrains hold. Let's still consider the following model,

$$\begin{cases} \max f(\boldsymbol{x},\boldsymbol{\xi}) \\ \text{s.t.} \begin{cases} g_j(\boldsymbol{x},\boldsymbol{\xi}) \leq 0,\ j = 1,2,\cdots,p, \\ \boldsymbol{x} \in X, \end{cases} \end{cases}$$

where $f(\boldsymbol{x},\boldsymbol{\xi})$ and $g_j(\boldsymbol{x},\boldsymbol{\xi}), j = 1,2\cdots,p$ are continuous functions in X and $\boldsymbol{\xi} = (\xi_1,\xi_2,\cdots,\xi_n)$ is a random vector on probability space $(\Omega,\mathscr{A},Pr)$. Based on the chance-constraint operator, the random chance-constrained model (CCM):

$$\begin{cases} \max \bar{f} \\ \text{s.t.} \begin{cases} Pr\{f(\boldsymbol{x},\boldsymbol{\xi}) \geq \bar{f}\} \geq \beta, \\ Pr\{g_j(\boldsymbol{x},\boldsymbol{\xi}) \leq 0\} \geq \alpha_j,\ j = 1,2,\cdots,p, \\ \boldsymbol{x} \in X, \end{cases} \end{cases} \tag{2.22}$$

where β and α_j are the predetermined confidence levels, $\bar{f}$ is the critical value which needs to determine.

Definition 2.9. A solution $\boldsymbol{x} \in X$ is said to be the feasible solution of problem (2.22) if and only if $Pr\{f(\boldsymbol{x},\boldsymbol{\xi}) \geq \bar{f}\} \geq \beta$ and $Pr\{g_j(\boldsymbol{x},\boldsymbol{\xi}) \leq 0\} \geq \alpha_j$ hold for all j. For any feasible $\boldsymbol{x}$, if there is a solution $\boldsymbol{x}^*$ such $\bar{f}^* > \bar{f}$, then $\boldsymbol{x}^*$ is called the optimal solution.

If the objective is to be minimized (for example, the objective is a cost function), the CCM should be as follows,

$$\begin{cases} \min \bar{f} \\ \text{s.t.} \begin{cases} Pr\{f(\boldsymbol{x},\boldsymbol{\xi}) \leq \bar{f}\} \geq \beta, \\ Pr\{g_j(\boldsymbol{x},\boldsymbol{\xi}) \leq 0\} \geq \alpha_j,\ j = 1,2,\cdots,p, \\ \boldsymbol{x} \in X, \end{cases} \end{cases} \tag{2.23}$$

where β and α_j are the predetermined confidence levels. Similarly, we have the following definition,

Definition 2.10. A solution $\boldsymbol{x} \in X$ is called the feasible solution of problem (2.23) if and only if $Pr\{f(\boldsymbol{x},\boldsymbol{\xi}) \leq \bar{f}\} \geq \beta$ and $Pr\{g_j(\boldsymbol{x},\boldsymbol{\xi}) \leq 0\} \geq \alpha_j$ holds for all j. For any feasible $\boldsymbol{x}$, if there is a solution $\boldsymbol{x}^*$ such $\bar{f}^* < \bar{f}$, then $\boldsymbol{x}^*$ is called the optimal solution.

In practice, the real-life problems are more complex, and there usually exist multiple objectives which need decision makers to decide. Thus we have to employ the following multi-objective CCM:

$$\begin{cases} \max[\bar{f}_1,\bar{f}_2,\cdots,\bar{f}_m] \\ \text{s.t.} \begin{cases} Pr\{f_i(\boldsymbol{x},\boldsymbol{\xi}) \geq \bar{f}_i\} \geq \beta_i,\ i = 1,2,\cdots,m, \\ Pr\{g_j(\boldsymbol{x},\boldsymbol{\xi}) \leq 0\} \geq \alpha_j,\ j = 1,2,\cdots,p, \\ \boldsymbol{x} \in X, \end{cases} \end{cases} \tag{2.24}$$

where β_i and α_j are the predetermined confidence levels, $\bar{f}_i$ are critical values which need to be determined.

Definition 2.11. $x \in X$ is called the Pareto solution of problem (2.24) if there doesn't exist a feasible x such that

$$\bar{f_i} \geq \bar{f_i}^*, \quad i = 1, 2, \cdots, m \tag{2.25}$$

and there at least exists one $j (j = 1, 2, \cdots, m)$ such that $\bar{f_j} > \bar{f_j}^*$.

Let's consider the typical bi-level single objective with random parameters, i.e, Eq. (2.19). Based on the chance-constraint operator, the random CCM:

$$\begin{cases} \max\limits_{x} \bar{F} \\ \text{s.t.} \begin{cases} Pr\{F(\boldsymbol{x},\boldsymbol{y},\boldsymbol{\xi}) \geq \bar{F}\} \geq \beta_1, \\ Pr\{G_i(\boldsymbol{x},\boldsymbol{y},\boldsymbol{\xi}) \leq 0\} \geq \alpha_i, \ i = 1, 2, \cdots, q \\ \text{where } \boldsymbol{y} \text{ solves:} \\ \max\limits_{y} \bar{f} \\ \text{s.t.} \begin{cases} Pr\{f(\boldsymbol{x},\boldsymbol{y},\boldsymbol{\xi}) \geq \bar{f}\} \geq \beta_2, \\ Pr\{g_j(\boldsymbol{x},\boldsymbol{y},\boldsymbol{\xi}) \leq 0\} \geq \alpha_j, \ j = 1, 2, \cdots, p \\ \boldsymbol{x} \in X, \boldsymbol{y} \in Y, \end{cases} \end{cases} \end{cases} \tag{2.26}$$

where β_1, β_2 and α_i, α_j are the predetermined confidence levels, $\bar{F}$ and $\bar{f}$ is the critical value which needs to determine.

Definition 2.12. A solution $(\boldsymbol{x},\boldsymbol{y})$ is said to be a feasible solution of problem (2.26) if and only if $Pr\{F(\boldsymbol{x},\boldsymbol{y},\boldsymbol{\xi}) \geq \bar{F}\} \geq \beta_1, Pr\{G_i(\boldsymbol{x},\boldsymbol{y},\boldsymbol{\xi}) \leq 0\} \geq \alpha_i$ hold for all i and $Pr\{f(\boldsymbol{x},\boldsymbol{y},\boldsymbol{\xi}) \geq \bar{f}\} \geq \beta_2, Pr\{g_j(\boldsymbol{x},\boldsymbol{y},\boldsymbol{\xi}) \leq 0\} \geq \alpha_j$ hold for all j. A feasible solution (x^*, y^*) is called optimal, if $\bar{F}^*$ is unique and for (x^*, y^*) such $\bar{F}^* > \bar{F}$, then (x^*, y^*) is called the optimal solution.

If the objective is to be minimized (for example, the objective is a cost function), the CCM should be as follows,

$$\begin{cases} \min\limits_{x} \bar{F} \\ \text{s.t.} \begin{cases} Pr\{F(\boldsymbol{x},\boldsymbol{y},\boldsymbol{\xi}) \leq \bar{F}\} \geq \beta_1, \\ Pr\{G_i(\boldsymbol{x},\boldsymbol{y},\boldsymbol{\xi}) \leq 0\} \geq \alpha_i, \ i = 1, 2, \cdots, q \\ \text{where } \boldsymbol{y} \text{ solves:} \\ \min\limits_{y} \bar{f} \\ \text{s.t.} \begin{cases} Pr\{f(\boldsymbol{x},\boldsymbol{y},\boldsymbol{\xi}) \leq \bar{f}\} \geq \beta_2, \\ Pr\{g_j(\boldsymbol{x},\boldsymbol{y},\boldsymbol{\xi}) \leq 0\} \geq \alpha_j, \ j = 1, 2, \cdots, p \\ \boldsymbol{x} \in X, \boldsymbol{y} \in Y, \end{cases} \end{cases} \end{cases} \tag{2.27}$$

where β_1, β_2 and α_i, α_j are the predetermined confidence levels, $\bar{F}$ and $\bar{f}$ is the critical value which needs to determine.

Definition 2.13. A solution $(\boldsymbol{x},\boldsymbol{y})$ is said to be a feasible solution of problem (2.27) if and only if $Pr\{F(\boldsymbol{x},\boldsymbol{y},\boldsymbol{\xi}) \leq \bar{F}\} \geq \beta_1, Pr\{G_i(\boldsymbol{x},\boldsymbol{y},\boldsymbol{\xi}) \leq 0\} \geq \alpha_i$ hold for all i and $Pr\{f(\boldsymbol{x},\boldsymbol{y},\boldsymbol{\xi}) \leq \bar{f}\} \geq \beta_2, Pr\{g_j(\boldsymbol{x},\boldsymbol{y},\boldsymbol{\xi}) \leq 0\} \geq \alpha_j$ hold for all j. A feasible solution (x^*, y^*) is called optimal, if $\bar{F}^*$ is unique and for (x^*, y^*) such $\bar{F}^* < \bar{F}$, then (x^*, y^*) is called the optimal solution.

If there exist multiple objectives in bi-level programming, we have to employ the following multi-objective CCM for bi-level programming:

$$\begin{cases} \max\limits_{\boldsymbol{x}}[\bar{F}_1, \bar{F}_2, \cdots, \bar{F}_U] \\ \text{s.t.} \begin{cases} Pr\{F_u(\boldsymbol{x},\boldsymbol{y},\boldsymbol{\xi}) \geq \bar{F}_u\} \geq \beta_u, u = 1,2,\cdots,U \\ Pr\{G_i(\boldsymbol{x},\boldsymbol{y},\boldsymbol{\xi}) \leq 0\} \geq \alpha_i,\ i = 1,2,\cdots,q \\ \text{where } \boldsymbol{y} \text{ solves:} \\ \max\limits_{\boldsymbol{y}}[\bar{f}_1, \bar{f}_2, \cdots, \bar{f}_L] \\ \text{s.t.} \begin{cases} Pr\{f_l(\boldsymbol{x},\boldsymbol{y},\boldsymbol{\xi}) \geq \bar{f}_l\} \geq \beta_l, l = 1,2,\cdots,L \\ Pr\{g_j(\boldsymbol{x},\boldsymbol{y},\boldsymbol{\xi}) \leq 0\} \geq \alpha_j,\ j = 1,2,\cdots,p \\ \boldsymbol{x} \in X, \boldsymbol{y} \in Y, \end{cases} \end{cases} \end{cases} \tag{2.28}$$

where β_u, β_l and α_i, α_j are the predetermined confidence levels, $\bar{F}_u$ and $\bar{f}_l$ are the critical values which needs to determine.

Definition 2.14. $(\boldsymbol{x}^*, \boldsymbol{y}^*)$ is a Pareto solution of problem (2.28), if and only if there doesn't exist feasible solutions $(\boldsymbol{x},\boldsymbol{y})$ such that

$$\bar{F}_u(\boldsymbol{x},\boldsymbol{y},\boldsymbol{\xi}) \geq \bar{F}_u(\boldsymbol{x}^*,\boldsymbol{y}^*,\boldsymbol{\xi}), u = (1,2,\cdots,U),$$

and there exists at least one $u(u = 1,2,\cdots,U)$ such that $\bar{F}_u(\boldsymbol{x},\boldsymbol{y},\boldsymbol{\xi}) > \bar{F}_u(\boldsymbol{x}^*,\boldsymbol{y}^*,\boldsymbol{\xi})$.

3. Random Dependent-Chance Operator

In practice, there usually exist multiple events in a complex random decision system. Sometimes, the decision-maker wishes to maximize the chance functions of these events (i.e. the probabilities of satisfying the events). In order to model this type of random decision system, Schneider [113] developed another technique, called probability maximization model (also called dependent-chance model (abbr. DCM) by some scholars, we will use this name in this book), in which the underlying philosophy is based on selecting the decision with maximal chance to meet the event. Then [64] discussed some coverage probability maximization problems. From then on, it was widely used in many fields to solve some realistic problems [47, 110].

DCM theory breaks the concept of feasible set and replaces it with uncertain environment. Roughly speaking, DCM involves maximizing chance functions of events in an uncertain environment. In deterministic model, expected value model (EVM), and chance-constrained programming (CCM), the feasible set is essentially assumed to be deterministic after the real problem is modeled. That is, an optimal solution is given regardless of whether it can be performed in practice. However, the given solution may be impossible to perform if the realization of uncertain parameter is unfavorable. Thus DCM theory never assumes that the feasible set is deterministic. In fact, DCM is constructed in an uncertain environment. This special feature of DCM is very different from the other existing types of random programming.

Let's consider the following typical DCM model,

$$\begin{cases} \max Pr\{f(\boldsymbol{x}, \boldsymbol{\xi}) \geq f\} \\ \text{s.t.} \begin{cases} g_j(\boldsymbol{x}, \boldsymbol{\xi}) \leq 0, j = 1, 2, \cdots, p, \\ \boldsymbol{x} \geq 0. \end{cases} \end{cases} \tag{2.29}$$

Since a complex decision system usually undertakes multiple events, there undoubtedly exist multiple potential objectives (some of them are chance functions) in a decision process. A typical formulation of dependent-chance multi-objective programming model (DCM) is represented as maximizing multiple chance functions subject to an uncertain environment,

$$\begin{cases} \max \begin{bmatrix} Pr\{h_1(\boldsymbol{x}, \boldsymbol{\xi}) \leq 0\} \\ Pr\{h_2(\boldsymbol{x}, \boldsymbol{\xi}) \leq 0\} \\ \cdots \\ Pr\{h_m(\boldsymbol{x}, \boldsymbol{\xi}) \leq 0\} \end{bmatrix} \\ \text{s.t.} \quad g_j(\boldsymbol{x}, \boldsymbol{\xi}) \leq 0, j = 1, 2, \cdots, p, \end{cases} \tag{2.30}$$

where $h_i(\boldsymbol{x}, \boldsymbol{\xi}) \leq 0$ are represent events ε_i for $i = 1, 2, \cdots, m$, respectively.

It follows from the principle of uncertainty that we can construct a relationship between decision vectors and chance function, thus calculating the chance functions by stochastic simulations or traditional methods. Then we can solve DCM by utility theory if complete information of the preference function is given by the decision-maker or search for all of the efficient solutions if no information is available. In practice, the decision maker can provide only partial information. In this case, we have to employ the interactive methods.

Sometimes, the objective function may minimize the deviations, positive, negative, or both, with a certain priority structure set by the decision maker. Then we may formulate the stochastic decision system as the following model,

$$\begin{cases} \min \sum_{j=1}^{l} P_j \sum_{i=1}^{m} (u_{ij} d_i^+ + v_{ij} d_i^-) \\ \text{s.t.} \begin{cases} Pr\left\{ h_i(\boldsymbol{x}, \boldsymbol{\xi}) \leq 0 \right\} + d_i^- - d_i^+ = b_i, & i = 1, 2, \cdots, m, \\ g_j(\boldsymbol{x}, \boldsymbol{\xi}) \leq 0, & j = 1, 2, \cdots, p, \\ d_i^-, d_i^+ \geq 0, & i = 1, 2, \cdots, m, \end{cases} \end{cases} \tag{2.31}$$

where P_j is the preemptive priority factor which express the relative importance of various goals, $P_j >> P_{j+1}$, for all j, u_{ij} is the weighting factor corresponding to positive deviation for goal i with priority j assigned, v_{ij} is the weighting factor corresponding to negative deviation for goal i with priority j assigned, d_i^+ is the positive deviation from the target value according to goal i, d_i^- is the negative deviation from the target of goal i, b_i is the target value according to goal i, l is the number of priorities, and m is the number of goal constraints.

Let's consider the typical bi-level single objective with random parameters, i.e, Eq. (2.19). Based on the dependent-chance operator, the random DCM:

$$\begin{cases} \max\limits_{\boldsymbol{x}} Pr\{F(\boldsymbol{x},\boldsymbol{y},\boldsymbol{\xi}) \le 0 \\ \text{s.t.} \begin{cases} G_i(\boldsymbol{x},\boldsymbol{y},\boldsymbol{\xi}) \le 0,\ i = 1,2,\cdots,q \\ \text{where } \boldsymbol{y} \text{ solves:} \\ \max\limits_{\boldsymbol{y}} Pr\{f(\boldsymbol{x},\boldsymbol{y},\boldsymbol{\xi}) \le 0\} \\ \text{s.t.} \begin{cases} g_j(\boldsymbol{x},\boldsymbol{y},\boldsymbol{\xi}) \le 0,\ j = 1,2,\cdots,p \\ \boldsymbol{x} \in X, \boldsymbol{y} \in Y. \end{cases} \end{cases} \end{cases} \tag{2.32}$$

If there exist multiple objectives in both upper- and lower-level models, based on the DCM operator, we have:

$$\begin{cases} \max\limits_{\boldsymbol{x}} \begin{bmatrix} Pr\{F_1(\boldsymbol{x},\boldsymbol{y},\boldsymbol{\xi}) \le 0\} \\ Pr\{F_2(\boldsymbol{x},\boldsymbol{y},\boldsymbol{\xi}) \le 0\} \\ \cdots \\ Pr\{F_U(\boldsymbol{x},\boldsymbol{y},\boldsymbol{\xi}) \le 0\} \end{bmatrix} \\ \text{s.t.} \begin{cases} G_i(\boldsymbol{x},\boldsymbol{y},\boldsymbol{\xi}) \le 0,\ i = 1,2,\cdots,q \\ \text{where } \boldsymbol{y} \text{ solves:} \\ \max\limits_{\boldsymbol{y}} \begin{bmatrix} Pr\{f_1(\boldsymbol{x},\boldsymbol{y},\boldsymbol{\xi}) \le 0\} \\ Pr\{f_2(\boldsymbol{x},\boldsymbol{y},\boldsymbol{\xi}) \le 0\} \\ \cdots \\ Pr\{f_L(\boldsymbol{x},\boldsymbol{y},\boldsymbol{\xi}) \le 0\} \end{bmatrix} \\ \text{s.t.} \begin{cases} g_j(\boldsymbol{x},\boldsymbol{y},\boldsymbol{\xi}) \le 0,\ j = 1,2,\cdots,p \\ \boldsymbol{x} \in X, \boldsymbol{y} \in Y. \end{cases} \end{cases} \end{cases} \tag{2.33}$$

2.3 Random EEEE Model

If the decision makers in both levels tend to optimize the mean values of their objectives, as well as make the constraints in both of levels satisfied with mean values, then the EEEE model is adopted to converts the stochastic model into a deterministic one. As mentioned above, EEEE denotes that expected value operator is used to deal with the randomness in upper-level model and lower-level model.

2.3.1 General Form of Random EEEE Model

By applying EEEE pattern into models (2.1), (2.2) and (2.3), the corresponding EEEE models are

$$\begin{cases} \max\limits_{x \in R^{n_1}} E[F(x,y,\xi)] \\ \text{s.t.} \begin{cases} E[G_i(x,y,\xi)] \leq 0, i = 1,2,\cdots,q \\ \text{where } y \text{ solves:} \\ \begin{cases} \max\limits_{y \in R^{n_2}} E[f(x,y,\xi)] \\ \text{s.t. } E[g_j(x,y,\xi)] \leq 0, j = 1,2,\cdots,p, \end{cases} \end{cases} \end{cases} \tag{2.34}$$

$$\begin{cases} \min\limits_{x \in R^{n_1}} [E[F_1(x,y,\xi)], E[F_2(x,y,\xi)], \cdots, E[F_m(x,y,\xi)]] \\ \text{s.t.} \begin{cases} E[G_i(x,y,\xi)] \leq 0, i = 1,2,\cdots,q \\ \text{where } y \text{ solves:} \\ \begin{cases} \min\limits_{y \in R^{n_2}} [E[f_1(x,y,\xi)], E[f_2(x,y,\xi)], \cdots, E[f_m(x,y,\xi)]] \\ \text{s.t. } E[g_j(x,y,\xi)] \leq 0, j = 1,2,\cdots,p \end{cases} \end{cases} \end{cases} \tag{2.35}$$

and

$$\begin{cases} \min\limits_{x \in R^{n_1}} \left[E[\tilde{a}_1^T x + \tilde{b}_1^T y], E[\tilde{a}_2^T x + \tilde{b}_2^T y], \cdots, E[\tilde{a}_m^T x + \tilde{b}_m^T y]\right] \\ \text{s.t.} \begin{cases} E[\tilde{c}_i^{1T} x + \tilde{c}_i^{2T} y] \leq E[\tilde{d}_i], i = 1,2,\cdots,q \\ \text{where } y \text{ solves:} \\ \min\limits_{y \in R^{n_2}} \left[E[\tilde{e}_1^T x + \tilde{f}_1^T y], E[\tilde{e}_2^T x + \tilde{f}_2^T y], \cdots, E[\tilde{e}_m^T x + \tilde{f}_m^T y]\right] \\ \text{s.t.} \begin{cases} E[\tilde{g}_j^{1T} x + \tilde{g}_j^{2T} y] \leq E[\tilde{h}_j], j = 1,2,\cdots,p \\ x > 0, y > 0. \end{cases} \end{cases} \end{cases} \tag{2.36}$$

It is noted that models (2.34), (2.35) and (2.36) have strictly mathematical meaning, different from models (2.1). Similarly, the EEEE models of (2.4), (2.5) and (2.6) ban be easily developed.

Since this book fucus on bi-level multi-objective problems, we concentrate on model (2.35) and (2.36). Sone relevant concepts of (2.35) are defined as follows.

Definition 2.15. The *constraint region* S of model (2.35) is defined by

$$S = \{((x,y)|E[G_i(x,y,\xi)] \leq 0, i = 1,2,\cdots,q; E[g_j(x,y,\xi)] \leq 0, j = 1,2,\cdots,p\}. \tag{2.37}$$

Definition 2.16. For fixed $\hat{x}$ of leader, the *feasible region of the follower*, denoted by $S(x)$, is defined by

$$S(\hat{x}) = \{y|E[G_i(\hat{x},y,\xi)] \leq 0, i = 1,2,\cdots,q; E[g_j(\hat{x},y,\xi)] \leq 0, j = 1,2,\cdots,p\}. \tag{2.38}$$

Definition 2.17. For fixed $\hat{x}$ of leader, if y^* is a Pareto-optimal (noninferior, efficient) solution of the lower model of (2.35), i.e., there exists no other $y \in S(\hat{x})$ such that $E[F_i(\hat{x}, y, \xi)] \le E[F_i(\hat{x}, y^*, \xi)]$ for all $i = 1, 2, \cdots, q$ with strict inequality for at least one i, then y^* is said to be a *rational response with respect to* $\hat{x}$. The set of all the rational responses with respect to $\hat{x}$ is denoted by $R(\hat{x})$, i.e.,

$$R(\hat{x}) = \{y \in S(\hat{x}) | y \text{ is a Pareto-optimal solution of the lower model of (2.35)}\}. \tag{2.39}$$

Remark 2.1. Each $y \in R(\hat{x})$ is also said to be a Pareto optimal response.

Remark 2.2. If there is only one objective for the lower level model, i.e., $m_2 = 1$, then the ration response set is defined by

$$R(\hat{x}) = \{y | y \in \arg\min_{y \in S(\hat{x})} E[f(x, y, \xi)]\}. \tag{2.40}$$

Definition 2.18. The *inducible region IR* of (2.35) is defined by

$$IR = \{(x, y) | (x, y) \in S, y \in R(x)\}. \tag{2.41}$$

Obviously, it follows from above definitions that

$$R(\hat{x}) \subset S(\hat{x}) \tag{2.42}$$

and

$$IR \subset S. \tag{2.43}$$

Remark 2.3. The inducible region IR of model (2.35) plays the role of feasible region in single-level model. Actually, although a solution $(x, y)(\in S)$ satisfies all the constraints, it is not acceptable if y is not a rational response since it damages the follower's interests. Thus, it is not actually feasible. Only the solution belongs to inducible set can be acceptable by both of decision makers.

The final decision step is to search Pareto optimal solution in the inducible region IR of model.

Definition 2.19. (x^*, y^*) is said to be a *Pareto optimal solution* of model (2.35) if and only if there exists no other $x, y \in IR$ such that $E[F_i(x, y, \xi)] \ge E[F_i(x^*, y^*, \xi^*)]$ for all $i = 1, 2, \cdots, m_1$ with strict inequality for at least one i.

Remark 2.4. If the leader optimizes only one objective, i.e., $m_1 = 1$, the concepts of Pareto optimal solution is substituted by the concept of Pareto optimal solution defined by

$$\{(x, y) | (x, y) \in \arg\min_{(x,y) \in IR} E[F(x, y, \xi)]\}. \tag{2.44}$$

In general, the leader does not know which response is chosen from the set of Pareto optimal responses by the follower. However, with some subjective anticipation or belief, we can provide reasonable decisions of the leader. Such anticipation or belief can be considered in three cases: (i) the leader anticipates that the follower will take a decision desirable for the leader (optimistic anticipation); (ii) the leader anticipates that the follower will take a decision undesirable for the leader (pessimistic anticipation); and (iii) the leader knows the preference of the follower. When the leader does not know the preference of the follower, the leader would make a decision by consulting outcomes yielded from the optimistic anticipation and the pessimistic anticipation.

The following example illustrates some concepts defined above.

Example 2.1. Consider the following random EEEE model problem, which has a single objective function and three objective functions of the follower:

$$\begin{cases} \min\limits_{x\in\Re} E[\tilde{a}_1 x + \tilde{a}_2 y_1 + \tilde{a}_3 y_2] \\ \text{s.t.} \begin{cases} \text{where } y_1, y_2 \text{ solves:} \\ f_1 = \min\limits_{y_1,y_2\in\Re} E\left[\tilde{b}_1 x + \tilde{b}_2 y_1 + \tilde{b}_3 y_2\right] \\ f_2 = \min\limits_{y_1,y_2\in\Re} E[\tilde{c}_1 x + \tilde{c}_2 y_1 + \tilde{c}_3 y_2] \\ \text{s.t.} \begin{cases} 50 \le x + y_1 \le 170 \\ -60 \le x - y_1 \le 60 \\ 50 \le x + y_2 \le 170 \\ -60 \le x - y_2 \le 60 \\ -170 \le y_1 + y_2 \le 170 \\ -60 \le y_1 - y_2 \le 60 \\ 90 \le x + y_1 + y_2 \le 240 \\ -20 \le x - y_1 + y_2 \le 70 \\ -20 \le x + y_1 - y_2 \le 130 \\ -130 \le x - y_1 - y_2 \le 20 \\ 10 \le x \le 100 \\ 10 \le y_1 \le 100 \\ 10 \le y_2 \le 100, \end{cases} \end{cases} \end{cases} \tag{2.45}$$

where $a_i, b_i, c_i (i = 1, 2, 3)$ are independent random variables characterized as

$$\tilde{a}_1 \sim \mathscr{U}(-2, 0), \tilde{a}_2 \sim \mathscr{U}(-3, -1), \tilde{a}_3 \sim \mathscr{U}(2, 6),$$

$$\tilde{b}_1 \sim \exp(1), \tilde{b}_2 \sim \exp(2), \tilde{b}_3 \sim \exp(-1),$$

$$\tilde{c}_1 \sim \mathscr{N}(2, 2), \tilde{c}_2 \sim \mathscr{N}(-2, 2), \tilde{c}_3 \sim \mathscr{N}(-1, 1),$$

where $\mathscr{U}(\cdot,\cdot)$, $\exp(\cdot)$ and $\mathscr{N}(\cdot,\cdot)$ denote uniform distribution, exponential distribution and normal distribution, respectively. The feasible region is shown by Fig. 2.3.

To examine rational responses of the follower, suppose that the leader has chosen a decision $x = 70$. Then, the feasible region of the follower $S(x)$ is an octagonal area

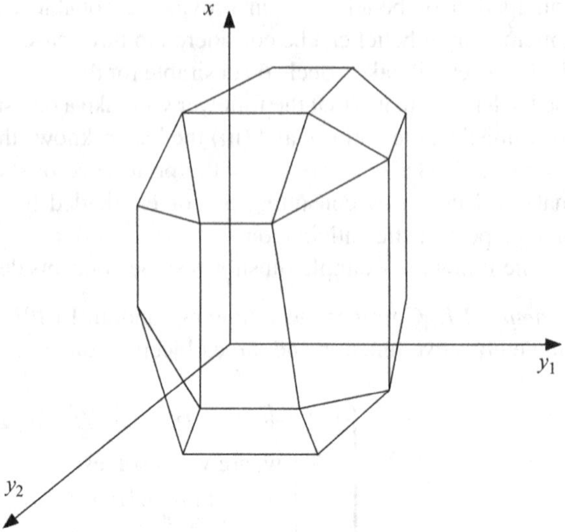

Fig. 2.3 The feasible region of model (2.45)

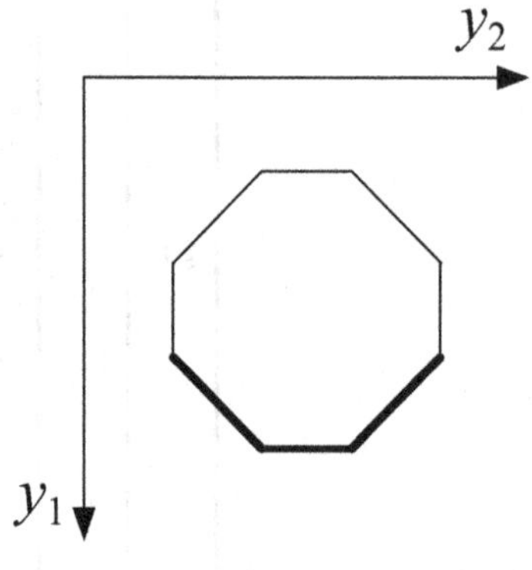

Fig. 2.4 A rational response set

in the $y_1 - y_2$ plane as shown in Fig. 2.4. The follower can choose a decision in the region $S(x)$ and must choose it in the set of Pareto optimal responses $P(x)$, which is depicted as thick lines and is derived by considering the preference cone of the follower.

2.3.2 Reference Point Method for Linear Models

In what follows, we focus on the linear random EEEE models (2.36). Although it is difficult to deal with general linear random EEEE models due to the complexity of randomness, bi-level structure and multiple objectives, some of linear random EEEE models with special forms can be transformed into equivalent models and then solved by the existed methods for deterministic bi-level models.

Theorem 2.1. *Assume that random vectors* $\tilde{a}_i = (\tilde{a}_{i1}, \tilde{a}_{i2}, \cdots, \tilde{a}_{in})^T$ *is a normally distributed with mean vectors* $\mu_i^a = (\mu_{i1}^a, \mu_{i2}^a, \cdots, \mu_{in}^a)^T$ *and positive definite covariance matrix* V_i^a*, written as* $\tilde{a}_i \sim \mathcal{N}(\mu_i^a, V_i^a)$*. At the same time,* $\tilde{b}_i \sim \mathcal{N}(\mu_i^b, V_i^b), \tilde{c}_i^1 \sim \mathcal{N}(\mu_i^{c^1}, V_i^{c^1}), \tilde{c}_i^2 \sim \mathcal{N}(\mu_i^{c^2}, V_i^{c^2}), \tilde{e}_i \sim \mathcal{N}(\mu_i^e, V_i^e), \tilde{f}_i \sim \mathcal{N}(\mu_i^f, V_i^f), \tilde{g}_i^1 \sim \mathcal{N}(\mu_i^{g^1}, V_i^{g^1}), \tilde{g}_i^2 \sim \mathcal{N}(\mu_i^{g^2}, V_i^{g^2})$*.* $\tilde{d}_i, \tilde{h}_j$ *are random variables characterized by*

$$\tilde{d}_i \sim \mathcal{N}(\mu_i^d, (\sigma_i^d)^2), \tilde{h}_j \sim \mathcal{N}(\mu_j^h, (\sigma_j^h)^2).$$

In addition, all the random variables are assumed to be independent of each other. Then problem (2.36) is equivalent to

$$\begin{cases} \min\left[\mu_1^{aT}\boldsymbol{x} + \mu_1^{bT}\boldsymbol{y}, \mu_2^{aT}\boldsymbol{x} + \mu_2^{bT}\boldsymbol{y}, \cdots, \mu_m^{aT}\boldsymbol{x} + \mu_m^{bT}\boldsymbol{y}\right] \\ \text{s.t.} \begin{cases} \mu_r^{c^1T}\boldsymbol{x} + \mu_r^{c^2T}\boldsymbol{y} \le \mu_r^d \\ \textit{where } y \textit{ solves:} \\ \min\left[\mu_1^{eT}\boldsymbol{x} + \mu_1^{fT}\boldsymbol{y}, \mu_2^{eT}\boldsymbol{x} + \mu_2^{fT}\boldsymbol{y}, \cdots, \mu_m^{eT}\boldsymbol{x} + \mu_m^{fT}\boldsymbol{y}\right] \\ \text{s.t.} \begin{cases} \mu_r^{g^1T}\boldsymbol{x} + \mu_r^{g^2T}\boldsymbol{y} \le \mu_r^h \\ \boldsymbol{x} \ge 0, \boldsymbol{y} \ge 0. \end{cases} \end{cases} \end{cases} \tag{2.46}$$

Proof. It follows from the nonnegativity of decision variables and linearity of the random expected value operator that the first objective for the leader $E[\tilde{a}_1^T\boldsymbol{x} + \tilde{b}_1^T\boldsymbol{y}]$ is equivalent to

$$\mu_1^{aT}\boldsymbol{x} + \mu_1^{bT}\boldsymbol{y}.$$

By the same way, the objectives and constrains can be converted into their crisp equivalent forms. Thus the theorem is proved. □

As linear random EEEE models (2.36) can be transformed into its equivalent crisp model, it is a general multi-objective bi-level model. After the leader has chosen a decision $\hat{\boldsymbol{x}}$, the follower chooses a decision $\boldsymbol{y}$ with full knowledge of the decision $\hat{\boldsymbol{x}}$ of the leader by solving the following multi-objective programming problem:

$$\begin{cases} \min\limits_{y \in R^{n_2}} \left[E[\tilde{e}_1^T\hat{\boldsymbol{x}} + \tilde{f}_1^T\boldsymbol{y}], E[\tilde{e}_2^T\hat{\boldsymbol{x}} + \tilde{f}_2^T\boldsymbol{y}], \cdots, E[\tilde{e}_m^T\hat{\boldsymbol{x}} + \tilde{f}_m^T\boldsymbol{y}]\right] \\ \text{s.t.} \begin{cases} E[\tilde{c}_i^{1T}\hat{\boldsymbol{x}} + \tilde{c}_i^{2T}\boldsymbol{y}] \le E[\tilde{d}_i], i = 1, 2, \cdots, p_1 \\ E[\tilde{g}_j^{1T}\hat{\boldsymbol{x}} + \tilde{g}_j^{2T}\boldsymbol{y}] \le E[\tilde{h}_j], j = 1, 2, \cdots, p_2 \\ \boldsymbol{x} \ge 0, \boldsymbol{y} \ge 0. \end{cases} \end{cases} \tag{2.47}$$

Denote the set of Pareto optimal responses (solutions) of the multi-objective programming problem (2.47) by $P(\tilde{\boldsymbol{x}})$ and feasible solution region of $\{(\boldsymbol{x}, \boldsymbol{y}) | E[\tilde{c}_i^{1T}\tilde{\boldsymbol{x}} + \tilde{c}_i^{2T}\boldsymbol{y}] \le E[\tilde{d}_i], i = 1, 2, \cdots, p_1; E[\tilde{g}_j^{1T}\tilde{\boldsymbol{x}} + \tilde{g}_j^{2T}\boldsymbol{y}] \le E[\tilde{h}_j], j = 1, 2, \cdots, p_2; \boldsymbol{x} \ge 0, \boldsymbol{y} \ge 0\}$ by S.

Hereafter, we call a Stackelberg solution based on the optimistic anticipation the optimistic Stackelberg solution simply. The optimistic Stackelberg solution is an optimal solution to the following problem if the utility function U_1 of the leader can be identified:

$$\begin{cases} \max_{x \in R^{n_1}} \max_{y \in R^{n_2}} U_1(E[\tilde{a}_1^T x + \tilde{b}_1^T y], E[\tilde{a}_2^T x + \tilde{b}_2^T y], \cdots, E[\tilde{a}_m^T x + \tilde{b}_m^T y]) \\ \text{s.t.} \begin{cases} (x, y) \in S \\ y \in P(x). \end{cases} \end{cases} \tag{2.48}$$

Problem (2.48) expresses a situation where the leader chooses a decision x so as to maximize the leader's utility function $U_1(E[\tilde{a}_1^T x + \tilde{b}_1^T y], E[\tilde{a}_2^T x + \tilde{b}_2^T y], \cdots, E[\tilde{a}_m^T x + \tilde{b}_m^T y])$ on the assumption (anticipation) that, for the given decision x of the leader, the follower takes a decision y in the set of Pareto optimal responses $P(x)$ such that the leaderąŕs utility function $U_1(E[\tilde{a}_1^T x + \tilde{b}_1^T y], E[\tilde{a}_2^T x + \tilde{b}_2^T y], \cdots, E[\tilde{a}_m^T x + \tilde{b}_m^T y])$ is maximized.

It can be observed that, in most real-world situations, identifying the utility function U_1 even by the leader in person is difficult; even if possible, the function might be a nonlinear function. Because of the difficulty of identifying the utility function and the difficulty in computing solutions, it would be appropriate to use an interactive method in which a decision maker can learn and realize a local preference around solutions derived by solving some problem at each iteration. Here, we employ the reference point method by Wierzbicki [130]. Let an achievement function be

$$\max_{i=1,2,\cdots,m_1} \{E[\tilde{a}_i^T x + \tilde{b}_i^T y] - \bar{F}_i\} + \rho \sum_{i=1}^{m_1} (E[\tilde{a}_i^T x + \tilde{b}_i^T y] - \bar{F}_i), \tag{2.49}$$

where $(\bar{F}_1, \bar{F}_2, \cdots, \bar{F}_{m_1})$ is a reference point specified by the leader, and ρ is a small positive scalar value. The function (2.49) is called the augmented Tchebyshev scalarizing function, which has some desirable properties.

Hence, the optimistic Stackelberg solutions can be obtained by interactively solving the following problem and updating the reference points:

$$\begin{cases} \min_x \min_y \left\{ \max_{i=1,2,\cdots,m_1} \{E[\tilde{a}_i^T x + \tilde{b}_i^T y] - \bar{F}_i\} + \rho \sum_{i=1}^{m_1} (E[\tilde{a}_i^T x + \tilde{b}_i^T y] - \bar{F}_i) \right\} \\ \text{s.t.} \begin{cases} (x, y) \in S \\ y \in P(x). \end{cases} \end{cases} \tag{2.50}$$

Problem (2.50) can be transformed into the following equivalent problem by introducing an auxiliary variable σ

$$\begin{cases} \min\limits_{x,y,\sigma} \sigma + \rho \sum\limits_{i=1}^{m_1} (E[\tilde{a}_i^T x + \tilde{b}_i^T y] - \bar{F}_i) \\ \text{s.t.} \begin{cases} E[\tilde{a}_i^T x + \tilde{b}_i^T y] - \bar{F}_i \le \sigma \\ (x, y) \in S \\ y \in P(x). \end{cases} \end{cases} \tag{2.51}$$

To solve problem (2.51), we apply the idea of the k-th best method [11]. We first start out by solving the following problem without the constraint $y \in P(x)$ of the Pareto optimality

$$\begin{cases} \min\limits_{x,y,\sigma} \sigma + \rho \sum\limits_{i=1}^{m_1} (E[\tilde{a}_i^T x + \tilde{b}_i^T y] - \bar{F}_i) \\ \text{s.t.} \begin{cases} E[\tilde{a}_i^T x + \tilde{b}_i^T y] - \bar{F}_i \le \sigma \\ (x, y) \in S. \end{cases} \end{cases} \tag{2.52}$$

Let $(\hat{x}^1, \hat{x}^1, \hat{\sigma}^1)$ denote an optimal solution to problem (2.52), and let $(\hat{x}^2, \hat{x}^2, \hat{\sigma}^2)$, $\cdots$, $(\hat{x}^N, \hat{x}^N, \hat{\sigma}^N)$ be the rest of $N-1$ basic feasible solutions such that

$$\sigma^j + \rho \sum_{i=1}^{k_1} (E[\tilde{a}_i^T x^j + \tilde{b}_i^T y^j] - \bar{F}_i) \le \sigma^{j+1} + \rho \sum_{i=1}^{k_1} (E[\tilde{a}_i^T x^{j+1} + \tilde{b}_i^T y^{j+1}] - \bar{F}_i),$$

$j = 1, 2, \cdots, N-1.$

To verify if the response $\hat{y}^j$ of the follower belongs to the set of Pareto optimal solutions with respect to the decision $\hat{x}^j$ of the leader, solve the linear programming problem

$$\begin{cases} \min\limits_{y,\varepsilon} \nu = -\sum\limits_{i=1}^{k_2} \varepsilon_i \\ \text{s.t.} \begin{cases} E[\tilde{f}_i^T y] + \varepsilon_i = E[\tilde{f}_i^T \hat{y}^j], i = 1, 2, \cdots, k_2 \\ \varepsilon_i \ge 0, i = 1, 2, \cdots, k_2 \\ y \in P(\hat{x}^j), \end{cases} \end{cases} \tag{2.53}$$

where $\varepsilon = (\varepsilon_1, \varepsilon_2, \cdots, \varepsilon_{k_2})^T$. If an optimal value is equal to 0, i.e., $\nu = 0$, the response $\hat{y}^j$ satisfies Pareto optimality for the decision $\hat{x}^j$. Thus, solving problem (2.51) is equivalent to finding the minimal index j such that $\nu = 0$. Starting from the point $(\hat{x}^j, \hat{y}^j, \hat{\sigma}^j)$, we examine adjacent points in turn. A procedure for obtaining the optimistic Stackelberg solution is summarized as follows:

Step 0. The leader specifies an initial reference point $(\bar{F}_1, \bar{F}_2, \cdots, \bar{F}_{k_2})$.

Step 1. Let $j = 1$. Solve the linear programming problem (2.52), and let $(\hat{x}^1, \hat{y}^1, \hat{\sigma}^1)$ denote an optimal solution to problem (2.52). Set $W := (\hat{x}^1, \hat{y}^1, \hat{\sigma}^1)$ and $T := \Phi$.

Step 2. Solve the linear programming problem (2.53). If an optimal value is equal to 0, i.e., $\nu = 0$, then $(\hat{x}^j, \hat{y}^j)$ is the optimistic Stackelberg solution with respect to the reference point $(\bar{F}_1, \bar{F}_2, \cdots, \bar{F}_{k_1})$. If the leader is satisfied with the solution, the interactive procedure stops. Otherwise, update the reference point $(\bar{F}_1, \bar{F}_2, \cdots, \bar{F}_{k_2})$ and return to Step 1. If an optimal value to problem (2.53) is not equal to 0, i.e., $\nu \neq 0$, then go to Step 3.

Step 3. Let W_j be a set of extreme points of problem (2.52) which is adjacent to $(\hat{x}^j, \hat{y}^j, \hat{y}^j, \hat{\sigma}^j)$ and satisfies

$$\sigma^j + \rho \sum_{i=1}^{k_1} (E[\tilde{a}_i^T x^j + \tilde{b}_i^T y^j] - \bar{F}_i) \leq \sigma^{j+1} + \rho \sum_{i=1}^{k_1} (E[\tilde{a}_i^T x^{j+1} + \tilde{b}_i^T y^{j+1}] - \bar{F}_i)$$

and let $T := T \cup (\hat{x}^j, \hat{y}^j, \hat{y}^j, \hat{\sigma}^j)$ and $W := (W \cup W^j) \perp T$.

Step 4. Let $j := j + 1$. Choose an extreme point $(\hat{x}^j, \hat{y}^j, \hat{y}^j, \hat{\sigma}^j)$ such that

$$\sigma^j + \rho \sum_{i=1}^{k_1} (E[\tilde{a}_i^T x^j + \tilde{b}_i^T y^j] - \bar{F}_i) = \min_{(x,y,\sigma) \in W} \left\{ \sigma^{j+1} + \rho \sum_{i=1}^{k_1} (E[\tilde{a}_i^T x^{j+1} + \tilde{b}_i^T y^{j+1}] - \bar{F}_i) \right\},$$

and return to Step 2.

Hereafter, we call a Stackelberg solutions based on the pessimistic anticipation as the pessimistic Stackelberg solutions. Let $(\bar{F}_1, \bar{F}_2, \cdots, \bar{F}_{k_1})$ be the reference point, and let Eq. (2.49) be employed as an achievement function. Then, the pessimistic Stackelberg solution can be obtain by iteratively solving the following problem and updating the reference point

$$\begin{cases} \min_{x} \max_{y} \left\{ \max_{i=1,2,\cdots,m_1} \{E[\tilde{a}_i^T x + \tilde{b}_i^T y] - \bar{F}_i\} + \rho \sum_{i=1}^{m_1} (E[\tilde{a}_i^T x + \tilde{b}_i^T y] - \bar{F}_i) \right\} \\ \text{s.t.} \begin{cases} (x, y) \in S \\ y \in P(x). \end{cases} \end{cases} \tag{2.54}$$

Problem (2.54) can be transformed into the following equivalent problem by introducing an auxiliary variable σ

$$\begin{cases} \max_{x} \min_{y,\sigma} \sigma + \rho \sum_{i=1}^{m_1} (E[\tilde{a}_i^T x + \tilde{b}_i^T y] - \bar{F}_i) \\ \text{s.t.} \begin{cases} E[\tilde{a}_i^T x + \tilde{b}_i^T y] - \bar{F}_i \leq \sigma \\ (x, y) \in S \\ y \in P(x). \end{cases} \end{cases} \tag{2.55}$$

Suppose that the leader makes a decision $\hat{x}^j$ and anticipates that the follower chooses a decision undesirable for the leader in the set of Pareto optimal responses.

Namely, we assume that the leader judges that the follower takes an optimal solution $y(\hat{x}^j)$ to the following problem as a response to the decision $\hat{x}^j$

$$\begin{cases} \max\limits_{y,\sigma} \sigma + \rho \sum\limits_{i=1}^{m_1} (E[\tilde{a}_i^T \hat{x}^j + \tilde{b}_i^T y] - \bar{F}_i) \\ \text{s.t.} \begin{cases} E[\tilde{a}_i^T \hat{x}^j + \tilde{b}_i^T y] - \bar{F}_i \le \sigma \\ (\hat{x}^j, y) \in S \\ y \in P(\hat{x}^j). \end{cases} \end{cases} \tag{2.56}$$

Turning our attention to the similarity between problems (2.56) and (2.51), we find that problem (2.56) can be solved by a procedure similar to that for solving problem (2.51). Let $(\hat{y}^l(\hat{x}^j), \sigma^l(\hat{x}^j))$ be a basic feasible solution which yields the lth largest value of the objective function of the following linear programming problem:

$$\begin{cases} \max\limits_{y,\sigma} \sigma + \rho \sum\limits_{i=1}^{m_1} (E[\tilde{a}_i^T \hat{x}^j + \tilde{b}_i^T y] - \bar{F}_i) \\ \text{s.t.} \begin{cases} E[\tilde{a}_i^T \hat{x}^j + \tilde{b}_i^T y] - \bar{F}_i \le \sigma \\ (\hat{x}^j, y) \in S. \end{cases} \end{cases} \tag{2.57}$$

To examine if the response $\hat{y}^l(\hat{x}^j)$ of the follower belongs to the set of Pareto optimal responses $P(\hat{x}^j)$ of problem (2.47), we solve the following linear programming problem:

$$\begin{cases} \min\limits_{y,\varepsilon} w = -\sum\limits_{i=1}^{m_2} \varepsilon_i \\ \text{s.t.} \begin{cases} E[\tilde{f}_i^T y] + \varepsilon_i = E[\tilde{f}_i^T \hat{y}^l], i = 1, 2, \cdots, m_2 \\ \varepsilon_i \ge 0, i = 1, 2, \cdots, m_2 \\ y \in P(\hat{x}^j). \end{cases} \end{cases} \tag{2.58}$$

If an optimal value is equal to 0, i.e., $w = 0$, the response $\hat{y}^l(\hat{x}^j)$ satisfies the Pareto optimality for the decision $\hat{x}^j$. Thus, solving problem (2.56) is equivalent to finding the minimal index l such that $w = 0$. Let l^* be such a minimal index. Then, the response $\hat{y}^{l^*}(\hat{x}^j)$ which is not desirable for the leader in the set of Pareto optimal responses is determined. Namely, solving problem (2.55) is equivalent to finding the minimal index j such that $\hat{y}^j$ is equal to $\hat{y}^{l^*}(\hat{x}^j)$. First, pick up an extreme point of the following problem (2.59) in nondecreasing order of the value of the objective function from $(\hat{x}^l, \hat{y}^l)$ to $(\hat{x}^N, \hat{y}^N)$ these points satisfy

$$\sigma^j + \rho \sum_{i=1}^{k_1} (E[\tilde{a}_i^T x^j + \tilde{b}_i^T y^j] - \bar{F}_i) \le \sigma^{j+1} + \rho \sum_{i=1}^{k_1} (E[\tilde{a}_i^T x^{j+1} + \tilde{b}_i^T y^{j+1}] - \bar{F}_i),$$

$j = 1, 2, \cdots, N-1.$

$$\begin{cases} \min_{x,y,\sigma} \sigma + \rho \sum_{i=1}^{m_1} (E[\tilde{a}_i^T x + \tilde{b}_i^T y] - \bar{F}_i) \\ \text{s.t.} \begin{cases} E[\tilde{a}_i^T x + \tilde{b}_i^T y] - \bar{F}_i \leq \sigma \\ (x, y) \in S. \end{cases} \end{cases} \tag{2.59}$$

Second, find $\hat{y}^{l^*}(\hat{x}^j)$ by the above mentioned procedure. Third, check whether $\hat{y}^j$ is equal to $\hat{y}^{l^*}(\hat{x}^j)$. A procedure for obtaining the pessimistic Stackelberg solution is summarized as follows:

Step 1. The leader specifies an initial reference point $(\bar{F}_1, \bar{F}_2, \cdots, \bar{F}_{m_1})$.

Step 2. Let $j = 1$. Solve the linear programming problem (2.59), and let $(\hat{x}^1, \hat{y}^1, \hat{\sigma}^1)$ denote an optimal solution to problem (2.59). Set $W^1 := (\hat{x}^1, \hat{y}^1, \hat{\sigma}^1)$ and $T^1 := \Phi$.

Step 3. Let $l := 1$. Solve the linear programming problem (2.57). and let $\hat{y}^l(\hat{x}^j)$ denote an optimal solution to problem (2.57). Let $W^2 := \{\hat{y}^l(\hat{x}^j), \hat{\sigma}^1\}$ and $T^2 := \emptyset$.

Step 4. Solve the linear programming problem (2.58). If an optimal value is equal to 0, i.e., $w = 0$ and $\hat{y}^j$ is equal to $\hat{y}^{l^*}(\hat{x}^j)$, then $(\hat{x}^j, (\hat{y}^j))$ is the pessimistic Stackelberg solution with respect to the reference point $(\bar{F}_1, \bar{F}_2, \cdots, \bar{F}_{m_1})$. If the leader is satisfied with the solution, the interactive procedure stops. Otherwise, update the reference point $(\bar{F}_1, \bar{F}_2, \cdots, \bar{F}_{m_1})$, and return to Step 2. If an optimal value to problem (2.58) is not equal to 1, i.e., $w \neq 0$, then go to Step 5. If $w = 0$ and $\hat{y}^j \neq \hat{y}^{l^*}(\hat{x}^j)$, then go to Step 7.

Step 5. Let W^{2l} be a set of extreme points of problem (2.57) which is adjacent to $\hat{y}^l(\hat{x}^j)$ and satisfies:

$$\hat{\sigma}^l + \rho \sum_{i=1}^{m_1} (E[\tilde{a}_i^T x^j + \tilde{b}_i^T y^j] - \bar{F}_i) \geq \sigma^{j+1} + \rho \sum_{i=1}^{m_1} (E[\tilde{a}_i^T x^{j+1} + \tilde{b}_i^T y^{j+1}] - \bar{F}_i),$$

and let $T^2 := T^2 \cup (\hat{y}^l(\hat{x}^j), \hat{\sigma}^l)$ and $W^2 := (W^2 \cup W^{2l}) \perp T^2$.

Step 6. Let $l := l + 1$. Choose an extreme point $(\hat{y}^l(\hat{x}^j), \hat{\sigma}^l)$ such that:

$$\hat{\sigma}^l + \rho \sum_{i=1}^{m_1} (E[\tilde{a}_i^T x^j + \tilde{b}_i^T y^j] - \bar{F}_i) = \max_{(y,\sigma) \in W^2} \sigma^{j+1} + \rho \sum_{i=1}^{m_1} (E[\tilde{a}_i^T x^{j+1} + \tilde{b}_i^T y^{j+1}] - \bar{F}_i),$$

and return to Step 4.

Step 7. Let W^{1j} be a set of extreme points of problem (2.59) which is adjacent to $\hat{x}^j, \hat{y}^j, \hat{\sigma}^j$ and satisfies:

$$\hat{\sigma}^j + \rho \sum_{i=1}^{m_1} (E[\tilde{a}_i^T x^j + \tilde{b}_i^T y^j] - \bar{F}_i) \leq \sigma + \rho \sum_{i=1}^{m_1} (E[\tilde{a}_i^T x + \tilde{b}_i^T y] - \bar{F}_i),$$

and let $T^1 := T^1 \cup (\hat{x}^j, \hat{y}^j, \hat{\sigma}^j)$ and $W^1 := (W^1 \cup W^{1j}) \perp T^1$.

Step 8. Let $j := j + 1$. Choose an extreme point $\hat{x}^j, \hat{y}^j, \hat{\sigma}^j$ such that:

$$\hat{\sigma}^j + \rho \sum_{i=1}^{m_1} (E[\tilde{a}_i^T x^j + \tilde{b}_i^T y^j] - \bar{F}_i) = \min_{(x,y,\sigma)\in W^1} \sigma + \rho \sum_{i=1}^{m_1} (E[\tilde{a}_i^T x + \tilde{b}_i^T y] - \bar{F}_i),$$

and return to Step 3.

We consider an algorithm based on an interactive method similar to the methods shown in the above because it can be expected that the decision maker derives the satisfactory solutions by learning and recognizing local preferences during an interactive process.

Suppose situations where the leader and the follower have been confronted often with the decision making problem represented by (2.36); in such a case, it is natural to consider that the leader knows or can estimate the preference of the follower. We assume that the leader can learn and recognize the local preference of the leader by an interactive process, but cannot learn and recognize that of the follower; therefore, the followerąŕs reference point is estimated only once by the leader.

Let $(\bar{F}_1, \bar{F}_2, \cdots, \bar{F}_{m_1})$ denote the follower's reference point specified by the leader. Then, it follows that the follower makes a decision $\hat{y}$ by solving the following problem with respect to a decision $\hat{x}$ of the leader

$$\begin{cases} \min_{y} \left\{ \max_{i=1,2,\cdots,m_2} \{E[\tilde{e}_i^T x + \tilde{f}_i^T y] - \bar{f}_i\} + \rho \sum_{i=1}^{m_1} (E[\tilde{f}_i^T x + \tilde{g}_i^T y] - \bar{f}_i) \right\} \\ \text{s.t. } (\hat{x}, y) \in S. \end{cases} \tag{2.60}$$

Problem (2.60) can be transformed into the following equivalent problem by introducing an auxiliary variable η

$$\begin{cases} \min_{y,\eta} \eta + \rho \sum_{i=1}^{m_2} (E[\tilde{e}_i^T \hat{x} + \tilde{f}_i^T y] - \bar{f}_i) \\ \text{s.t.} \begin{cases} E[\tilde{f}_i^T \hat{x} + \tilde{f}_i^T y] - \bar{f}_i \leq \eta, i = 1, , \cdots, m \\ (\hat{x}, y) \in S. \end{cases} \end{cases} \tag{2.61}$$

Given a decision $\hat{x}$ of the leader, a set of rational responses $R(\hat{x})$ of the follower is a set of optimal solutions to problem (2.61). We assume that the set of rational responses $R(\hat{x})$ is a singleton for any Ĺx, i.e., $R(\hat{x})$ is a function of $\hat{x}$.

By introducing an auxiliary variable σ, problem (2.61) can be transformed into the following equivalent problem

$$
\begin{cases}
\min_{x,\sigma} \sigma + \rho \sum_{i=1}^{m_1} (E[\tilde{a}_i^T x + \tilde{b}_i^T y] - \bar{F}_i) \\
\text{s.t.} \begin{cases} y = R(x) \\ E[\tilde{a}_i^T x + \tilde{b}_i^T y] - \bar{F}_i \leq \sigma \\ (x, y) \in S. \end{cases}
\end{cases}
\tag{2.62}
$$

Because problem (2.60) is a linear programming problem and problem (2.62) is also linear with the exception of the constraint $y = R(x)$, problem (2.62) can be reduced to a conventional two-level linear programming problem, and it can be solved by using one of the developed algorithms [7–9, 11, 17, 129].

Example 2.2. We consider the following random bi-level multiobjective linear programming problem , which has a single objective function and three objective functions of the follower:

$$
\begin{cases}
\min_{x_1,x_2,x_3} F = \xi_{11}x_1 + \xi_{21}x_2 + \xi_{31}x_3 + 2\xi_{41}y_1 + \xi_{51}y_2 + \xi_{61}y_3 \\
\text{s.t.} \begin{cases}
\text{where } (y_1, y_2, y_3) \text{ solves} \\
f_1 = \min_{y_1,y_2,y_3} \xi_{21}x_1 + \xi_{22}x_2 + \xi_{23}x_3 + \xi_{24}y_1 + \xi_{25}y_2 + \xi_{26}y_3 \\
f_2 = \min_{y_1,y_2,y_3} \xi_{31}x_1 + \xi_{32}x_2 + \xi_{33}x_3 + \xi_{34}y_1 + \xi_{35}y_2 + \xi_{36}y_3 \\
f_3 = \min_{y_1,y_2,y_3} \xi_{41}x_1 + \xi_{42}x_2 + \xi_{43}x_3 + \xi_{44}y_1 + \xi_{45}y_2 + \xi_{46}y_3 \\
\text{s.t.} \begin{cases}
15x_1 - x_2 + 3x_3 + 2y_1 - y_2 + 2y_3 \leq 200 \\
7x_1 + 7x_2 + 6x_3 + y_1 + 13y_2 + y_3 \leq 140 \\
2x_1 + 2x_2 - x_3 + 14y_1 + 2y_2 + 2y_3 \leq 240 \\
-3x_1 + 6x_2 + 12x_3 + 4y_1 - 8y_2 + y_3 \leq 140 \\
4x_1 - 7x_2 + 7x_3 + 2y_1 + 4y\_7y_3 \leq 45 \\
4x_1 + 5x_2 + x_3 - 7y_1 - 6y_2 + y_3 \leq 800 \\
x_1, x_2, x_3, y_1, y_2, y_3 \geq 0,
\end{cases}
\end{cases}
\end{cases}
\tag{2.63}
$$

where $\xi_{ij}, i = 1,2,3,4; j = 1,2,\cdots,6$ are normal distributed random variables characterized as:

$$
\begin{aligned}
&\xi_{11} \sim \mathcal{N}(-14,2), \xi_{11} \sim \mathcal{N}(11,2), \xi_{13} \sim \mathcal{N}(8,2), \xi_{14} \sim \mathcal{N}(-15,2),\\
&\xi_{15} \sim \mathcal{N}(-3,2), \xi_{16} \sim \mathcal{N}(4,2), \xi_{21} \sim \mathcal{N}(6,2), \xi_{22} \sim \mathcal{N}(-2,1),\\
&\xi_{23} \sim \mathcal{N}(4,2), \xi_{24} \sim \mathcal{N}(-4,3), \xi_{25} \sim \mathcal{N}(7,2), \xi_{26} \sim \mathcal{N}(-7,1),\\
&\xi_{31} \sim \mathcal{N}(-1,0), \xi_{32} \sim \mathcal{N}(-13,3), \xi_{33} \sim \mathcal{N}(-3,2), \xi_{34} \sim \mathcal{N}(4,1),\\
&\xi_{35} \sim \mathcal{N}(2,0), \xi_{36} \sim \mathcal{N}(4,3), \xi_{41} \sim \mathcal{N}(-1,0), \xi_{42} \sim \mathcal{N}(-2,1),\\
&\xi_{43} \sim \mathcal{N}(-18,8), \xi_{44} \sim \mathcal{N}(3,3), \xi_{45} \sim \mathcal{N}(-9,2), \xi_{34} \sim \mathcal{N}(8,1).
\end{aligned}
$$

The EEEE model of (2.63) is:

$$\left\{\begin{array}{l}\min\limits_{x_1,x_2,x_3} E[\xi_{11}x_1+\xi_{21}x_2+\xi_{31}x_3+2\xi_{41}y_1+\xi_{51}y_2+\xi_{61}y_3]\\ \text{s.t.}\left\{\begin{array}{l}\text{where } (y_1,y_2,y_3) \text{ solves}\\ f_1=\min\limits_{y_1,y_2,y_3} E[\xi_{21}x_1+\xi_{22}x_2+\xi_{23}x_3+\xi_{24}y_1+\xi_{25}y_2+\xi_{26}y_3]\\ f_2=\min\limits_{y_1,y_2,y_3} E[\xi_{31}x_1+\xi_{32}x_2+\xi_{33}x_3+\xi_{34}y_1+\xi_{35}y_2+\xi_{36}y_3]\\ f_3=\min\limits_{y_1,y_2,y_3} E[\xi_{41}x_1+\xi_{42}x_2+\xi_{43}x_3+\xi_{44}y_1+\xi_{45}y_2+\xi_{46}y_3]\\ \text{s.t.}\left\{\begin{array}{l}15x_1-x_2+3x_3+2y_1-y_2+2y_3\le 200\\ 7x_1+7x_2+6x_3+y_1+13y_2+y_3\le 140\\ 2x_1+2x_2-x_3+14y_1+2y_2+2y_3\le 240\\ -3x_1+6x_2+12x_3+4y_1-8y_2+y_3\le 140\\ 4x_1-7x_2+7x_3+2y_1+4y\_7y_3\le 45\\ 4x_1+5x_2+x_3-7y_1-6y_2+y_3\le 800\\ x_1,x_2,x_3,y_1,y_2,y_3\ge 0.\end{array}\right.\end{array}\right.\end{array}\right. \tag{2.64}$$

It follows from Theorem 2.1 that model is equivalent to:

$$\left\{\begin{array}{l}\min\limits_{x_1,x_2,x_3} -14x_1+11x_2+8x_3-15y_1-3y_2+4y_3\\ \text{s.t.}\left\{\begin{array}{l}\text{where } (y_1,y_2,y_3) \text{ solves}\\ \min\limits_{y_1,y_2,y_3} 6x_1-2x_2+4x_3-4y_1+7y_2-7y_3\\ \min\limits_{y_1,y_2,y_3} -x_1-13x_2-3x_3+4y_1+2y_2+4y_3\\ \min\limits_{y_1,y_2,y_3} -x_1-2x_2-18x_3+3y_1-9y_2+8y_3]\\ \text{s.t.}\left\{\begin{array}{l}15x_1-x_2+3x_3+2y_1-y_2+2y_3\le 200\\ 7x_1+7x_2+6x_3+y_1+13y_2+y_3\le 140\\ 2x_1+2x_2-x_3+14y_1+2y_2+2y_3\le 240\\ -3x_1+6x_2+12x_3+4y_1-8y_2+y_3\le 140\\ 4x_1-7x_2+7x_3+2y_1+4y\_7y_3\le 45\\ 4x_1+5x_2+x_3-7y_1-6y_2+y_3\le 800\\ x_1,x_2,x_3,y_1,y_2,y_3\ge 0.\end{array}\right.\end{array}\right.\end{array}\right. \tag{2.65}$$

The decision variable vectors (x_1, x_2, x_3) and (y_1, y_2, y_3) of the leader and the follower, respectively, are three-dimensional vectors. This problem is larger than Example 2.1 cannot be solved by the graphical solution procedure. Thus, we must use the algorithms described above to obtain the optimistic and the pessimistic Stackelberg solutions to this problem. The solutions are shown in Table 2.2.

Table 2.2 Stackelberg solutions and objective function values for the problem

	Optimistic solution	Pessimistic solution
(x_1, x_2, x_3)	(11.94, 0.00, 0.00)	(14.22, 0.00, 0.00)
(y_1, y_2, y_3)	(14.18, 2.79, 6.04)	(0.00, 2:86, 3.33)
F	−364.01	−194.35
f_1	−7.83	82.01
f_2	74.49	4.81
f_3	53.81	−13.28

2.3.3 *Genetic Approach for Nonlinear Models*

If model (2.35) is a nonlinear model, i.e., some functions in model (2.35) is nonlinear, the random features make it impossible to be converted into its crisp equivalent model. In order solve nonlinear model (2.35), three issues are needed to solve: the trade-off of the multiple objectives, the randomness and the bi-level structure. In this section, we design a hybrid method

consisting of simulation technique, goal programming method and genetic algorithm, where they are deal with random expected operators, multiple objectives and bi-level structure, respectively.

Random Simulation 1 for Expected Value

Let ξ be an n-dimensional random vector defined on the probability space $(\Omega, \mathscr{A}, Pr)$ (equivalently, it is characterized by a probability distribution $F(\cdot)$), and $f : \mathbf{R}^n \to \mathbf{R}$ a measurable function. Then $f(\xi)$ is a random variable. In order to calculate the expected value $E[f(\xi)]$, we generate ω_k from Ω according to the probability measure Pr, and write $\xi_k = \xi(\omega_k)$ for $k = 1, 2, \cdots, N$. Equivalently, we generate random vectors $\xi_k, k = 1, 2, \cdots, N$ according to the probability distribution $F(\cdot)$. It follows from the strong law of large numbers that:

$$\frac{\sum\limits_{k=1}^{N} f(\xi_k)}{N} \to E[f(\xi)], \tag{2.66}$$

as $N \to \infty$. Therefore, the value $E[f(\xi)]$ can be estimated by $\frac{1}{N}\sum_{k=1}^{N} f(\xi_k)$ provided that N is sufficiently large.

The process of random simulation 1 for expected value can be summarized as follows:

Step 1. Set $L = 0$.
Step 2. Generate ω from Ω according to the probability measure Pr.
Step 3. $L \leftarrow L + f(\xi(\omega))$.
Step 4. Repeat the second and third steps N times.
Step 5. $E[f(\xi)] = L/N$.

Example 2.3. Let ξ_1 be an exponentially distributed variable $exp(2)$, ξ_2 a normally distributed variable $\mathcal{N}(4,1)$, and ξ_3 a uniformly distributed variable $\mathcal{U}(4,8)$. A run of random simulation with 10000 cycles shows that

$$E\left[\sqrt{\xi_1^3+\xi_2^2+\xi_3}\right] = 5.3698.$$

Goal Programming Method

The goal programming method was initialized by Charnes and Cooper [20] in 1961. After that, Ijiri [60], Lee [77], Kendall and Lee [67], Ignizio [59], Narasimhan [99], Dyer [35], Lee and Kim [76], Freed and Glover [39], Karsak et al. [65], Schniederjans [114], Buffa and Jackson [14], Barichard [10], Calvete et al. [18], Chu [24] deeply researched and widely developed it. When dealing with many multiple objective decision-making problems, the goal programming method is widely applied since it could provide with a technique that is accepted by many decision makers, that is, it could point out the preference information and harmoniously inosculate it into the model.

The basic idea of the goal programming method is that, for the objective function $\boldsymbol{f}(\boldsymbol{x}) = (f_1(\boldsymbol{x}),f_2(\boldsymbol{x}),\cdots,f_m(\boldsymbol{x}))^T$, decision makers give a goal value $\boldsymbol{f}^o = (f_1^o,f_2^o,\cdots,f_m^o)^T$ such that every objective function $f_i(\boldsymbol{x})$ approximates the goal value f_i^o as closely as possible. Let $d_p(\boldsymbol{f}(\boldsymbol{x}),\boldsymbol{f}^o) \in \mathbf{R}^m$ be the deviation between $\boldsymbol{f}(\boldsymbol{x})$ and $\boldsymbol{f}^o$, and then consider the following problem:

$$\min_{\boldsymbol{x}\in X} d_p(\boldsymbol{f}(\boldsymbol{x}),\boldsymbol{f}^o), \tag{2.67}$$

where the goal value $\boldsymbol{f}^o$ and the weight vector $\mathbf{w}$ are predetermined by the decision maker. The weight w_i expresses the important factor that the objective function $f_i(\boldsymbol{x})$ $(i = 1,2,\cdots,m)$ approximates the goal value f_i^o, $1 \le p \le \infty$.

When $p = 1$, it is recalled the simple goal programming method, which is most widely used. Then we have:

$$d_p(\boldsymbol{f}(\boldsymbol{x}),\boldsymbol{f}^o) = \sum_{i=1}^{m} w_i|\boldsymbol{f}(\boldsymbol{x}),\boldsymbol{f}^o)|.$$

Since there is the notation $|\cdot|$ in $d_p(\boldsymbol{f}(\boldsymbol{x}),\boldsymbol{f}^o)$, it is not a differentiable function anymore. Therefore, denote that $d_i^+ = \frac{1}{2}(|f_i(\boldsymbol{x}) - f_i^o| + (f_i(\boldsymbol{x}) - f_i^o))$ and $d_i^- = \frac{1}{2}(|f_i(\boldsymbol{x}) - f_i^o| - (f_i(\boldsymbol{x}) - f_i^o))$, where d_i^+ expresses the quantity that $f_i(\boldsymbol{x})$ exceeds f_i^o, and d_i^- expresses the quantity that $f_i(\boldsymbol{x})$ is less than f_i^o. It is easy to prove that, for any $i = 1,2,\cdots,m$,

$$d_i^+ + d_i^- = |f_i(\boldsymbol{x}) - f_i^o|, d_i^+ - d_i^- = f_i(\boldsymbol{x}) - f_i^o, d_i^+ d_i^- = 0,\ d_i^+, d_i^- \ge 0. \tag{2.68}$$

When $p = 1$, problem (2.67), can be rewritten as:

$$\begin{cases} \min \sum_{i=1}^{m} w_i(d_i^+ + d_i^-) \\ \text{s.t.} \begin{cases} f_i(\boldsymbol{x}) + d_i^+ - d_i^- = f_i^o \\ d_i^+ d_i^- = 0 \\ d_i^+, d_i^- \geq 0 \\ i = 1, 2, \cdots, m \\ \boldsymbol{x} \in X. \end{cases} \end{cases} \tag{2.69}$$

In order to easily solve problem (2.69), abandon the constraint $d_i^+ d_i^- = 0$ $(i = 1, 2, \cdots, m)$, and we have:

$$\begin{cases} \min \sum_{i=1}^{m} w_i(d_i^+ + d_i^-) \\ \text{s.t.} \begin{cases} f_i(\boldsymbol{x}) + d_i^+ - d_i^- = f_i^o \\ d_i^+, d_i^- \geq 0 \\ i = 1, 2, \cdots, m \\ \boldsymbol{x} \in X. \end{cases} \end{cases} \tag{2.70}$$

Theorem 2.2. *If $(\boldsymbol{x}, \bar{\boldsymbol{d}}^+, \bar{\boldsymbol{d}}^-)$ is the optimal solution to problem (2.70), then $\bar{\boldsymbol{x}}$ is doubtlessly the optimal solution to problem (2.67), where $\bar{\boldsymbol{d}}^+ = (\bar{d}_1^+, \bar{d}_2^+, \cdots, \bar{d}_m^+)$ and $\bar{\boldsymbol{d}}^- = (\bar{d}_1^-, \bar{d}_2^-, \cdots, \bar{d}_m^-)$.*

Proof. Since $(\boldsymbol{x}, \bar{\boldsymbol{d}}^+, \bar{\boldsymbol{d}}^-)$ is the optimal solution to problem (2.70), we have $\boldsymbol{x} \in X$, $\bar{\boldsymbol{d}}^+ \geq 0, \bar{\boldsymbol{d}}^- \geq 0$ and

$$f_i(\boldsymbol{x}) + \bar{d}_i^+ - \bar{d}_i^- = f_i^o,\ i = 1, 2, \cdots, m. \tag{2.71}$$

(1) If $\bar{d}_i^+ = \bar{d}_i^- = 0$, we have $f_i(\boldsymbol{x}) = f_i^o$, which means $\boldsymbol{x}$ is the optimal solution problem (2.67).
(2) If there exists $i_0 \in \{1, 2, \cdots, m\}$ such that $f_i(\boldsymbol{x}) \neq f_i^o$, $\bar{d}_i^+ \bar{d}_i^- = 0$ doubtlessly holds. If not, we have $\bar{d}_i^+ > 0$ and $\bar{d}_i^- > 0$. We discuss them respectively, as follows:

(a) If $\bar{d}_i^+ - \bar{d}_i^- > 0$, for $i \in \{1, 2, \cdots, m\}$, let

$$\tilde{d}_i^+ = \begin{cases} \bar{d}_i^+ - \bar{d}_i^-, & \text{if } i = i_0 \\ \bar{d}_i^+, & \text{if } i \neq i_0 \end{cases} \text{ and } \tilde{d}_i^- = \begin{cases} 0, & \text{if } i = i_0 \\ \bar{d}_i^-, & \text{if } i \neq i_0. \end{cases} \tag{2.72}$$

Thus, $\tilde{d}_{i_0}^+ < \bar{d}_{i_0}^+$ and $\tilde{d}_{i_0}^- < \bar{d}_{i_0}^-$ both hold. It follows from equations. (2.71) and (2.72) that

$$f_i(\boldsymbol{x}) + \tilde{d}_i^+ - \tilde{d}_i^- = \begin{cases} f_i(\boldsymbol{x}) + 0 - (\bar{d}_i^+ - \bar{d}_i^-) = f_i^o, & i = i_0 \\ f_i(\boldsymbol{x}) + \bar{d}_i^+ - \bar{d}_i^- = f_i^o, & i \neq i_0. \end{cases}$$

We also know that $\boldsymbol{x} \in X$, $\tilde{d}_i^+ \geq 0$ and $\tilde{d}_i^- \geq 0$. Denote $\tilde{\boldsymbol{d}}^+ = (\tilde{d}_1^+, \tilde{d}_2^+, \cdots, \tilde{d}_m^+)$ and $\tilde{\boldsymbol{d}}^- = (\tilde{d}_1^-, \tilde{d}_2^-, \cdots, \tilde{d}_m^-)$, and then we have $(\boldsymbol{x}, \tilde{\boldsymbol{d}}^+, \tilde{\boldsymbol{d}}^-)$ as a feasible solution to problem (2.70). It follows from $\tilde{d}_{i_0}^+ < \bar{d}_{i_0}^+$ and $\tilde{d}_{i_0}^- < \bar{d}_{i_0}^-$ that

$$\sum_{i=1}^m (\tilde{d}_{i_0}^+ + \tilde{d}_{i_0}) < \sum_{i=1}^m (\bar{d}_{i_0}^+ + \bar{d}_{i_0}). \tag{2.73}$$

This conflicts with the assumption that $(\boldsymbol{x}, \bar{\boldsymbol{d}}^+, \bar{\boldsymbol{d}}^-)$ is the optimal solution to problem (2.70).

(b) If $\bar{d}_i^+ - \bar{d}_i^- < 0$, for $i \in \{1, 2, \cdots, m\}$, let

$$\tilde{d}_i^+ = \begin{cases} 0, & \text{if } i = i_0 \\ \bar{d}_i^+, & \text{if } i \neq i_0 \end{cases} \text{ and } \tilde{d}_i^- = \begin{cases} -(\bar{d}_i^+ - \bar{d}_i^-), & \text{if } i = i_0 \\ \bar{d}_i^-, & \text{if } i \neq i_0. \end{cases} \tag{2.74}$$

We can similarly prove that this conflicts with the assumption that $(\boldsymbol{x}, \bar{\boldsymbol{d}}^+, \bar{\boldsymbol{d}}^-)$ is the optimal solution to problem (2.70).

So far we have proved that $(\boldsymbol{x}, \bar{\boldsymbol{d}}^+, \bar{\boldsymbol{d}}^-)$ is the optimal solution to problem (2.69). Since the feasible region of problem (2.69) is included in problem (2.70), $(\boldsymbol{x}, \bar{\boldsymbol{d}}^+, \bar{\boldsymbol{d}}^-)$ is the optimal solution to problem (2.70). Next, we will prove that $(\boldsymbol{x}, \bar{\boldsymbol{d}}^+, \bar{\boldsymbol{d}}^-)$ is the optimal solution to problem (2.67). For any feasible solution $(\boldsymbol{x}, \bar{\boldsymbol{d}}^+, \bar{\boldsymbol{d}}^-)$, it follows from (2.68) that

$$|f_i(\boldsymbol{x}) - f_i^o| = d_i^+ + d_i^-, \ |f_i(\bar{\boldsymbol{x}}) - f_i^o| = \bar{d}_i^+ + \bar{d}_i^-, \ i = 1, 2, \cdots, m.$$

For any $\boldsymbol{x} \in X$, since

$$\sum_{i=1}^m |f_i(\bar{\boldsymbol{x}}) - f_i^o| = \sum_{i=1}^m (\bar{d}_i^+ + \bar{d}_i^-) \leq \sum_{i=1}^m (d_i^+ + d_i^-) = \sum_{i=1}^m |f_i(\boldsymbol{x}) - f_i^o|,$$

it means that $\bar{\boldsymbol{x}}$ is the optimal solution to problem (2.67).

RWGA

Since the 1960s, there has been increasing interest in developing powerful algorithms for difficult optimization problems that imitate the decision making of people. A term now in common use for such techniques is evolutionary computation. The best known algorithms in this class are Genetic Algorithms (GAs) developed by Holland [55], evolutionary strategies developed by Rechenberg and Eigen [106], and Schwefel [115], and genetic programming developed by Koza [70]. There are also many hybrid versions that incorporate various features of these foregoing paradigms. State-of-the-art overviews in the field of evolutional computation can be found in Bäck et al. [6], and Michalewicz [88].

Genetic algorithms, perhaps the most widely known evolutionary computation methods, are powerful and broadly applicable stochastic search and optimization techniques. In the past few years, the genetic algorithm community has turned much of its attention to optimization problems in industrial engineering, resulting in a fresh body of research and applications [13, 27, 78]. A bibliography on genetic algorithms can be found in Alander [2].

In general, a genetic algorithm has five basic components as summarized by Michalewic [88]:

(1) A genetic representation of solutions to the problem.
(2) A method for the creation of an initial population of solutions.
(3) An evaluation function which rates the solutions in terms of their fitness.
(4) Genetic operators that alter the genetic composition of the children during reproduction.
(5) Parameter values for the genetic algorithms.

Ishibuchi et al. [62] proposed a weight-sum based fitness assignment method, called the random weighted genetic algorithm (RWGA), to determine the variable search direction toward the Pareto frontier. This weighted-sum approach is an extension of the methods used in conventional approaches to GA multiple objective optimization through the assignment of weights to each objective function and the combination of the weighted objectives into a single objective function. Typically, there are two types of search behaviors in the objective space: a fixed direction search and a multiple direction search, as seen in Figs. 2.5 and 2.6. The random-weight approach allows the genetic algorithms to have a variable search direction tendency which uniformly samples the area over the entire frontier.

To solve multiple objective problems, this section applies rough simulation and makes use of the random-weight genetic algorithm to convert uncertain multi-objective problems into deterministic problems. For the following model:

$$\begin{cases} \max[E[f_1(\boldsymbol{x},\boldsymbol{\xi})], E[f_2(\boldsymbol{x},\boldsymbol{\xi})], \cdots, E[f_m(\boldsymbol{x},\boldsymbol{\xi})]] \\ \text{s.t.} \begin{cases} E[g_r(\boldsymbol{x},\boldsymbol{\xi})] \leq 0, r = 1, 2, \cdots, p \\ x_j \geq 0, j = 1, 2, \cdots, n. \end{cases} \end{cases}$$

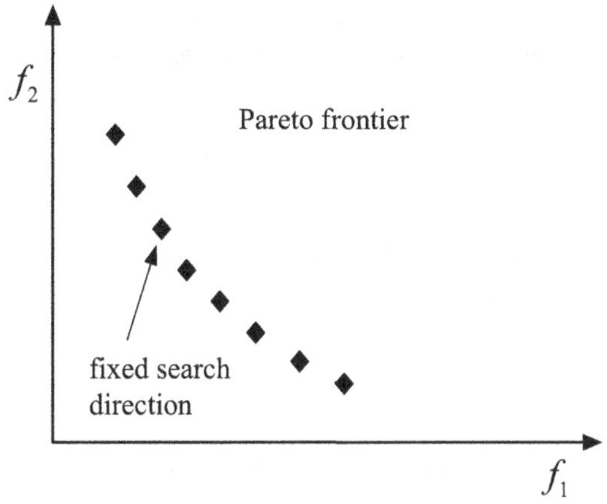

Fig. 2.5 Search in a fixed direction in criterion space

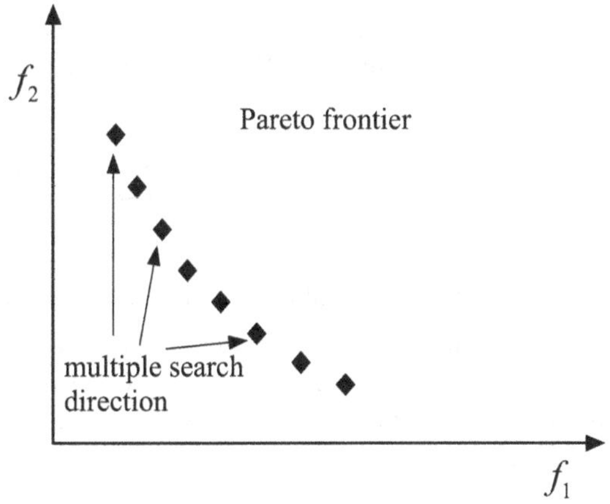

Fig. 2.6 Search in multiple directions in criterion space

We first simulate the expected value using random rough simulation and then apply the genetic algorithm to solve the multiple objective programming problem. This process can be summarized as follows:

Representation. A vector $\boldsymbol{x}$ is chosen as the chromosome to represent a possible solution to the optimization problem.

Handling the objective and constraint function. To derive the determined multiple objective programming problem, we apply a rough simulation technique.

Initialization. Suppose a decision maker is able to predetermine a region which contains the feasible set. Generate a random vector x from this region until a feasible chromosome is accepted. Repeat the above process $N_{popsize}$ times, until the initial feasible chromosomes $\boldsymbol{x}^1, \boldsymbol{x}^2, \cdots, \boldsymbol{x}^{N_{popsize}}$ are found.

Evaluation function. The decision maker's aim is to obtain the maximum expected value for each goal. Suppose $E[f(\boldsymbol{x}, \boldsymbol{\xi})] = \sum_{i=1}^{m} E[f_i(\boldsymbol{x}, \boldsymbol{\xi})]$, where the weight coefficient w_i expresses the importance of $E[f_i(\boldsymbol{x}, \xi)]$ to the decision maker. Then the evaluation function can be written as follows:

$$eval(\boldsymbol{x}) = \sum_{i=1}^{m} E[f_i(\boldsymbol{x}, \xi)],$$

where the random weight is generated as

$$w_i = \frac{r_i}{\sum_{i=1}^{m} r_i}, i = 1, 2, \cdots, m,$$

where r_i are nonnegative random numbers.

Selection process. The roulette wheel method is applied to develop the selection process, so at each turn of the wheel, a single chromosome for a new population is selected, as shown in the following: Compute the total probability q,

$$q = \sum_{j=1}^{N_{popsize}} eval(\boldsymbol{x}^j).$$

Then compute the probability of the ith chromosome $q_i, q_i = \frac{eval(\boldsymbol{x}^i)}{q}$. Generate a random number r in [0,1] and select the ith chromosome $\boldsymbol{x}^i$ such that $q_{i-1} < r \leq q_i, 1 \leq i \leq N_{popsize}$. Repeat the above process $N_{pop-size}$ times and $N_{pop-size}$ copies of chromosomes are determined. Selection probability is computed using the following function:

$$p_i = \frac{eval(\boldsymbol{x}^i) - eval(x)_{min}}{\sum_{j=1}^{popsize} eval(\boldsymbol{x}^i) - eval(\boldsymbol{x})_{min}},$$

where $eval(\boldsymbol{x})_{min}$ is the minimum fitness value for the current population.

Crossover operation. Generate two random numbers λ_1, λ_2 from the open interval $(0, 1)$ that satisfy $\lambda_1 + \lambda_2 = 1$, and select the chromosome $\boldsymbol{x}^i$ as a parent provided that $\lambda_i < P_{\lambda_i}$, where parameter P_{λ_i} is the probability of the crossover operation. Repeat this process $N_{popsize}$ times with $P_{\lambda_i} \cdot N_{popsize}$ chromosomes selected to undergo the crossover operation. The crossover operator on x^1 and x^2 produces two children y^1 and y^2 as follows:

$$\mathbf{y}^1 = \lambda_1 \boldsymbol{x}^1 + \lambda_2 \boldsymbol{x}^2, \mathbf{y}^2 = \lambda_1 \boldsymbol{x}^2 + \lambda_2 \boldsymbol{x}^1.$$

If both children are feasible, then the parents are replaced, otherwise, the feasible child is used and the above operation is repeated until two feasible children are obtained or a given number of cycles are completed.

Mutation operation. Similar to the crossover process, chromosome x^i is selected as the parent to undergo the mutation operation provided that the random number $m < P_m$, where parameter P_m is the probability of the mutation operation. After the process is repeated $N_{popsize}$ times, $P_{\lambda_i} \cdot N_{popsize}$ are the expected selection. Suppose x^1 is chosen as a parent. A mutation direction $\mathbf{d} \in \mathbf{R^n}$ is chosen randomly. Replace x with $\boldsymbol{x} + M \cdot \mathbf{d}$ if $\boldsymbol{x} + M \cdot \mathbf{d}$ is feasible; otherwise set M as a random between 0 and M until it is feasible or a given number of cycles are completed. Here, M is a sufficiently large positive number.

We illustrate the random simulation-based random weight genetic algorithm procedure as follows:

Step 1. Initialize $N_{popsize}$ chromosomes, the feasibility of which can be checked through the fuzzy random simulation.
Step 2. Update the chromosomes using crossover and mutation operations and then use rough simulation to check the feasibility of the offspring. Compute the fitness of each chromosome based on the weighted-sum objective.
Step 3. Select chromosomes by spinning the roulette wheel.
Step 4. Conduct the crossover operation.
Step 5. Conduct the mutation operation for the chromosomes generated in the crossover operation.
Step 6. Repeat the second to fourth steps for a given number of cycles.
Step 7. Report the best chromosome as the optimal solution.

Numerical Example

In what follows, a numerical example is presented to illustrate the usage of the proposed solution method to the nonlinear random bi-level EEEE model.

Example 2.4. Consider the problem

$$\begin{cases} \max\limits_{x_1,x_2,x_3} F_1 = 2\xi_1^2 x_1 + \sqrt{\xi_2 x_2 + \xi_3 x_3} + \xi_4 y_1 + \sqrt{\xi_5} y_2 + \sqrt{\xi_6} y_3 \\ \max\limits_{x_1,x_2,x_3} F_2 = x_1 + x_2 + x_3 + y_1 + y_2 + y_3 \\ \text{s.t.} \begin{cases} \text{where } (y_1, y_2, y_3) \text{ solves} \\ \max\limits_{y_1,y_2,y_3} f_1 = \sqrt{7\xi_1 x_1 + 3\xi_2 x_2 + 8\xi_3 x_3} + (12\xi_4 y_1 + 11\xi_5 y_2 + 5\xi_6 y_3)^2 \\ \max\limits_{y_1,y_2,y_3} f_2 = x_1 + 5x_2 + 7x_3 + y_1 + 11y_2 - 8y_3 \\ \text{s.t.} \begin{cases} \sqrt{x_1\xi_1 + x_2\xi_2} + x_3^2\xi_3 + \xi_4 y_2 + 17y_3\xi_6 \le 1540 \\ 3x_1 + 6x_2 + 12x_3 + 14y_1 + 19y_2 \le 2400 \\ x_1 + x_2 + x_3 + y_1 + y_2 - y_3 \ge 150 \\ x_1 + x_2 + x_3 + y_1 + y_2 + y_3 \le 200 \\ x_1, x_2, x_3, y_1, y_2, y_3 \ge 0, \end{cases} \end{cases} \end{cases} \quad (2.75)$$

where $\xi_j (j = 1, 2, \cdots, 6)$ are random variables characterized as

$$\xi_1 \sim \mathcal{N}(8, 2), \quad \xi_2 \sim \mathcal{N}(16, 1), \quad \xi_3 \sim \mathcal{N}(10, 2),$$
$$\xi_4 \sim \mathcal{N}(10, 2), \; \xi_5 \sim \mathcal{N}(-20, 2), \; \xi_6 \sim \mathcal{N}(16, 1).$$

In order to solve it, we use the expected operator to deal with objectives and constraints, then we can obtain the model

$$\begin{cases} \max\limits_{x_1,x_2,x_3} E[2\xi_1^2 x_1 + \sqrt{\xi_2 x_2 + \xi_3 x_3} + \xi_4 y_1 + \sqrt{\xi_5 y_2} + \sqrt{\xi_6 y_3}] \\ \max\limits_{x_1,x_2,x_3} x_1 + x_2 + x_3 + y_1 + y_2 + y_3 \\ \text{s.t.} \begin{cases} \text{where } (y_1, y_2, y_3) \text{ solves} \\ \max\limits_{y_1,y_2,y_3} E[\sqrt{7\xi_1 x_1 + 3\xi_2 x_2 + 8\xi_3 x_3} + (12\xi_4 y_1 + 11\xi_5 y_2 + 5\xi_6 y_3)^2] \\ \max\limits_{y_1,y_2,y_3} \{x_1 + 5x_2 + 7x_3 + y_1 + 11y_2 - 8y_3\} \\ \text{s.t.} \begin{cases} E[\sqrt{x_1\xi_1 + x_2\xi_2} + x_3^2\xi_3 + \xi_4 y_2 + 17y_3\xi_6] \le 1540 \\ 3x_1 + 6x_2 + 12x_3 + 14y_1 + 19y_2 \le 2400 \\ x_1 + x_2 + x_3 + y_1 + y_2 - y_3 \ge 150 \\ x_1 + x_2 + x_3 + y_1 + y_2 + y_3 \le 200 \\ x_1, x_2, x_3, y_1, y_2, y_3 \ge 0. \end{cases} \end{cases} \end{cases} \tag{2.76}$$

After running the hybrid algorithm in this section, we obtain $(x_1^*, x_2^*, x_3^*, y_1^*, y_2^*, y_3^*) = (12.3, 16.7, 12.4, 20.5, 20.7, 15.3)$ and $(E([F_1^*], E[F_2^*]) = (106.9, 118.9)$, $(E[f_1^*], E[f_2^*]) = (106.5, 78.3)$.

2.4 Random CECC Model

CECC denotes that expected value operator is used to deal with the randomness in the constraints of upper-level model. At the same time, chance constraint operators are used to deal with the randomness in the upper-level objectives and lower-level model.

2.4.1 General Form of Random CECC Model

Let's consider the typical bi-level single-objective model with random parameters, i.e. model (2.1). The CECC version of model (2.1) is developed as follows.

$$\left\{\begin{array}{l}\min\limits_{x\in R^{n_1}} \bar{F}\\ \text{s.t.}\left\{\begin{array}{l}Pr\{F(\boldsymbol{x},\boldsymbol{y},\boldsymbol{\xi})\leq \bar{F}\}\geq \beta_1\\ E[G_i(\boldsymbol{x},\boldsymbol{y},\boldsymbol{\xi})]\leq 0, i=1,2,\cdots,q\\ \text{where } y \text{ solves:}\\ \min\limits_{y\in R^{n_2}} \bar{f}\\ \text{s.t.}\left\{\begin{array}{l}Pr\{f(\boldsymbol{x},\boldsymbol{y},\boldsymbol{\xi})\leq \bar{f}\}\geq \beta_2\\ Pr\{g_j(\boldsymbol{x},\boldsymbol{y},\boldsymbol{\xi})\leq 0\}\geq \theta_j, j=1,2,\cdots,p,\end{array}\right.\end{array}\right.\end{array}\right. \tag{2.77}$$

where β_1 is the predetermined confidence levels in the upper-level model β_2, θ_j are the predetermined confidence levels in the lower-level model. Similarly, if decision makers want to maximize the objective values, then,

$$\left\{\begin{array}{l}\max\limits_{x\in R^{n_1}} \bar{F}\\ \text{s.t.}\left\{\begin{array}{l}Pr\{F(\boldsymbol{x},\boldsymbol{y},\boldsymbol{\xi})\geq \bar{F}\}\geq \beta_1\\ E[G_i(\boldsymbol{x},\boldsymbol{y},\boldsymbol{\xi})]\leq 0, i=1,2,\cdots,q\\ \text{where } y \text{ solves:}\\ \max\limits_{y\in R^{n_2}} \bar{f}\\ \text{s.t.}\left\{\begin{array}{l}Pr\{f(\boldsymbol{x},\boldsymbol{y},\boldsymbol{\xi})\geq \bar{f}\}\geq \beta_2\\ Pr\{g_j(\boldsymbol{x},\boldsymbol{y},\boldsymbol{\xi})\leq 0\}\geq \theta_j, j=1,2,\cdots,p,\end{array}\right.\end{array}\right.\end{array}\right. \tag{2.78}$$

where β_1 is the predetermined confidence levels in the upper-level programming, β_2, θ_j are the predetermined confidence levels in the lower-level programming.

Definition 2.20. A solution $\boldsymbol{x} \in R^{n_1}$ is said to be the upper-level feasible solution of problem (2.77) if and only if $Pr\{F(\boldsymbol{x},\boldsymbol{y},\boldsymbol{\xi}) \leq \bar{F}\} \geq \beta_1$ and $E\{G_i(\boldsymbol{x},\boldsymbol{y},\boldsymbol{\xi}) \leq 0\}$ for all i. For any feasible $\boldsymbol{x}$, if there is a solution $\boldsymbol{x}^*$ such that $\bar{F}^* < \bar{F}$, then $\boldsymbol{x}^*$ is called the upper-level optimal solution.

Definition 2.21. A solution $\boldsymbol{y} \in R^{n_2}$ is said to be the lower-level feasible solution of problem (2.77) if and only if $Pr\{f(\boldsymbol{x},\boldsymbol{y},\boldsymbol{\xi}) \leq \bar{f}\} \geq \beta_2$ and $Pr\{g_j(\boldsymbol{x},\boldsymbol{y},\boldsymbol{\xi}) \leq 0\} \geq \theta_j$ for all j. For any feasible $\boldsymbol{y}$, if there is a solution $\boldsymbol{y}^*$ such that $\bar{f}^* < \bar{f}$, then $\boldsymbol{y}^*$ is called the lower-level optimal solution.

Consider the following bi-level multi-objective programming problem with random coefficients,

$$\left\{\begin{array}{l}\max\limits_{x\in R^{n_1}}[F_1(\boldsymbol{x},\boldsymbol{y},\boldsymbol{\xi}),F_2(\boldsymbol{x},\boldsymbol{y},\boldsymbol{\xi}),\cdots,F_m(\boldsymbol{x},\boldsymbol{y},\boldsymbol{\xi})]\\ \text{s.t.}\left\{\begin{array}{l}G_i(\boldsymbol{x},\boldsymbol{y},\boldsymbol{\xi})\leq 0, i=1,2,\cdots,q\\ \text{where } y \text{ solves:}\\ \left\{\begin{array}{l}\max\limits_{y\in R^{n_2}}[f_1(\boldsymbol{x},\boldsymbol{y},\boldsymbol{\xi}),f_2(\boldsymbol{x},\boldsymbol{y},\boldsymbol{\xi}),\cdots,f_r(\boldsymbol{x},\boldsymbol{y},\boldsymbol{\xi})]\\ \text{s.t. } g_j(x,y,\xi)\leq 0, j=1,2,\cdots,p,\end{array}\right.\end{array}\right.\end{array}\right. \tag{2.79}$$

where $\boldsymbol{x} = (x_1, x_2, \cdots, x_n)^T$ is an n-dimensional upper-level decision vector; $\boldsymbol{y} = (y_1, y_2, \cdots, y_n)^T$ is an n-dimensional lower-level decision vector; $\boldsymbol{\xi} = (\xi_1, \xi_2, \cdots, \xi_n)$ is a random vector; $F_u(\boldsymbol{x}, \boldsymbol{y}, \boldsymbol{\xi})$ are upper-level objective functions, $u = 1, 2, \cdots, m$; $f_l(\boldsymbol{x}, \boldsymbol{y}, \boldsymbol{\xi})$ are lower-level objective functions, $l = 1, 2, \cdots, r$; $G_i(\boldsymbol{x}, \boldsymbol{\xi}) \leq 0$ and $g_j(x, y, \xi) \leq 0$ are random constraints. For a fixed decision vector $\boldsymbol{x}$, it is meaningless to maximize the objectives $F(\boldsymbol{x}, \boldsymbol{y}, \boldsymbol{\xi})$, before we know the exact value of the random vector $\boldsymbol{\xi}$, just as we can not maximize a random function in stochastic programming. Also, we can not judge weather or not a decision $\boldsymbol{x}$ is feasible before we know the value of $\boldsymbol{\xi}$. Hence, both the objectives and constraints in problem (2.79) are ill-defined. For presenting a mathematically meaningful random programming, we build a new class of random programming to model random decision problems via chance measure which was proposed above. We present the bi-level multi-objective programming with CECC operator as follows,

$$\begin{cases} \max\limits_{x \in R^{n_1}} [F_1, F_2, \cdots, F_m] \\ \text{s.t.} \begin{cases} Pr\{F_u(\boldsymbol{x}, \boldsymbol{y}, \boldsymbol{\xi}) \geq F_u\} \geq \beta_u, u = 1, 2, \cdots, m, \\ E[G_i(\boldsymbol{x}, \boldsymbol{y}, \boldsymbol{\xi})] \leq 0, i = 1, 2, \cdots, q \\ \text{where } y \text{ solves:} \\ \max\limits_{y \in R^{n_2}} [f_1, f_2, \cdots, f_m] \\ \text{s.t.} \begin{cases} Pr\{f_l(\boldsymbol{x}, \boldsymbol{y}, \boldsymbol{\xi}) \geq f_l\} \geq \beta_l, l = 1, 2, \cdots, r, \\ Pr\{g_j(\boldsymbol{x}, \boldsymbol{y}, \boldsymbol{\xi}) \leq 0\} \geq \theta_j, j = 1, 2, \cdots, p, \end{cases} \end{cases} \end{cases} \tag{2.80}$$

where β_u is predetermined confidence levels in upper-level programming, $u = 1, 2, \cdots, m$; β_l and θ_j are predetermined confidence levels in upper-level programming, $l = 1, 2, \cdots, r, j = 1, 2, \cdots, p$ are predetermined confidence levels in lower-level programming. If the objectives is to minimize the cost, then problem (2.80) should be formulated as follows,

$$\begin{cases} \min\limits_{x \in R^{n_1}} [F_1, F_2, \cdots, F_m] \\ \text{s.t.} \begin{cases} Pr\{F_u(\boldsymbol{x}, \boldsymbol{y}, \boldsymbol{\xi}) \leq F_u\} \geq \beta_u, u = 1, 2, \cdots, m, \\ E[G_i(\boldsymbol{x}, \boldsymbol{y}, \boldsymbol{\xi})] \leq 0, i = 1, 2, \cdots, q \\ \text{where } y \text{ solves:} \\ \min\limits_{y \in R^{n_2}} [f_1, f_2, \cdots, f_m] \\ \text{s.t.} \begin{cases} Pr\{f_l(\boldsymbol{x}, \boldsymbol{y}, \boldsymbol{\xi}) \leq f_l\} \geq \beta_l, l = 1, 2, \cdots, r, \\ Pr\{g_j(\boldsymbol{x}, \boldsymbol{y}, \boldsymbol{\xi}) \leq 0\} \geq \theta_j, j = 1, 2, \cdots, p. \end{cases} \end{cases} \end{cases} \tag{2.81}$$

Definition 2.22. Suppose for any given $\boldsymbol{y}$, a upper-level programming feasible solution $\boldsymbol{x}^*$ of problem (2.80) satisfies

$$Pr\{F_u(\boldsymbol{x}^*, \boldsymbol{y}, \boldsymbol{\xi}) \geq F_u(\boldsymbol{x}^*)\} \geq \beta_u, u = 1, 2, \cdots, m,$$

where confidence levels $\beta_u \in [0,1]$. $\boldsymbol{x}^*$ is a random efficient solution to problem (2.80) if and only if there exists no other feasible solution $\boldsymbol{x}$ such that

$$Pr\{F_u(\boldsymbol{x},\boldsymbol{y},\boldsymbol{\xi}) \geq F_u(\boldsymbol{x})\} \geq \beta_u, u = 1,2,\cdots,m,$$

$F_u(\boldsymbol{x}) \geq F_u(\boldsymbol{x}^*)$ for all u and $F_{u_0}(\boldsymbol{x}) > F_{u_0}(\boldsymbol{x}^*)$ for at least one $u_0 \in \{1,2,\cdots,m\}$.

Definition 2.23. Suppose for any given $\boldsymbol{x}$, a lower-level programming feasible solution $\boldsymbol{y}^*$ of problem (2.80) satisfies

$$Pr\{f_l(\boldsymbol{x},\boldsymbol{y}^*,\boldsymbol{\xi}) \geq f_l(\boldsymbol{y}^*)\} \geq \beta_l, l = 1,2,\cdots,r,$$

where confidence levels $\alpha_l, \beta_l \in [0,1]$. $\boldsymbol{y}^*$ is a random efficient solution to problem (2.80) if and only if there exists no other feasible solution $\boldsymbol{y}$ such that

$$Pr\{f_l(\boldsymbol{x},\boldsymbol{y},\boldsymbol{\xi}) \geq f_l(\boldsymbol{x})\} \geq \beta_l, l = 1,2,\cdots,r,$$

$f_l(\boldsymbol{y}) \geq F_l(\boldsymbol{y}^*)$ for all l and $F_{l_0}(\boldsymbol{y}) > f_{l_0}(\boldsymbol{y}^*)$ for at least one $l_0 \in \{1,2,\cdots,r\}$.

For the random expected operator, we have discussed above. In what follows, we recall some properties for the probability constraints. Without loss of generality, the probability constraint is denote by $Pr\{g(\boldsymbol{x},\boldsymbol{\xi})\}$.

Theorem 2.3. *Assume that the random vector* $\boldsymbol{\xi}$ *degenerates to a random variable* ξ *with distribution function* Φ*, and the function* $g(\boldsymbol{x},\xi)$ *has the form* $g(x,\xi) = h(\boldsymbol{x})-\xi$*. Then* $Pr\{g(\boldsymbol{x},\xi)\} \geq \alpha$ *if and only if* $h(\boldsymbol{x}) \leq K_\alpha$*, where* $K_\alpha = \sup\{K|K = \Phi^{-1}(1-\alpha)\}$.

Proof. The assumption implies that $Pr\{g(\boldsymbol{x},\xi) \leq 0\} \geq \alpha$ can be written in the following form:

$$Pr\{h(\boldsymbol{x}) \leq \xi\} \geq \alpha. \tag{2.82}$$

It is clear that, for each given confidence level $\alpha (0 < \alpha < 1)$, there exists a number K_α (may be multiple or ∞) such that

$$Pr\{K_\alpha \leq \xi\} = \alpha, \tag{2.83}$$

and the probability $Pr\{\alpha \leq \xi\}$ will increase if K_α is replaced with a smaller number. Hence, $Pr\{h(\boldsymbol{x}) \leq \xi\} \geq \alpha$ if and only if $h(\boldsymbol{x}) \leq K_\alpha$.

Notice that the equation $Pr\{\alpha \leq \xi\} = 1 - \Phi(K_\alpha)$ always holds, and we have

$$K_\alpha = \Phi^{-1}(1-\alpha),$$

where Φ^{-1} is the inverse function of Φ. Sometimes the solution to (2.83) is not unique. Equivalently, the function Φ^{-1} is multi-valued. For this case, we should choose it as the largest one, that is,

$$K_\alpha = sup\{K|K = \Phi^{-1}(1-\alpha)\}.$$

Thus the deterministic equivalent is $h(\boldsymbol{x}) \leq K_\alpha$. The theorem is proved. □

Theorem 2.4. *Let the random vector* $\boldsymbol{\xi} = (a_1, a_2, \cdots, a_n, b)$*, and the function* $g(\boldsymbol{x}, \boldsymbol{\xi})$ *has the form* $g(\boldsymbol{x}, \boldsymbol{\xi}) = a_1x_1 + a_2x_2 + \cdots + a_nx_n - b$*. If* a_i *and* b *are assumed to be independently normally distributed variables, then* $Pr\{g(\boldsymbol{x}, \boldsymbol{\xi}) \leq 0\} \geq \alpha$ *if and only if*

$$\sum_{i=1}^{n} E[a_i]x_i + \Phi^{-1}(\alpha)\sqrt{\sum_{i=1}^{n} V[a_i]x_i^2 + V[b]} \leq E[b], \tag{2.84}$$

where Φ *is the standardized normal distribution.*

Proof. The chance constraint $Pr\{g(\boldsymbol{x}, \boldsymbol{\xi}) \leq 0\} \geq \alpha$ can be written in the following form:

$$Pr\left\{\sum_{i=1}^{n} a_ix_i \leq b\right\} \geq \alpha. \tag{2.85}$$

Since a_i and b are assumed to be independently normally distributed variables, the function

$$y(\boldsymbol{x}) = \sum_{i=1}^{n} a_ix_i - b,$$

is also normally distributed with the following expected value and variance:

$$E[y(\boldsymbol{x})] = \sum_{i=1}^{n} E[a_i]x_i - E[b],$$

$$V[y(\boldsymbol{x})] = \sum_{i=1}^{n} V[a_i]x_i^2 + V[b].$$

We note that

$$\frac{\sum_{i=1}^{n} a_ix_i - b - \left(\sum_{i=1}^{n} E[a_i]x_i - E[b]\right)}{\sqrt{\sum_{i=1}^{n}\sum_{i=1}^{n} V[a_i]x_i^2 + V[b]}},$$

must be standardized normally distributed. Since the inequality $\sum_{i=1}^{n} a_i x_i \le b$ is equivalent to

$$\frac{\sum_{i=1}^{n} a_i x_i - b - \left(\sum_{i=1}^{n} E[a_i]x_i - E[b]\right)}{\sqrt{\sum_{i=1}^{n}\sum_{i=1}^{n} V[a_i]x_i^2 + V[b]}} \le -\frac{\sum_{i=1}^{n} E[a_i]x_i - E[b]}{\sqrt{\sum_{i=1}^{n}\sum_{i=1}^{n} V[a_i]x_i^2 + V[b]}},$$

the chance constraint (2.85) is equivalent to

$$Pr\left\{\eta \le -\frac{\sum_{i=1}^{n} E[a_i]x_i - E[b]}{\sqrt{\sum_{i=1}^{n}\sum_{i=1}^{n} V[a_i]x_i^2 + V[b]}}\right\} \ge \alpha, \tag{2.86}$$

where η is the standardized normally distributed variable. Then the chance constraint (2.86) holds if and only if

$$\Phi^{-1}(\alpha) \le -\frac{\sum_{i=1}^{n} E[a_i]x_i - E[b]}{\sqrt{\sum_{i=1}^{n}\sum_{i=1}^{n} V[a_i]x_i^2 + V[b]}}\alpha. \tag{2.87}$$

That is, the deterministic equivalent of chance constraint is (2.84). The theorem is proved. □

2.4.2 KKT Method for Linear Model

In what follows we focus on the linear version of CECC model as follows.

$$\begin{cases} \max\limits_{x \in R^{n_1}} [\overline{F}_1, \overline{F}_2, \cdots, \overline{F}_{m_1}] \\ \text{s.t.} \begin{cases} Pr\{\widetilde{\boldsymbol{C}}_i^T \boldsymbol{x} + \widetilde{\boldsymbol{D}}_i^T \boldsymbol{y} \ge \overline{F}_i\} \ge \alpha_i^U, i = 1, 2, \cdots, m_1 \\ E[\widetilde{\boldsymbol{A}}_i^T \boldsymbol{x} + \widetilde{\boldsymbol{B}}_i^T \boldsymbol{y}] \le 0, i = 1, 2, \cdots, p_1 \\ \text{where } \boldsymbol{y} \text{ solves:} \\ \max\limits_{y \in R^{n_2}} [\alpha_1^L, \alpha_2^L, \cdots, \alpha_{m_2}^L] \\ \text{s.t.} \begin{cases} Pr\{\tilde{\boldsymbol{c}}_i^T \boldsymbol{x} + \tilde{\boldsymbol{d}}_i^T \boldsymbol{y} \ge \bar{f}_i\} \ge \alpha_i^L, i = 1, 2, \cdots, m_2 \\ Pr\{\tilde{\boldsymbol{a}}_i^T \boldsymbol{x} + \tilde{\boldsymbol{b}}_i^T \boldsymbol{y} \le 0\} \ge \beta_i^L, i = 1, 2, \cdots, p_2, \end{cases} \end{cases} \end{cases} \tag{2.88}$$

where α_i^U, α_i^L and β_i^Ląi are predetermined confidence levels by decision makers.

Since we have discussed the methods to deal with random expected value, it is enough to consider the lower model of (2.88).

Theorem 2.5. *Assume that $\tilde{\boldsymbol{m}}_i$ is normally distributed with mean vector μ_i^c and positive definite covariance matrix V_i^c, written as $\tilde{\boldsymbol{m}}_i \sim \mathcal{N}(\mu_i^c, V_i^c)$. Similarly, $\tilde{d}_i \sim \mathcal{N}(\mu_i^d, V_i^d)$. All the random variables are independent with each other. Then $Pr\{\tilde{\boldsymbol{c}}_i^T\boldsymbol{x} + \tilde{\boldsymbol{d}}_i^T\boldsymbol{y} \geq \bar{f}_i\} \geq \alpha_i^L$ is equivalent to*

$$\bar{f}_i \leq \Phi^{-1}(1-\alpha_i^L)\sqrt{\boldsymbol{x}^T V_i^c \boldsymbol{x} + \boldsymbol{y}^T V_i^d \boldsymbol{y}} + \mu_i^{cT}\boldsymbol{x} + \mu_i^{dT}\boldsymbol{y}. \tag{2.89}$$

Proof. It follows from the assumptions that $\tilde{\boldsymbol{c}}_i^T$ and $\tilde{\boldsymbol{d}}_i^T$ is normally distributed with mean vector $\mu_i^{cT}\boldsymbol{x} + \mu_i^{dT}\boldsymbol{y}$ and positive definite covariance matrix $\boldsymbol{x}^T V_i^c \boldsymbol{x} + \boldsymbol{y}^T V_i^d \boldsymbol{y}$. So

$$\frac{\tilde{\boldsymbol{c}}_i^T\boldsymbol{x} + \tilde{\boldsymbol{d}}_i^T\boldsymbol{y} - (\mu_i^{cT}\boldsymbol{x} + \mu_i^{dT}\boldsymbol{y})}{\sqrt{\boldsymbol{x}^T V_i^c \boldsymbol{x} + \boldsymbol{y}^T V_i^d \boldsymbol{y}}},$$

must be standardized normally distributed. Thus we have

$$\begin{aligned}
& Pr\{\tilde{\boldsymbol{c}}_i^T\boldsymbol{x} + \tilde{\boldsymbol{d}}_i^T\boldsymbol{y} \geq \bar{f}_i\} \geq \alpha_i^L \\
\Leftrightarrow\ & \alpha_i^L \leq Pr\left\{\frac{\tilde{\boldsymbol{c}}_i^T\boldsymbol{x} + \tilde{\boldsymbol{d}}_i^T\boldsymbol{y} - (\mu_i^{cT}\boldsymbol{x} + \mu_i^{dT}\boldsymbol{y})}{\sqrt{\boldsymbol{x}^T V_i^c \boldsymbol{x} + \boldsymbol{y}^T V_i^d \boldsymbol{y}}} \geq \frac{\bar{f}_i - (\mu_i^{cT}\boldsymbol{x} + \mu_i^{dT}\boldsymbol{y})}{\sqrt{\boldsymbol{x}^T V_i^c \boldsymbol{x} + \boldsymbol{y}^T V_i^d \boldsymbol{y}}}\right\} \\
\Leftrightarrow\ & \alpha_i^L \leq 1 - Pr\left\{\frac{\tilde{\boldsymbol{c}}_i^T\boldsymbol{x} + \tilde{\boldsymbol{d}}_i^T\boldsymbol{y} - \mu_i^{cT}\boldsymbol{x} + \mu_i^{dT}\boldsymbol{y}}{\sqrt{\boldsymbol{x}^T V_i^c \boldsymbol{x} + \boldsymbol{y}^T V_i^d \boldsymbol{y}}} \leq \frac{\bar{f}_i - (\mu_i^{cT}\boldsymbol{x} + \mu_i^{dT}\boldsymbol{y})}{\sqrt{\boldsymbol{x}^T V_i^c \boldsymbol{x} + \boldsymbol{y}^T V_i^d \boldsymbol{y}}}\right\} \\
\Leftrightarrow\ & \alpha_i^L \leq 1 - \Phi\left(\frac{\bar{f}_i - (\mu_i^{cT}\boldsymbol{x} + \mu_i^{dT}\boldsymbol{y})}{\sqrt{\boldsymbol{x}^T V_i^c \boldsymbol{x} + \boldsymbol{y}^T V_i^d \boldsymbol{y}}}\right) \\
\Leftrightarrow\ & \bar{f}_i \leq \Phi^{-1}(1-\alpha_i^L)\sqrt{\boldsymbol{x}^T V_i^c \boldsymbol{x}\boldsymbol{y}^T V_i^d \boldsymbol{y}} + \mu_i^{cT}\boldsymbol{x} + \mu_i^{dT}\boldsymbol{y},
\end{aligned}$$

where Φ is the standardized normal distribution. This completes the proof. □

Theorem 2.6. *Assume that $\tilde{\boldsymbol{a}}_i$ is normally distributed with mean vector μ_i^a and positive definite covariance matrix V_i^a, written as $\tilde{\boldsymbol{a}}_i \sim \mathcal{N}(\mu_i^a, V_i^a)$. Similarly, $\tilde{b}_i \sim \mathcal{N}(\mu_i^b, V_i^b)$. All the random variables are independent with each other. Then $Pr\{\tilde{\boldsymbol{a}}_i^T\boldsymbol{x} + \tilde{\boldsymbol{b}}_i^T\boldsymbol{y} \leq 0\} \geq \beta_i^L$ is equivalent to*

$$\bar{f}_i \leq \Phi^{-1}(1-\alpha_i^L)\sqrt{\boldsymbol{x}^T V_i^c \boldsymbol{x}} + \mu_i^{cT}\boldsymbol{x} + \mu_i^{dT}\boldsymbol{y}. \tag{2.90}$$

Proof. It follows from the assumptions that $\tilde{\boldsymbol{a}}_i^T$ and $\tilde{\boldsymbol{b}}_i^T$ is normally distributed with mean vector $\mu_i^{aT}\boldsymbol{x}+\mu_i^{bT}\boldsymbol{y}$ and positive definite covariance matrix $\boldsymbol{x}^T V_i^a \boldsymbol{x}+\boldsymbol{y}^T V_i^b \boldsymbol{y}$. So

$$\frac{\tilde{\boldsymbol{a}}_i^T\boldsymbol{x} + \tilde{\boldsymbol{b}}_i^T\boldsymbol{y} - (\mu_i^{aT}\boldsymbol{x} + \mu_i^{bT}\boldsymbol{y})}{\sqrt{\boldsymbol{x}^T V_i^a \boldsymbol{x} + \boldsymbol{y}^T V_i^b \boldsymbol{y}}},$$

must be standardized normally distributed. Thus we have

$$\begin{aligned}
& Pr\{\tilde{\boldsymbol{a}}_i^T\boldsymbol{x} + \tilde{\boldsymbol{b}}_i^T\boldsymbol{y} \le 0\} \ge \beta_i^L \\
\Leftrightarrow\ & \beta_i^L \le Pr\left\{\frac{\tilde{\boldsymbol{a}}_i^T\boldsymbol{x} + \tilde{\boldsymbol{b}}_i^T\boldsymbol{y} - (\mu_i^{aT}\boldsymbol{x} + \mu_i^{bT}\boldsymbol{y})}{\sqrt{\boldsymbol{x}^T V_i^a \boldsymbol{x} + \boldsymbol{y}^T V_i^b \boldsymbol{y}}} \le \frac{-(\mu_i^{aT}\boldsymbol{x} + \mu_i^{bT}\boldsymbol{y})}{\sqrt{\boldsymbol{x}^T V_i^a \boldsymbol{x} + \boldsymbol{y}^T V_i^b \boldsymbol{y}}}\right\} \\
\Leftrightarrow\ & \beta_i^L \le \Phi\left(\frac{-(\mu_i^{aT}\boldsymbol{x} + \mu_i^{bT}\boldsymbol{y})}{\sqrt{\boldsymbol{x}^T V_i^a \boldsymbol{x} + \boldsymbol{y}^T V_i^b \boldsymbol{y}}}\right) \\
\Leftrightarrow\ & \Phi^{-1}(\beta_i^L)\sqrt{\boldsymbol{x}^T V_i^a \boldsymbol{x} + \boldsymbol{y}^T V_i^b \boldsymbol{y}} + \mu_i^{cT}\boldsymbol{x} + \mu_i^{dT}\boldsymbol{y} \le 0.
\end{aligned}$$

By using Theorem 2.5 and Theorem 2.6, linear CECC model (2.88) is converted into a crisp equivalent model.

KKT Method

As model (2.88) can be converted into its crisp equivalent, then we can use KKT methods, proposed by Dempe and Zemkoho [30].

Consider the following optimistic bi-level programming problem to

$$\min F(x, y) \text{ s.t. } x \in\subset \Re^n, y \in S(x) \tag{2.91}$$

also called the upper level problem, where $F : \Re^n \times \Re^m \to \Re$ is a continuously differentiable function and the set-valued mapping $S : \Re^n \rightrightarrows \Re^m$, describes the solution set of the following parametric optimization problem also known as the lower level problem

$$\min f(x, y) \text{ s.t. } y \in K(x), \tag{2.92}$$

where $K(x)$ is a closed subset of $\Re^m$, for all $x \in X$, and the function $f : \Re^n \times \Re^m \to \Re$ is twice continuously differentiable. We assume that the upper and lower level feasible sets are given as

$$X := \{\boldsymbol{x} \in \Re^n | G(x) \le 0\}, K(\boldsymbol{x}) = \{y \in \Re^m | g(x, y) \le 0\}, \forall x \in X, \tag{2.93}$$

respectively; the functions $G : \Re^n \to \Re^k$ and $g : \Re^n \times \Re^m \to \Re^p$ being continuously and twice continuously differentiable, respectively. Also, unless otherwise stated, the functions $f(x, \cdot)$ and $g_i(x, \cdot), i = 1, \cdots, p$ are assumed to be convex for all $x \in X$. It is well known from convex optimization that the lower level problem would be equivalent to the parametric generalized equation:

$$0 \in \nabla_y f(x, y) + N_{K(x)}(y), \tag{2.94}$$

where $N_{K(x)}(y)$ denotes the normal cone (in the sense of convex analysis) to $K(x)$ at y, provided $y \in K(x)$, and $N_{K(x)}(y) := \emptyset$ otherwise.

Hence, the following one level reformulation of the bi-level program that can be called primal KKT reformulation (this terminology may be justified by the fact that the above generalized equation can be considered as a compact form of the KKT conditions of the lower level problem) [30]:

$$\begin{cases} \min F(x, y) \\ \text{s.t.} \begin{cases} G(x) \le 0 \\ 0 \in \nabla_y f(x, y) + N_{K(x)}(y). \end{cases} \end{cases} \tag{2.95}$$

Theorem 2.7 ([98]). *The point $(\bar{x}, \bar{y})$ is a local (resp. global) optimal solution of (2.91) if and only if $(\bar{x}, \bar{y})$ is a local (resp. global) optimal solution of (2.95).*

This complete equivalence between problem (2.91) and its primal KKT reformulation (2.95) is lost if one considers the following detailed form of the normal cone to $K(x)$ at y:

$$N_{(x)}(y) = \{\nabla_y g(x, y)^T u | u \ge 0, u^T g(x, y) = 0\},$$

which holds under a certain constraint qualification (CQ) [107], In fact, the resulting problem is the so-called KKT reformulation of the bi-level optimization problem:

$$\begin{cases} \min F(x, y) \\ \text{s.t.} \begin{cases} G(x) \le 0, L(x, y, u) = 0 \\ u \ge 0, g(x, y) \le 0, u^T g(x, y) = 0, \end{cases} \end{cases} \tag{2.96}$$

where $L(x, y, u) :=$ ąÔ$\nabla_y f(x, y) + \nabla_y g(x, y)^T u$. The relationship between the latter problem, that we call classical KKT reformulation in the sequel, and the bi-level program, in terms of optimal solutions, has recently been studied in [31]. This link can be summarized in the following result where the fulfillment of a CQ at $(\bar{x}, \bar{y})$, say

$$[\nabla_y g(\bar{x}, \bar{y})^T \beta = 0, \beta \ge 0, \beta^T g(\bar{x}, \bar{y}) = 0] \Rightarrow \beta = 0, \tag{2.97}$$

is necessary. Furthermore, $\Lambda(\bar{x}, \bar{y}) = 0$ will denote the set of Lagrange multipliers of the lower level problem, i.e. the set of all the vectors u satisfying: $u \geq 0, u^T g(\bar{x}, \bar{y}) = 0$ and $L(\bar{x}, \bar{y}, u) = 0$ [30].

Theorem 2.8 ([30]). *Let $(\bar{x}, \bar{y})$ be a global (resp. local) optimal solution of (2.91) and assume that CQ (2.97) is satisfied at $(\bar{x}, \bar{y})$. Then, for each $\bar{u} \in \Lambda(\bar{x}, \bar{y})$, the point $(\bar{x}, \bar{y}, \bar{u})$ is a global (resp. local) optimal solution of (2.96). Conversely, let CQ (2.97) be satisfied at (x, y), for all $y \in S(x), x \in X$ (resp. at $(\bar{x}, \bar{y})$. Assume that $(\bar{x}, \bar{y}, \bar{u})$ is a global optimal solution (resp. local optimal solution for all $\bar{u} \in \Lambda(\bar{x}, \bar{y})$ of (2.96). Then, $(\bar{x}, \bar{y})$ is a global (resp. local) optimal solution of (2.91).*

Clearly, for $(\bar{x}, \bar{y})$ to be a local optimal solution of (2.91), one needs to make sure that $(\bar{x}, \bar{y}, \bar{u})$ is a local optimal solution of problem (2.96), for all $\bar{u} \in \Lambda(\bar{x}, \bar{y})$. In fact, an example of bi-level program was provided in [31], where $(\bar{x}, \bar{y})$ fails to solve (1.1) locally, whereas $\bar{u} \in \Lambda(\bar{x}, \bar{y})$ is a local solution of (2.96), for all but one $\bar{u} \in \Lambda(\bar{x}, \bar{y})$. This fact has motivated the following definition for the notion of optimality conditions for the bi-level optimization problem from the perspective of the KKT reformulation. As usually done in the literature on MPCCs (mathematical programs with complementarity constraints), we partition the set of indices of the functions involved in the complementarity slackness as [30]:

$$\eta := \eta(\bar{x}, \bar{y}, \bar{u}) := \{i | \bar{u}_i = 0, g_i(\bar{x}, \bar{y}) < 0\},$$

$$\mu := \mu(\bar{x}, \bar{y}, \bar{u}) := \{i | \bar{u}_i = 0, g_i(\bar{x}, \bar{y}) = 0\},$$

$$\nu := \mu(\bar{x}, \bar{y}, \bar{u}) := \{i | \bar{u}_i > 0, g_i(\bar{x}, \bar{y}) = 0\}.$$

Definition 2.24 ([30]). A point $(\bar{x}, \bar{y})$ will be said to be M-stationary for the bi-level optimization problem (2.91) if there exists $(\alpha, \beta, \gamma) \in \Re^{k+p+m}$ such that $\forall \bar{u} \in \Lambda(\bar{x}, \bar{y})$:

$$\nabla_x F(\bar{x}, \bar{y}) + \nabla G(\bar{x})^T \alpha + \nabla_x g(\bar{x}, \bar{y})^T \beta + \nabla_x L(\bar{x}, \bar{y}, \bar{u})^T \gamma = 0, \tag{2.98}$$

$$\nabla_y F(\bar{x}, \bar{y}) + \nabla_x g(\bar{x}, \bar{y})^T \beta + \nabla_x L(\bar{x}, \bar{y}, \bar{u})^T \gamma = 0, \tag{2.99}$$

$$\alpha \geq 0, \alpha^T G(\bar{x}) = 0, \tag{2.100}$$

$$\nabla_y g_\nu(\bar{x}, \bar{y}) \gamma = 0, \beta_\eta = 0, \tag{2.101}$$

$$\forall i \in \mu, (\beta_i > 0 \wedge \nabla_y g_i(\bar{x}, \bar{y})^T \gamma) = 0. \tag{2.102}$$

Conditions (2.98), (2.99), (2.100), (2.101), (2.102) are called the M-stationarity conditions for problem (2.91).

Definition 2.25 ([30]). A point $(\bar{x}, \bar{y})$ will be said to be S-stationary for the bi-level optimization problem (2.91) if there exists $(\alpha, \beta, \gamma) \in \Re^{k+p+m}$ such that $\forall \bar{u} \in \Lambda(\bar{x}, \bar{y})$: (2.98), 2.99), 2.100), (2.101) and

$$\forall \bar{u}, \beta_i \geq 0 \wedge \nabla_y g_i(\bar{x}, \bar{y}) \gamma \geq 0. \tag{2.103}$$

Conditions (2.98), (2.99), (2.100), (2.101) and (2.103) are called the S-stationarity conditions for (2.91).

Similarly, surrogates of other well-known types of stationarity concepts could also be defined for the bi-level optimization problem. In this chapter though, we will focus our attention only on the above M- and S-types, since they are the most important ones.

A third possibility to write problem (2.91) as a one level optimization problem is the following optimal value reformulation:

$$\begin{cases} \min F(x, y) \\ \text{s.t.} \begin{cases} f(x, y) \leq \phi(x) \\ G(x) \leq 0, g(x, y) \leq 0, \end{cases} \end{cases} \tag{2.104}$$

where ϕ is the optimal value function of the lower level problem, defined as

$$\phi(x) := \min\{f(x, y)|y \in K(x)\}.$$

The latter problem is globally and locally equivalent to the bi-level programming problem (2.91) [30].

Some notations are introduced as following that will simplify the presentation of the section. Let $\mathbb{N}^*$ be the set of positive integers. For $l \in \mathbb{N}^*, \Re^{n_1+\cdots+n_l} = \Re^{n_1} \times R^{n_2} \times \Re^{n_l}$ and $0_{n_1+\cdots+n_l}$ is the origin of the space $\Re^{n_1+\cdots+n_l}$. For vectors $a^i \in \Re^{n_i}, i = 1, \cdots, l$, the joint vector $(a^1, \cdots, a^{nl})$ may be used instead of its transposed vector $(a^1, \cdots, a^{nl})^T$. Let $a^i \in \Re^{n_i}, i = 1, 2; (a^1)^T a^2 = (a_1, a_2)$. is used for the inner product of a^1 and a^2. For two properties a and b, property $a \vee b$ means that either a or b is satisfied, whereas $a \wedge b$ denotes the fulfillment of a and b simultaneously. For $A \subseteq \Re^l$, and A denotes the topological boundary of A, while $d_A(a) = d(a, A)$ is the distance from the point a to A [30].

First consider the Fréchet normal cone to a closed set $A \subseteq \Re^l$ at some point $\bar{z} \in A$:

$$\widehat{N}_A\bar{z} := \{z^* \in \Re^l|\langle z^*, z - \bar{z}\rangle \leq o(||z - \bar{z}||)\forall \bar{z} \in A\}.$$

This cone is known to be the polar of the Bouligand tangent cone [30]:

$$T_A(\bar{z} := \{z^* \in \Re^l|\exists t_k \downarrow 0, z_k \to z : \bar{z} + t_k z_k \in A\}).$$

The Mordukhovich normal cone to A at z is the Kuratowski-Painlevé upper limit [108] of the Fréchet normal cone, i.e.

$$N_A(\bar{z}) := \{z^* \in \Re^l|\exists z_k^* \to z^*, z_k \to \bar{z}(z_k \in A) : z_k^* \in \widehat{N}_A(z_k)\}.$$

The Mordukhovich subdifferential can also be defined from the Fréchet subdifferential as in the case of the normal cone. But in order to save some space, we simply consider the well-known interplay between most of the normal cone and subdifferential objects in the literature. That is, for a lower semicontinuous function $f : \Re^l \to \overline{\Re}$, the Mordukhovich subdifferential of f at some point $\bar{z} \in dom f$ is [30]:

$$\partial f(\bar{z}) = \{z^* \in \Re^l | (z^*, l) \in N_{epif}(\bar{z}, f(\bar{z}))\},$$

where *epif* denotes the epigraph of f. It is important to mention that if f is a continuously differentiable function, then ∂f coincides with the gradient of f. Considering two functions, the sum and chain rules are obtained respectively as:

Theorem 2.9 ([95]). *Let the functions $f, g : \Re^l \to \Re$ be locally Lipschitz continuous around $\bar{z}$. Then $\partial(f + g)(\bar{z}) \subseteq \partial f(\bar{z}) + \partial g(\bar{z})$. Equality holds if f or g is continuously differentiable.*

Theorem 2.10 ([93]). *Let $f : \Re^{l_1} \to \Re^{l_2}$ be Lipschitz continuous around $\bar{z}$, and $g : \Re^{l_2} \to \Re$ Lipschitz continuous around $\bar{w} = f(\bar{z}) \in dom g$. Then*

$$\partial(g \circ f)(\bar{z}) \subseteq [\partial \langle w^*, f(\bar{z}\rangle : w^* \in \partial g(\bar{w})].$$

In the next result, recall the necessary optimality condition for a Lipschitz optimization problem with geometric constraint.

Theorem 2.11 ([108]). *We let $f : \Re^l \to \Re$ be a locally Lipschitz continuous function and $A \subseteq \Re^l$, a closed set. For $\bar{z}$ to be a local minimizer of f on A, it is necessary that*

$$0 \in \partial f(\bar{z}) + N_A(\bar{z}).$$

For a set-valued mapping $\Phi : \Re^{l_1} \rightrightarrows \Re^{l_2}$, a derivative-like object, called coderivative, and introduced by Mordukhovich (see [46]), can also be defined. Let $(\bar{u}, \bar{z}) \in gph\Phi$, the coderivative of Φ at $(\bar{u}, \bar{z})$ is a positively homogeneous set-valued mapping $D^*\Phi(\bar{u}, \bar{z}) : \Re^{l_2} \rightrightarrows \Re^l$, such that for any $z^* \in \Re^{l_2}$, we have [30]

$$D^*\Phi(\bar{u}, \bar{z})(z^*) := \{u^* \in \Re^{l_1} (u^*, -z^*) \in N_{gph\Phi}(\bar{u}, \bar{z})\}.$$

If Φ reduces to a single-valued Lipschitz continuous function, then $\bar{z}$ can be omitted and the coderivative of Φ reduces to $D^*\Phi(\bar{u}, \bar{z})(z^*) = \partial\langle z^*, \Phi\rangle(\bar{u})$, for all $z^* \in \Re^{l_2}$ with $\langle z^*, \Phi\rangle(\bar{u}) = \langle z^*, \Phi(\bar{u})\rangle$ and ∂ being the basic subdifferential defined above. It clearly follows that, in case Φ is single-valued and continuously differentiable, then $D^*\Phi(\bar{u})(z^*) = \{\nabla\Phi(\bar{u})^T z^*\}$, for all $z^* \in \Re^{l_2}$, where $\nabla\Phi$ denotes the Jacobian matrix of ϕ [30].

The set-valued mapping Φ is said to be Lipschitz-like at $(\bar{u}, \bar{z}) \in gph\Phi$, if there exist neighborhoods U of $\bar{u}$, V of $\bar{z}$, and a number $L > 0$ such that

$$\Phi(u) \cap V \subseteq \Phi(u') + L||u - u'||\mathbb{B}, \forall u, u' \in U. \tag{2.105}$$

This property, often called the Aubin property, was introduced by Aubin [5]. It is worth mentioning that the Aubin property is a natural extension of the Lipschitz continuity known for a single-valued function. If we fix $u' := \bar{u}$ in (2.105), then we obtain the following inclusion

$$\Phi(u) \cap V \subseteq \Phi(\bar{u}) + L||u - \bar{u}||\mathbb{B}, \forall u \in U, \tag{2.106}$$

which defines the calmness or upper pseudo-Lipschitz continuity of the set-valued mapping Φ. Hence, it is obvious that the Aubin property implies the calmness property. As shown in the next theorem, the calmness property may be very useful in computing the normal cone to a subset of $\Re^l$ defined by finitely many inequalities and equalities [30].

Theorem 2.12 ([51]). *Consider the set*

$$A := \{z \in \Re^l | g(z) \leq 0, h(z) = 0\},$$

where $g : \Re^l \to \Re^{l_1}$ *and* $h : \Re^l \to \Re^{l_2}$ *are continuously differentiable functions. Then*

$$N_A(\bar{z})\{\nabla g(\bar{z})^T \lambda + \nabla h(\bar{z})^T \mu | \lambda \geq 0, \lambda^T g(\bar{z}) = 0\}$$

for $\bar{z} \in A$*, provided the following set-valued mapping is calm at* $(0, 0, \bar{z})$

$$M(t_1, t_2) = \{z \in \Re^l | g(z) + t_1 \leq 0, h(z) + t_2 = 0\}.$$

We consider the optimistic bi-level programming problem to

$$\min F(x, y) \text{ suject to } xinX \subseteq \mathbb{R}^n, y \in S(x), \tag{2.107}$$

also called the upper level problem, where $F : \mathbb{R}^n \times \mathbb{R}^m \to \mathbb{R}$ is a continuously differentiable function and the set-valued mapping $S : \mathbb{R}^n \rightrightarrows \mathbb{R}^m$, describes the solution set of the following parametric optimization problem also known as the lower level problem:

$$\min f(x, y) \text{ suject to } y \in K(x), \tag{2.108}$$

where $K(x)$ is a closed subset of $\mathbb{R}^m$, for all $x \in X$, and the function $f : \mathbb{R}^n \times \mathbb{R}^m \to \mathbb{R}^p$ is twice continuously differentiable. We assume that the upper and lower level feasible sets are given as

$$X := \{x \in \mathbb{R}^n | G(x) \leq 0\} \text{ and } K(x) := \{y \in \mathbb{R}^m | g(x, y) \leq 0\} \text{ for all } x \in X, \tag{2.109}$$

respectively; the functions $G : \mathbb{R}^n \to \mathbb{R}^k$ and $g : \mathbb{R}^n \times \mathbb{R}^m \to \mathbb{R}^p$ being continuously and twice continuously differentiable, respectively. Also, unless otherwise stated, the functions $f(x,\cdot)$ and $g_i(x,\cdot)$, $i = 1, \cdots, p$ are assumed to be convex for all $x \in X$. It is well known from convex optimization that the lower level problem would be equivalent to the parametric generalized equation:

$$0 \in \nabla_y f(x,y) + N_{k(x)}(y), \tag{2.110}$$

where $N_{k(x)}(y)$ denotes the normal cone (in the sense of convex analysis) to $K(x)$ at y, provided $y \in K(x)$, and $+N_{k(x)}(y) := \emptyset$, otherwise [30].

Hence, the following one level reformulation of the bi-level program that we call primal KKT reformulation (this terminology may be justified by the fact that the above generalized equation can be considered as a compact form of the KKT conditions of the lower level problem) [30]:

$$\begin{cases} \min F(x,y) \\ \text{s.t.} \begin{cases} G(x) \le 0 \\ 0 \in \nabla_y f(x,y) + N_{Kx}(y). \end{cases} \end{cases} \tag{2.111}$$

This problem, also corresponding to an optimization problem with (parametric) generalized equation constraint (OPEC), where K is a moving set, has recently been studied in [97].

The optimization problem with operator constraint, which may be seen as a special optimization problem with geometric constraint is

$$\min F(z) \text{ subject to } z \in \Omega \cap \Psi^{-1}(\Lambda), \tag{2.112}$$

where $F : \mathbb{R}^l \to \mathbb{R}$ and $\Psi : \mathbb{R}^l \to \mathbb{R}^{l_1}$ are locally Lipschitz continuous functions, and the sets $\Omega \sqsubseteq \mathbb{R}^l$, $\Lambda \subseteq \mathbb{R}^{l_1}$ are closed. To derive KKT-type dual optimality conditions for problem (2.112), we consider the basic CQ. Let $\bar{z}$ be a feasible point of problem (2.112); the basic CQ, which was introduced in [92] is said to be satisfied at $\bar{z}$ if

$$\left.\begin{array}{l} 0 \in \partial < u^*, \Psi > (\bar{z}) + N_\Omega(\bar{z}) \\ u^* \in N_\Lambda(\Psi(\bar{z})) \end{array}\right\} \Rightarrow u^* = 0. \tag{2.113}$$

Some CQs closely related to the basic CQ [63].

Remark 2.5 ([30]). If $\Omega := \mathbb{R}^l, \Lambda := \mathbb{R}^{l_2} \times \{0_{l_1-l_2}\}$ and Ψ is a continuously differentiable function, then the basic CQ coincides with the dual form of the well-known Mangasarian Fromovitz Constraints Qualification (MFCQ). Hence, the basic CQ is a generalization of the MFCQ to the optimization problem with operator constraint.

We consider the following perturbation map of the operator constraint:

$$\Psi(u) := \{z \in \Omega | \Psi(z) + u \in \Lambda\}. \tag{2.114}$$

The next lemma, which is a consequence of [95] (Theorem 6.10) and [94] (Corollary 4.2), shows that this set-valued mapping is Lipschitz-like at $(0, \bar{z}) \in \Psi$, if the basic CQ is satisfied at $\bar{z}$.

Lemma 2.4. *Assume that $\bar{z} \in \Psi(0)$. Then*

$$D^*\Psi(0, \bar{z})(z^*) \subseteq \{u^* \in N_\Lambda(\Psi(\bar{z}))| - z^* \in \partial < u^*, \Psi > (\bar{z}) + N_\Omega(\bar{z})\}.$$

If in addition, the basic CQ is satisfied at $\bar{z}$, then Ψ is Lipschitz-like at $(0, \bar{z})$.

Since we are only interested in designing optimality conditions for local optimal solutions, it is necessary to show that the Lipschitz-like and calmness properties defined in the previous section are locally preserved for the set-valued mapping Ψ.

Lemma 2.5 ([30]). *Let $\bar{z} \in \Psi(\bar{u})$ and let V be a neighborhood of $\bar{z}$. If Ψ is calm (resp. Lipschitz-like) at $(\bar{u}, \bar{z})$, then the set-valued mapping*

$$\Psi_v(u) := \{z \in \Omega \cap V | \Psi(z) + u \in \Lambda\},$$

is also calm (resp. Lipschitz-like) at $(\bar{u}, \bar{z})$.

We are now ready to state a KKT-type optimality condition for problem (2.112) under the basic CQ [30].

Proposition 2.1 ([30]). *Let $\bar{z}$ be a local optimal solution of problem (2.112). Assume that the basic CQ is satisfied at $\bar{z}$. Then, there exists $u > 0$ such that for any $r \geq u$, one can find $u^* \in \mathbb{B} \cap N_\Lambda(\Psi(\bar{z}))$ such that*

$$0 \in \partial F(\bar{z}) + \partial < u^*, \Psi > (\bar{z}) + N_\Omega(\bar{z}). \tag{2.115}$$

Remark 2.6 ([30]). Under the setting of Remark 2.5, the basic CQ coincides with the usual MFCQ. Further, it is well known that under the MFCQ, the set of Lagrange multipliers is bounded. But the bound is not usually provided with the classical technique to derive KKT conditions via the MFCQ. Hence, the interesting feature of the approach in Proposition 2.1.

Remark 2.7 ([30]). One can easily check that the above result remains valid if the basic CQ is replaced by the weaker calmness of the set-valued mapping Ψ. The optimality condition (2.115) also follows from [135], Theorem 3.1], where one has to set $\emptyset(z) := \Psi(z) + \Lambda$. However, the approach in [135] does not allow us to detect the fact that u^* also belongs to $N_\Lambda(\Psi(\bar{z}))$, which is an important component of Proposition 2.1, regarding the structure of problem (2.112). Furthermore, the inclusion $u^* \in N_\Lambda(\Psi(\bar{z}))$ plays an important role in the applications.

For the rest of this section, we assume that Ψ is a real-valued function and $\Lambda = \mathbb{R}$. Then the optimization problem with operator constraint takes the form

$$\min F(z) \text{ st. } z \in \Omega, \Psi(z) \leq 0. \tag{2.116}$$

Here, $\Omega := \{z|g(z) \leq 0, h(z) = 0\}$, with g and h being some given continuous functions. Next, we recall the definition of the concept of partial calmness for problem (2.116), as introduced in [134] in the framework of the optimal value reformulation of the bi-level program. The term partial as opposed to the stronger notion of calmness by Clarke [25] refers to the fact that only some of the constraints (Ψ in our case) are perturbed. Since we will be dealing with the classical and primal KKT reformulations of (2.112), we will use the terminology of Ψ-partial calmness in order to differentiate between the two values to be taken by Ψ in the corresponding reformulation.

Definition 2.26 ([30]). Let $\bar{z}$ be a local optimal solution of problem (2.116). Problem (2.116) is Ψ-partially calm at $\bar{z}$ if there is a neighborhood U of $(0, \bar{z})$ and a number $\lambda > 0$ such that

$$F(z) - F(\bar{z}) + \lambda|t| \geq 0 \quad \forall (t, z) \in U : z \in \Omega, \Psi(z) + t \leq 0.$$

Broadly speaking, the latter concept was tailored to move a disturbing constraint (in the sense of leading to the failure of a CQ) to the objective function. This corresponds in the case of problem (2.116), to the following exact penalization, where $z \to \Psi(z)_+$ (with $\Psi(z)_+ = \max\{0, \Psi(z)\}$) represents the penalty function and λ the penalty coefficient.

Theorem 2.13 ([136], Proposition 2.2). *Let $\bar{z}$ be a local optimal solution of problem (2.116). Problem (2.116) is Ψ-partially calm at $\bar{z}$ if and only if there exists a number $\lambda > 0$ such that $\bar{z}$ is a local optimal solution of the problem to*

$$\min F(z) + \lambda \Psi(z)_+ \ st. \ z \in \Omega.$$

To conclude this section, let us mention a sufficient condition for problem (2.116) to be Ψ-partially calm. The proof can be found in [25].

Proposition 2.2 ([30]). *Let $\bar{z}$ be a local optimal solution of problem (2.116) such that the set-valued mapping Ψ in (2.114) (with Ω and Λ defined in (2.116)) is calm at $(0, \bar{z})$. Then, problem (2.116) is Ψ-partially calm at $\bar{z}$.*

Deriving the classical KKT reformulation from the generalized equation, it is clear that if $K(x) := \mathbb{R}^m$, the MFCQ remains applicable. Otherwise, if the following problem:

$$\begin{cases} \min F(x,y) \\ \text{s.t.} \begin{cases} G(x) \le 0, \mathbb{L}(x,y,u) = 0 \\ u \ge 0, g(x,y) \le 0, u^\top g(x,y) = 0, \end{cases} \end{cases} \tag{2.117}$$

is considered as a usual nonlinear optimization problem, it would fail at any feasible point; cf. [23, 112, 136]. However, the basic CQ, which can be seen as a generalization of the MFCQ (cf. Remark 2.5) may well still be applied provided the feasible set is written differently. To motivate our discussion, we recall that the failure of the MFCQ is due to the following complementarity system

$$u \ge 0, g(x,y) \le 0, u^\top g(x,y) = 0. \tag{2.118}$$

Next example shows that the basic CQ is applicable to problem (2.117), if we assume that the function g is linear in (x,y), and the feasible set is reformulated as an operator constraint; with $\Psi(x,y,u) = (G(x), \mathscr{L}(x,y,u)), \Lambda = \mathbb{R}^k \times \{0_m\}$ and Ω denoting the set of (x,y,u) solving the complementarity problem (2.118).

Example 2.5 ([30]). We consider the bi-level optimization problem to

$$\min x^2 + y^2 \text{ subject to } x \ge 0, y \in S(x) := \mathit{argmin}\{xy + y | y \ge 0\}.$$

One can easily check that $(0,0)$ is the optimal solution of the above problem. The classical KKT reformulation of this problem is:

$$\begin{cases} \min x^2 + y^2 \\ \text{s.t.} \begin{cases} x \ge 0, x - u + 1 = 0 \\ u \ge 0, y \ge 0, uy = 0. \end{cases} \end{cases}$$

It is obvious that the lower level multiplier corresponding to the optimal solution is $\bar{u} = 1$; and hence that the MFCQ fails to hold at $(0,0,1)$. We are now going to show that the basic CQ is satisfied if we set $\Psi(x,y,u) = (-x, x-u+1), \Lambda = \mathbb{R} \times \{0\}$ and $\Omega = \{(x,y,u) \in \mathbb{R}^3 | y \ge 0, u \ge 0, yu = 0\}$. For some point $(\alpha,\beta) \in N_\Lambda(\Psi(0,0,1))$, i.e. $(\alpha,\beta) \in \mathbb{R}_+ \times \mathbb{R}$, $(0,0,0) \in < \nabla\Psi(0,0,1), (\alpha,\beta) > + N_\Omega(0,0,1)$ if and only if $\alpha - \beta = 0$ and $(0,-\beta) \in N_\Theta(0,1)$ (with $\Theta := \{(y,u) \in \mathbb{R}^2 | y \ge 0, u \ge 0, yu = 0\}$). If follows from following Lemma 2.6 that $\beta = 0$ and hence that $\alpha = 0$. This shows that the basic CQ holds at $(0,0,1)$.

By setting $v := -g(x,y)$ and hence introducing a new (dummy) variable in the problem, the idea in the above example can be extended to the more general problem (2.117). The technicality behind this is that the new constraint $g(x,y) + v = 0$ is moved to the function Ψ nd thus allowing just the computation of the normal cone to the polyhedral set

$$\Theta := \{(u,v) \in \mathbb{R}^{2p} | u \ge 0, v \ge 0, u^\top v = 0\}, \tag{2.119}$$

which is possible without any qualification condition.

Lemma 2.6 ([38], Proposition 2.1). *Let* $(\bar{u}, \bar{v}) \in \Theta$*, then*

$$N_\Theta(\bar{u}, \bar{v}) = \left\{ (u^*, v^*) \in \mathbb{R}^{2p} : \begin{array}{ll} u_i^* = 0 & \forall i : \bar{u}_i > 0 = \bar{v}_i \\ v_i^* = 0 & \forall i : \bar{u}_i = 0 < \bar{v}_i \\ (u_i^* < 0 \wedge v_i^* < 0) \vee u_i^* v_i^* = 0 & \forall i : \bar{u}_i = 0 = \bar{v}_i \end{array} \right\}$$

Thanks to the aforementioned transformation, M-type stationarity conditions can be derived for problem (2.117).

Definition 2.27 ([30]). A point $(\bar{x}, \bar{y})$ will be said to be M-stationary for the bi-level optimization problem (2.107) if there exists $(\alpha, \beta, \gamma) \in \mathbb{R}^{k+p+m}$ such that $\forall \bar{u} \in \Lambda(\bar{x}, \bar{y})$:

$$\nabla_x F(\bar{x}, \bar{y}) + \nabla G(\bar{x})^\top \alpha + \nabla_x g(\bar{x}, \bar{y})^\top \beta + \nabla_x \mathscr{L}(\bar{x}, \bar{y}, \bar{u})^\top \gamma = 0, \tag{2.120}$$

$$\nabla_y F(\bar{x}, \bar{y}) + \nabla_y g(\bar{x}, \bar{y})^\top \beta + \nabla_y \mathscr{L}(\bar{x}, \bar{y}, \bar{u})^\top \gamma = 0, \tag{2.121}$$

$$\alpha \geq 0, \alpha^\top G(\bar{x}) = 0, \tag{2.122}$$

$$\nabla_y g_v(\bar{x}, \bar{y})\gamma = 0, \beta_\eta = 0, \tag{2.123}$$

$$\forall i \in \mu, (\beta_i > 0 \wedge \nabla_y g_i(\bar{x}, \bar{y})\gamma > 0) \vee \beta_i(\nabla_y g_i(\bar{x}, \bar{y})\gamma) = 0. \tag{2.124}$$

Conditions (2.120), (2.121), (2.122), (2.123), (2.124) are called the M-stationarity conditions for problem (2.107).

Definition 2.28 ([30]). A point $(\bar{x}, \bar{y})$ will be said to be S-stationary for the bi-level optimization problem (2.107) if there exists $(\alpha, \beta, \gamma) \in \mathbb{R}^{k+p+m}$ such that $\forall \bar{u} \in \Lambda(\bar{x}, \bar{y})$: (2.120), (2.121), (2.122), (2.123), (2.124) and

$$\forall i \in \mu, \beta_i \geq 0. \tag{2.125}$$

Conditions (2.120), (2.121), (2.122), (2.123) and (2.125) are called the S-stationarity conditions for (2.107).

Similarly, surrogates of other well-known types of stationarity concepts could also be defined for the bi-level optimization problem. In this chapter though, we will focus our attention only on the above M- and S- types, since they are the most important ones. For the other ones, the interested reader is referred, for example, refer to [37]. A third possibility to write problem (2.107) as a one level optimization problem is the following optimal value reformulation:

$$\begin{array}{l} \min F(x, y) \\ \text{s.t.} \begin{cases} f(x, y) \leq \varphi(x) \\ G(x) \leq 0, g(x, y) \leq 0, \end{cases} \end{array} \tag{2.126}$$

where φ is the optimal value function of the lower level problem, defined as

$$\varphi(x) := \min\{f(x, y) | y \in K(x)\}.$$

Theorem 2.14 ([30]). *Let $(\bar{x}, \bar{y}, \bar{u})$ be a local optimal solution of problem (2.117) and assume that the following CQ holds at $(\bar{x}, \bar{y}, \bar{u})$:*

$$\left.\begin{array}{l} \nabla G(\bar{x})^\top \alpha + \nabla_x g(\bar{x}, \bar{y})^\top \beta + \nabla_x \mathscr{L}(\bar{x}, \bar{y}, \bar{u})^\top \gamma = 0 \\ \nabla_y g(\bar{x}, \bar{y})^\top \beta + \nabla_y \mathscr{L}(\bar{x}, \bar{y}, \bar{u})^\top \gamma = 0 \\ \alpha \geq 0, \alpha^\top G(\bar{x}) = 0 \\ \beta_v = 0, \nabla_y g_\eta(\bar{x}, \bar{y})\gamma = 0 \\ \forall i \in u, (\beta_i > 0) \vee \beta_i(\nabla_y g_i(\bar{x}, \bar{y})\gamma) = 0 \end{array}\right\} \Rightarrow \left\{\begin{array}{l} \alpha = 0 \\ \beta = 0 \\ \gamma = 0. \end{array}\right. \tag{2.127}$$

Then, there exists $(\alpha, \beta, \gamma) \in \mathbb{R}^{k+p+m}$, with $\| (\alpha, \beta, \gamma) \| \leq r$ (for some $\gamma > 0$) such that the M-stationarity conditions are satisfied.

Proof ([30]). Let us set $\Phi(x, y, u, v) = (G(x), g(x, y) + v, \mathscr{L}(x, y, u), \Lambda = \mathbb{R}^k_- \times \{0_{p+m}\})$ and $\Omega = \mathbb{R}^{n+m} \times \Theta$. Let $(\bar{x}, \bar{y}, \bar{u})$ be a local optimal solution of problem (2.117). One can easily verify that there is a vector $\bar{v}$ such that $(\bar{x}, \bar{y}, \bar{u}, \bar{v})$ is a local optimal solution of the problem to

$$\min F(x, y) \text{ suject to } (x, y, u, v) \in \Omega \cap \Psi^{-1}(\Lambda). \tag{2.128}$$

We have

$$N_\Omega(\bar{x}, \bar{y}, \bar{u}, \bar{v}) = \{0_{n+m}\} \times N_\Theta(\bar{u}, \bar{v}), \tag{2.129}$$

$$N_\Lambda(\Psi(\bar{x}, \bar{y}, \bar{u}, \bar{v})) = \{(\alpha, \beta, \gamma) | \alpha \geq 0, \alpha^\top G(\bar{x}) = 0\}, \tag{2.130}$$

$$\nabla \Psi(x, y, u, v)^\top (\alpha, \beta, \gamma) = \begin{bmatrix} A(\alpha, \beta, \gamma) \\ \beta \end{bmatrix} \tag{2.131}$$

where

$$A(\alpha, \beta, \gamma) = \begin{bmatrix} \nabla G(\bar{x})^\top \alpha + \nabla_x g(\bar{x}, \bar{y})^\top \beta + \nabla_x \mathscr{L}(\bar{x}, \bar{y}, \bar{u})^\top \gamma \\ \nabla_y g(\bar{x}, \bar{y})^\top \beta + \nabla_y \mathscr{L}(\bar{x}, \bar{y}, \bar{u})^\top \gamma \\ \nabla_y g(\bar{x}, \bar{y})\gamma \end{bmatrix} \tag{2.132}$$

It follows from equalities (2.129), (2.130) and (2.131) that the basic CQ applied to problem (2.128) at $(\bar{x}, \bar{y}, \bar{u}, \bar{v})$ can equivalently be formulated as follows: there is no nonzero vector $(\alpha, \beta, \gamma) \in \mathbb{R}^{k+p+m}$ such that

$$\nabla G(\bar{x})^\top \alpha + \nabla_x g(\bar{x}, \bar{y})^\top \beta + \nabla_x \mathscr{L}(\bar{x}, \bar{y}, \bar{u})^\top \gamma = 0, \tag{2.133}$$

$$\nabla_y g(\bar{x}, \bar{y})^\top \beta + \nabla_y \mathscr{L}(\bar{x}, \bar{y}, \bar{u})^\top \gamma = 0, \tag{2.134}$$

$$\alpha \geq 0, \alpha^\top G(\bar{x}) = 0, \tag{2.135}$$

$$(-\nabla_y g(\bar{x}, \bar{y})\gamma, -\beta) \in N_\Theta(\bar{u}, \bar{v}). \tag{2.136}$$

By noting that $\bar{v}_i = -g_i(\bar{x}, \bar{y})$, for $i := 1, \cdots, p$, it follows from Lemma 2.6 that the basic CQ applied to problem (2.128) is equivalent to CQ (2.127). Hence, from Proposition 2.1 there exists $(\alpha, \beta, \gamma) \in \mathbb{R}^{k+p+m}$, with $\| (\alpha, \beta, \gamma) \| \leq r$ (for some $r > 0$) such that (2.120), (2.121), (2.122) and (2.136) are satisfied, given that the objective function of problem (2.128) is independent of (u, v). The result then follows by interpreting inclusion (2.136), as already made above. □

The bound on the multiplier vector, usually neglected for MPCCs, can be explicitly given in terms of problem data; see the proof of Proposition 2.1. It may be important to mention that this bound can be very useful in developing an effective algorithm for problem (2.117), and hence for the bi-level optimization problem.

The technique used in the proof of Theorem 2.14, i.e. to transform the nonlinear complementarity problem in (2.118) into a linear one, has been used in various occasions, for the MPCC [38]. One can easily check that the M-stationarity conditions obtained here are identical to those in [135] or [137] under various CQs, among which CQ (b) of [135] (Theorem 4.1) or (b) of [137] (Theorem 5.1) coincides with the CQ in Theorem 2.14. But, it should be mentioned that in the latter case, this CQ is recovered from a perspective different from that of [135, 137], where an enhanced generalized equation formulation of the KKT conditions of the lower level problem was used to design the CQ.

The following part introduces a different way to choose Ψ, Ω and Λ; that would lead to a new and weaker CQ allowing us to obtain the same optimality conditions as in Theorem 2.14. To proceed, let us recall that the complementarity system (2.118) is equivalent to

$$u_i \geq 0, g_i(x, y) \leq 0, u_i g_i(x, y) = 0, i = 1, \cdots, p,$$

meaning that

$$(u_i, -g_i(x, y)) \in \Lambda_i := \{(a, b) \in \mathbb{R}^2 | a \geq 0, b \geq 0, ab = 0\}, i = 1, \cdots, p.$$

Theorem 2.15 ([30]). *Let $(\bar{x}, \bar{y}, \bar{u})$ be a local optimal solution of problem (2.117) and assume that the following assertions are satisfied:*

1. *The following set-valued mapping is calm at $(0, 0, \bar{x}, \bar{y}, \bar{u})$*

$$\mathcal{M}_1(t_{1,2}) := \{(\bar{x}, \bar{y}, \bar{u}) | G(x) + t_1 \leq 0, \mathcal{L}(\bar{x}, \bar{y}, \bar{u}) + t_2\},$$

2. *The following implication holds at* $(0, 0, \bar{x}, \bar{y}, \bar{u})$:

$$\left.\begin{array}{l}\nabla G(\bar{x})^\top \alpha + \nabla_x g(\bar{x}, \bar{y})^\top \beta + \nabla_x \mathscr{L}(\bar{x}, \bar{y}, \bar{u})^\top \gamma = 0 \\ \nabla_y g(\bar{x}, \bar{y})^\top \beta + \nabla_y \mathscr{L}(\bar{x}, \bar{y}, \bar{u})^\top \gamma = 0 \\ \alpha \geq 0, \alpha^\top G(\bar{x}) = 0 \\ \beta_\upsilon = 0, \nabla_y g_\eta(\bar{x}, \bar{y})\gamma = 0 \\ \forall i \in u, (\beta_i > 0 \wedge \nabla_y g_i(\bar{x}, \bar{y})\gamma > 0) \vee \beta_i(\nabla_y g_i(\bar{x}, \bar{y})\gamma) = 0\end{array}\right\} \Rightarrow \left\{\begin{array}{l}\beta = 0 \\ \nabla_y g(\bar{x}, \bar{y})\gamma = 0.\end{array}\right. \tag{2.137}$$

Then, there exists $(\alpha, \beta, \gamma) \in \mathbb{R}^{k+p+m}$, with $\| \beta \| \leq r$ (for some $r > 0$) such that the M-stationarity conditions are satisfied.

Proof ([30]). We consider the set $\Omega = \{(x, y, u) | G(x) \leq 0, \mathscr{L}(x, y, u) = 0\}$ and the function $\psi(x, y, u) = (\mu_u, -g_i(x, y))_{i=1,\cdots,p}$. If $(\bar{x}, \bar{y}, \bar{u})$ is a local optimal solution of problem (2.117), it means, in other words, that $(\bar{x}, \bar{y}, \bar{u})$ is a local optimal solution of the problem to

$$\min F(x, y) \text{ st. } (x, y, u) \in \Omega \cap \Psi^{-1}(\Lambda), \tag{2.138}$$

where $\Lambda = \Lambda_1, \cdots, \Lambda_p$, with $\Lambda_i = \{(a, b) \in \mathbb{R}^2 | a \geq 0, b \geq 0, ab = 0\}$ for $i = 1, \cdots, p$.

Applying Proposition 2.1 to (2.138), there exists a vector $(\delta, \beta) \in \mathbb{R}^{2p}$ with $\| (\delta, \beta) \| \leq r$ (for some $r > 0$) such that

$$(\delta_i, \beta_i) \in N_{\Lambda_i}(\Psi_i(\bar{x}, \bar{y}, \bar{u})), i := 1, \cdots, p, \tag{2.139}$$

$$(0, 0) \in \begin{bmatrix} \nabla F(\bar{x}, \bar{y}) \\ 0 \end{bmatrix} + \begin{bmatrix} -\nabla g(\bar{x}, \bar{y})^\top \beta \\ \delta \end{bmatrix} + N_\Omega(\bar{x}, \bar{y}, \bar{u}), \tag{2.140}$$

provided there is no nonzero vector $(\delta, \beta) \in \mathbb{R}^{2p}$ such that

$$(\delta_i, \beta_i) \in N_{\Lambda_i}(\Psi_i(\bar{x}, \bar{y}, \bar{u})), i := 1, \cdots, p, \tag{2.141}$$

$$(0, 0) \in \begin{bmatrix} -\nabla g(\bar{x}, \bar{y})^\top \beta \\ \delta \end{bmatrix} + N_\Omega(\bar{x}, \bar{y}, \bar{u}). \tag{2.142}$$

It follows, under assumption 1 (see Theorem 2.19), that

$$N_\Omega(\bar{x}, \bar{y}, \bar{u}) = \{A(\alpha, \beta, \gamma) - (\nabla_x g(\bar{x}, \bar{y})^\top, \beta, \nabla_y g(\bar{x}, \bar{y})^\top \beta, 0)^\top | \alpha \geq 0, \alpha^\top G(\bar{x}) = 0\},$$

where $A(\alpha, \beta, \gamma)$ denotes the matrix given in (2.132). Hence, either from (2.140) or from (2.142), it follows that there exists $\gamma \in \mathbb{R}^m$ such that $\delta = -\nabla_y g(\bar{x}, \bar{y})\gamma$; à fortiori, either (2.139) or (2.141) implies that there exists $\gamma \in \mathbb{R}^m$ such that $(-\nabla_y g_i(\bar{x}, \bar{y})\gamma, \beta_i) \in N_{\Lambda_i}(\Psi_i(\bar{x}, \bar{y}, \bar{u})), i := 1, \cdots, p$. The result then follows by noting that $N_{\Lambda_i}(\Psi_i(\bar{x}, \bar{y}, \bar{u}))$ is obtained from Lemma 2.6. □

The approach in the above result is similar to the one used in [53], for a mathematical program with vanishing constraints.

Remark 2.8 ([30]).

(i) Assumptions 1 and 2 in the previous result are satisfied, provided CQ (2.127) holds at $(\bar{x}, \bar{y}, \bar{u})$. In fact, it is obvious that CQ (2.127) implies assumption 2. On the other hand, CQ (2.127) can equivalently be written as

$$\begin{aligned} A(\bar{x}, \bar{y}, \bar{u}) := \big\{ &(\alpha, \beta, \gamma) | \nabla G(\bar{x})^\top \alpha + \nabla_x g(\bar{x}, \bar{y})^\top \beta + \nabla_x \mathscr{L}(\bar{x}, \bar{y}, \bar{u})^\top \gamma = 0 \\ &\nabla_y g(\bar{x}, \bar{y})^\top \beta + \nabla_y \mathscr{L}(\bar{x}, \bar{y}, \bar{u})^\top \gamma = 0, \alpha \geq 0, \alpha^\top G(\bar{x}) = 0 \\ &\beta_v = 0, \nabla_y g_\eta(\bar{x}, \bar{y})\gamma = 0, \forall i \in u, (\beta_i > 0 \wedge \nabla_y g_i(\bar{x}, \bar{y})\gamma > 0) \\ &\vee \beta_i(\nabla_y g_i(\bar{x}, \bar{y})\gamma) = 0 \big\} \\ = &\{(0, 0, 0)\}. \end{aligned}$$

Furthermore, one has

$$\begin{aligned} &A(\bar{x}, \bar{y}, \bar{u}) \supseteq \{(\alpha, 0, \gamma) | \nabla G(\bar{x})^\top \alpha + \nabla_x \mathscr{L}(\bar{x}, \bar{y}, \bar{u})^\top \gamma = 0, \nabla_y \mathscr{L}(\bar{x}, \bar{y}, \bar{u})^\top \gamma = 0, \\ &\alpha \geq 0, \alpha^\top G(\bar{x}) = 0\} := B(\bar{x}, \bar{y}, \bar{u}), \end{aligned}$$

which means that CQ (2.127) is also a sufficient condition for $B(\bar{x}, \bar{y}, \bar{u}) = \{(0, 0, 0)\}$. Following Lemma 2.4, the latter equality implies the fulfillment of assumption 1.

(ii) A second possibility to recover CQ (2.127) in the above theorem is to move the constraints defining Ω to the function Ψ, i.e. to set $\Omega := \mathbb{R}^n \times \mathbb{R}^m$ and

$$\Psi(x, y) := [G(x), \mathscr{L}(x, y, u), (u_i, -g_i(x, y))_{i=1,\cdots,p}].$$

This would also help recover the bound on all the multipliers.

To conclude this subsection, we now deduce the M-type optimality conditions for the bi-level optimization (2.107), from the above developments on the classical KKT reformulation (2.117).

Corollary 2.1 ([30]). *Let $(\bar{x}, \bar{y})$ be a local optimal solution of problem (2.107). Assume that the following assertions hold:*

1. CQ (2.143) holds at $(\bar{x}, \bar{y})$

$$[\nabla_y g(\bar{x}, \bar{y})^\top \beta = 0, \beta \geq 0, \beta^\top g(\bar{x}, \bar{y}) = 0] \Rightarrow \beta = 0. \tag{2.143}$$

2. CQ (2.127) holds at $(\bar{x}, \bar{y}, \bar{u})$, for all $\bar{u} \in \Lambda(\bar{x}, \bar{y})$

Then, $(\bar{x}, \bar{y})$ is M-stationary, with $\| (\alpha, \beta, \gamma) \| \leq r$ (for some $r > 0$).

This result can be restated with CQ (2.127) of assumption 2 replaced by the CQs in Theorem 2.15.

In the framework of MPCCs, the Guignard CQ has been shown to be one of the few CQs to be directly applicable to (2.117) considered as a usual nonlinear optimization problem; cf. [37]. In the next result, we derive the S-type stationarity conditions for problem (2.117) under the Guignard CQ, which holds at a point $(\bar{x}, \bar{y}, \bar{u}) \in \mathscr{C}$ ($\mathscr{C}$ being the feasible set of problem (2.117)) if the Fréchet normal cone to $\mathscr{C}$ at $\bar{z}$ takes from form

$$\hat{N}_{\mathscr{C}}(\bar{x}, \bar{y}, \bar{u}) = \left\{ d := (d_x, d_y, d_u) : \begin{array}{ll} \nabla G_i(\bar{x})^\top d_x \geq 0, & \forall i : G_i(\bar{x}) = 0 \\ \nabla g_i(\bar{x}, \bar{y})^\top (d_x, d_y) \geq 0, & \forall i : g_i(\bar{x}, \bar{y}) = 0 \\ \nabla \mathscr{L}_i(\bar{x}, \bar{y}, \bar{u})^\top d = 0, & \forall i : i = 1, \cdots, m \\ \nabla \sigma(\bar{x}, \bar{y}, \bar{u})^\top d = 0, & \end{array} \right\} \tag{2.144}$$

where σ denotes the function defined as $\sigma(x, y, u) := -u^\top g(x, y)$.

Theorem 2.16 ([30]). *Let $(\bar{x}, \bar{y}, \bar{u})$ be a local optimal solution of problem (2.117) and assume that the Guignard CQ is satisfied at $(\bar{x}, \bar{y}, \bar{u})$. Then, there exists $(\alpha, \beta, \gamma) \in \mathbb{R}^{k+p+m}$ such that the S-stationarity conditions are satisfied.*

Proof ([30]). According to [31] (Theorem 3.5), under the Guignard CQ, there exists $(\alpha, \beta, \gamma, \lambda) \in \mathbb{R}^{k+p+m+1}$ such that the following optimality conditions are satisfied:

$$\nabla_x F(\bar{x}, \bar{y}) + \nabla G(\bar{x})^\top \alpha + \nabla_x g(\bar{x}, \bar{y})^\top (\beta - \lambda \bar{u}) + \nabla_x \mathscr{L}(\bar{x}, \bar{y}, \bar{u})^\top \gamma = 0, \tag{2.145}$$

$$\nabla_y F(\bar{x}, \bar{y}) + \nabla_y g(\bar{x}, \bar{y})^\top (\beta - \lambda \bar{u}) + \nabla_y \mathscr{L}(\bar{x}, \bar{y}, \bar{u})^\top \gamma = 0, \tag{2.146}$$

$$\alpha \geq 0, \alpha^\top G(\bar{x}) = 0, \tag{2.147}$$

$$\beta \geq 0, \beta^\top g(\bar{x}, \bar{y}) = 0, \tag{2.148}$$

$$\nabla_y g(\bar{x}, \bar{y})\gamma - \lambda g(\bar{x}, \bar{y}) \geq 0, \bar{u}^\top (\nabla_y g(\bar{x}, \bar{y})\gamma) = 0. \tag{2.149}$$

It suffices now to show that these conditions are equivalent to the S-stationarity conditions in the sense of Definition 2.28, i.e. for a Lagrange multiplier vector $\bar{v} = (\alpha, \beta, \gamma, \lambda)$, the triple $(\bar{x}, \bar{y}, \bar{v})$ satisfies (2.145), (2.146), (2.147), (2.148), and (2.149) if and only if there exists $\bar{v}^* = (\alpha^*, \beta^*, \gamma^*)$ such that $(\bar{x}, \bar{y}, \bar{v}^*)$ satisfies (2.120), (2.121), (2.122), (2.123) and (2.125).

An example of bi-level optimization problem for which the Guignard CQ is satisfied can be found in [31]. For more on this CQ and its application to MPCCs, refer to [37].

For the rest of this section, we focus our attention to the concept of partial calmness. Precisely, we start by showing how a combination of partial calmness and basic CQ could lead to the S -stationarity conditions. The principle of this result is very simple. In fact, since the failure of the MFCQ is due to the complementarity system, then moving the function ęÒ to the upper level objective function paves the way to the application of the same CQ.

Theorem 2.17 ([30]). *Let $(\bar{x}, \bar{y}, \bar{u})$ be a local optimal solution of problem (2.117) and assume that the following assertions are satisfied:*

1. *Problem (2.117) is σ-partially calm at $(\bar{x}, \bar{y}, \bar{u})$.*
2. *The following implication holds at $(\bar{x}, \bar{y}, \bar{u})$:*

$$\left.\begin{array}{l} \nabla G(\bar{x})^\top \alpha + \nabla_x g(\bar{x}, \bar{y})^\top \beta + \nabla_x \mathcal{L}(\bar{x}, \bar{y}, \bar{u})^\top \gamma = 0 \\ \nabla_y g(\bar{x}, \bar{y})^\top \beta + \nabla_y \mathcal{L}(\bar{x}, \bar{y}, \bar{u})^\top \gamma = 0 \\ \alpha \geq 0, \alpha^\top G(\bar{x}) = 0 \\ \beta \geq 0, \beta^\top g(\bar{x}, \bar{y}) = 0 \\ \nabla_y g(\bar{x}, \bar{y})\gamma \geq 0, \bar{u}^\top (\nabla_y g(\bar{x}, \bar{y})\gamma) = 0 \end{array}\right\} \Rightarrow \left\{\begin{array}{l} \alpha = 0 \\ \beta = 0 \\ \gamma = 0. \end{array}\right. \tag{2.150}$$

Then, there exists $(\alpha, \beta, \gamma) \in \mathbb{R}^{k+p+m}$, with $\| (\alpha, \beta, \gamma) \| \leq r$ (for some $r > 0$) such that the S-stationarity conditions are satisfied.

Proof ([30]). Let $(\bar{x}, \bar{y}, \bar{u})$ be a local optimal solution of problem (2.117) and let assumption 1 of the theorem be satisfied. Then, it follows from Theorem 2.13 that, there exists $\lambda > 0$ such that $(\bar{x}, \bar{y}, \bar{u})$ is also a local optimal solution of the problem

$$\min F(x, y) - \lambda u^\top g(x, y) \text{ subject to } (x, y, u) \in \Omega \cap \Psi^{-1}(\Lambda), \tag{2.151}$$

where $\Omega = \mathbb{R}^{n+m} \times \mathbb{R}^p_+$, $\Lambda = \mathbb{R}^{k+p}_- \times \{0_m\}$ and $\Psi(x, y, u) = (G(x), g(x, y), \mathcal{L}(x, y, u))$. One can easily check that:

$$N_\Omega(\bar{x}, \bar{y}, \bar{u}) = \{0_{n+m}\} \times \{\eta \in \mathbb{R}^p | \eta \leq 0, \eta^\top \bar{u} = 0\}, \tag{2.152}$$

$$N_\Lambda(\Psi(\bar{x}, \bar{y}, \bar{u})) = \{(\alpha, \beta, \gamma) | \alpha \geq 0, \alpha^\top G(\bar{x}) = 0, \beta \geq 0, \beta^\top g(\bar{x}, \bar{y}) = 0\}, \tag{2.153}$$

$$\nabla \Psi(x, y, u)^\top (\alpha, \beta, \gamma) = A(\alpha, \beta, \gamma), \tag{2.154}$$

where $A(\alpha, \beta, \gamma)$ is the matrix in (2.132). It follows from equalities (2.152), (2.153) and (2.154) that the basic CQ applied to problem (2.151) at $(\bar{x}, \bar{y}, \bar{u})$ can equivalently be formulated as: there is no nonzero vector $(\alpha, \beta, \gamma) \in \mathbb{R}^{k+p+m}$ and a vector $\eta \in \mathbb{R}^p$ (dummy multiplier) such that the first four lines of the left hand side of implication (2.150) and the system

$$\nabla_y g(\bar{x}, \bar{y})\gamma + \eta = 0, \eta \leq 0, \eta^\top \bar{u} = 0, \tag{2.155}$$

are satisfied, respectively. It clearly follows that assumption 2 corresponds to the basic CQ applied to problem (2.151), where the last line of the system in the left hand side of implication (2.150) is recovered from (2.155). Hence, applying

Proposition 2.1 to problem (2.151), it also follows from (2.152), (2.153) and (2.154) that there exists $(\alpha, \beta, \gamma) \in \mathbb{R}^{k+p+m}$, with $\| (\alpha, \beta, \gamma) \| \leq r$ (for some $r > 0$) and $\lambda > 0$ such that (2.145), (2.146), (2.147), (2.148) and

$$-\lambda g(\bar{x}, \bar{y}) + \nabla_y g(\bar{x}, \bar{y})\gamma + \eta = 0, \eta \leq 0, \eta^\top \bar{u} = 0 \tag{2.156}$$

hold. Hence, (2.149) is regained from system (2.156) while noting that the feasibility of $(\bar{x}, \bar{y}, \bar{u})$ implies $\bar{u}^\top g(\bar{x}, \bar{y}) = 0$.

We have shown that there exists $(\alpha, \beta, \gamma, \lambda)$, with $\lambda > 0$, such that (2.145), (2.146), (2.147), (2.148) and (2.149). The S-stationarity conditions (2.120), (2.121), (2.122), (2.123) and (2.125) are then obtained as in the previous result. □

In the next result, we show that the CQ in assumption 2 of the previous theorem can be weakened, if the perturbation map of the joint upper and lower level feasible set is calm.

Theorem 2.18 ([30]). *Let $(\bar{x}, \bar{y}, \bar{u})$ be a local optimal solution of problem (2.117) and assume that the following assertions are satisfied:*

1. *Problem (2.117) is σ-partially calm at $(\bar{x}, \bar{y}, \bar{u})$.*
2. *The following set-valued mapping is calm at $(0, 0, \bar{x}, \bar{y})$*

$$M_2(t_1, t_2) := \{(x, y) | G(x) + t_1 \leq 1, g(x, y) + t_2 \leq 0\}.$$

3. *The following implication holds at $(\bar{x}, \bar{y}, \bar{u})$:*

$$\left.\begin{array}{l} \nabla G(\bar{x})^\top \alpha + \nabla_x g(\bar{x}, \bar{y})^\top \beta + \nabla_x \mathscr{L}(\bar{x}, \bar{y}, \bar{u})^\top \gamma = 0 \\ \nabla_y g(\bar{x}, \bar{y})^\top \beta + \nabla_y \mathscr{L}(\bar{x}, \bar{y}, \bar{u})^\top \gamma = 0 \\ \alpha \geq 0, \alpha^\top G(\bar{x}) = 0 \\ \beta \geq 0, \beta^\top g(\bar{x}, \bar{y}) = 0 \\ \nabla_y g(\bar{x}, \bar{y})\gamma \geq 0, \bar{u}^\top (\nabla_y g(\bar{x}, \bar{y})\gamma) = 0 \end{array}\right\} \Rightarrow \gamma = 0. \tag{2.157}$$

Then, there exists $(\alpha, \beta, \gamma) \in \mathbb{R}^{k+p+m}$, with $\| \gamma \| \leq r$ (for some $r > 0$) such that the S-stationarity conditions are satisfied.

Proof ([30]). Set $\Omega = \{(x, y, u) | u \geq 0, G(x) \leq 0, g(x, y) \leq 0\}$, and let $(\bar{x}, \bar{y}, \bar{u})$ be a local optimal solution of problem (2.117). Then, under assumption 1, there exists $\lambda > 0$ such that $(\bar{x}, \bar{y}, \bar{u})$ is also a local optimal solution of

$$\min F(x, y) - \lambda u^\top g(x, y) \text{ subject to } (x, y, u) \in \Omega \cap \mathscr{L}^{-1}(0).$$

Hence, it follows from Proposition 2.1 that if

$$[0 \in \nabla \mathscr{L}(\bar{x}, \bar{y}, \bar{u})^\top \gamma + N_\Omega(\bar{x}, \bar{y}, \bar{u}), \gamma \in \mathbb{R}^m] \Rightarrow \gamma = 0, \tag{2.158}$$

then there exists $\gamma \in \mathbb{R}^m$ with $\| \gamma \| \leq r$ (for some $r > 0$) such that

$$0 \in \begin{bmatrix} \nabla F(\bar{x}, \bar{y}) \\ -\lambda g(\bar{x}, \bar{y}) \end{bmatrix} + \nabla \mathscr{L}(\bar{x}, \bar{y}, \bar{u})^\top \gamma + N_\Omega(\bar{x}, \bar{y}, \bar{u}). \tag{2.159}$$

Now, observe that $\Omega = \Omega' \times \mathbb{R}^p_+$ with $\Omega' := \{(X, Y) | G(x) \leq 0, g(x, y) \leq 0\}$.

According to [51] (Theorem 5), we have following theorem:

Theorem 2.19 ([30]). *We consider the set* $A := \{z \in \mathbb{R}^l | g(z) \leq 0, h(z) = 0\}$, *where* $g : \mathbb{R}^l \to \mathbb{R}^{l_1}$ *and* $h : \mathbb{R}^l \to \mathbb{R}^{l_2}$ *are continuously differentiable functions. Then*

$$N_A(\bar{z}) = \{\nabla g(\bar{z})^\top \lambda + \nabla h(\bar{z})^\top \mu | \lambda \geq 0, \lambda^\top g(\bar{z}) = 0\},$$

for $\bar{z} \in A$, *provided the following set-valued mapping is calm at* $(0, 0, \bar{z})$

$$\mathscr{M}(t_1, t_2) := \{z \in \mathbb{R}^l | g(z) + t_1 \leq 0, h(z) + t_2 = 0\}.$$

It follows from Theorem 2.19 that

$$N_{\Omega'}(\bar{x}, \bar{y}) = \{(\nabla G(\bar{x})^\top \alpha + \nabla_x g(\bar{x}, \bar{y})^\top \beta, \nabla_y g(\bar{x}, \bar{y})^\top \beta)^\top | (2.147) \text{ and } (2.148)\}$$

under assumption 2. Hence, we regain assumption 3 and the desired optimality conditions by applying the last equality to (2.158) and (2.159), while noting that $u^* \in N_{\mathbb{R}^p_+}(\bar{u})$ if and only if $u^* \leq 0$ and $\bar{u}^\top u^* = 0$. □

Proceeding as in Remark 2.8, one can easily check that assumption 2 in Theorem 2.17 implies the fulfillment of both assumption 2 and 3 in Theorem 2.18.

To conclude this section, it seems important to mention some sufficient conditions for the partial calmness used in Theorems 2.17 and 2.18. For this purpose, a slightly modified notion of uniform weak sharp minimum was introduced in [29]. We recall that the initial definition first appeared in [134].

Definition 2.29 ([30]). The family $\{(2.108) | x \in X\}$ is said to have a uniformly weak sharp minimum if there exist $c > 0$ and a neighborhood $\mathscr{N}(x)$ of $S(x), x \in X$ such that

$$f(x, y) - \varphi(x) \geq cd(y, S(x)), \forall y \in K(x) \cap \mathscr{N}(x), \forall x \in X.$$

If we set $\mathscr{N}(x) = \mathbb{R}^m$, the definition can be found in [134]. A uniform weak sharp minimum, as given in the above definition, was shown to exist under the uniform calmness [29].

Theorem 2.20 ([30]). *Let* $(\bar{x}, \bar{y}, \bar{u})$ *be a local optimal solution of problem (2.117). Then, problem (2.117) is* σ*-partially calm at* $(\bar{x}, \bar{y}, \bar{u})$, *provided one of the following assumptions hold:*

1. *The family* $\{(2.108)|x \in X\}$ *has a uniform weak sharp minimum.*
2. *The set* $\mathscr{A} := \{(x, y, u)|G(x) \leq 0, \mathscr{L}(x, y, u) = 0, g(x, y) \leq 0, u \geq 0\}$ *is semismooth and* $-(\bar{u}^\top \nabla g(\bar{x}, \bar{y}), g(\bar{x}, \bar{y})) \notin bdN_{\mathscr{A}}(\bar{x}, \bar{y}, \bar{u})$.

Proof. Under assumption 1, the result follows from [136]. Under assumption 2, it follows from [50] that the set-valued mapping $\mathscr{M}_3(t) := \{(x, y, u) \in \mathscr{A}| - u^\top g(x, y) + t \leq 0\}$ is calm at $(0, \bar{x}, \bar{y}, \bar{u})$. Hence, the result follows from Proposition 2.2. □

For the definition of the semismoothness [50]. This is automatically satisfied for our set $\mathscr{A}$, if G is convex, and $(x, y) \mapsto \nabla_y f(x, y)$ and g are affine linear.

Finally, the S-stationarity conditions for the bi-level program (2.107) can be obtained via its classical KKT reformulation as follows.

Corollary 2.2 ([30]). *Let* $(\bar{x}, \bar{y})$ *be a local optimal solution of problem (2.107). Assume that the following assertions hold:*

1. *CQ (2.143) holds at* $(\bar{x}, \bar{y})$.
2. *Guignard CQ holds at* $(\bar{x}, \bar{y}, \bar{u})$, *for all* $\bar{u} \in \Lambda(\bar{x}, \bar{y})$.

Then, $(\bar{x}, \bar{y})$ *is S-stationary in the sense of Definition 2.28.*

Similarly to the M-stationarity case, this corollary can be restated with the Guignard CQ of assumption 2 replaced by the CQs in Theorem 2.17 or Theorem 2.18.

Consider the set-valued mapping Q defined as $Q(x, y) := N_{k(x)}(y)$, for $y \in K(x)$, and $Q(x, y) := \emptyset$, otherwise. To obtain the closedness of gph Q , necessary in order to apply the basic CQ to problem (2.111), we introduce the concept of inner semicontinuity for a set-valued mapping; see [96] for more details.

A set-valued mapping Φ is said to be inner semicontinuous at $(x, y) \in \text{gph}\Phi$ if for every sequence $x^k \to x$ there is a sequence $y^k \in \Phi(x^k)$ such that $y^k \to y$. Φ will be said to be inner semicontinuous if it is inner semicontinuous at every point of its graph. In the following result, we show that gph Q is closed if the lower level feasible set mapping K is inner semicontinuous.

Proposition 2.3 ([30]). *Assume that* K *is inner semicontinuous. Then, gph* Q *is closed as a subset of gph* $K \times \mathbb{R}^m$, *i.e. if* $(x^k, y^k) \to (x, y)((x^k, y^k) \in gphK)$ *and* $z^k \in z$ *with* $z^k \in Q(x^k, y^k)$, *then* $z \in Q(x, y)$.

Proof. Let $(x^k, y^k) \to (x, y)((x^k, y^k) \in \text{gph}K)$ and $z^k \in z$ with $z^k \in Q(x^k, y^k)$. Since $K(x^k)$ is assumed to be convex for all k, we have

$$< z^k, u^k - y^k > \leq 0, \forall u^k in K(x^k), k \in \mathbb{N}. \tag{2.160}$$

Given that K is inner semicontinuous, then for an arbitrary $v \in K(x)$, there exists $v^k \in K(x^k)$ such that $v^k \to v$. It follows from (2.160) that $< z^k, u^k - y^k > \leq 0$, $\forall k \in \mathbb{N}$. This implies that $< z, v - y > \leq 0$, $\forall v \in K(x)$, which concludes that proof. □

This result can be seen as an extension to parametric sets, of the result stated in [107] (Proposition 3.3), with the difference that our cones are defined in the sense of convex analysis. But, it may well be extended to the case where $N_{K(x)}(y)$ is the more general Mordukhovich normal cone used in [107], when $K(x) := K$ for all x.

Considering the case where the set-valued mapping K is defined as in (2.109), the concept of inner semicontinuity can be brought to usual terms through the following well-known result; see e.g. [28].

Lemma 2.7 ([30]). *If CQ (2.143) holds at* $(\bar{x}, \bar{y})$*, then K is inner semicontinuous near* $(\bar{x}, \bar{y})$*. The normal cone to the graph of* $N_{\mathbb{R}^p_-}$*, which is also useful in this section, can be obtained from the normal cone to* Θ *(defined in (2.119)) as follows.*

Proposition 2.4 ([30]). *Let* $(\bar{\chi}, \bar{v}) \in gphN_{\mathbb{R}^p_-}$*, then*

$$N_{gph\mathbb{R}^p_-}(\bar{\chi}, \bar{v}) = \{(-\chi^*, v^*) \in \mathbb{R}^{2p} | (-\chi^*, v^*) \in N_{\Theta}(\bar{\chi}, \bar{v})\}.$$

Proof ([30]). The proof starts by noting that

$$\begin{aligned} \text{gph}N_{\mathbb{R}^p_-} &= \{(\chi, v) \in \mathbb{R}^{2p} | \chi \leq 0, v \geq 0, \chi^\top v = 0\} \\ &= \{(\chi, v) \in \mathbb{R}^{2p} | (-\chi, v) \in \Theta\}. \end{aligned}$$

This means that $\text{gph}N_{\mathbb{R}^p_-} = \vartheta(\chi, v) := (-\chi, v)$ and for $(\bar{\chi}, \bar{v}) \in \mathbb{R}^{2p}$, one obviously has

$$\nabla\vartheta(\bar{\chi}, \bar{v}) = \begin{pmatrix} -I_p & O \\ O & I_p \end{pmatrix}$$

with I_p and O denoting the $p \times p$ identity and zero matrix, respectively. Hence, the Jacobian matrix $\nabla\vartheta(\bar{\chi}, \bar{v})$ is quadratic and nonsingular and it follows from [93] (Corollary 2.12) that

$$N_{\text{gph}\mathbb{R}^p_-}(\bar{\chi}, \bar{v}) = \nabla\vartheta(\bar{\chi}, \bar{v})^\top N_{\Theta}(\vartheta(\bar{\chi}, \bar{v})),$$

given that $\text{gph}N_{\mathbb{R}^p_-}$ and Θ are closed sets. The result then follows. □

Exploiting the polyhedrality of $\text{gph}N_{\mathbb{R}^p_-}$ and Θ, the equality in the above result can also be proven, at least in the case where $\bar{\chi} = 0$, by using a combination of [104] (Proposition 1) and [51] (Theorem 5).

In the next theorem, a slightly modified version of Theorem 6.1 in [97] is presented.

Theorem 2.21 ([30]). *Let* $(\bar{x}, \bar{y})$ *be a local optimal solution of problem (2.111) (i.e. of (2.107)). Assume that the following assertions hold:*

1. *CQ (2.143) holds at* $(\bar{x}, \bar{y})$
2. *The following set-valued mapping is calm at* $(0, \bar{x}, \bar{y}, \bar{u})$*, for all* $\bar{u} \in \Lambda(\bar{x}, \bar{y})$

$$M(\vartheta) := \left\{ (x, y, u) \middle| \begin{bmatrix} g(x, y) \\ u \end{bmatrix} + \vartheta \in gphN_{\mathbb{R}^p_-} \right\}.$$

3. The following set-valued mapping is calm at $(0, 0, \bar{x}, \bar{y}, \bar{u})$, *for all* $\bar{u} \in \Lambda(\bar{x}, \bar{y})$

$$P(z, \vartheta) := \left\{ (x, y, u) \middle| \begin{bmatrix} G(x) \\ \mathscr{L}(x, y, u) \end{bmatrix} + z \in \mathbb{R}^k_- \times \{0_m\} \right\} \cap M(\vartheta).$$

Then, there exist $(\alpha, \beta, \gamma) \in \mathbb{R}^{k+p+m}$ *and* $\bar{u} \in \Lambda(\bar{x}, \bar{y})$ *such that the M-stationarity conditions (2.120), (2.121), (2.122), (2.123) and (2.124) are satisfied, with* $\| (\alpha, \gamma) \| \leq r$ *(for some* $r > 0$*).*

Proof ([30]). We organize the proof in three steps in order to simplify further reference.

Step 1. Consider the following values for ψ and Λ, respectively:

$$\psi(x, y) := [G(x), x, y, -\nabla_y f(x, y)], \text{ and } \Lambda := \mathbb{R}^k_- \times \text{gph}Q.$$

Then it follows from Proposition 2.1 that there exists $(\alpha, \gamma) \in \mathbb{R}^{k+m}$, with $\| (\alpha, \gamma) \| \leq r$ (for some $r > 0$) such that

$$0 \in \nabla F(\bar{x}, \bar{y}) + \begin{bmatrix} \nabla G(\bar{x})^\top \alpha + \nabla^2_{xy} f(\bar{x}, \bar{y})^\top \gamma \\ \nabla^2_{yy} f(\bar{x}, \bar{y})^\top \gamma \end{bmatrix} + D^* Q((\bar{x}, \bar{y})| - \nabla_y f(\bar{x}, \bar{y}))(\gamma). \tag{2.161}$$

provided Λ is closed and the set-valued mapping Ψ in (2.114) (with $\Omega := \mathbb{R} \times \mathbb{R}^m$, ψ and Λ given as in the beginning of this proof) is calm at $(0, \bar{x}, \bar{y})$. Obviously, the closeness of Λ is ensured by assumption 1, by a combination of Proposition 2.3 and Lemma 2.7. As for the calmness of Ψ, it is obtained by assumption 3. In fact the proof of the latter claim can be adapted from the proof of Theorem 4.3 in [97].

Step 2. Under assumption 1 and 2, an upper bound for the coderivative of Q at $(\bar{x}, \bar{y}, -\nabla_y f(\bar{x}, \bar{y}))$ is derived from Theorem 3.1 in [97]:

$$D^* Q((\bar{x}, \bar{y})| - \nabla_y f(\bar{x}, \bar{y}))(\gamma) \subseteq \bigcup_{\bar{u} \in \Lambda(\bar{x}, \bar{y})} \{ (\nabla(\nabla_y g(\bar{x}, \bar{y})^\top \bar{u}))^\top \gamma$$
$$+ \nabla g(\bar{x}, \bar{y})^\top D^* N_{\mathbb{R}^p_-}(g(\bar{x}, \bar{y}), \bar{u})(\nabla_y g(\bar{x}, \bar{y}) \gamma) \}.$$

Step 3. To conclude the proof, note that $\beta \in D^* N_{\mathbb{R}^p_-}(g(\bar{x}, \bar{y}), \bar{u})(\nabla_y g(\bar{x}, \bar{y}) \gamma)$ if and only if $(\beta, -\nabla_y g(\bar{x}, \bar{y}) \gamma) \in N_{\text{gph}N_{\mathbb{R}^p_-}}(g(\bar{x}, \bar{y}), \bar{u})$. Hence, the result follows by considering the equality in Proposition 2.4. □

In the case, constraint $G(x) \leq 0$ is included in ψ whereas it is part of Ω in [97]. The reason for this is to get a close link between CQ (2.127) and assumptions

2 and 3. Also, $\lambda(\bar{x}, \bar{y})$ is not a singleton as in [97]. To obtain this, at the place of assumption 1, it is required in [97] that $\nabla_y g(\bar{x}, \bar{y})$ have full rank. Finally, in contrary to Theorem 6.1 of [97], the multipliers α and γ are bounded in Theorem 2.21 by a known number, something which can be useful when constructing an algorithm for the bi-level program.

There are two equivalent ways to interpret the coderivative term in the right hand side of the inclusion in step 2 of the proof of Theorem 2.21. The first one used in [120] consists in writing it directly in terms of $g(\bar{x}, \bar{y}), \bar{u}$ and $\nabla_y g(\bar{x}, \bar{y})\gamma$. It should however be mentioned that in [120], g does not depend on the parameter x. The second one that we have used here consists in first computing the normal cone to the graph of $N_{\mathbb{R}^p}$ at $(g(\bar{x}, \bar{y}), \bar{u})$. Then, translating inclusion $(\beta, -\nabla_y g(\bar{x}, \bar{y})\gamma) \in N_{\mathrm{gph} N_{\mathbb{R}^p}}(g(\bar{x}, \bar{y}), \bar{u})$, directly leads to the M-stationarity conditions in the sense of Definition 2.27.

At first view, it is not apparent that the optimality conditions in [120] are in fact equivalent to the M-stationarity conditions in Definition 2.27. Moreover, a condition was later suggested in [52], in order to obtain S -type optimality conditions for an OPEC from the M-ones. By the way, let us mention that in the case of our problem, the S-stationarity conditions defined in [52] correspond to those of Definition 2.28. Hence, a CQ similar to the one suggested in [138] and called Partial MPEC LICQ can also lead from M- to S -type optimality conditions for OPEC. For problem (2.107), one can easily check that the Partial MPEC LICQ takes the form:

$$\left.\begin{array}{l} \nabla G(\bar{x})^\top \alpha + \nabla_x g(\bar{x}, \bar{y})^\top \beta + \nabla_x \mathcal{L}(\bar{x}, \bar{y}, \bar{u})^\top \gamma = 0 \\ \nabla_y g(\bar{x}, \bar{y})^\top \beta + \nabla_y \mathcal{L}(\bar{x}, \bar{y}, \bar{u})^\top \gamma = 0 \\ \beta_v = 0, \nabla_y g_\eta(\bar{x}, \bar{y})\gamma = 0 \end{array}\right\} \Rightarrow \beta_\mu = 0, \nabla_y g_\mu(\bar{x}, \bar{y})\gamma = 0.$$

For a more clear comparison between the approach in the previous section and the current one, one should note that assumption 2 and 3 of Theorem 2.21 are satisfied, provided that CQ (2.127) holds at $(\bar{x}, \bar{y}, \bar{u})$, for all $\bar{u} \in \Lambda(\bar{x}, \bar{y})$. This follows similarly as in Remark 2.8. Hence, the following corollary of the last theorem [30].

Corollary 2.3 ([30]). *Let $(\bar{x}, \bar{y})$ be a local optimal solution of problem (2.107). Assume that the following assertions hold:*

1. *CQ (2.143) holds at $(\bar{x}, \bar{y})$.*
2. *CQ (2.127) holds at $(\bar{x}, \bar{y}, \bar{u})$, for all $\bar{u} \in \Lambda(\bar{x}, \bar{y})$.*

Then, there exist $(\alpha, \beta, \gamma) \in \mathbb{R}^{k+p+m}$ and $\bar{u} \in \Lambda \bar{x}, \bar{y}$ such that the M-stationarity conditions (2.120), (2.121), (2.122), (2.123) and (2.124) are satisfied, with $\| (\alpha, \gamma) \| \leq r$ (for some $r > 0$).

If we neglect the bounds on the multipliers, the only difference between Corollaries 2.1 and 2.3 is that for the former, the M-stationarity conditions (2.120), (2.121), (2.122), (2.123) and (2.124) have to be satisfied for all $\bar{u} \in \Lambda(\bar{x}, \bar{y})$, whereas for the latter, they have to hold for some $u \in \Lambda(x, y)$. It cannot be otherwise in the case of Corollary 2.3, if one considers the inclusion in Step 2 of the proof of Theorem 2.21. This means that, for the bi-level optimization problem (2.91), if we adopt the definition of the M-stationarity of a local optimal solution as in Definition 2.27, the primal KKT reformulation leads us to weaker conditions. Hence, one could say that the gain we have with the primal KKT reformulation in terms of local optimal solution, is lost when considering M-stationarity. From the view point of CQs, obviously, the same effort (i.e. in terms of CQs) leads to M-type optimality conditions for the primal KKT reformulation, which are weaker than those obtained via the classical KKT reformulation.

Moreover, the classical KKT reformulation provides a much bigger flexibility in designing surrogates for the other optimality conditions known for MPCCs; see for example the S -type stationarity conditions obtained above. However, it may be very difficult to derive S-type optimality conditions for an OPEC. The reason for this is that the Fréchet normal cone does not have as good calculus rules as that of Mordukhovich.

For the last part of this section, we now introduce a concept of partial calmness/exact penalization for the primal KKT reformulation of the bi-level programming problem, that will allow us to substantially weaken CQ (2.127), while still being able to obtain the optimality conditions in Theorem 2.21. To bring the concept of partial calmness in Definition 2.26 to the primal KKT reformulation (2.111), one possibility is to observe that:

$$\psi(x,y) \in \Lambda \Leftrightarrow \rho(x,y) := d_\Lambda 0 \psi(x,y) = 0,$$

where $\Lambda := \mathrm{gph}Q, \psi(x,y) := (x, y, -\nabla_y f(x,y))$ and d_Λ denotes the distance function. Hence, the primal KKT reformulation can equivalently be written as:

$$\min F(x,y) \text{ subject to } (x,y) \in X \times \mathbb{R}^m, \rho(x,y) = 0. \tag{2.162}$$

Such a transformation for an OPEC has already been suggested in [50], but with no further details. We start by showing in the next result that problem (2.162) leads to the same optimality conditions as in Theorem 2.21. Second, we show that the latter reformulation of an OPEC induces new but not so fruitful ideas for CQs. $\bar{x} \in X$ will be said to be upper level regular if there exists no nonzero vector $\alpha \geq 0; \alpha^\top G(\bar{x}) = 0$ and $\nabla G(\bar{x})^\top \alpha = 0$.

Theorem 2.22 ([95] Corollary 4.6). *Let the functions $f, g : \mathbb{R}^{l_1} \to \mathbb{R}$ be locally Lipschitz continuous around $\bar{z}$. Then*

$$\partial(f+g)(\bar{z}) \subseteq \partial f(\bar{z}) + \partial g(\bar{z}).$$

Equality holds iff or g is continuously differentiable.

Theorem 2.23 ([93] Proposition 2.10). *Letf : $\mathbb{R}^{l_1} \to \mathbb{R}^{l_2}$ be Lipschitz continuous around $\bar{z}$, and $g : \mathbb{R}^{l_2} \to \bar{\mathbb{R}}$ Lipschitz continuous around $\bar{\omega} = f(\bar{z}) \in$ dom g. Then*

$$\partial(gof)(\bar{z}) \subseteq \bigcup [\partial < \omega^*, f(\bar{z}) >: \omega^* \in \partial g(\bar{\omega})].$$

Theorem 2.24 ([30]). *Let $(\bar{x}, \bar{y})$ be a local optimal solution of (2.162) (i.e. of problem (2.107)). Assume that the following assertions hold:*

1. *$\bar{x}$ is upper level regular,*
2. *assumption 1 and 2 in Theorem 2.21,*
3. *ρ-partial calmness at $(\bar{x}, \bar{y})$.*

Then, there exist $(\alpha, \beta, \gamma) \in \mathbb{R}^{k+P+m}$ and $\bar{u} \in \Lambda(\bar{x}, \bar{y})$ such that the M-stationarity conditions (2.120), (2.121), (2.122), (2.123) and (2.124) are satisfied, with $\| (\alpha, \gamma) \| \leq r$ (for some $r > 0$).

Proof ([30]). Under assumption 3, it follows from Theorem 2.13, that there exists $r_1 > 0$ such that $(\bar{x}, \bar{y})$ is a local optimal solution of

$$\min F(x, y) + r_1 \rho(x, y) \text{ subject to } G(x) \leq 0.$$

Hence, from Theorems 2.22 and 2.14, there exists α, with $\| \alpha \| \leq r_2$, for some $r_2 > 0$ such that (2.122) and the following inclusion hold

$$0 \in \nabla F(\bar{x}, \bar{y}) + (\nabla G(\bar{x})^\top \alpha, 0)^\top + r_1 \partial \rho(\bar{x}, \bar{y}). \tag{2.163}$$

Applying Theorem 2.23 to ρ, it follows that

$$\partial \rho(\bar{x}, \bar{y}) \subseteq \bigcup [\partial < u^*, \psi > (\bar{x}, \bar{y}), u^* \in \mathbb{B} \cap N_\Lambda(\psi(\bar{x}, \bar{y}))], \tag{2.164}$$

given that Λ is (locally) closed and hence, $\partial d_\Lambda(a) = \mathbb{B} \cap N_\Lambda(a)$; cf. [108] (Example 8.53). Thus, the combination of (2.163) and (2.164) implies that there exists $\gamma \in \mathbb{R}^m$, with $\| \gamma \| \leq r_1$, such that inclusion (2.161) holds.

The rest of the proof then follows as that of Theorem 2.21. In this case, r can be chosen as $r = \min r_1, r_2$. □

Remark 2.9 ([30]). The need of the partial calmness in order to have ρ (composition of ψ and the distance function on Λ) as an exact penalization term can be avoided by considering a classical result of Clarke [22] (Proposition 2.4.3), which amounts to saying that the distance function is automatically an exact penalty term provided the objective function (F in our case) is Lipschitz continuous. To proceed, one should observe that $\psi(x, y) \in \Lambda$ is also equivalent to $d_{\psi^{-1}(\Lambda)}(x, y) = 0$. Hence, $(\bar{x}, \bar{y})$ is a local optimal solution of problem (2.111) implies that, there exists $r > 0$ such that $(\bar{x}, \bar{y})$ is a local optimal solution of the problem to

$$\min F(x, y) + r d_{\psi^{-1}(\Lambda)}(x, y) \text{ subject to } (x, y) \in X \times \mathbb{R}^m,$$

without any CQ. In exchange though, computing the basic subdifferential of the distance function $d_{\psi^{-1}(\Lambda)}$ would then require an assumption closely related to the ρ-partial calmness, i.e. the calmness of a certain set-valued mapping.

Consider the set-valued mapping $\Psi(\vartheta) := \{(x, y) \in X \times \mathbb{R}^m | \psi(x, y) + \vartheta \in \Lambda\}$, where $\psi(x, y) := (x, y, -\nabla_y f(x, y))$ and $\Lambda :=$ gph$Q$. The next result helps to show that the ρ-partial calmness is closely related to the CQ in assumption 3 of Theorem 2.21.

Theorem 2.25 ([30]). *Let $(\bar{x}, \bar{y}) \in \Psi(0)$. Then, Ψ is calm at $(0, \bar{x}, \bar{y})$ if and only if the following set-valued mapping is calm at $(0, \bar{x}, \bar{y})$*

$$\tilde{\Psi}(t) := \{(x, y) \in X \times \mathbb{R}^m | \rho(x, y) \leq t\}. \tag{2.165}$$

This result established in [51], for the case where $X \times \mathbb{R}^m$ corresponds to a normed space, remains valid in our setting. Additionally, one can easily check that the calmness of $\tilde{\Psi}$ is equivalent to the calmness of a set-valued mapping obtained by replacing $\rho(x, y) \leq t$ in (2.165) by $\rho(x, y) + t \leq 0$. Hence, by the combination of Proposition 2.2 and Theorem 2.25, it is clear that the calmness of

$$\bar{P}(z, \vartheta) := \{(x, y, u) \in X \times \mathbb{R}^m \times \mathbb{R}^p | \mathscr{L}(x, y, u) + z = 0\} \cap M(\vartheta)$$

at $(0, 0, \bar{x}, \bar{y}, \bar{u})$, for all $\bar{u} \in \Lambda(\bar{x}, \bar{y})$, is a sufficient condition for the ρ-partial calmness to hold. Here, Mis defined as in assumption 2 of Theorem 2.21. Hence, a dual condition, similar to (and weaker than) CQ (2.127), sufficient for the ρ-partial calmness, can be stated.

Moreover, applying Proposition 2.2, the ρ-partial calmness also inspires a different kind of CQ in the dual form. In fact, it follows from Proposition 2.2, that a sufficient condition for the ρ-partial calmness to hold is that:

$$\partial\rho(\bar{x}, \bar{y}) \cap -N_{X \times \mathbb{R}^m}(\bar{x}, \bar{y}) = \emptyset. \tag{2.166}$$

Proposition 2.5 ([30]). *If equality holds in (2.164), then CQ (2.166) fails at $(\bar{x}, \bar{y})$.*

Proof. If equality holds in (2.164), then $0 \in \partial\rho(\bar{x}, \bar{y})$ since as a normal cone, $N_\Lambda(\psi(\bar{x}, \bar{y}))$ always contains the origin point. For the latter reason, we also have $0 \in N_{X \times \mathbb{R}^m}(\bar{x}, \bar{y})$. Hence, the result. □

This behavior of CQ (2.166) is close to that of a similar CQ considered in [29] for the optimal value reformulation (2.126), where instead of $\rho(x, y)$ one has $f(x, y) - \varphi(x)$. In the case of problem (2.126), the corresponding CQ was shown to automatically fail, provided the value function is locally Lipschitz continuous. Even though one can easily construct examples where equality holds in (2.164), the generalization of this fact would generally require the set Λ to be normally regular at $\psi(\bar{x}, \bar{y})$, which may not be easy to have given that Λ is the graph of a normal cone mapping. Moreover, if we assume X to be convex, passing to the boundary of $N_{X \times \mathbb{R}^m}$ generates a CQ, analogous to the one also considered in [29] for (2.126), that may have more chances to be satisfied [30]:

$$\partial\rho(\bar{x}, \bar{y}) \cap bdN_{X \times \mathbb{R}^m}(\bar{x}, \bar{y}) = \emptyset. \tag{2.167}$$

Considering the abstract nature of CQ (2.167) (which is weaker than CQ (2.166)), an immediate attempt to write it in terms of problem data, i.e. in a verifiable form, produces the following condition

$$\nabla g(\bar{x},\bar{y})^\top \beta + \nabla_{x,y}\mathcal{L}(\bar{x},\bar{y},\bar{u})^\top \gamma \notin bdN_{X\times\mathbb{R}^m}(\bar{x},\bar{y})\forall \bar{u} \in \Lambda(\bar{x},\bar{y}), \forall \gamma \in \mathbb{R}^m, \tag{2.168}$$

$$\forall \beta \in D^*N_{\mathbb{R}_p}(g(\bar{x},\bar{y}),\bar{u})(\nabla_y g(\bar{x},\bar{y})\gamma), \tag{2.169}$$

provided inclusion (2.164) holds, while noting that for $A, B, C \subseteq \mathbb{R}^l, A \subseteq B$ and $B \cap C = \emptyset$ imply $A \cap C = \emptyset$. It should however be mentioned that CQ (2.168) also fails in many cases, in particular, if $0 \in bdN_{X\times\mathbb{R}^m}(\bar{x},\bar{y})$ [30].

Numerical Example

In what follows, a numerical example is presented to illustrate the effectiveness of the proposed solution method to the linear RCECC. model.

Example 2.6. We consider the following random bi-level multiobjective linear programming problem, which has two objective functions of the leader and two objective functions of the follower:

$$\begin{cases} \min\limits_{x_1,x_2} F_1 = \xi_{11}x_1 + \xi_{12}x_2 + \xi_{13}y_1 + \xi_{14}y_2 \\ \min\limits_{x_1,x_2} F_2 = \xi_{21}x_1 + \xi_{22}x_2 + \xi_{23}y_1 + \xi_{24}y_2 \\ \text{s.t.} \begin{cases} \text{where } (y_1,y_2) \text{ solves} \\ \min\limits_{y_1,y_2} f_1 = \eta_{11}x_1 + \eta_{12}x_2 + \eta_{13}y_1 + \eta_{14}y_2 \\ \min\limits_{y_1,y_2} f_2 = \eta_{21}x_1 + \eta_{22}x_2 + \eta_{23}y_1 + \eta_{24}y_2 \\ \text{s.t.} \begin{cases} \zeta_1 x_1 + \zeta_2 x_2 + \zeta_3 y_1 + \zeta_4 y_2 \le 200 \\ 2x_1 - x_2 + 3y_1 + y_2 \le 100 \\ x_1, x_2, y_1, y_2 \ge 0, \end{cases} \end{cases} \end{cases} \tag{2.170}$$

where $\xi_{ij}, \eta_{ij}, \zeta_j (i = 1, 2; j = 1, 2, 4)$ are normal distributed random variables characterized as:

$$\xi_{11} \sim \mathcal{N}(6,2), \xi_{12} \sim \mathcal{N}(10,2), \xi_{13} \sim \mathcal{N}(8,2), \xi_{14} \sim \mathcal{N}(12,2),$$

$$\xi_{21} \sim \mathcal{N}(3,2), \xi_{22} \sim \mathcal{N}(-5,2), \xi_{23} \sim \mathcal{N}(7,2), \xi_{24} \sim \mathcal{N}(-9,2),$$

$$\eta_{11} \sim \mathcal{N}(12,1), \eta_{12} \sim \mathcal{N}(10,1), \eta_{13} \sim \mathcal{N}(8,1), \eta_{24} \sim \mathcal{N}(6,1),$$

$$\eta_1 \sim \mathcal{N}(12,1), \eta_2 \sim \mathcal{N}(5,1), \eta_3 \sim \mathcal{N}(18,1), \eta_4 \sim \mathcal{N}(20,1),$$

$$\zeta_1 \sim \mathcal{N}(6,1), \zeta_2 \sim \mathcal{N}(12,1), \zeta_3 \sim \mathcal{N}(18,1), \zeta_4 \sim \mathcal{N}(24,1).$$

If the confidence levels are given by the decision makers, the CECC model of (2.170) is:

$$\begin{cases} \min\limits_{x_1,x_2}\{\overline{F}_1,\overline{F}_2\} \\ \text{s.t.} \begin{cases} Pr\{\xi_{11}x_1+\xi_{12}x_2+\xi_{13}y_1+\xi_{14}y_2\leq \overline{F}_1\}\geq 0.8 \\ Pr\{\xi_{21}x_1+\xi_{22}x_2+\xi_{23}y_1+\xi_{24}y_2\leq \overline{F}_2\}\geq 0.8 \\ \text{where } (y_1,y_2) \text{ solves} \\ \min\limits_{y_1,y_2}\{\bar{f}_1,\bar{f}_2\} \min\limits_{y_1,y_2} f_1=\eta_{11}x_1+\eta_{12}x_2+\eta_{13}y_1+\eta_{14}y_2 \\ \min\limits_{y_1,y_2} f_2=\eta_{21}x_1+\eta_{22}x_2+\eta_{23}y_1+\eta_{24}y_2 \\ \text{s.t.} \begin{cases} Pr\{\eta_{11}x_1+\eta_{12}x_2+\eta_{13}y_1+\eta_{14}y_2\leq \bar{f}_1\}\geq 0.8 \\ Pr\{\eta_{21}x_1+\eta_{22}x_2+\eta_{23}y_1+\eta_{24}y_2\leq \bar{f}_2\}\geq 0.8 \\ Pr\{\zeta_1x_1+\zeta_2x_2+\zeta_3y_1+\zeta_4y_2\leq 200\}\geq 0.8 \\ 2x_1-x_2+3y_1+y_2\leq 100 \\ x_1,x_2,y_1,y_2\geq 0, \end{cases} \end{cases} \end{cases} \tag{2.171}$$

$$\begin{cases} \min\limits_{x_1,x_2}\{\overline{F}_1,\overline{F}_2\} \\ \text{s.t.} \begin{cases} \overline{F}_1\geq 16x_1+10x_2+8y_1+12y_2+0.84\sqrt{2(x_1^2+x_2^2+y_1^2+y_2^2)} \\ \overline{F}_1\geq 3x_1-5x_2+7y_1-9y_2+0.84\sqrt{2(x_1^2+x_2^2+y_1^2+y_2^2)} \\ \text{where } (y_1,y_2) \text{ solves} \\ \min\limits_{y_1,y_2}\{\bar{f}_1,\bar{f}_2\} \\ \text{s.t.} \begin{cases} \bar{f}_1\geq 12x_1+10x_2+8y_1+6y_2+0.84\sqrt{x_1^2+x_2^2+y_1^2+y_2^2} \\ \bar{f}_2\geq 12x_1+5x_2+18y_1+20y_2+0.84\sqrt{x_1^2+x_2^2+y_1^2+y_2^2} \\ 200\geq 6x_1+12x_2+18y_1+24y_2+0.84\sqrt{x_1^2+x_2^2+y_1^2+y_2^2} \\ 2x_1-x_2+3y_1+y_2\leq 100 \\ x_1,x_2,y_1,y_2\geq 0. \end{cases} \end{cases} \end{cases} \tag{2.172}$$

Using the KKT method mentioned above, the solutions are $(x_1^*,x_2^*,y_1^*,y_2^*) = (23.6,45.9,0,42.7)$, the corresponding objective function values are $(f_1^*,f_2^*,F_1^*,F_2^*) = (320.9,114.7,35.7,12.6)$.

2.4.3 Genetic Algorithm for Nonlinear Model

As (2.80) is a nonlinear model, i.e., some functions in model (2.80) is nonlinear, the randomness can not be converted into its crisp equivalent model. In order solve nonlinear (2.80), three issues are needed to solve: the trade-off of the multiple objectives, the random parameters and the bi-level structure. In this section, we design a hybrid method consisting of ε-constraint method, random simulations for critical values, and EBS-based genetic algorithm.

ε-Constraint Method

The ε-constraint method (also be called the reference objective method) was presented by Haimes et al. [19]. The main idea of the ε-constraint method is to select one primary objective $f_{i_0}(x)$ to be minimization (if the problem is to minimize the objectives) and other objectives are converted into constraints. If the multiobjective problem is

$$(P)\ \min_{x \in X}[f_1(\boldsymbol{x}), f_2(\boldsymbol{x}), \cdots, f_m(\boldsymbol{x})],$$

then the ε-constraint method for the problem may be denoted as follows:

$$(P_{i_0}(\varepsilon)) \begin{cases} \min f_{i_0}(\boldsymbol{x}) \\ \text{s.t.} \begin{cases} f_i(\boldsymbol{x}) \le \varepsilon_i,\ i = 1, 2, \cdots, m, i \ne i_0 \\ \boldsymbol{x} \in X, \end{cases} \end{cases}$$

where parameters $\varepsilon_i (i = 1, 2, \cdots, m, i \ne i_0)$ are predetermined by the decision maker, which represent the decision maker's tolerant threshold for objective i.

Remark 2.10. Consider the problem the maximization of multi-objectives

$$(P')\ \max_{x \in X}[f_1(\boldsymbol{x}), f_2(\boldsymbol{x}), \cdots, f_m(\boldsymbol{x})].$$

Then the ε-constraint method for the problem may be denoted as follows:

$$(P'_{i_0}(\varepsilon)) \begin{cases} \max f_{i_0}(\boldsymbol{x}) \\ \text{s.t.} \begin{cases} f_i(\boldsymbol{x}) \ge \varepsilon_i,\ i = 1, 2, \cdots, m, i \ne i_0 \\ \boldsymbol{x} \in X. \end{cases} \end{cases}$$

The relationship between the efficient solution to problem (P) and the optimal solution to problem $(P_{i_0}(\varepsilon))$ is given by the following Lemma.

Lemma 2.8 (Chakong and Haimes [19]).

1. *$\boldsymbol{x}^*$ is an efficient solution to problem (P) if and only if $\boldsymbol{x}^*$ solves $P_i(\varepsilon^*)$ for every $i = 1, 2, \cdots, m$.*
2. *If $\boldsymbol{x}^*$ solves $P_i(\varepsilon^*)$ for some i, and the solution is unique, then $\boldsymbol{x}^*$ is an efficient solution to problem (P).*

Lemma 2.9 (Li [79]). *Assume that x^* is an efficient solution of (P). Then there exists $\varepsilon_i (i = 1, 2, \cdots, m, i \neq i_0)$ such that x^* is a optimal solution of $(P_{i_0}(\varepsilon))$.*

For the ε-constraint method, there are two advantages as follows:

(1) Guaranteeing the i_0th optimization and considering other objectives simultaneously, which is preferred by the decision maker in practice decision making.
(2) The trade-off rate can be obtained by the Kuhn-Tuker operator of the point x^*, which help decision maker find a more preferred decision.

We have noted that how to decide the proper value of ε. If the value of every ε_i is too small, it is possible that problem $(P_{Pi_0}(\varepsilon))$ has no feasible solution. Conversely, If the value of ε_i is too large, the loss of the other objective f_i may be great. Usually, the analyst can provide $f_i^0 = \min_{x \in X} f_i(x) (i = 1, 2, \cdots, m)$ and objective values $[f_1(\boldsymbol{x}), f_2(\boldsymbol{x}), \cdots, f_m(\boldsymbol{x})]$ of some feasible solution x. Then, the decision maker determines the values of ε_i according to the experience and some specific requirements. For more details, please refer to [19].

Random Simulation for Critical Values

Suppose that ξ is an n-dimensional random vector defined on the probability space $(\Omega, \mathscr{A}, Pos)$, and $f : \mathbf{R}^n \to \mathbf{R}$ is a measurable function. The problem is to determine the maximal value $\bar{f}$ such that

$$Pr\{f(\xi) \geq \bar{f}\} \geq \alpha, \tag{2.173}$$

where α is a predetermined confidence level with $0 < \alpha < 1$. We generate ω_k from Ω according to the probability measure Pr and write $\xi_k = \xi(\omega_k)$ for $k = 1, 2, \cdots, N$. Now we define

$$h(\xi_k) = \begin{cases} 1, \text{ if } f(\xi_k) \geq \bar{f} \\ 0, \text{ otherwise,} \end{cases} \tag{2.174}$$

for $k = 1, 2, \cdots, N$, which are a sequence of random variables, and $E[h(\xi_k)] = \alpha$ for all k. By the strong law of large numbers, we obtain

$$\frac{\sum_{k=1}^{N} h(\xi_k)}{N} \to \alpha,$$

as $N \to \infty$. Note that the sum $\sum_{k=1}^{N} h(\xi_k)$ is just the number of ξ_k satisfying $f(\xi_k) \geq \bar{f}$ for $k = 1, 2, \cdots, N$. Thus the value $\bar{f}$ can be taken as the N'th largest element in the sequence $\{f(\xi_1), f(\xi_2), \cdots, f(\xi_N)\}$, where N' is the integer part of αN. The procedure of random simulation for CCM are summarized as follows:

Step 1. Set N' as the integer part of αN.
Step 2. Generate $\omega_1, \omega_2, \cdots, \omega_N$ from Ω according to the probability measure Pr.
Step 3. Return the N'th largest element in $\{f(\xi_1), f(\xi_2), \cdots, f(\xi_N)\}$.

Example 2.7. Let us employ the stochastic simulation to search for the maximal $\bar{f}$ such that

$$Pr\left\{\sqrt{\xi_1^2 + \xi_2^2 + \xi_3^2} \geq \bar{f}\right\} \geq 0.8,$$

where $\xi_1 \sim exp(1)$ is an exponentially distributed variable, $\xi_2 \sim \mathcal{N}(3, 1)$ is a normally distributed variable, and $\xi_3 \sim \mathcal{U}(0, 1)$ is a uniformly distributed variable. A run of stochastic simulation with 10000 cycles shows that $\bar{f} = 4.2086$.

EBS-Based GA

After $\varepsilon-$constraint method and random simulation technique are applied, we focus on the method to deal with bi-level structure. In this section, an improved GA, i.e., EBS-based GA, is proposed in the follows.

Introduced by Holland [54], a genetic algorithm is a metaheuristic belonging to the class of evolutionary approaches. Metaheuristics [45] are devised to address complex optimization problems for which conventional optimization methods are unable to be either effective or efficient. Among the various metaheuristics available, genetic algorithms are recognized as practical solution approaches for many real-world problems. Several factors have boosted the appeal of genetic algorithms as global optimization approaches: (1) The need for more sophisticated models makes it difficult or even impossible the use of traditional optimization techniques. These models typically result in complex optimization problems with many local optima and little inherent structure to guide the search. (2) Genetic algorithms are generally quite effective for rapid global search of large solution spaces. As a result, near-optimal solutions are likely to be attained in reasonable computation times. (3) Genetic algorithms operate on a pool of individuals, thus multiple solutions are suggested. (4) The search mechanism is intrinsically parallel, thus lending itself to a parallel implementation with the potential reduction in the computational requirement.

Genetic algorithms have also been successfully applied to several instances of bi-level programming [16, 49, 71, 89]. Coevolutionary algorithms consists in associating several evolutionary algorithms and applying transformations, such as mutation and cross-over, to distinct populations. A coevolution operator is then regularly applied between sub-populations to keep a global view on the whole problem. Oduguwa and Roy described Bi-level Genetic Algorithm (BiGA) [96], a coevolutionary algorithm to solve bi-level problems.

Procedure BiGA

Data: initial population pop;
$\text{pop}_l \leftarrow \text{selection}_l\mathit{ower}$ (pop);
$\text{pop}_u \leftarrow \text{selection}_u\mathit{pper}$ (pop);
Coevolution (pop_u, popu_l);
while Stopping criterion not met **do**;
    Crossover (pop_u), crossover (pop_l);
    Mutation (pop_u), mutation (pop_l);
    Evaluation (popu), evaluation (pop_l);
    Elitist coevolution (pop_u, pop_l);
    Evaluation (pop_u), evaluation (pop_l);
    Archiving (pop_u), archiving (pop_l);
end;
return archive

The genetic operators include selection, crossover and mutation. In the EBS-based GA, the entropy-Boltzmann selection is employed, and changes are made in crossover and mutation. The modified crossover and mutation operators are carefully designed for preserving constraints.

Entropy-Boltzmann Selection: The Boltzmann selection has been used successfully in a variety of water resources applications [33, 116] and represents an interesting mixture of concepts from EA literature with the basic sampling scheme used in SA [69]. To assure population variety, the entropy-Boltzmann selection mechanism [73] is employed in the EBS-based GA. Instead of utilizing an importance sampling method individually in the traditional Boltzmann selection, the entropy-Boltzmann selection adopts entropy sampling [75] and importance sampling [12] methods simultaneously in the Monte Carlo simulation to partially overcome premature convergence problems. Entropy sampling directs the evolution to select configurations of lower entropy. In this way, the rate of selection of rare configurations is higher than that of abundant ones. As a result, when a configuration falls into a local optimum and many same or similar configurations pile up, the rate of configurations acceptance concentrated near the local optimum is highly suppressed, which enables the system to escape easily from the local optimum [73]. Let $\mathbf{z}_d = (\mathbf{x}_d, \mathbf{y}_d, \lambda_d)$, then the probability of occurrence of a configuration $\mathbf{z}_d$ with energy $E(\mathbf{z}_d)$ has the form,

$$P_{eB}(\mathbf{z}_d) = Ce^{-J(E(\mathbf{z}_d))}, \tag{2.175}$$

where l represents the individual index of configuration, $P_{eB}(\mathbf{z}_d)$ is the probability of entropy-Boltzmann selection for $\mathbf{z}_d$, C is the normalization factor given by $C = (\sum_{\mathbf{z}_d} e^{-J(E(\mathbf{z}_d))})^{-1}$, $J(E(\mathbf{z}_d))$ is set to be $J(E(\mathbf{z}_d)) = S(E(\mathbf{z}_d)) + \beta E(\mathbf{z}_d)$, the entropy $S(E(\mathbf{z}_d))$ of the system having the energy $E(\mathbf{z}_d)$ is defined as $S(E(\mathbf{z}_d)) = k \ln \Lambda(E(\mathbf{z}_d))$, here $\Lambda(E(\mathbf{z}_d))$ is the number of configurations with the energy $E(\mathbf{z}_d)$, and $\beta = (kT_\tau)^{-1}$ with the inverse of temperature T_τ and the Boltzmann constant k, here τ is the iteration number. The Boltzmann constant k is set to $k = 1$ for convenience. In order to get a desired probability distribution for the case of the entropy-Boltzmann sampling, the Metropolis algorithm [87] is adopted to produce new configurations of a desired probability distribution.

The temperature, T_τ, which represents the amount of selective pressure in the algorithm and indirectly its skill in finding good solutions, is updated using the expression $T_{\tau+1} = T_\tau * \gamma$ at the end of τth iteration, with the algorithm starting with the highest temperature (i.e., T_0). There has been much work done on the choice of the initial constant temperature T_0 [81, 90]. However, it is still difficult to determine T_0 because it is dependent on the strategies used for different problems. In general, T_0 is the same magnitude as the objective function value and can be expressed as a function of $f_0^{\max}$ and $f_0^{\min}$ [58], which represent the maximum and minimum objective function values of the initial population, respectively. In this chapter, a carefully designed computational experiment was implemented based on a series of linear combinations of $f_0^{\max}$ and $f_0^{\min}$. Finally, it was found that T_0 could be chosen as $T_0 = f_0^{\max}$ because of its sufficient convergence. In order to guarantee a sufficiently slow search process, the cooling rate γ was set at $\gamma = 0.96$ which is close to 1, so that after I generations, the final temperature becomes $T_M = \gamma^I T_0$.

The corresponding fitness function is calculated as:

$$F(f_\tau(\mathbf{z}_d)) = e^{-\beta(f_\tau^{\max} - f_\tau(\mathbf{z}_d))}, \tag{2.176}$$

where $F(f_\tau(\mathbf{z}_d))$ is the fitness value of configuration $\mathbf{z}_d$ in the τth iteration, $f_\tau(\mathbf{z}_d)$ is the objective function of configuration $\mathbf{z}_d$ in the τth iteration, $f_\tau^{\max}$ represents the maximum objective function value of the τth population.

Crossover Operation with Three Segmental Crossover Probabilities: To fully search the feasible region for all possible solutions, the crossover operation is conducted in the three sections of the chromosomes respectively. Suppose the probability of the crossover operation in every section is respectively p_a, p_b, p_c, called segmental crossover probabilities. Three random numbers a, b, c are randomly generated from $(0, 1)$, and $a < P_a$, $b < P_b$, $c < P_c$. Suppose that $\mathbf{s}^1 = (\mathbf{x}^1, \mathbf{y}^1, \lambda^1)$ and $\mathbf{s}^2 = (\mathbf{x}^2, \mathbf{y}^2, \lambda^2)$ are selected as a pair of chromosomes that conduct the crossover operation, $\mathbf{s}^{1'}$ and $\mathbf{s}^{2'}$ are the offspring. The crossover operation is as follows:

$$\mathbf{s}^{1'} = a\begin{Bmatrix} \mathbf{x}^1 \\ 0 \\ 0 \end{Bmatrix} + (1-a)\begin{Bmatrix} \mathbf{x}^2 \\ 0 \\ 0 \end{Bmatrix} + b\begin{Bmatrix} 0 \\ \mathbf{y}^1 \\ 0 \end{Bmatrix} + (1-b)\begin{Bmatrix} 0 \\ \mathbf{y}^2 \\ 0 \end{Bmatrix} + c\begin{Bmatrix} 0 \\ 0 \\ \lambda^1 \end{Bmatrix}$$
$$+(1-c)\begin{Bmatrix} 0 \\ 0 \\ \lambda^2 \end{Bmatrix}, \tag{2.177}$$

$$\mathbf{s}^{2'} = (1-a)\begin{Bmatrix} \mathbf{x}^1 \\ 0 \\ 0 \end{Bmatrix} + a\begin{Bmatrix} \mathbf{x}^2 \\ 0 \\ 0 \end{Bmatrix} + (1-b)\begin{Bmatrix} 0 \\ \mathbf{y}^1 \\ 0 \end{Bmatrix} + b\begin{Bmatrix} 0 \\ \mathbf{y}^2 \\ 0 \end{Bmatrix} + (1-c)\begin{Bmatrix} 0 \\ 0 \\ \lambda^1 \end{Bmatrix}$$
$$+c\begin{Bmatrix} 0 \\ 0 \\ \lambda^2 \end{Bmatrix}. \tag{2.178}$$

Adaptive Mutation: The mutation probability is fixed for the entire optimization process for the domain in question, which results in local optimization and increases the search time. The adaptive mutation probability method advised by [82] is provided to improve the convergence rapidity of the genetic algorithm.

At the beginning of the mutation process, the mutation probability $p_m(\tau)$ is supposed to be a larger value to spur the individual mutation, whose fitness value is small. The $P_m(\tau)$ value is decreased to restrain the individual mutation to improve the computing speed and widen the searching scale when the result is near the optimum. On the other hand, if the mutation probability has the same value for all the solutions of the population, that is solutions with high fitness values as well as those with low fitness values are subjected to the same level of mutation, this will certainly deteriorate the performance of GAs. The adaptive strategy for updating mutation probability is computed as:

$$p_m^d(\tau) = \begin{cases} p_{m0}, & F(f_\tau(\mathbf{z}_d)) \geq F_\tau^a \\ p_{m0}\left(1 + \exp\left(\eta \frac{F_\tau^a - F(f_\tau(\mathbf{z}_d))}{F_\tau^a}\right)\exp(-\tau)\right), & \text{other,} \end{cases} \tag{2.179}$$

where $p_m^d(\tau)$ is the mutation probability of the dth individual at the τ iteration, F_τ^a is the average fitness value, p_{m0} is the initialization of mutation probability, and η is a constant.

Numerical Example

In this section, a numerical example is given to illustrate the effectiveness of the application of the models and algorithms proposed in above section.

Example 2.8. Consider the problem

$$\begin{cases} \max\limits_{x_1,x_2,x_3} F_1 = 2\xi_1^2 x_1 + \sqrt{\xi_2 x_2 + \xi_3 x_3} + \xi_4 y_1 + \sqrt{\xi_5} y_2 + \sqrt{\xi_6} y_3 \\ \max\limits_{x_1,x_2,x_3} F_2 = x_1 + x_2 + x_3 + y_1 + y_2 + y_3 \\ \text{s.t.} \begin{cases} \text{where } (y_1, y_2, y_3) \text{ solves} \\ \max\limits_{y_1,y_2,y_3} f_1 = \sqrt{7\xi_1 x_1 + 3\xi_2 x_2 + 8\xi_3 x_3} + (12\xi_4 y_1 + 11\xi_5 y_2 + 5\xi_6 y_3)^2 \\ \max\limits_{y_1,y_2,y_3} f_2 = x_1 + 5x_2 + 7x_3 + y_1 + 11y_2 - 8y_3 \\ \text{s.t.} \begin{cases} \sqrt{x_1\xi_1 + x_2\xi_2} + x_3^2\xi_3 + \xi_4 y_2 + 17 y_3 \xi_6 \leq 1540 \\ 3x_1 + 6x_2 + 12x_3 + 14y_1 + 19y_2 \leq 2400 \\ x_1 + x_2 + x_3 + y_1 + y_2 - y_3 \geq 150 \\ x_1 + x_2 + x_3 + y_1 + y_2 + y_3 \leq 200 \\ x_1, x_2, x_3, y_1, y_2, y_3 \geq 0, \end{cases} \end{cases} \end{cases} \tag{2.180}$$

where $\xi_j, j = 1, 2, \cdots, 6$ are random variables as

$$\xi_1 \sim \mathcal{N}(10, 2),\ \xi_2 \sim \mathcal{N}(8, 2),\ \ \xi_3 \sim \mathcal{N}(12, 2),$$
$$\xi_4 \sim \mathcal{N}(20, 2),\ \xi_5 \sim \mathcal{N}(16, 2),\ \xi_6 \sim \mathcal{N}(8, 2).$$

The CECC model of (2.180) is

$$\begin{cases} \max\limits_{x_1,x_2,x_3} \{\bar{F}_1, x_1 + x_2 + x_3 + y_1 + y_2 + y_3\} \\ \text{s.t.} \begin{cases} Pr\left\{\xi_1^2 x_1 + \sqrt{\xi_2 x_2 + \xi_3 x_3} + \xi_4 y_1 + \sqrt{\xi_5} y_2 + \sqrt{\xi_6} y_3 \geq \bar{F}_1\right\} \geq 0.8 \\ \text{where } (y_1, y_2, y_3) \text{ solves} \\ \max\limits_{y_1,y_2,y_3} [\bar{f}_1, x_1 + 5x_2 + 7x_3 + y_1 + 11y_2 - 8y_3] \\ \text{s.t.} \begin{cases} Pr\left\{\sqrt{7\xi_1 x_1 + 3\xi_2 x_2 + 8\xi_3 x_3} + (12\xi_4 y_1 + 11\xi_5 y_2 + 5\xi_6 y_3)^2 \geq \bar{f}_1\right\} \\ \geq 0.8 \\ Pr\{\sqrt{x_1\xi_1 + x_2\xi_2} + x_3^2\xi_3 + \xi_4 y_2 + 17 y_3 \xi_6 \leq 1400\} \geq 0.6 \\ 3x_1 + 6x_2 + 12x_3 + 14y_1 + 19y_2 \leq 2400 \\ x_1 + x_2 + x_3 + y_1 + y_2 - y_3 \geq 150 \\ x_1 + x_2 + x_3 + y_1 + y_2 + y_3 \leq 200 \\ x_1, x_2, x_3, y_1, y_2, y_3 \geq 0. \end{cases} \end{cases} \end{cases} \tag{2.181}$$

After running of the above algorithm, we obtain $(x_1^*, x_2^*, x_3^*, y_1^*, y_2^*, y_3^*) = (11.3, 6.5, 11.8, 20.3, 10.9, 25.2)$ and $(F_1^*, F_2^*) = (134.5, 155.4)$, $(f_1^*, f_2^*) = (103.4, 76.3)$.

2.5 Random DEDC Model

DEDC denotes that expected value operator is used to deal with the randomness in the upper-level constraints, dependent-chance operator is used to deal with the upper-level objectives and lower-level constraints. At the same time, chance constraint operator is used to deal with the randomness in the lower-level constraints.

2.5.1 General Form of Random DEDC Model

Let's consider the typical bi-level single objective model with random parameters, i.e. model (2.1). It follows from the DEDC operator that:

$$\begin{cases} \min\limits_{x \in R^{n_1}} Pr\{F(x, y, \xi) \leq \bar{F}\} \\ \text{s.t.} \begin{cases} E\{G_i(x, y, \xi) \leq 0\}, i = 1, 2, \cdots, q \\ \text{where } y \text{ solves:} \\ \begin{cases} \min\limits_{y \in R^{n_2}} Pr\{f(x, y, \xi) \leq \bar{f}\} \\ Pr\{g_j(x, y, \xi) \leq 0\} \geq \alpha_j, j = 1, 2, \cdots, p, \end{cases} \end{cases} \end{cases} \tag{2.182}$$

where α_j is the predetermined confidence levels in the lower-level model, $\bar{f}$ is a predetermined objective value. ξ is a random vector. Similarly, if decision makers want to maximize the objective value, then

$$\begin{cases} \max\limits_{x \in R^{n_1}} Pr\{F(x, y, \xi) \geq \bar{F}\} \\ \text{s.t.} \begin{cases} E\{G_i(x, y, \xi) \leq 0\}, i = 1, 2, \cdots, q \\ \text{where } y \text{ solves:} \\ \begin{cases} \max\limits_{y \in R^{n_2}} Pr\{f(x, y, \xi) \geq \bar{f}\} \\ Pr\{g_j(x, y, \xi) \leq 0\} \geq \theta_j, j = 1, 2, \cdots, p, \end{cases} \end{cases} \end{cases} \tag{2.183}$$

where θ_j is the predetermined confidence levels in the lower-level programming, ξ is a random vector.

Definition 2.30. A solution $x \in R^{n_1}$ is said to be the upper-level feasible solution of problem (2.182) if and only if $E\{G_i(x, y, \xi) \leq 0\}$ for all i. For any feasible x, if there is a solution x^* such that $\bar{F}^* < \bar{F}$, then x^* is called the upper-level optimal solution.

Definition 2.31. A solution $y \in R^{n_2}$ is said to be the lower-level feasible solution of problem (2.182) if and only if $Pr\{g_j(x, y, \xi) \leq 0\} \geq \theta_j$ for all j. For any feasible y, if there is a solution y^* such that $\bar{f}^* < \bar{f}$, then y^* is called the lower-level optimal solution.

Consider the bi-level multi-objective programming model (2.79), we present the bi-level multi-objective programming with DEDC operator as follows,

$$\begin{cases} \max\limits_{x\in R^{n_1}} \begin{bmatrix} Pr\{F_1(\boldsymbol{x},\boldsymbol{y},\boldsymbol{\xi}) \ge F_u\} \\ Pr\{F_2(\boldsymbol{x},\boldsymbol{y},\boldsymbol{\xi}) \ge F_u\} \\ \cdots \\ Pr\{F_u(\boldsymbol{x},\boldsymbol{y},\boldsymbol{\xi}) \ge F_u\} \end{bmatrix} \\ \text{s.t.} \begin{cases} E\{G_i(\boldsymbol{x},\boldsymbol{y},\boldsymbol{\xi}) \le 0\}, i = 1,2,\cdots,q \\ \text{where } y \text{ solves:} \\ \max\limits_{y\in R^{n_2}} \begin{bmatrix} Pr\{f_1(\boldsymbol{x},\boldsymbol{y},\boldsymbol{\xi}) \ge f_u\} \\ Pr\{f_2(\boldsymbol{x},\boldsymbol{y},\boldsymbol{\xi}) \ge f_u\} \\ \cdots \\ Pr\{f_u(\boldsymbol{x},\boldsymbol{y},\boldsymbol{\xi}) \ge f_u\} \end{bmatrix} \\ Pr\{g_j(\boldsymbol{x},\boldsymbol{y},\boldsymbol{\xi}) \le 0\} \ge \alpha_j, j = 1,2,\cdots,p, \end{cases} \end{cases} \quad (2.184)$$

where α_j is predetermined confidence levels in lower-level programming. If the objectives is to minimize the cost, then problem (2.184) should be formulated as follows,

$$\begin{cases} \min\limits_{x\in R^{n_1}} \begin{bmatrix} Pr\{F_1(\boldsymbol{x},\boldsymbol{y},\boldsymbol{\xi}) \le F_u\} \\ Pr\{F_2(\boldsymbol{x},\boldsymbol{y},\boldsymbol{\xi}) \le F_u\} \\ \cdots \\ Pr\{F_u(\boldsymbol{x},\boldsymbol{y},\boldsymbol{\xi}) \le F_u\} \end{bmatrix} \\ \text{s.t.} \begin{cases} E\{G_i(\boldsymbol{x},\boldsymbol{y},\boldsymbol{\xi}) \le 0\}, i = 1,2,\cdots,q \\ \text{where } y \text{ solves:} \\ \min\limits_{y\in R^{n_2}} \begin{bmatrix} Pr\{f_1(\boldsymbol{x},\boldsymbol{y},\boldsymbol{\xi}) \le f_u\} \\ Pr\{f_2(\boldsymbol{x},\boldsymbol{y},\boldsymbol{\xi}) \le f_u\} \\ \cdots \\ Pr\{f_l(\boldsymbol{x},\boldsymbol{y},\boldsymbol{\xi}) \le f_u\} \end{bmatrix} \\ Pr\{g_j(\boldsymbol{x},\boldsymbol{y},\boldsymbol{\xi}) \le 0\} \ge \theta_j, j = 1,2,\cdots,p, \end{cases} \end{cases} \quad (2.185)$$

where θ_j is predetermined confidence levels in lower-level programming.

Definition 2.32. Suppose for any given $\boldsymbol{y}$, a upper-level programming feasible solution $\boldsymbol{x}^*$ of problem (2.184) satisfies

$$E\{G_i(\boldsymbol{x}^*,\boldsymbol{y},\boldsymbol{\xi}) \le 0\}, i = 1,2,\cdots,q,$$

$\boldsymbol{x}^*$ is a random efficient solution to problem (2.184) if and only if there exists no other feasible solution $\boldsymbol{x}$ such that

$$E\{G_i(\boldsymbol{x},\boldsymbol{y},\boldsymbol{\xi}) \le 0\}, i = 1,2,\cdots,q,$$

$F_u(\boldsymbol{x}) \ge F_u(\boldsymbol{x}^*)$ for all u and $F_{u_0}(\boldsymbol{x}) > F_{u_0}(\boldsymbol{x}^*)$ for at least one $u_0 \in \{1,2,\cdots,m\}$.

Definition 2.33. Suppose for any given x, a lower-level programming feasible solution y^* of problem (2.184) satisfies

$$Pr\{g_j(x^*, y, \xi) \leq 0\} \geq \theta_j, j = 1, 2, \cdots, p,$$

where y^* is a random efficient solution to problem (2.184) if and only if there exists no other feasible solution y such that

$$Pr\{f_l(x, y, \xi) \geq f_l(x)\} \geq \beta_l, l = 1, 2, \cdots, r,$$

$f_l(y) \geq F_l(y^*)$ for all l and $F_{l_0}(y) > f_{l_0}(y^*)$ for at least one $l_0 \in \{1, 2, \cdots, r\}$.

2.5.2 Interactive Fuzzy Goal Programming for Linear Model

In this section, we concentrate on the DEDC version of the bi-level multi-objective linear programming problem with random coefficients as follows:

$$\begin{cases} \max\limits_{x \in R^{n_1}} [Pr\{\widetilde{C}_1^T x + \widetilde{D}_2^T y \geq \overline{F}_1\}, Pr\{\widetilde{C}_2^T x + \widetilde{D}_2^T y \geq \overline{F}_2\}, \cdots, Pr\{\widetilde{C}_{m_1}^T x + \widetilde{D}_{m_1}^T y \geq \overline{F}_{m_1}\}] \\ \text{s.t.} \begin{cases} E[\widetilde{A}_i^T x + \widetilde{B}_i^T y] \leq 0, i = 1, 2, \cdots, p_1 \\ \text{where } y \text{ solves:} \\ \max\limits_{y \in R^{n_2}} [Pr\{\tilde{c}_i^T x + \tilde{d}_1^T y \geq \bar{f}_1\}, Pr\{\tilde{c}_2^T x + \tilde{d}_2^T y \geq \bar{f}_2\}, \cdots, Pr\{\tilde{c}_{m_2}^T x + \tilde{d}_{m_2}^T y \geq \bar{f}_{m_2}\}] \\ \text{s.t.} Pr\{\tilde{a}_i^T x + \tilde{b}_i^T y \leq 0\} \geq \beta_i^L, i = 1, 2, \cdots, p_2, \end{cases} \end{cases} \tag{2.186}$$

where β_i^L are predetermined confidence levels by the follower, $\overline{F}_i$ and $\bar{f}_i$ are the predetermined objectives' values.

Remark 2.11. If the random variables in model (2.186) are normal distributed and independent with each other, then we can obtain its crisp equivalent model from Theorem 2.1 and Theorem 2.5.

Interactive Fuzzy Goal Programming Approach

As model (2.186) is converted into its crisp equivalent model, multiple objective technique is used to deal with objectives for both makers. Next, we focus on how to deal with the bi-level structure.

In this section, we use the Interactive fuzzy goal programming (FGP) approach proposed by Arora and Gupta [4]. The method has been successfully applied to many bi-level or multi-objective problems [119, 123].

In the bi-level programming problem under consideration, in order to arrive at a solution which is acceptable to both the decision makers they would have to cooperate with each other to make a balance of decision powers. For this they may compromise by giving a possible relaxation of their individual optimal decision. In such a case, the objective functions F_1 and F_2 and the decision vector X_1 are required to be transformed into fuzzy goals by means of assigning an imprecise aspiration level to each of them. The fuzzy goals are then characterized by the membership functions for achieving their respective aspired level.

Construction of Membership Functions

In a hierarchical decision making situation, since each of the decision maker's aim is to maximize his/her own objective function over the same feasible region, optimal solution for each of them, when calculated in isolation, can be considered as the aspiration level for the associated fuzzy goal. To build membership functions, goal and tolerances should be determined first. However, they could hardly be determined without the meaningful supporting data. We first obtain the individual best solution for each objective.

Let $(X_1^{h_1}, X_2^{h_1}, F_2^{h_1})$ and $(X_1^{l_1}, X_2^{l_1}, F_2^{l_1})$ be the best solutions when the leader and the follower independently solves the problems, i.e., $F_1^{h_1} = \max X_1, X_2 \in SF_1(X_1, X_2)$ and $F_2^{h_1} = \max X_1, X_2 \in SF_2(X_1, X_2)$, where S is the feasible region.

Then the fuzzy objective goals for the leader and the follower takes the form

$$F_1 \gtrsim F_1^{h_1}, F_2 \gtrsim F_2^{l_1}. \tag{2.187}$$

Similarly, the fuzzy goals for the control vectors of the leader appear as

$$X_1 \gtrsim X_1^{h_1}. \tag{2.188}$$

Now, in the decision making context, the above individual best solutions are usually different, because of the conflicting nature of the objectives at both the levels. However, in the hierarchical decision process, since execution of decision is sequential from the leader to the follower, the leader must give some relaxation to his decision for the benefit of the follower. So it can reasonably be assumed that the value $F_1^{h_2} = \min X_1, X_2 \in SF_1(X_1, X_2)$ be the lower tolerance limit of the fuzzy goals of the leader. Let the worst possible solution for the followerąŕs objective function be considered as the lower tolerance limit of the fuzzy goal of the follower, i.e. $F_2^{h_2} = \min X_1, X_2 \in SF_2(X_1, X_2)$. Again, the decision $X_1^{h_1}$ and a decision which gives the objective value higher than $F_1^{h_1}$ are absolutely acceptable to the leader. The leader knows that using the optimal decision $X_1^{h_1}$ as a control factor for the follower is not practical. It is more reasonable to have some tolerance that gives the follower a feasible region to search for his/her optimal solution and also reduce searching time or interaction.

The range of the decision variable should be "around $X_1^{h_1}$ with its maximum tolerance t_1".

The membership functions for the defined fuzzy goals can be expressed algebraically as:

$$\mu_{F_1}(F_1(X_1,X_2)) = \begin{cases} 1 & \text{if } F_1(X_1,X_2) \geq F_1^{h_1} \\ \frac{F_1(X_1,X_2)-F_1^{h_2}}{F_1^{h_1}-F_1^{h_2}} & \text{if } F_1^{h_2} \leq F_1(X_1,X_2) \geq F_1^{h_1} \\ 0 & \text{if } F_1(X_1,X_2) < F_1^{h_1}, \end{cases} \tag{2.189}$$

$$\mu_{F_2}(F_2(X_1,X_2)) = \begin{cases} 1 & \text{if } F_2(X_1,X_2) \geq F_2^{h_1} \\ \frac{F_2(X_1,X_2)-F_2^{h_2}}{F_2^{h_1}-F_2^{h_2}} & \text{if } F_2^{h_2} \leq F_2(X_1,X_2) \geq F_2^{h_1} \\ 0 & \text{if } F_2(X_1,X_2) < F_2^{h_1}, \end{cases} \tag{2.190}$$

$$\mu_{x_1} = \begin{cases} \frac{X_1-(X_1^{h_1}-t)}{t} & \text{if } X_1^{h_1}-t \leq X_1 \leq X_1^{h_1} \\ \frac{(X_1^{h_1}+t)-X_1}{t} & \text{if } X_1^{h_1} \leq X_1 \leq X_1^{h_1}+t \\ 0 & \text{otherwise}, \end{cases} \tag{2.191}$$

where $X_1^{h_1}$ is the most preferred solution, $X_1^{h_1}-t$ and $X_1^{h_1}+t$ are the worst acceptable decision and the satisfaction is linearly increasing within the interval $[X_1^{h_1}-t, X_1^{h_1}]$ and linearly decreasing within the interval $[X_1^{h_1}-t, X_1^{h_1}]$ and the other decisions are not acceptable. The membership function is a triangular fuzzy number.

The negative tolerance $X_1^{h_1}-t$ and positive tolerance $X_1^{h_1}+t$ are not necessarily same. X_1 generally lies between $X_1^{h_1}-t$ and $X_1^{h_1}+t$. But the decision maker may desire to shift the range of X1 altogether which may not be around $X_1^{h_1}$. The desired range can be shifted to the right or left of $X_1^{h_1}$ depending on the discretion of the decision maker.

Now, in a fuzzy decision making situation, it is worth mentioning that the achievement of highest membership value to the extent of a fuzzy goal is always desired by a decision maker. Regarding this aspect of BLPP, FGP approach seems to be the most appropriate to make a proper distribution of decision powers to the decision makers. The membership function of X_1 is shown in Fig. 2.7.

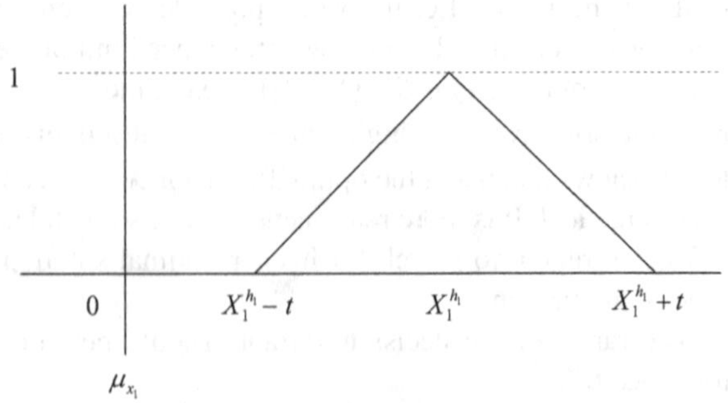

Fig. 2.7 Membership function of X_1

Formulation of FGP Model

The direct use of conventional goal programming (GP) to multi-objective fuzzy programming (FP) was first proposed by Mohamed [91] in 1997. Mohamed solution concept has been further extended by Pal et al. [102].

As a robust tool for solving MODM problems, preemptive based GP [59, 100] approach to non-linear FP has also been studied by Pal and Moitra [103].

In a fuzzy decision environment, achievement of a fuzzy goal to its aspired level means achievement of the associated membership function as 1, the decision makers' problem is actually to achieve the aspired level 1 of the membership goals to the extent possible by minimizing the under-deviational variable from it. However, the leader can specify their minimum satisfactory level say δ.

Now, for the defined membership functions in (2.189), (2.190) and (2.191), the FGP model can be presented as

$$\begin{cases} \min Z = \sum_{i=1}^{n_1} w_{1i} d_{1i}^{i} + w_2 d_2^- + w_3 d_3^- \\ s.t. \begin{cases} \frac{X_1 - X_1^t}{X_1^{h_1} - X_1^t} + d_1^- - d_1^+ = I_1 \\ \frac{F_1(X_1, X_2) - F_1^{h_2}}{F_1^{h_1} - F_1^{h_2}} + d_1^- - d_1^+ = 1 \\ \frac{F_2(X_1, X_2) - F_2^{h_2}}{F_2^{h_1} - F_2^{h_2}} + d_1^- - d_1^+ = 1 \\ X \in S \\ 0 \le \mu_{X_1}, \mu_{F_1}, \mu_{F_2} \le 1 \\ d_{1i}^-, d_{1i}^+ \ge 0, \text{ with } d_{1i}^- \cdot d_{1i}^+ = 0, i = 1, 2, \cdots, n_1 \\ d_i^-, d_i^+ \ge 0, \text{ with } d_i^- \cdot d_i^+ = 0, i = 2, 3, \end{cases} \end{cases} \tag{2.192}$$

where d_1^-, dt_1 represent the vectors of deviational variables associated with the membership goals for the decision vector X1, I1 is the column vector with all elements equal to 1 and the dimension of it depends on X_1 and $d_i^-, d_i^+ (i = 2, 3)$ represent under and over-deviational variables, respectively from the aspired levels of the respective fuzzy goals. w1i $(i = 1, 2, \cdots, n_1), w_2, w_3$ are numerical weights associated with the deviational variables $d_{1i}^- (i = 1, 2, \cdots, n_1), d_2^-$ and d_3^-, respectively.

The values of w_1, w_2 are determined as

$$w_1 = \frac{1}{X_1^{h_1} - X_1^t}, \quad w_2 = \frac{1}{F_1^{h_1} - F_1^{h_2}}, \quad w_3 = \frac{F_2^{l_1}}{F_2^{l_2}}. \tag{2.193}$$

To formulate the decision making model of the problem (2.192), the problem under consideration can be divided into two phases: (i) at the first phase the feasible region of the state variables are determined. (ii) at the second phase the attainment of the objective goals of both the decision makers and the decision vector controlled by the leader are obtained by performing recursive analysis.

The First Phase Problem

The attainment of the objective goals to their target levels depend on $X \in S$, where the feasible region S corresponds to the structural constraints. Since the membership functions represent the degree of satisfaction, so we introduce the additional constraints $0 \leq \mu_{X_1}, \mu_{F_1}, \mu_{F_2} \leq 1$ Then the feasible region of the corresponding problem is called the extended feasible region.

The phase I method can be used to find all the extreme points. The optimal solution of the problem (2.192) may correspond to any one of the extreme point or nearer to any of them in the extended feasible region S. Due to the convexity property of the stage returns in dynamic programming (DP) [100], the feasible region of the decision variables associated with the respective stages can be found as follows.

Let $X^v(= x_1^v, x_2^v, \cdots, x_n^v)$ be the extreme point solution corresponding to the mth iteration step in solving the problem. The convex combination of all such extreme point solutions can be presented as $\sum_{v=1}^{V} \lambda_v X^v \in S$, where $\sum_{v=1}^{V} \lambda_v, 0 \leq \lambda_v \leq 1$ and λ_v are scalars.

Let $X^w (w = 1, \cdots, W), W \geq V$, be the different solutions obtained for different sets of values of $\lambda_v (v = 1, 2, \cdots, V)$ in the augmented feasible region S, which may then be called the effective set of solutions towards achieving the respective target levels of the objective goals. Then, the feasible regions of the decision variables appear as $X_j = (x_j^1, x_j^2, \cdots, x_j^W) \in S, X_j \neq \emptyset$ and $j = 1, 2, \cdots, n$.

Now, the feasible regions of the state variables can be determined as

$$y_{j1,i} = T_{ji}(y_{ji}, x_j), j = 1, 2, \cdots, n; i = 1, 2, \cdots, m, \tag{2.194}$$

where y_{ji} and T_{ji} are the jth state variable and jth state transformations respectively, which corresponds to the ith system constraint.

Then for the defined constraint in (2.194), the problem in (2.192) can be transformed into equivalent problem involving m state variables.

The Second Phase Problem

Now, in making stage-wise decision on the basis of the assigned weight to the membership goals of the problem (2.194), the DP formulation can be presented as: Evaluate $F_j = \min_{d_j^-} Q_j(y_{ji}, d_j^-)$ where

$$Q_j(y_{ji}, d_j^-) = R_j(y_{ji}, d_j^-), j = 1; Q_j(y_{ji}, d_j^-) = R_j(y_{ji}, d_j^-) + F_{j-1}(y_{j-1,i}),$$
$$j = 2, 3, \cdots, n; \tag{2.195}$$

subject to the membership goals in (2.192) and $y_{j-1,i} = T_{ji}(y_{ji}, x_j), x_j \in X_j, x_j \geq 0$ and $X_j \neq \emptyset$ where $F_j(\cdot)$ is the optimal value of the achievement function composed of under-deviational variables and $R_j(\cdot)$ is the return function at stage j.

The problem in (2.195) can be solved recursively. If both the decision makers are satisfied by the achieved solution (X^*), then it is taken as the solution of the system. However, the decision makers are not always satisfied by the solution obtained. Then the ratio of satisfactory degree between both the levels

$$\Delta = \frac{\mu_{F_2}(X^*)}{\mu_{F_1}(X^*)},$$

which is defined by Lai [72] is useful. If $\Delta > 1$, i.e. $\mu_{F_2}(X^*) > \mu_{F_1}(X^*)$, then the follower updates the aspiration level δ_{F_2} by decreasing the value of δ_{F_2}. Receiving the updated level $\hat{\delta}_{F_2}$ the problem (2.195) is solved with $\hat{\delta}_{F_2}$ and then the leader obtains a larger satisfactory degree and the follower accepts a smaller satisfactory degree. Conversely, if $\Delta < 1$, i.e. $\mu_{F_2}(X^*) < \mu_{F_1}(X^*)$, then the leader updates the aspiration level δ_{F_1} or δ_{X_1} or both by decreasing their values and the leader accepts a smaller satisfactory degree and the follower obtains a larger satisfactory degree.

At an iteration k, let $\mu_{F_i}^k, i = 1, 2$ denote satisfactory degrees of decision maker $i, i = 1, 2$ and let

$$\Delta^k = \frac{\mu_{F_2}^k(X^*)}{\mu_{F_1}^k(X^*)},$$

denote the satisfactory degrees of the upper and the lower levels. Let a corresponding solution be X^{*k}. When the ratio D^k of satisfactory degrees is in the closed interval, the lower and the upper bounds specified by the leader, then the leader concludes the solution as a satisfactory solution and the iterative interactive process terminates.

Termination condition of the interactive process for bi-level programming problems

The ratio D^k of satisfactory degrees is in the closed interval, the lower and the upper bounds specified by the leader.

This condition is provided in order to keep overall satisfactory balance between both the levels.

Procedure for Updating the Minimal Satisfactory Level δ

If the ratio Δ_k exceeds its upper bound, the follower decreases its minimal satisfactory level. Conversely, if the ratio Δ_k is below its lower bound, then the leader decreases its minimal satisfactory level.

The problem in (2.195) can be solved recursively. If both the leader and the follower are satisfied by the achieved solution, then it is taken as the solution of the system. But if they are not satisfied then either or both of them can change their aspiration levels and again solve the problem. The process continues until the satisfactory solution is attained.

Algorithm and Flowchart

The outline of the above procedure is summarized in the following algorithm:

Step 0. (Pre-process) Leader specifies the lower and the upper bounds of the ratio of satisfactory degrees Δ.

Step 1. Formulation of FGP model (1a) Determine the fuzzy goals for the objective functions of the leader and the follower and the decision variable controlled by the leader as in (2.187) and (2.188). (1b) Construct the membership functions for the objective functions of both the decision makers and the decision vector controlled by the leader by determining the goal and tolerances as given in (2.189), (2.190) and (2.191). (1c) Determine the weights associated with underdeviationals variables as in (2.193). (1d) Set the aspiration level for the objective functions of the leader and follower as $\delta_{F_1}, \delta_{F_2}$. Also set the aspiration level vector for the decision vector X1 of the leader as unity vector i.e. $\delta_{X_1} = I_1$. (1e) By incorporating the concept of membership functions and weights, formulate the FGP model of the BLPP as in (2.192).

Step 2. Solve by DP (2a) Find the feasible regions of the decision variables x_j and then the feasible regions of the state variables y_{ji} for $j = 1, \cdots, n; i = 1, \cdots, m$. Set $k = 1$. (2b) Solve the transformed problem recursively by using DP.

Step 3. Termination process If the solution obtained satisfies the termination condition, the leader concludes the solution as a satisfactory solution and the algorithm stops. Otherwise $k = k + 1$.

Step 4. Updating aspiration level The decision makers updates the satisfactory level or aspiration levels in accordance with the procedure of updating minimal satisfactory level. Go to step 2(b).

The flow chart of the algorithm is as follows (see Fig. 2.8).

Numerical Example

In this section, a numerical example is given to illustrate the effectiveness of the application of the models and algorithms proposed earlier.

Example 2.9. Consider the problem

$$\begin{cases} \max\limits_{x_1,x_2} F_1 = 2\xi_1 x_1 + \xi_2 x_2 + 2\xi_3 y_1 + \xi_4 y_2 \\ \max\limits_{x_1,x_2} F_2 = \xi_1 x_1 + 2\xi_2 x_2 + 2\xi_3 y_1 + 4\xi_4 y_2 \\ \text{s.t.} \begin{cases} \text{where } (y_1, y_2) \text{ solves} \\ \max\limits_{y_1,y_2} f_1 = \xi_1 x_1 + 2\xi_2 x_2 + \xi_3 y_1 + 2\xi_4 y_2 \\ \max\limits_{y_1,y_2} f_2 = 2\xi_1 x_1 + \xi_2 x_2 + 2\xi_3 y_1 + 2\xi_4 y_2 \\ \text{s.t.} \begin{cases} x_1 + x_2 + y_1 + y_2 \geq 100 \\ x_1 + x_2 + y_1 + y_2 \leq 200 \\ x_1, x_2, y_1, y_2 \geq 0, \end{cases} \end{cases} \end{cases} \tag{2.196}$$

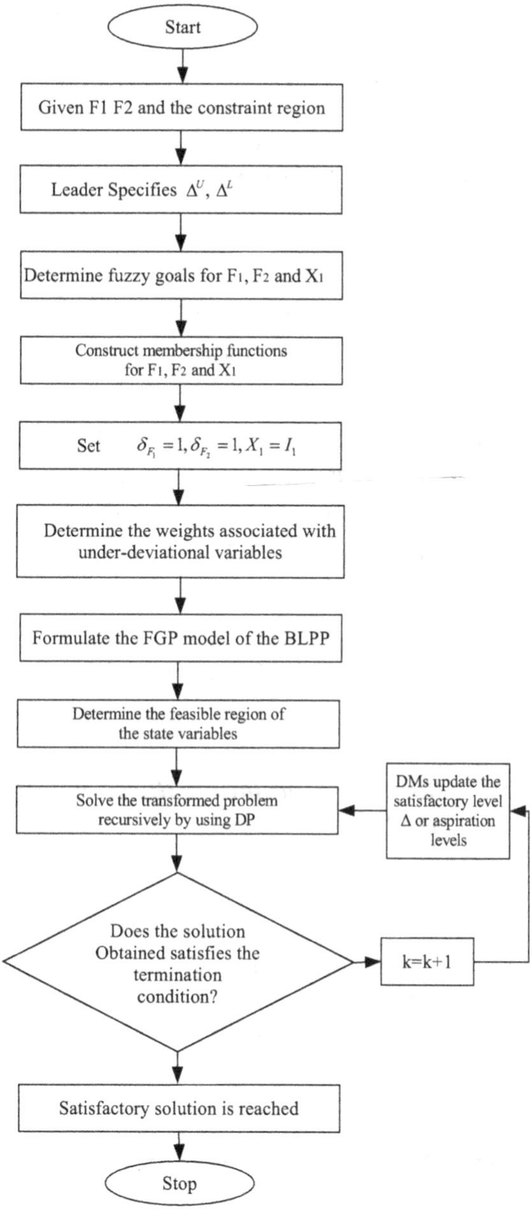

Fig. 2.8 Flow chart of the algorithm

where $\xi_j (j = 1, 2, 3, 4)$ are random variables characterized as

$$\xi_1 \sim \mathcal{N}(4, 1), \xi_2 \sim \mathcal{N}(5, 1), \xi_3 \sim \mathcal{N}(6, 1), \xi_4 \sim \mathcal{N}(7, 1).$$

In order to solve it, we use the chance operator to deal with rough objectives and rough constraints, and then we can obtain the DEDC

$$\begin{cases} \max\limits_{x_1, x_2} [\alpha_1, \alpha_2] \\ \text{s.t.} \begin{cases} Pr\{2\xi_1 x_1 + \xi_2 x_2 + 2\xi_3 y_1 + \xi_4 y_2 \geq 1500\} \geq \alpha_1 \\ Pr\{\xi_1 x_1 + 2\xi_2 x_2 + 2\xi_3 y_1 + 4\xi_4 y_2 \geq 3000\} \geq \alpha_2 \\ \text{where } (y_1, y_2) \text{ solves} \\ \max\limits_{y_1, y_2} [\beta_1, \beta_2] \\ \text{s.t.} \begin{cases} Pr\{\xi_1 x_1 + 2\xi_2 x_2 + \xi_3 y_1 + 2\xi_4 y_2 \geq 1800\} \geq \beta_1 \\ Pr\{2\xi_1 x_1 + \xi_2 x_2 + 2\xi_3 y_1 + 2\xi_4 y_2 \geq 1800\} \geq \beta_2 \\ x_1 + x_2 + y_1 + y_2 \geq 100 \\ x_1 + x_2 + y_1 + y_2 \leq 200 \\ x_1, x_2, y_1, y_2 \geq 0. \end{cases} \end{cases} \end{cases} \tag{2.197}$$

After running of interactive method, we obtain $(x_1^*, x_2^*, y_1^*, y_2^*) = (30.3, 32.4, 31.5, 83.8)$ and $(\alpha_1^*, \alpha_2^*) = (0.72, 0.93)$, $(\beta_1^*, \beta_2^*) = (0.69, 0.89)$.

2.5.3 Random Simulation-Based BPMOGA for Nonlinear Model

As (2.184) is a nonlinear model, i.e., some functions in model (2.184) is nonlinear, the randomness can not be converted into its crisp equivalent model. To solve nonlinear (2.184), three issues are needed to solve: the trade-off of the multiple objectives, the random parameters and the bi-level structure. In this section, we design a hybrid method consisting of two-stage method, random simulations for values, and BPMOGA.

Random Simulation for Probability

Let ξ be an n-dimensional random vector defined on the probability space $(\Omega, \mathcal{A}, Pr)$, and $f : \mathbf{R}^n \to \mathbf{R}$ a measurable function. In order to obtain the probability

$$L = Pr\{f(\xi) \leq 0\}. \tag{2.198}$$

We generate ω_k from Ω according to the probability measure Pr, and write $\xi_k = \xi(\omega_k)$ for $k = 1, 2, \cdots, N$. Let N' denote the number of occasions on which $f(\xi_k) \leq 0$ for $k = 1, 2, \cdots, N$ (i.e., the number of random vectors satisfying the system of inequalities). Let us define

$$h(\xi_k) = \begin{cases} 1, & \text{if } f(\xi_k) \leq 0 \\ 0, & \text{otherwise.} \end{cases}$$

Then we have $E[h(\xi_k)] = L$ for all k, and $N' = \sum_{k=1}^{N} h(\xi_k)$. It follows from the strong law of large numbers that

$$\frac{N'}{N} = \frac{\sum\limits_{k=1}^{N} h(\xi_k)}{N},$$

converges to L. Thus the probability L can be estimated by N'/N provided that N is sufficiently large. The process of random simulation 3 for probability can be summarized as follows:

Step 1. Set $N' = 0$.
Step 2. Generate ω from Ω according to the probability measure Pr.
Step 3. If $f(\xi(\omega)) \leq 0$, then $N' + +$.
Step 4. Repeat the second and third steps N times.
Step 5. Return $L = N'/N$.

Example 2.10. Let $\xi_1 \sim exp(6)$ be an exponentially distributed variable, $\xi_2 \sim \mathcal{N}(4, 1)$ a normally distributed variable, and $\xi_3 \sim \mathcal{U}(0, 2)$ a uniformly distributed variable. A run of stochastic simulation with 10000 cycles shows that

$$Pr\left\{ \sqrt{\xi_1^2 + \xi_2^2 + \xi_3^2} \leq 8 \right\} = 0.6749.$$

Two-Stage Method

In this section, we will use the two-stage method to seek the efficient solution to the crisp multiobjective decision-making problem

$$\begin{cases} \max[H_1(\boldsymbol{x}), H_2(\boldsymbol{x}), \cdots, H_m(\boldsymbol{x})] \\ \text{s.t. } \boldsymbol{x} \in X. \end{cases} \tag{2.199}$$

The two-stage method is proposed by Li [80] on the basis of the maximin method proposed by Zimmermann [140].

Step 1. Apply Zimmermann's minimum operator to obtain the maximal satisfying degree α^0 of the objective set and the related feasible solution x^0, that is,

$$\begin{cases} \max \alpha \\ \text{s.t.} \begin{cases} \mu_k(\boldsymbol{x}) = \frac{H_k(\boldsymbol{x})-H_k'}{H_k^*-H_k'} \geq \alpha, k = 1, 2, \cdots, m \\ \boldsymbol{x} \in X. \end{cases} \end{cases} \tag{2.200}$$

Assume that the optimal solution to problem (2.200) is $(\boldsymbol{x}^0, \alpha^0)$, where α^0 is the optimal satisfying degree of the whole objective sets. If the optimal solution to problem (2.200) is unique, $\boldsymbol{x}^0$ is the efficient solution to problem (2.199). However, as we cannot usually know if the optimal solution to problem (2.200) is unique, then the efficiency of $\boldsymbol{x}^0$ must be checked by the following stage.

Step 2. Check the efficiency of $\boldsymbol{x}^0$ or seek the new efficient solution $\boldsymbol{x}^1$. Construct a new model whose objective function is to maximize the average satisfying degree of all objects subject to the additional constraint $\alpha_k \geq \alpha^0 (k = 1, 2, \cdots, m)$. Since the compensatory of the arithmetic mean operator, the solution obtained in the second stage is efficient. The existence of the constraint $\alpha_k \geq \alpha^0 (k = 1, 2, \cdots, m)$ guarantees the mutual equilibrium of every objective function.

$$\begin{cases} \max \frac{1}{m} \sum_{k=1}^{m} \alpha_k \\ \text{s.t.} \begin{cases} \alpha^0 \leq \alpha_k \leq \mu_k(x), k = 1, 2, \cdots, m \\ 0 \leq \alpha_k \leq 1 \\ x \in X. \end{cases} \end{cases} \tag{2.201}$$

Assume that the optimal solution to problem (2.201) is $\boldsymbol{x}^1$. It is easy to prove that $\boldsymbol{x}^1$ is also the solution to problem (2.200), thus we have $\boldsymbol{x}^1 = \boldsymbol{x}^0$ if the solution to problem (2.200) is unique. However, if the solution to problem (2.200) is not unique, $\boldsymbol{x}^0$ may or may not be an efficient solution, so we need to guarantee that $\boldsymbol{x}^1$ is definitely efficient. In any case, the two-stage method can find an efficient solution in the second stage.

BPMOGA

The bi-level structure and the multiple objectives greatly increase computing complexity and solution speed. In the bi-level modeling, the lower-level modeling acts as a constraint on the upper-level modeling. GAs are known to offer significant advantages over conventional algorithmic approaches, as they include population-wide searches, a continuous balance between convergence and diversity, and the principles of a building-block combination [139]. Because of the difficulty in balancing exploitation and exploration in the search space, the design strategy for the GA parameters: population size (*Pop*), crossover rate (p_c), and mutation rate (p_m); is one of the most critical issues. In this section we use a fuzzy logic controller (FLC) to dynamically tune the crossover rate and mutation rate during the optimization process. A bi-level decision procedure (BP) is also necessary to deal with the bi-level modeling. Therefore, in this section, a bi-level multi-objective genetic algorithm (BPMOGA) with FLC is proposed to solve the bi-level multi-objective optimization modeling.

Bi-level decision procedure. BP technology allows for a hierarchical structure of two levels. The basic theory behind BP is that leader makes decisions to fulfill their goals, which requires the follower to supply the optimal solutions in isolation. Then, the follower's decisions are amended subject to the global benefits. The process is continued until a satisfactory solution is obtained. Therefore, we can say that the BP is an order of two optimization problems, in which the constraints of the leader are decided by the solutions of the follower. In the BPMOGA, the bi-level multi-objective model is solved by interactively applying the upper-level GA and the lower-level GA.

Individual representation. For the GA, the decision variables can be coded as chromosomes. The gene representation can be sequence-based encoding, 0-1 encoding, Gray encoding or hybrid representation according to the nature of the decision variables.

Figure 2.9 shows a hybrid individual representation for a bi-level facility layout and hazardous-material transportation problem. The representation is combined using sequence-based encoding in the upper-level and a 0-1 representation for the lower-level programming. For example, on the upper level, all facilities are in accordance with an order marked as $(1, 2, 3, \cdots, M)$ and are established as a sequence S, where i denotes facility x is allocated to location i, $i \in \{1, N\}$. Different sequences lead to different layout solutions and any changes in sequence S correspond to a new layout plan.

Sequence crossover operator. The sequence crossover operator [121] is different from other order crossover approaches. In this approach, to generate an offspring, some positions are stochastically selected from one parent, while the same elements of the first parent are imposed on the sequence of movements in the homologous positions in the other parent. Assume that positions $7, 6, , 3$ have been selected for the operation. Offspring inherits elements $7, 6, 3$ from Parent$_1$. Offspring $=$ $(, , 7, 6, , 3, , , ,)$ and in the positions $7, 6, , 3$ in Parent$_2$, the elements are $4, 6, , 9$ with the positions in Parent$_1$ being $4, 6, , 9$. So Offspring $=$ $(4, 7, 6, , 3, , 9)$. The other elements are then copied from Parent$_1$ and the Offspring becomes $(3, 2, 7, 1, 5, 6, 8, 9, 4)$, as shown in Fig. 2.10.

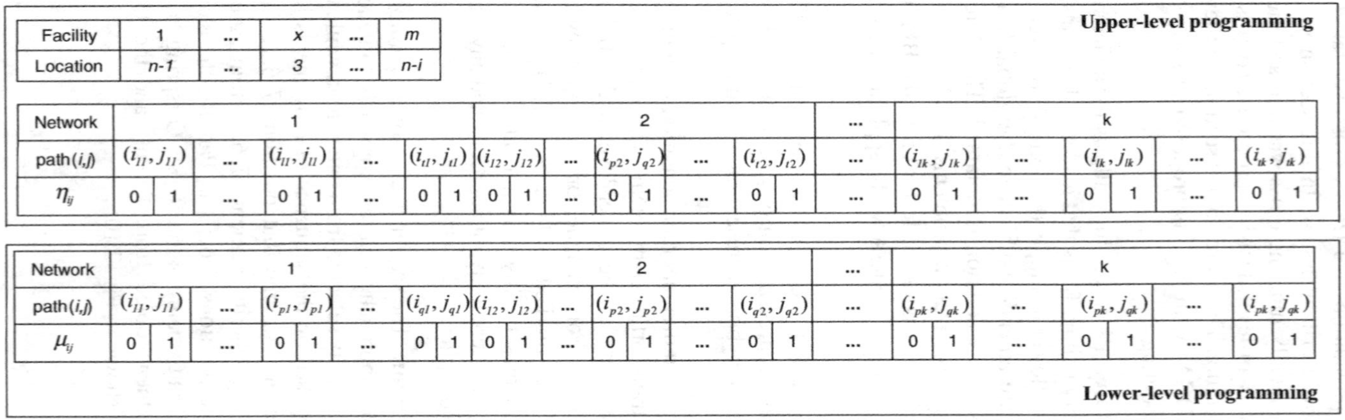
Upper-level programming
Facility | 1 | ... | x | ... | m
Location | n-1 | ... | 3 | ... | n-i
Network | 1 | 2 | ... | k
path(i,j) | (i_{11}, j_{11}) | ... | (i_{l1}, j_{l1}) | ... | (i_{t1}, j_{t1}) | (i_{12}, j_{12}) | ... | (i_{p2}, j_{q2}) | ... | (i_{t2}, j_{t2}) | ... | (i_{1k}, j_{1k}) | ... | (i_{lk}, j_{lk}) | ... | (i_{tk}, j_{tk})
η_{ij} | 0 | 1 | ... | 0 | 1 | ... | 0 | 1 | 0 | 1 | ... | 0 | 1 | ... | 0 | 1 | ... | 0 | 1 | ... | 0 | 1 | ... | 0 | 1
Network | 1 | 2 | ... | k
path(i,j) | (i_{11}, j_{11}) | ... | (i_{p1}, j_{p1}) | ... | (i_{q1}, j_{q1}) | (i_{12}, j_{12}) | ... | (i_{p2}, j_{p2}) | ... | (i_{q2}, j_{q2}) | ... | (i_{pk}, j_{qk}) | ... | (i_{pk}, j_{qk}) | ... | (i_{pk}, j_{qk})
μ_{ij} | 0 | 1 | ... | 0 | 1 | ... | 0 | 1 | 0 | 1 | ... | 0 | 1 | ... | 0 | 1 | ... | 0 | 1 | ... | 0 | 1 | ... | 0 | 1
Lower-level programming

Fig. 2.9 Hybrid individual representation for the bi-level facility layout and hazardous-material transportation problem

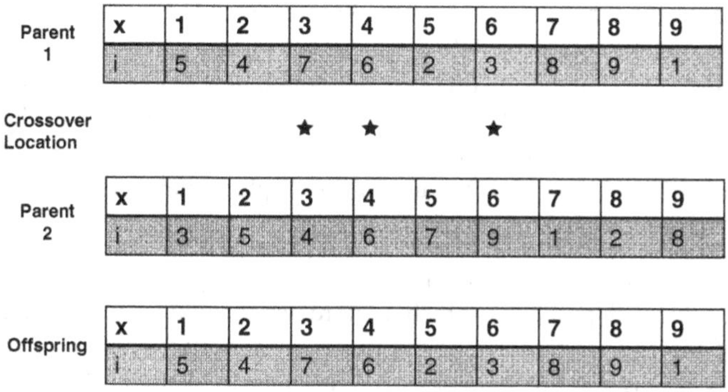

Fig. 2.10 Sequence crossover operator

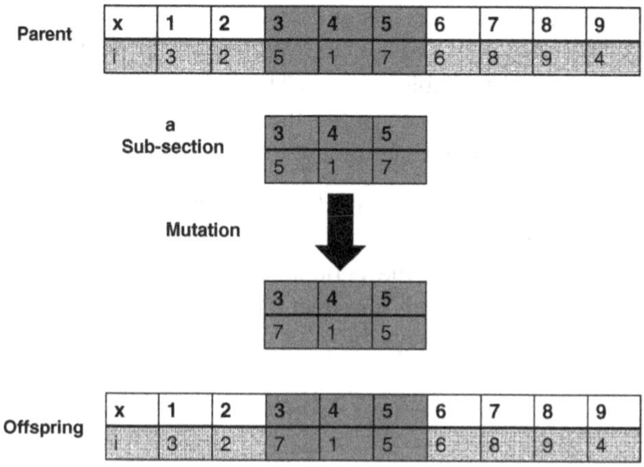

Fig. 2.11 Scramble sub-list mutation operator

Scramble sub-list mutation operator. A mutation operator called a scramble sub-list mutation [43] is included in the sequence based GA. In this method, a sub-list of the section from a parent chromosome is selected and the operator alters these in the offspring, leaving the remainder of the chromosome as it was in the parent. There are no parameters in a scramble sub-list mutation; however, the length of the chromosome section to be scrambled can be limited. An example is shown in Fig. 2.11, in which a subsection containing $5, 1, 7$ is selected and a permuted sequence $7, 1, 5$ is produced and selected. The remaining chromosomes do not change.

Roulette wheel selection with elitist strategy. Roulette wheel selection, the most common GA selection operation, is a type of proportional selection mechanism. Basically, the function determines an option possibility and a survival possibility proportional to the fitness value of every chromosome. These possibilities are then put into the roulette wheel modeling, where the option process relies on rotating the wheel the same number of times as the population size, and on each rotation selecting a chromosome for the new generation. The characteristic of the wheel is its random sampling selection process. To protect the best chromosomes in the next generation, elitist selection is often used as a supplementary method if this is not chosen in the proportional option procedure.

Fuzzy logic controller. In general, an FLC is indispensable when seeking to resolve complicated problems which require large complex computations. The function of the FLC is to enhance the precision and effectiveness of the GA by dynamically tuning the GA parameters. According to Wang [127], there are two main FLCs; the crossover FLC and the mutation FLC; which can be employed to dynamically regulate the crossover and mutation ratio during the optimization process. Therefore, there can be a significant difference between a GA which considers a FLC and a GA that does not. To overcome the difficulties in the proposed modeling, we adopt FLC technology here. The heuristic updating theory for the crossover ratio and mutation ratio considers the changes in the average fitness of the population over two continuous generations.

Multi-objective operation. In general, if the multiple objectives are of the same dimension; for instance, one objective (Obj_1) is to minimize the cost, and another objective (Obj_2) is to minimize the economic loss; we can integrate these two objectives using weighting methods and treat the aggregated objective as the fitness value for the BPMOGA, namely:

$$\text{eval} = (w_1 Obj_1 + w_2 Obj_2), \tag{2.202}$$

in which w_1 and w_2 are the two weights for the decision makers' two objectives. Obviously, $w_1 + w_2 = 1$. In addition, eval is the fitness value for the BPMOGA.

If the multiple objectives do not have the same dimension, the Pareto method can be adopted for the multi-objective operation.

Overall BLMOGA procedure. At the start of the algorithm, some feasible solutions for the upper level decision variables which meet the constraints of upper level modeling are generated. Then the solutions are introduced into the lower-level modeling, and the lower-level solutions determined by running the genetic operations. The multi-objective operation is applied to handle the multiple objectives in either the upper or lower levels. Then, new solutions are generated through an update to the solutions. This program continues until the stop condition is met. The overall procedure for BLMOGA is shown in Fig. 2.12.

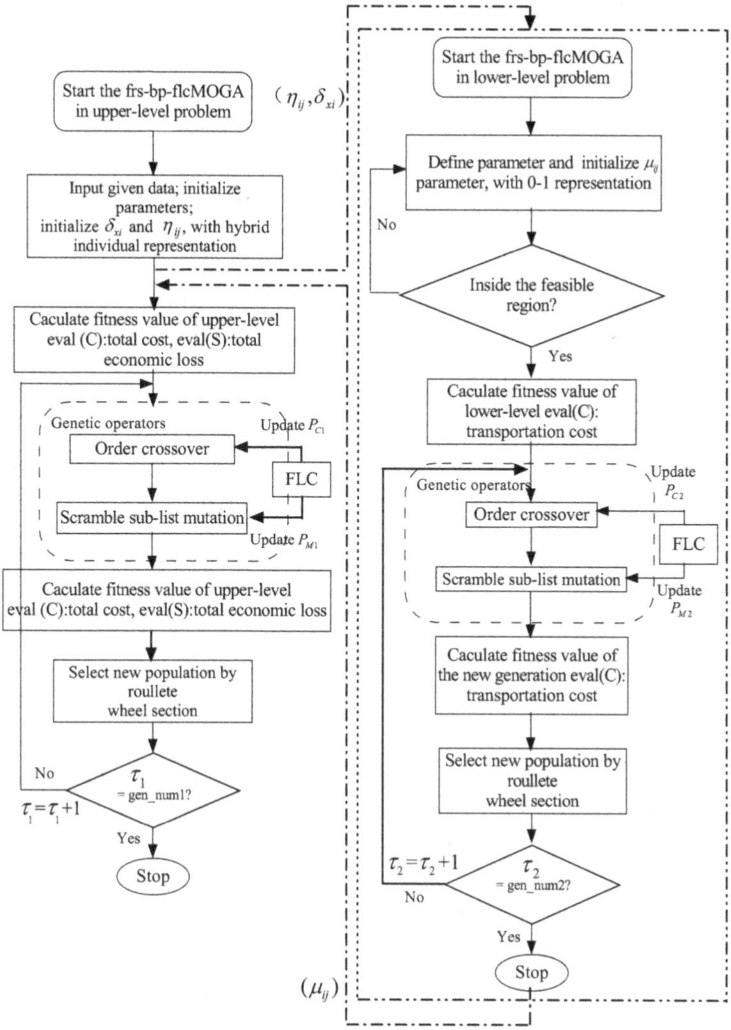

Fig. 2.12 The overall procedure of BLMOGA

Numerical Example

In this section, a numerical example is given to illustrate the effectiveness of the application of the models and algorithms proposed earlier.

Example 2.11. Consider the problem

$$\begin{cases} \max\limits_{x_1,x_2,x_3} F_1 = 2\xi_1^2 x_1 + \sqrt{\xi_2 x_2 + \xi_3 x_3} + \xi_4 y_1 + \sqrt{\xi_5} y_2 + \sqrt{\xi_6} y_3 \\ \max\limits_{x_1,x_2,x_3} F_2 = \xi_1^2 x_1 + \xi_2^2 x_2 + x_3 + y_1 + y_2 + y_3 \\ \text{where } (y_1, y_2, y_3) \text{ solves} \\ \begin{cases} \max\limits_{y_1,y_2,y_3} f_1 = \sqrt{7\xi_1 x_1 + 3\xi_2 x_2 + 8\xi_3 x_3} + (12\xi_4 y_1 + 11\xi_5 y_2 + 5\xi_6 y_3)^2 \\ \max\limits_{y_1,y_2,y_3} f_2 = x_1 + 5x_2 + 7x_3 + y_1 + 11y_2 - 8\xi_6^3 y_3 \\ \text{s.t.} \begin{cases} \sqrt{x_1\xi_1 + x_2\xi_2} + x_3^2\xi_3 + \xi_4 y_2 + 17y_3\xi_6 \le 1540 \\ 3x_1 + 6x_2 + 12x_3 + 14y_1 + 19y_2 \le 2400 \\ x_1 + x_2 + x_3 + y_1 + y_2 - y_3 \ge 150 \\ x_1 + x_2 + x_3 + y_1 + y_2 + y_3 \le 200 \\ x_1, x_2, x_3, y_1, y_2, y_3 \ge 0, \end{cases} \end{cases} \end{cases} \tag{2.203}$$

where $\xi_j, j = 1, 2, \cdots, 6$ are random variables characterized as

$$\begin{array}{l} \xi_1 \sim \mathcal{N}(8,1), \quad \xi_2 \sim \mathcal{N}(10,1), \ \xi_3 \sim \mathcal{N}(12,1), \\ \xi_4 \sim \mathcal{N}(14,1), \ \xi_5 \sim \mathcal{N}(16,1), \ \xi_6 \sim \mathcal{N}(18,1). \end{array}$$

We can obtain the DEDC model as follows

$$\begin{cases} \max\limits_{x_1,x_2,x_3} [\alpha_1, \alpha_2] \\ \text{s.t.} \begin{cases} Pr\{\xi_1^2 x_1 + \sqrt{\xi_2 x_2 + \xi_3 x_3} + \xi_4 y_1 + \sqrt{\xi_5} y_2 + \sqrt{\xi_6} y_3 \ge \bar{F}_1\} \ge \alpha_1 \\ Pr\{x_1 + 5x_2 + 7x_3 + y_1 + 11y_2 - 8\xi_6^3 y_3 \ge 680\} \ge \alpha_2 \\ \text{where } (y_1, y_2, y_3) \text{ solves} \\ \max\limits_{y_1,y_2,y_3} [\beta_1, \beta_2] \\ \text{s.t.} \begin{cases} Pr\{\sqrt{7\xi_1 x_1 + 3\xi_2 x_2 + 8\xi_3 x_3} + (12\xi_4 y_1 + 11\xi_5 y_2 + 5\xi_6 y_3)^2 \\ \ge 460\} \ge \beta_1 \\ Pr\{x_1 + 5x_2 + 7x_3 + y_1 + 11y_2 - 8\xi_6^3 y_3 \ge 390\} \ge \beta_2 \\ Pr\{\sqrt{x_1\xi_1 + x_2\xi_2} + x_3^2\xi_3 + \xi_4 y_2 + 17y_3\xi_6 \le 1400\} \ge 0.6 \\ 3x_1 + 6x_2 + 12x_3 + 14y_1 + 19y_2 \le 2400 \\ x_1 + x_2 + x_3 + y_1 + y_2 - y_3 \ge 150 \\ x_1 + x_2 + x_3 + y_1 + y_2 + y_3 \le 200 \\ x_1, x_2, x_3, y_1, y_2, y_3 \ge 0. \end{cases} \end{cases} \end{cases} \tag{2.204}$$

After running the algorithm above, we obtain $(x_1^*, x_2^*, x_3^*, y_1^*, y_2^*, y_3^*) = (10.3, 26.7, 12.4, 20.5, 10.7, 0)$ and $(\alpha_1^*, \alpha_2^*) = (0.64, 0.81)$, $(\beta_1^*, \beta_2^*) = (0.72, 0.54)$.

2.6 Regional Water Resources Allocation Problem in the Gan-Fu Plain

This part turns back to the problem background which is illustrated in the first section in this chapter and applies the above methodologies to RWRAP problem.

2.6.1 Modelling

In this section, a bi-level programming model for the RWRAP considering fuzziness and randomness is constructed. The mathematical description of the problem is given as follows.

Assumptions

Before constructing the model, the following assumptions are adopted:

(1) All available water for distribution is provided by a single river basin;
(2) The minimal volume of available water that should be allocated to each sub-area, the minimal ecological water requirement in the river basin, the maximal and minimal values for water withdrawal and waste water emission proportions for water users in each sub-area are known;
(3) The regional authority in the upper level and water manager of each sub-area in the lower level will make rational decisions cooperatively for avoiding unsatisfied solutions;
(4) If the actual water demand is greater than water withdrawal, the difference in value between them is dealt with by conserving water and improving water efficiency. When the water withdrawal of a sub-area is more than its assigned initial water right, it needs to buy water from the water market. Conversely, it can sell excess water in the water market; and
(5) The total volume of stream flow in the river basin and water demand are considered as random variables, with attributes determined using historical data and professional experience using statistics.

Lower Level Model for Water Allocation Problems in Sub-areas

The lower level is discussed first. When the mathematical description for the lower level is founded, it is easier to determine the mathematical description for the upper level.

1. Objective functions for the lower level

For the lower level of the bi-level programming model, firstly, water manager of each sub-area distributes water to industrial water user, the municipal water user, and agricultural water user respectively, which generates the economic benefit $\sum_{k=1}^{m} b_{ik}y_{ik}$. However, the promised water sometimes cannot be delivered due to insufficient supply, which leads to the need to acquire of water from higher-priced alternatives and/or negative consequences generated from the curbing of regional development plans [84]. For example, municipal residents may have to curtail the watering of lawns, industries may have to reduce production levels or increase water recycling rates, and farmers may not be able to use irrigation as planned. These actions will result in increased costs or decreased benefits for regional development, that is,

$$\sum_{k=1}^{m} c_{ik}\left(E[\tilde{\omega}_{\tilde{d}_{ik}(r_i,\sigma_i)}] - y_{ik}\right),$$

where $E[\tilde{\omega}_{\tilde{d}_{ik}(r_i,\sigma_i)}]$ is the expected value of (r,σ)-level trapezoidal fuzzy variable $\tilde{\omega}_{\tilde{d}_{ik}(r_i,\sigma_i)}$. As mentioned in the assumptions, when the water withdrawal of a sub-area is more than its initial water rights, the shortage needs to be bought from the water market. Conversely, when the initial water rights are more than the water withdrawal of the sub-area, water manager of this sub-area can sell the excess water in the water market, which generates a transaction price

$$\left(x_i - \sum_{k=1}^{m} y_{ik}\right)p(r).$$

Water managers of the sub-areas need to pay water charges for water withdrawal to regional authority, the cost of which is denoted as $p'_k \sum_{k=1}^{m} y_{ik}$. Further, the cost of treating waste water is also considered in the objective function as different water users in different sub-areas discharge different proportions of waste water. So the cost is given as $\sum_{k=1}^{m} p_w\lambda_{ik}y_{ik}$.

The lower objective function is to measure and maximize the economic benefit of each sub-area, thus benefits should be maximized and costs minimized, so the expected value of the objective function can be described as follows:

$$E[V_i] = \sum_{k=1}^{m}\left(b_{ik}y_{ik} - c_{ik}(E[\tilde{\omega}_{\tilde{d}_{ik}(r_i,\sigma_i)}] - y_{ik}) - p'_k y_{ik} - p_w\lambda_{ik}y_{ik}\right) + \left(x_i - \sum_{k=1}^{m} y_{ik}\right)p(r). \tag{2.205}$$

2. Constraints of the lower level

There are 4 kinds of constraints on the lower level, details of them are explained in the following.

(1) Constraints on water withdrawal

The water withdrawal of each water user needs to satisfy their minimal requirements in order to guarantee the basic water use. If the water price is sufficiently high in the water market, the minimal requirements can efficiently prevent the water manager in each sub-area to excessively sell water rights in the market regardless of the basic need of the water users. On the other hand, maximal allocation targets are also set in order to balance the development among different water users. It should be noted that the water rights (i.e., x_i) of each sub-area can be used to meet the water demands of the users as well as for trade in the water market. Water withdrawal (i.e., y_{ik}) is defined as the amount of water used to meet the demand of a user (i.e., $E[\tilde{\omega}_{\tilde{d}_{ik}(r_i,\sigma_i)}]$). Thus, y_{ik} should be not more than $E[\tilde{\omega}_{\tilde{d}_{ik}(r_i,\sigma_i)}]$. In order to make full use of the initial water rights of each sub-area, if the benefit of selling the water rights in the market is less than that of using them to meet the demand of the water users, the best strategy, when there is sufficient water supply, is to give priority to the water demand of users and then trade the excess water in the water market. These constraints are given as Eq. (2.206), which ensures that these relationships are not violated.

$$T_{ik\min} \leq y_{ik} \leq T_{ik\max}, \quad y_{ik} \leq E[\tilde{\omega}_{\tilde{d}_{ik}(r_i,\sigma_i)}], \quad \forall i \in \Psi,\ k \in \Phi. \tag{2.206}$$

(2) Constraints on waste water

As is generally known, production activities and the general society produce different degrees of waste water. From the perspective of environmental protection, the waste water of each sub-area cannot exceed the maximal volume. Thus the following constraints are employed:

$$\sum_{k=1}^{m} \lambda_{ik} y_{ik} \leq q_i, \quad \forall i \in \Psi. \tag{2.207}$$

(3) Water transaction price function

In water allocation under a water market, the water trade price is the basic term that determines the arrangement. Although the water market is not a completely competitive market, market prices are still influenced by the relationship between supply and demand, namely, the price rises when demand is more than supply. Conversely, the price drops. According to the theoretical analysis of the oligarch competition model [40], the water rights transaction price function in water market can be expressed as $p(r) = \kappa - \delta r > 0$ $(\kappa > 0, \delta > 0)$, where r represents the total available water for trading in the water market, which is denoted by

$$\sum_{i=1}^{n} \left(x_i - \sum_{k=1}^{m} y_{ik} \right).$$

So the water transaction price function is as follows:

$$p(r) = \kappa - \delta \left(\sum_{i=1}^{n} \left(x_i - \sum_{k=1}^{m} y_{ik} \right) \right) > 0, \quad (\kappa > 0,\ \delta > 0). \tag{2.208}$$

(4) Non-negative constraints

Because water withdrawal cannot be minus, the constraint is presented:

$$y_{ik} \geq 0, \quad \forall i \in \Psi,\ k \in \Phi. \tag{2.209}$$

Upper Level Model for Water Allocation Problem in the Overall Region

After the mathematical description of the lower level model is given, the objective function and constraints of the upper level can now be determined. In order to fulfill the three key principles of water allocation (i.e., equity, efficiency and sustainability), the objective function is established considering the environmental and economic benefit for meeting the sustainability and efficiency principles; in the constraint on ecological water requirement, a minimal ecological water supply in the river basin is required for avoiding unsustainable water supplies for the environment dimension; for ensuring the equity principle, a cooperative decision making process is considered for guaranteeing the satisfaction of both levels.

1. Objective function of the upper level

The upper objective function defines the total benefit to society for water resources allocation, so the regional authority needs to create a plan to effectively allocate the water supply to the sub-areas in order to maximize the total benefits to society which also need to consider the environmental dimension of the water resources management in order to avoid any unsustainable water supplies and allocation. The total benefits are comprised of three components: ecological benefit, charge of water withdrawal from the sub-areas and the whole economic benefit of water output to the sub-areas. So the upper level objective function can be represented by:

$$E[V_0] = h(w) + \sum_{i=1}^{n}\sum_{k=1}^{m} p'_k y_{ik} + \sum_{i=1}^{n} E[V_i],$$

where $h(w) = ew$, which means ecological benefit function of the river basin. Then, the expected value of the upper level objective function can be transformed into:

$$\begin{aligned} E[V_0] = ew + \sum_{i=1}^{n}\Bigg[\sum_{k=1}^{m}\Big(b_{ik}y_{ik} - c_{ik}(E[\tilde{\omega}_{\tilde{d}_{ik}(r_i,\sigma_i)}] - y_{ik}) - p_w\lambda_{ik}y_{ik}\Big) \\ + \left(x_i - \sum_{k=1}^{m} y_{ik}\right)p(r)\Bigg]. \end{aligned} \tag{2.210}$$

2. Constraints of the upper level

There are 3 kinds of constraints on the upper level, detail of them are explained as follows.

(1) Constraints on water supply

Regarding the constraints of the upper level, it is assumed that the water supply is limited as the total volume of stream flow of the river basin has a capacity limit. The actual water allocated to water users does not exceed this capacity, and the initial water rights of water users and public water rights are equal to the total volume of available water for allocation. Thus there is a restriction on capacity as follows:

$$\sum_{i=1}^{n} x_i + w = E[\tilde{Q}], \quad \sum_{i=1}^{n}\sum_{k=1}^{m} y_{ik} + w \leq E[\tilde{Q}], \tag{2.211}$$

where E is used to transform the random variable into a deterministic one, which is based on the theory proposed by [48].

(2) Constraint on ecological water requirement

As mentioned above, if the ecological water requirement is not met, the ecological environment may be damaged. Because this situation does not conform with the requirements of sustainable development, public water rights w should be not lower than the minimal ecological water requirement η in the river basin, therefore, then we have the following inequalities:

$$w \geq \eta, \ (\eta > 0). \tag{2.212}$$

(3) Constraints on initial water rights for the sub-areas

The initial water rights x_i must be more than the minimal volume that is allocated to sub-area i, thus we have:

$$x_i \geq \theta_i \ (\theta_i > 0), \quad \forall i \in \Psi. \tag{2.213}$$

Global Model

Based on the discussion above, by integrating (2.205) $\sim$ (2.213), the following global expected model of bi-level programming can now be formulated for the RWRAP.

$$
\max_{x_i} \; E[V_0] = ew + \sum_{i=1}^{n} \left[\sum_{k=1}^{m} \left(b_{ik}y_{ik} - c_{ik}(E[\tilde{\omega}_{\tilde{d}_{ik}(r_i,\sigma_i)}] - y_{ik}) - p_w \lambda_{ik} y_{ik} \right) + \left(x_i - \sum_{k=1}^{m} y_{ik} \right) p(r) \right] \tag{2.214}
$$

$$
\text{s.t.} \begin{cases}
\sum_{i=1}^{n} x_i + w = E[\tilde{Q}] \\
\sum_{i=1}^{n} \sum_{k=1}^{m} y_{ik} + w \leq E[\tilde{Q}] \\
x_i \geq \theta_i \ (\theta_i > 0), \quad \forall i \in \Psi \\
w \geq \eta \ (\eta > 0) \\
\max_{y_{ik}} \; E[V_i] = \sum_{k=1}^{m} \left(b_{ik}y_{ik} - c_{ik}(E[\tilde{\omega}_{\tilde{d}_{ik}(r_i,\sigma_i)}] - y_{ik}) - p_k' y_{ik} - p_w \lambda_{ik} y_{ik} \right) \\
\qquad\qquad + \left(x_i - \sum_{k=1}^{m} y_{ik} \right) p(r) \\
\text{s.t.} \begin{cases}
T_{ik\min} \leq y_{ik} \leq T_{ik\max}, \quad \forall i \in \Psi, k \in \Phi \\
y_{ik} \leq E[\tilde{\omega}_{\tilde{d}_{ik}(r_i,\sigma_i)}], \quad \forall i \in \Psi, \ k \in \Phi \\
\sum_{k=1}^{m} \lambda_{ik} y_{ik} \leq q_i, \quad \forall i \in \Psi \\
p(r) = \kappa - \delta \left(\sum_{i=1}^{n} \left(x_i - \sum_{k=1}^{m} y_{ik} \right) \right) > 0, \ (\kappa > 0, \delta > 0) \\
y_{ik} \geq 0, \quad \forall i \in \Psi, k \in \Phi,
\end{cases}
\end{cases} \tag{2.215}
$$

where E is used to transform the random variable into a deterministic one based on theory of [48].

2.6.2 *Presentation of the Case Problem*

The largest irrigation area south of the Yangtze lies on the Gan-Fu plain at the end of the Fu-He valley in China, with a dedicated irrigation area of 82,000 ha. At present, the irrigation projects in the district primarily supply water to industrial, municipal, and agricultural users, with secondary projects in other areas such as waterpower and environmental use, thereby benefitting both the environment and the economy. The Gan-Fu plain irrigation district is divided into four sub-areas (in this case study, $i = 1, 2, 3, 4$, with $i = 1$ representing Nanchang County, $i = 2$ representing Jinxian County, $i = 3$ representing Linchuan District, and $i = 4$ representing Fengcheng City). Considering the main tasks of in the Gan-Fu plain irrigation district, this chapter hopes to meet the water demand for three main water users ($k = 1, 2, 3$, with $k = 1$ for industrial users, $k = 2$ for municipal users, and $k = 3$ for agricultural users). The flow network and water users in the Gan-Fu plain irrigation district are shown in Fig. 2.13.

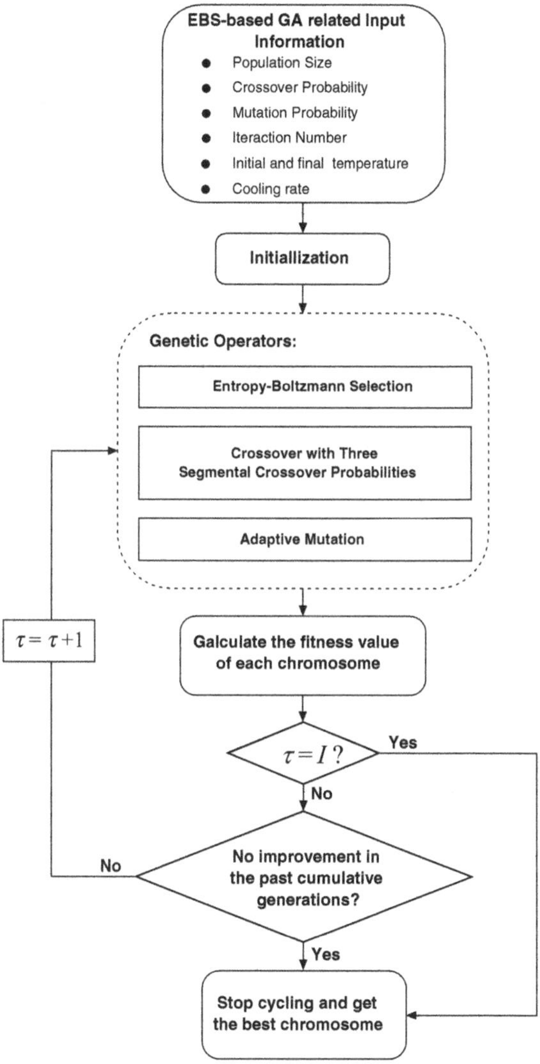

Fig. 2.13 Overall procedure of EBS-based GA for RWRAP

From water allocation system statistical data, the water price function is set at $p(r) = 6 - 1.2r$ ($r < 5.0$), the cost of treating the waste water is set at 0.80 RMB/m^3, the eco-efficiency is set at 25.8 RMB/m^3, and the minimal ecological water requirement is set at 3.50×10^6 m^3. Details and some other parameters are in Table 2.3, and the random data based on historical data and expert experience, are shown in Tables. 2.4 and 2.5. Random theory is a useful tool when seeking to deal with the uncertain water demand in a RWRAP under a random environment. However, some actions taken by water managers can be more deterministic; for example, if reservoir operations management plans are mandated by regulation or law, the managers are obliged to comply; so such decisions are, therefore, less

Table 2.3 Parameters in RWRAP

Parameter	Sub-area i											
	$i = 1$			$i = 2$			$i = 3$			$i = 4$		
$\theta_i (10^6\ m^3)$	25.50			42.80			51.30			305.60		
$q_i\ (10^6\ m^3)$	4.35			3.5			3.05			18.95		
r_i	0.85			0.9			0.88			0.95		
σ_i	0.05			0.08			0.03			0.01		
	$k = 1$	$k = 2$	$k = 3$	$k = 1$	$k = 2$	$k = 3$	$k = 1$	$k = 2$	$k = 3$	$k = 1$	$k = 2$	$k = 3$
b_{ik} (RMB/m^3)	55.5	71.4	23.5	—	—	26.4	—	68.2	25.8	58.2	67.8	24.3
c_{ik} (RMB/m^3)	11.5	12.8	3.4	—	—	2.7	—	12.2	3.1	10.6	13.4	3.8
λ_{ik}	70%	34%	5.2%	—	—	6.7%	—	41%	4.5%	71%	37%	5.3%
$T_{ik\min}\ (10^6\ m^3)$	3.00	2.75	15.50	—	—	36.80	—	1.05	43.00	1.85	3.60	235.50
$T_{ik\max}\ (10^6\ m^3)$	3.42	3.30	20.50	—	—	48.20	—	1.22	56.80	2.18	4.15	308.50
p'_k (RMB/m^3)	1.2	1	0.2	1.2	1	0.2	1.2	1	0.2	1.2	1	0.2

Table 2.4 Parameters of random water demand $\tilde{d}_{ik}$ (10^6 m^3)

Water user	Sub-area i			
	$i = 1$	$i = 2$	$i = 3$	$i = 4$
Industrial ($k = 1$)	$\mathcal{N}(3.40, 0.1^2)$	—	—	$\mathcal{N}(2.20, 0.01^2)$
Municipal ($k = 2$)	$\mathcal{N}(3.30, 0.08^2)$	—	$\mathcal{N}(1.20, 0.01^2)$	$\mathcal{N}(4.20, 0.02^2)$
Agricultural ($k = 3$)	$\mathcal{N}(23.38, 3^2)$	$\mathcal{N}(49.10, 4.5^2)$	$\mathcal{N}(58.53, 10^2)$	$\mathcal{N}(314.50, 45^2)$

Table 2.5 Parameters of random stream flow $\tilde{Q}$

Stream flow	Probability p_t	a_t (10^6 m^3)	b_t (10^6 m^3)	c_t (10^6 m^3)
Low ($t = 1$)	$p_1 = 0.3$	280	355	380
Medium ($t = 2$)	$p_2 = 0.5$	400	475	500
High ($t = 3$)	$p_3 = 0.2$	520	590	620

uncertain. In this special case, the deterministic water demand is regarded as a special random number, in which the left and right borders have the same value, and the most possible water demand degenerates to a deterministic single point value. Thus, the approach for dealing with random variables is also suitable for variables that are more deterministic.

To ensure equity between the lower level water managers, a lower satisfaction degree limit for each sub-area water managers is set at 0.70 (i.e., $\lambda_i = 0.70$, for $i = 1, 2, 3, 4$) to avoid dissatisfied water users. To guarantee equity between the regional authority on the upper level and each lower level sub-area water manager, the lower and upper bounds for the satisfaction degree ratio are set at 0.75 and 1.00 respectively (i.e., $[\Delta_l, \Delta_u] = [0.75, 1]$), which balances the satisfaction degrees on the two decision making levels.

2.6.3 Parameter Selection for EBS-Based GA

The EBS-based GA parameters are set from the results of preliminary experiments which were conducted to observe the behavior of the algorithm at different parameter settings. Many scholars have suggested how to determine the proper parameters [32, 66, 86, 131] . By comparing several sets of parameters, including population size (i.e., N), iteration number (i.e., I), crossover probability (i.e., p_a, p_b, p_c), and mutation probability initialization (i.e., p_{m0}), the most reasonable parameters are chosen. However, there is a trade-off between the time evaluation and population size. When a problem has nontrivial evaluation times, even a very limited number of search generations within a large population can take significant computational time [68].

The adopted parameters in the EBS-based GA are as follows: population size $N = 50$, iteration number $I = 200$, three segmental crossover probabilities $p_a = 0.5$, $p_b = 0.7$, $p_c = 0.6$, mutation probability initialization $p_{m0} = 0.05$.

Table 2.6 Pay-off table with 5 individual single-objective problems (10^6 RMB)

	Regional Authority	Nanchang County	Jinxian County	Linchuan District	Fengcheng City
Results	$E[V_0]$	$E[V_1]$	$E[V_2]$	$E[V_3]$	$E[V_4]$
$\max E[V_0]$	11651.43	839.4216	1225.442	1546.311	7646.958
$\max E[V_1]$	11617.30	981.7655	1225.442	1546.311	7530.930
$\max E[V_2]$	11445.21	756.6554	1257.842	1163.835	7487.579
$\max E[V_3]$	11450.34	711.0618	928.9334	1586.632	7530.930
$\max E[V_4]$	11391.53	862.6008	928.9937	1177.115	7687.662
Upper bound	11651.43	981.7655	1257.842	1586.632	7687.662
Lower bound	11391.53	711.0618	928.9334	1163.835	7487.579
Rate of difference	2.23%	27.57%	26.15%	26.65%	2.6%

2.6.4 Results and Discussion

To show the practicality and efficiency of the optimization method for the RWRAP, interactive fuzzy programming combined with EBS-based GA was conducted and run on Pentium 4, 2.40 GHz clock pulse with 1024 MB memory.

Table 2.6 shows the upper bound and lower bound of the objective function $E[V_i(\mathbf{x}, \mathbf{y})]$ (for $i = 0, 1, \cdots, 4$), which were calculated using the Zimmermann method previously proposed. From Table 2.6, it can be seen that the difference between the upper bound and lower bound for the economic benefit derived from the water use is relatively low for the sub-area that has a comparatively high economic benefit; i.e., Fengcheng City at only 2.6%. This indicates that further optimization using the EBS-based GA would take up less optimization space for Fengcheng City, while greater optimization space is needed in the other sub-areas. Although not conclusive, it seems that the optimization space size for subsequent optimization is determined by the related sub-area parameters.

Tables 2.7 and 2.8 list the detailed results of the 8 cases used in the sensitivity analyses. The results for each case were found after 50 runs of the EBS-based GA, in which the solution that had the maximal fitness value was selected, while locally trapped solutions were excluded. The eight candidate values for the upper level regional authorityąŕs lower satisfaction degree limit (i.e., λ_0) selected in the range [0.86, 1.00] were based on the decision makersąŕ opinions. The interval between each candidate value was set at 0.02. It can be concluded that the solutions were influenced significantly by the changes in candidate values. As the lower satisfaction degree limit for the upper level regional authority decreases (i.e., λ_0), there is an increasing trend in the initial water rights in each sub-area (i.e., $\sum_{i=1}^{4} x_i$), which would lead to a higher satisfaction degree for each lower level sub-area water manager. By comparing the variation tendencies in the available water rights for trade in the market (i.e., $\sum_{i=1}^{n}(x_i - \sum_{k=1}^{m} y_{ik})$), it appears that when the upper-level regional authority has a sufficiently high lower satisfactory degree limit (i.e., $\lambda_0 \geq 0.88$), the water rights in the market can be adequately traded (i.e.,

Table 2.7 Sensitivity analysis 1

Candidate	ω (10^6 m^3)	Sub-area i	x_i (10^6 m^3)	y_{i1} (10^6 m^3)	y_{i2} (10^6 m^3)	y_{i3} (10^6 m^3)	$\sum_{i=1}^{4}\left(x_i - \sum_{k=1}^{3} y_{ik}\right)$ (10^6 m^3)	$E[V_0]$ (10^6 RMB)	$E[V_i]$ (10^6 RMB)	$\mu_i(E[V_i])$	δ_i	G
Candidate 1	16.1421	$i = 1$	28.6019	3.4200	3.3000	16.0385	0	11651.43	806.0868	0.3510	0.3510	0.7404
($\lambda_0 = 1.00$)		$i = 2$	42.8000	—	—	48.2000			1225.443	0.9015	0.9015	
		$i = 3$	51.3000	—	1.2200	56.8000			1546.312	0.9046	0.9046	
		$i = 4$	311.1560	2.1800	4.1500	298.5494			7557.813	0.3510	0.3510	
Candidate 2	11.2392	$i = 1$	38.1554	3.3648	3.3000	16.7817	0	11646.23	875.4918	0.6074	0.6198	0.8310
($\lambda_0 = 0.98$)		$i = 2$	42.8000	—	—	48.2000			1225.442	0.9015	0.9200	
		$i = 3$	51.3000	—	1.2200	56.8000			1546.312	0.9046	0.9231	
		$i = 4$	306.5054	1.8500	4.1500	303.0943			7609.112	0.6074	0.6198	
Candidate 3	8.7848	$i = 1$	36.5135	3.1678	3.3000	19.4330	0	11641.03	908.7430	0.7302	0.7607	0.8681
($\lambda_0 = 0.96$)		$i = 2$	42.8000	—	—	48.2000			1225.443	0.9015	0.9391	
		$i = 3$	51.3000	—	1.2200	56.8000			1546.312	0.9046	0.9423	
		$i = 4$	310.0943	1.8500	4.1500	303.0943			7633.689	0.7302	0.7607	
Candidate 4	7.3249	$i = 1$	3 6.7195	3.0886	3.3000	20.5000	0	11635.84	927.3274	0.7989	0.8499	0.8836
($\lambda_0 = 0.94$)		$i = 2$	42.8000	—	—	48.2000			1225.442	0.9015	0.9590	
		$i = 3$	51.3000	—	1.2200	56.8000			1546.312	0.9046	0.9623	
		$i = 4$	311.8556	1.8500	4.0711	303.6455			7647.426	0.7989	0.8499	

Table 2.8 Sensitivity analysis 2

Candidate	ω (10^6 m^3)	Sub-area i	x_i (10^6 m^3)	y_{i1} (10^6 m^3)	y_{i2} (10^6 m^3)	y_{i3} (10^6 m^3)	$\sum_{i=1}^{4}\left(x_i - \sum_{k=1}^{3} y_{ik}\right)$ (10^6 m^3)	$E[V_0]$ (10^6 RMB)	$E[V_i]$ (10^6 RMB)	$\mu_i(E[V_i])$	δ_i	G
Candidate 5	6.5347	$i = 1$	38.1702	3.0886	3.3000	20.5000	0	11630.64	936.0316	0.8311	0.9033	0.8844
($\lambda_0 = 0.92$)		$i = 2$	42.8000	—	—	48.2000			1225.442	0.9015	0.9799	
		$i = 3$	51.3000	—	1.2200	56.8000			1546.311	0.9046	0.9833	
		$i = 4$	311.1951	1.8500	3.9389	304.5678			7653.859	0.8311	0.9033	
Candidate 6	5.7445	$i = 1$	39.6209	3.0886	3.3000	20.5000	0	11625.44	944.7359	0.8632	0.9591	0.8745
($\lambda_0 = 0.90$)		$i = 2$	42.8000	—	—	48.2000			1225.442	0.9015	1.0017	
		$i = 3$	51.3000	—	1.2200	56.8000			1546.311	0.9046	1.0051	
		$i = 4$	310.5347	1.8500	3.8068	305.4902			7660.294	0.8632	0.9591	
Candidate 7	4.9542	$i = 1$	41.0715	3.0886	3.3000	20.5000	0	11620.24	953.4401	0.8954	1.0175	0.8862
($\lambda_0 = 0.88$)		$i = 2$	42.8000	—	—	48.2000			1225.442	0.9015	1.0244	
		$i = 3$	51.3000	—	1.2200	56.8000			1546.311	0.9046	1.0280	
		$i = 4$	309.8742	1.8500	3.6747	306.4125			7666.726	0.8954	1.0175	
Candidate 8	4.3941	$i = 1$	41.9729	3.0886	3.3000	20.5000	0.1129	11615.04	956.7974	0.9078	1.0555	0.8791
($\lambda_0 = 0.86$)		$i = 2$	43.0264	—	—	48.2000			1227.504	0.9078	1.0555	
		$i = 3$	51.3694	—	1.2200	56.8000			1547.634	0.9078	1.0555	
		$i = 4$	309.2368	1.8500	3.6000	306.9340			7669.207	0.9078	1.0555	

$\sum_{i=1}^{n}(x_i - \sum_{k=1}^{m} y_{ik}) = 0$). Although the situation discussed in this chapter may not be analyzed as an application of the Coase Theorem [26] proposed by the Nobel Prize laureate Ronald H. Coase due to the fact that the water market is not a completely competitive market, the adequate trading of water rights in the market reflects that economic efficiency can be sufficiently achieved through the full allocation of water resources when $\lambda_0 \geq 0.88$.

When the lower satisfaction degree limit for each lower level sub-area water manager is set at 0.70 (i.e., $\lambda_i = 0.70$, for $i = 1, 2, 3, 4$), and the lower and upper bounds of the ratio of the satisfactory degree are set at 0.75 and 1.00 respectively (i.e., $[\Delta_l, \Delta_u] = [0.75, 1]$), Candidates 3–5 are the only feasible solutions that guarantee the RWRAP equity principle. To select a feasible solution that best fits the decision maker's preference, the following evaluation equation is introduced:

$$G = w_1 * \mu_0(E[V_0]) + w_2 * \min_i\{\mu_i(E[V_i])\}, \tag{2.216}$$

where w_1 and w_2 are the weights of the upper level satisfaction degree and the lower level minimal satisfaction degree respectively, both of which are determined by the decision makers; G is the evaluation value to find the optimal feasible solution that can best fit the decision makers' preferences. It should be noted that the satisfaction degree weights for the upper level and the lower level minimums in this case problem were set at $w_1 = 0.6$ and $w_2 = 0.4$. The evaluation value for each candidate was calculated as shown in Tables 2.7 and 2.8. As Candidate 5 (i.e., $\lambda_0 = 0.92$) has the highest evaluation value (i.e., $G = 0.8844$), it is selected as the optimal solution to this case problem. The Candidate 5 decision variable values indicate that Nanchang County (i.e., sub-area 1) and Fengcheng City (i.e., sub-area 4) water managers are able to sell excess water in the water market, while Jinxian County (i.e., sub-area 2) and Linchuan District (i.e., sub-area 3)water managers have to buy water from the water market. Figure 2.14 shows the detailed water allocation plan for the Gan-Fu plain irrigation district. This allocation plan allows decision makers on both levels to determine the appropriate water resources allocations to simultaneously achieve the equity, efficiency and sustainability principles. These results are quite useful and can serve as references for decision makers in guiding current practice.

Robustness analyses using different λ_0 are shown in Fig. 2.15. It can be concluded from the results that the minimal lower level satisfaction degree (i.e., λ) continuously and steadily decreases with an increase in the lower satisfaction degree limit of the upper level (i.e., λ_0). The perturbation of parameter λ_0 does not lead to an unstable tendency in the fitness value λ, which demonstrates the robustness of the results.

A traditional RWRAP only has an objective function focused on the maximization of the total benefits to society (i.e., the upper level objective), and seldom includes the users' objectives (i.e., the lower level objectives). This means that the regional authority satisfaction degree o is set at $\lambda_0 = 1$ (i.e., Candidate 1). It can be seen from Tables 2.7 and 2.8 that the lower satisfaction degree levels in sub-area

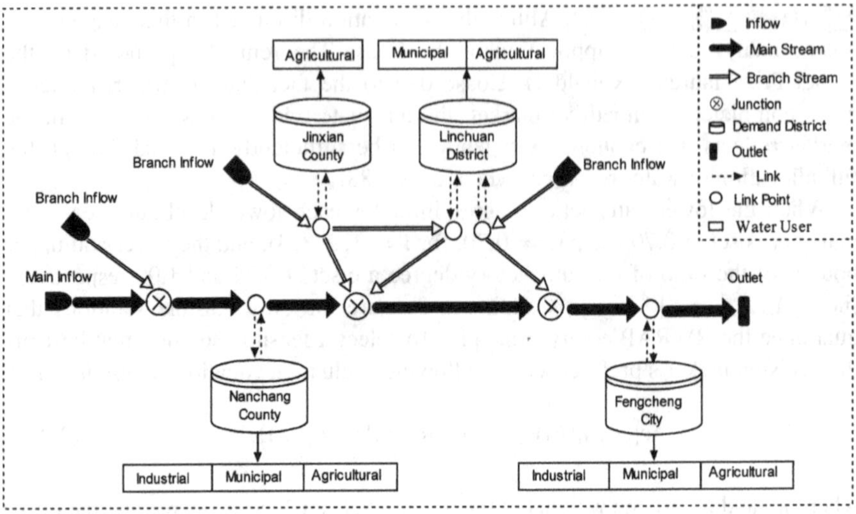

Fig. 2.14 Flow network and water users in Gan-Fu Plain irrigation district

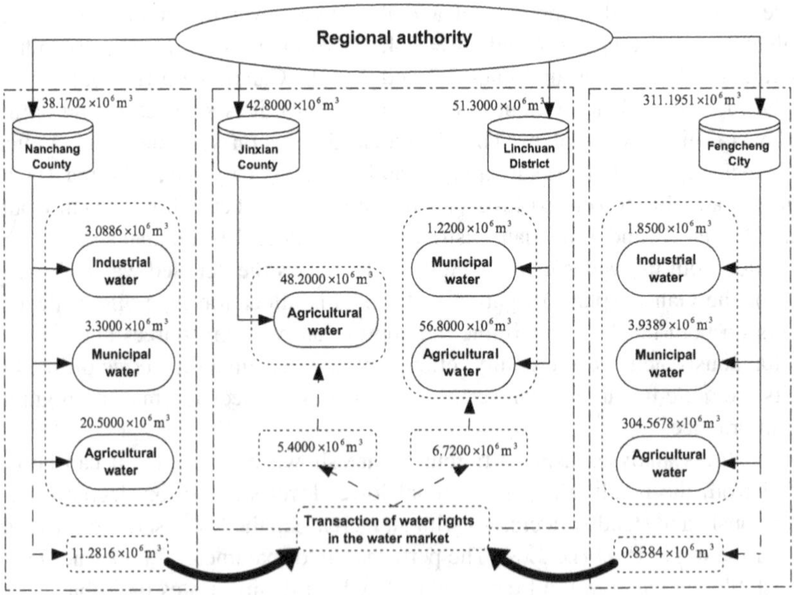

Fig. 2.15 The robust analysis with different λ_0

Table 2.9 Comparison results between EBS-based GA and SA

	Objective function value			
Algorithm	Best	Worst	Average	Variance of objective function value
EBS-based GA	0.8311	0.8225	0.8128	0.3408×10^{-2}
SA	0.8311	0.8203	0.8089	0.4213×10^{-2}
Algorithm	Average convergence iteration number			Average computation time (s)
EBS-based GA	97			23.3482
SA	165			49.6027

1 and 4 are too low, which could lead to unbalanced development across the whole society. However, the advantage of using bi-level optimization for the RWRAP is that it considers the overall satisfaction balance between the upper and lower levels, which can ensure water allocation equity.

To demonstrate the effectiveness and efficiency of the EBS-based GA, it was compared with a basic SA algorithm [69] based on the preceding irrigation district example, where the lower satisfaction degree limit on the upper level (i.e., λ_0) was set at 0.92. Since the performance of the SA is significantly impacted by the choice of the defined parameters (i.e., number of iterations, initial temperature, cooling rate, and final temperature), a range of values was used for the four variables (i.e., I, T_0, γ, and T_M). 12 different combinations were examined to allow the SA full opportunity to obtain a global solution. The parameters used for the SA were number of iterations 200, initial temperature 0.85, and cooling rate 0.97, and a final temperature $T_M = \gamma^I T_0$. To eliminate some of the deleterious effects of the random processes in the algorithms, comparison results between the two algorithms were given when averaged over 50 experimental runs (for the EBS-based GA, and the SA, respectively). The performance of the two algorithms is shown in Table 2.9. From this, it can be concluded that for practical problems, the EBS-based GA has a significant advantage in average computing time and average convergence iteration numbers than the SA. It was further demonstrated that the EBS-based GA for the RWRAP has the ability to better allocate the water than the SA, which took more than double the time to find the optimal solution.

From Table 2.9 it can be seen that: (1) both the EBS-based GA and the SA are able to obtain the best objective function value within 50 runs of trials but the SA may sometimes fall into a poorer local optimum; and (2) the EBS-based GA tends to be more stable than the SA when searching for the optima because the variance in the EBS-based GA objective function value is less than that of the SA. Because of the advanced operators, which are designed to avoid infeasible solutions and premature convergence, the proposed EBS-based GA demonstrates an improved search performance compared to the SA. Additionally, it has been proven that if the temperature is annealed sufficiently slowly, the SA can, with a high probability, approximately converge to a global optimal solution [105]. Yet, as is shown in the above comparison, since the EBS-based GA can arrive at the same best solution as the SA, it can be concluded that the EBS-based GA is able to produce sufficiently good solutions for the RWRAP.

References

1. Abolpour B, Javan M, Karamouz M (2007) Water allocation improvement in river basin using adaptive neural fuzzy reinforcement learning approach. Appl Soft Comput 7(1):265–285
2. Alander J (1994) An indexed bibliography of genetic algorithms: years 1957–1993. Art of CAD, Vaasa
3. Al-Radif A (1999) Integrated water resources management (IWRM): an approach to face the challenges of the next century and to avert future crises. Desalin 124(1):145–153
4. Arora SR, Gupta R (2009) Interactive fuzzy goal programming approach for bilevel programming problem. Eur J Oper Res 194(2):368–376
5. Aubin JP (1984) Lipschitz behavior of solutions to convex minimization problems. Math Oper Res 9(1):87–111
6. Bäck T, Hammel U, Schwefel H (1997) Evolutionary computation: comments on the history and current state. IEEE Trans Evol Comput 1(1):3–17
7. Bard J, Moore J (1990) A branch and bound algorithm for the bilevel programming problem. SIAM J Sci Stat Comput 11:281
8. Bard JF (1988) Convex two-level optimization. Math Program 40(1):15–27
9. Bard J, Falk J (1982) An explicit solution to the multi-level programming problem. Comput Oper Res 9(1):77–100
10. Barichard V, Ehrgott M, Gandibleux X, T'Kindt V (2009) Multiobjective programming and goal programming: theoretical results and practical applications. Lecture notes in economics and mathematical systems. Springer Science & Business Media, Berlin
11. Bialas W, Karwan M (1984) Two-level linear programming. Manag Sci 30:1004–1020
12. Binder K, Heermann DW (1988) Monte Carlo simulation in statistical physics: an introduction, vol 80. Springer, Berlin
13. Bortfeldt A (2006) A genetic algorithm for the two-dimensional strip packing problem with rectangular pieces. Eur J Oper Res 172(3):814–837
14. Buffa F, Jackson W (1983) A goal programming model for purchase planning. J Purch Mater Manag 19(3):27–34
15. Cai X, Lasdon L, Michelsen AM (2003) Group decision making in water resources planning using multiple objective analysis. J Water Resour Plann Manag 130(1):4–14
16. Calvete HI, Galé C, Mateo PM (2009) A genetic algorithm for solving linear fractional bilevel problems. Ann Oper Res 166:39–56
17. Candler W, Fortuny-Amat J, McCarl B (1981) The potential role of multilevel programming in agricultural economics. Am J Agric Econ 63(3):521–531
18. Calvete H, Gale C, Oliveros M, Sánchez-Valverde B (2007) A goal programming approach to vehicle routing problems with soft time windows. Eur J Oper Res 177(3):1720–1733
19. Chankong V, Haimes Y (1983) Multiobjective decision making: theory and methodology. North-Holland, New York
20. Charnes A, Cooper WW (1957) Management models and industrial applications of linear programming. Manag Sci 4(1):38–91
21. Charnes A, Cooper WW (1959) Chance-constrained programming. Manag Sci 6(1):73–79
22. Clarke FH (1990) Optimization and nonsmooth analysis. Society for Industrial Mathematics, Philadelphia
23. Chen Y, Florian M (1995) The nonlinear bilevel programming problem: formulations, regularity and optimality conditions. Optim 32(3):193–209
24. Chu S, Zhu M, Zhu, L (2010) Goal programming models and DSS for manpower planning of airport baggage service. Multiple criteria decision making for sustainable energy and transportation systems. Springer, Berlin pp 189–199
25. Clarke FH (1994) Optimization and Nonsmooth Analysis, In: SIAM classics in applied mathematics, vol 5. Wiley, New York. Reprint, Philadelphia
26. Coase RH (1960) The problem of social cost. J Law Econo 3(1):1–44

27. Coit D, Smith A (2005) Reliability optimization of series-parallel systems using a genetic algorithm. IEEE Trans Reliab 45(2):254–260
28. Dempe S, Dutta J, Mordukhovich BS (2007) Newnecessary optimality conditions in optimistic bilevel programming. Optimization 56(5–6):577–604
29. Dempe S, Zemkoho AB (2011) The generalized Mangasarian-Fromowitz constraint qualification and optimality conditions for bilevel programs. J Optim Theory Appl 148:46–68
30. Dempe S, Zemkoho A (2012) On the Karush-Kuhn-Tucker reformulation of the bilevel optimization problem. Nonlinear Anal 75(3):1202–1218
31. Dempe S, Dutta J (2012) Is bilevel programming a special case of a mathematical program with complementarity constraints? Math Program 131(1–2):37–48
32. Dimou CK, Koumousis VK (2003) Genetic algorithms in competitive environments. J Comput Civ Eng 17(3):142–149
33. Dougherty DE, Marryott RA (1991) Optimal groundwater management: 1. Simulated annealing. Water Resour Res 27(10):2493–2508
34. Durrett R, Durrett R (2010) Probability: theory and examples. Cambridge University Press, Cambridge
35. Dyer J (1972) Interactive goal programming. Manag Sci 19(1):62–70
36. Easter KW, Hearne R (1995) Water markets and decentralized water resources management: international problems and opportunities. JAWRA J Am Water Resour Assoc 31(1):9–20
37. Flegel ML (2005) Constraint qualification and stationarity concepts for mathematical programs with equilibrium constraints, Ph.D. Thesis, Institute of Applied Mathematics and Statistics. University of Würzburg, Würzburg
38. Flegel ML, Kanzow C (2006) A direct proof for M-stationarity under MPEC-GCQ for mathematical programs with equilibrium constraints. In: Dempe S, Kalashnikov VV (ed) Optimization with Multivalued Mappings. Springer, New York, pp 111–122
39. Freed N, Glover F (1981) Simple but powerful goal programming models for discriminant problems. Eur J Oper Res 7(1):44–60
40. Fudenberg D, Tirole J (1991) Game theory. MIT Press, Cambridge
41. Gan Q, Zhang FC, Zhang ZY (2015) Multi-objective optimal allocation for regional water resources based on ant colony optimization algorithm. Int J Smart Home 9(5):103–110
42. Garrido A (2000) A mathematical programming model applied to the study of water markets within the Spanish agricultural sector. Ann Oper Res 94(1):105–123
43. Davis L (1991) Handbook of genetic algorithms. Van Nostrand Reinhold, New York
44. Gendreau M, Marcotte P, Savard G (1996) A hybrid tabu-ascent algorithm for the linear bilevel programming problem. J Glob Optim 8(3):217–233
45. Glover F, Kochenberger GA (2003) Handbook of metaheuristics. Kluwer Academic, Dordrecht
46. Giannessi F (2006) Variational analysis and generalized differentiation. J Optim Theory Appl 131(2):313–314
47. Hasuike T, Katagiri H, Ishii H (2008) Probability maximization model of 0–1 knapsack problem with random fuzzy variables. In: IEEE International Conference on Fuzzy Systems (FUZZ-IEEE 2008) (IEEE World Congress on Computational Intelligence). IEEE, pp 548–554
48. Heilpern S (1992) The expected value of a fuzzy number. Fuzzy Set Syst 47(1):81–86
49. Hejazi SR, Memariani A, Jahanshahloo G, Sepehri MM (2002) Linear bilevel programming solution by genetic algorithm. Comput Oper Res 29(13):1913–1925
50. Henrion R, Outrata JV (2001) A subdifferential condition for calmness of multifunctions. J Math Anal Appl 258(1):110–130
51. Henrion R, Outrata JV (2005) Calmness of constraint systems with applications. Math Program 104(2–3):437–464
52. Henrion R, Outrata JV, Surowiec TM (2010) A note on the relation between strong and M-stationarity for a class of mathematical programs with equilibrium constraints. Kybernetika 46(3):423–434

53. Hoheisel T, Kanzow C, Outrata JV (2010) Exact penalty results for mathematical programs with vanishing constraints. Nonlinear Anal 72:2514–2526
54. Holland JH (1975) Adaptation in natural and artificial systems: an introductory analysis with applications to biology, control, and artificial intelligence. University of Michigan Press, Ann Arbor
55. Holland J (1992) Adaptation in Natural and Artificial Systems. MIT Press Cambridge, MA
56. Hu Z, Chen Y, Yao L et al (2016) Optimal allocation of regional water resources: from a perspective of equity–efficiency tradeoff. Resour Conserv Recycl 109:102–113
57. Huang GH, Loucks DP (2000) An inexact two-stage stochastic programming model for water resources management under uncertainty. Civ Eng Environ Syst 17(2):95–118
58. Hwang SF, He RS (2006) Improving real-parameter genetic algorithm with simulated annealing for engineering problems. Adv Eng Softw 37(6):406–418
59. Ignizio JP (1976) Goal programming and extensions. Lexington Books, Lexington
60. Ijiri Y (1965) Management Goals and Accounting for Control. North-Holland, Amsterdam
61. Isendahl N, Dewulf A, Pahl-Wostl C (2010) Making framing of uncertainty in water management practice explicit by using a participant-structured approach. J Environ Manag 91(4):844–851
62. Ishibuchi H, Murata T (1998) A multi-objective genetic local search algorithm and its application to flowshop scheduling. IEEE Trans Syst Man Cybern Part C Appl Rev 28(3):392–403
63. Jourani A (1994) Constraint qualifications and Lagrange multipliers in nondifferentiable programming problems. J Optim Theory Appl 81(3):533–548
64. Kabe D (1980) On some coverage probability maximization problems. Commun Stat-Simul Comput 9(1):73–79
65. Karsak E, Sozer S, Alptekin S (2003) Product planning in quality function deployment using a combined analytic network process and goal programming approach. Comput Ind Eng 44(1):171–190
66. Kavitha S, Thyagharajan KK (2016) Efficient DWT-based fusion techniques using genetic algorithm for optimal parameter estimation. Soft Comput 9:1–10
67. Kendall K, Lee S (1980) Formulating blood rotation policies with multiple objectives. Manag Sci 26(11):1145–1157
68. Kim JE, Jr RDE (2008) Permutation-based elitist genetic algorithm for optimization of large-sized resource-constrained project scheduling. J Constr Eng Manag 134(11):904–913
69. Kirkpatrick S, Gelatt CD, Vecchi MP (1983) Optimization by simulated annealing. Science 220(4598):671–680
70. Koza J (1992) Genetic programming: on the programming of computers by means of natural selection. MIT Press, Cambridge, MA
71. Kuo RJ, Lee YH, Zulvia FE et al (2015) Solving bi-level linear programming problem through hybrid of immune genetic algorithm and particle swarm optimization algorithm. Appl Math Comput 266(C):1013–1026
72. Lai YJ (1996) Hierarchical optimization: a satisfactory solution. Fuzzy Sets Syst 77(3):321–335
73. Lee CY (2003) Entropy-Boltzmann selection in the genetic algorithms. IEEE Trans Syst Man Cybern Part B 33(1):138–149
74. Lee ES (2001) Fuzzy multiple level programming. Appl Math Comput 120(1):79–90
75. Lee J (1993) New monte carlo algorithm: entropic sampling. Phys Rev Lett 71(2):211–214
76. Lee J, Kim S (2000) Using analytic network process and goal programming for interdependent information system project selection. Comput Oper Res 27(4):367–382
77. Lee S (1972) Goal programming for decision analysis. Auerbach, Philadelphia
78. Lee Z, Su S, Chuang C, Liu K (2008) Genetic algorithm with ant colony optimization (GA-ACO) for multiple sequence alignment. Appl Soft Comput 8(1):55–78
79. Li J, Xu J, Gen M (2006) A class of multiobjective linear programming model with fuzzy random coefficients. Math Comput Model 44(11–12):1097–1113

80. Li R (1990) Multiple Objective Decision Making in a Fuzzy Environment. Ph.D. thesis, Kansas State University
81. Lin FT, Kao CY, Hsu CC (1993) Applying the genetic approach to simulated annealing in solving some NP-hard problems. IEEE Trans Syst Man Cybern Part B 23(6):1752–1767
82. Liu M, Sun Z, Yan J, Kang J (2011) An adaptive annealing genetic algorithm for the job-shop planning and scheduling problem. Expert Syst Appl 38(8):9248–9255
83. Loucks DP, Beek EV, Stedinger JR, Dijkman, JPM, Villars MT (2005) Water resources systems planning and management: an introduction to methods, models and applications. UNESCO, Paris
84. Loucks DP, Stedinger JR, Douglas AH (1981) Water resource systems planning and analysis. Prentice-Hall, Englewood Cliffs
85. Lv Y, Huang GH, Yang ZF, Liu Y, Cheng GH (2010) Planning regional water resources system using an interval fuzzy bi-level programming method. J Environ Inform 16(2):43–56
86. Majumdar A, Ghosh D (2015) Genetic algorithm parameter optimization using Taguchi Robust design for multi-response optimization of experimental and historical data. Astrophysics 45(1):52–63
87. Metropolis N, Rosenbluth AW, Rosenbluth MN, Teller AH, Teller E (1953) Equation of state calculations for fast computing machines. J Chem Phys 21(6):1087–1092
88. Michalewicz Z (1996) Genetic Algorithms + Data Structures. Springer, New York
89. Ming CL, Hui MW, Wu S et al (2015) A bi-level inventory replenishment strategy using clustering genetic algorithm. Eur J Ind Eng 9(6):774
90. Mohanta DK, Sadhu PK, Chakrabarti R (2007) Deterministic and stochastic approach for safety and reliability optimization of captive power plant maintenance scheduling using GA/SA-based hybrid techniques: a comparison of results. Reliab Eng Syst Saf 92(2):187–199
91. Mohamed RH (1997) The relationship between goal programming and fuzzy programming. Fuzzy Sets Syst 89(2):215–222
92. Mordukhovich BS (1984) Nonsmooth analysis with nonconvex generalized differentials and conjugate mappings. Dokl Akad Nauk BSSR 28:976–979
93. Mordukhovich B (1994) Stability theory for parametric generalized equations and variational inequalities via nonsmooth analysis. Trans Am Math Soc 343(2):609–657
94. Mordukhovich BS (1994) Lipschitzian stability of constraint systems and generalized equations. Nonlinear Anal 22(2):173–206
95. Mordukhovich BS (1994) Generalized differential calculus for nonsmooth and set-valued mappings. J Math Anal Appl 183(1):250–288
96. Mordukhovich BS (2006) Variational analysis and generalized differentiation. I: basic theory. II: applications. Springer, Berlin
97. Mordukhovich BS, Outrata JV (2007) Coderivative analysis of quasi-variational inequalities with applications to stability and optimization. SIAM J Control Optim 18(2):389–412
98. Mordukhovich BS, Outrata JV (2007) Coderivative analysis of quasi-variational inequalities with applications to stability and optimization. SIAM J Optim 18(2):389–412
99. Narasimhan R (1980) Goal programming in a fuzzy environment. Decis Sci 11(2):325–336
100. Nemhauser GL (1967) Introduction to dynamic programming. Wiley, New York
101. Nicklow J, Reed P, Savic D, Dessalegne T et al (2010) State of the art for genetic algorithms and beyond in water resources planning and management. J Water Res Plann Manag 136(4):412–432
102. Pal B, Basu I (1995) A goal programming method for solving fractional programming problems via dynamic programming. Optim 35(2):145–157
103. Pal BB, Moitra BN (2003) A goal programming procedure for solving problems with multiple fuzzy goals using dynamic programming. Eur J Oper Res 144(3):480–491
104. Robinson SM (1981) Some continuity properties of polyhedral multifunctions. Math Program Stud 14:206–214
105. Rajasekaran S (2000) On simulated annealing and nested annealing. J Glob Optim 16(1): 43–56

106. Rechenberg I (1973) Evolutionsstrategie, Optimierung technischer Systeme nach Prinzipien der biologischen Evolution, Volume 15 von Reihe Problemata. F
107. Rockafellar RT (1993) Lagrange multipliers and optimality. SIAM Rev 35:183–238
108. Rockafellar RT, Wets, RJB (2009) Variational analysis, vol 317. Springer, Berlin/New York
109. Sahin KH, Cirit AR (1998) A dual temperature simulated annealing approach for solving bilevel programming problems. Comput Chem Eng 23(1):11–25
110. Sakawa M, Kato K, Katagiri H (2004) An interactive fuzzy satisficing method for multiobjective linear programming problems with random variable coefficients through a probability maximization model. Fuzzy Sets Syst 146(2):205–220
111. Sakawa M, Nishizaki I (2002) Interactive fuzzy programming for decentralized two-level linear programming problems. Fuzzy Set Syst 125(3):301–315
112. Scheel H, Scholtes S (2000) Mathematical programs with complementarity constraints: stationarity, optimality, and sensitivity. Math Oper Res 25(1):1–22
113. Schneider M (1973) Probability maximization in networks. In: Proceedings of international conference on transportation research, pp. 748–755
114. Schniederjans M (1984) Linear goal programming. Petrocelli Books, New York
115. Schwefel H (1995) Evolution and optimum seeking. Sixth-generation computer technology series. Wiley, New York
116. Shieh HJ, Peralta RC (2005) Optimal in situ bioremediation design by hybrid genetic algorithm-simulated annealing. J Water Res Plann Manag 131(1):67–78
117. Simaan M, Cruz JB (1973) On the Stackelberg strategy in nonzero-sum games. J Optim Theory Appl 11(5):533–555
118. Sinha S, Sinha SB (2002) KKT transformation approach for multi-objective multi-level linear programming problems. Eur J Oper Res 143(1):19–31
119. Singh P, Kumari S, Singh P (2016) Fuzzy efficient interactive goal programming approach for multi-objective transportation problems. Int J Appl Comput Math doi:10.1007/s40819-016-0155-x
120. Surowiec TM (2010) Explicit stationary conditions and solution characterization for equilibrium problems with equilibrium constraints. Ph.D. Thesis, Mathematisch-Naturwissentschaften Fakulät II, Humbolt Universität zu Berlin (2010)
121. Syswerda G, Palmucci J (1991) The application of Genetic Algorithms to resource scheduling. In: International conference on genetic algorithms, pp 502–508
122. Tiryaki F (2006) Interactive compensatory fuzzy programming for decentralized multi-level linear programming (DMLLP) problems. Fuzzy Set Syst 157(23):3072–3090
123. Toksari MD, Bilim T (2015) Interactive fuzzy goal programming based on Jacobian matrix to solve decentralized bi-level multi-objective fractional programming problems. Int J Fuzzy Syst 17(4):1–10
124. UNESCAP (United Nations Economic and Social Commission for Asia and the Pacific), Principles and practices of water allocation among water-use sectors (2000). ESCAP Water Resources Series No. 80, Bangkok
125. Viessman W, Schilling KE (2015) Social and environmental objectives in water resources planning and management. American Society of Civil Engineers, New York
126. Wang L, Fang L, Hipel KW (2008) Basin-wide cooperative water resources allocation. Eur J Oper Res 190(3):798–817
127. Wang P (1997) Speeding up the search process of genetic algorithm by fuzzy logic. In: Proceedings 5th European Congress on Intelligent Techniques and Soft Computing, pp 665–671
128. Wang X, Sun Y, Song L, Mei C (2009) An eco-environmental water demand based model for optimising water resources using hybrid genetic simulated annealing algorithms. Part I. Model development. J Environ Manag 90(8):2612–2619
129. White DJ, Anandalingam G (1993) A penalty function approach for solving bi-level linear programs. J Glob Optim 3(4):397–419
130. Wierzbicki A (1999) Reference Point Approaches multicriteria decision making. Adv MCDM Models Algorithms Appl 21(9):9–13

131. Windarto (2016) An implementation of continuous genetic algorithm in parameter estimation of predator-prey model. AIP Conf Proc 1718(1):3342–3345
132. Xu JP, Yao LM (2011) Random-like multiple objective decision making. Springer, Berlin/Heidelberg
133. Xu J, Zeng Z (2010) Applying optimal control model to dynamic equipment allocation problem: case study of concrete-faced rockfill dam construction project. J Constr Eng Manag 137(7):536–550
134. Ye JJ, Zhu DL (1995) Optimality conditions for bilevel programming problems. Optimization 33(1):9–7
135. Ye JJ, Ye XY (1997) Necessary optimality conditions for optimization problems with variational inequality constraints. Math Oper Res 22(4):977–997
136. Ye JJ, Zhu DL, Zhu QJ (1997) Exact penalization and necessary optimality conditions for generalized bilevel programming problems. SIAM J Control Optim 7(2):481–507
137. Ye JJ (2000) Constraint qualifications and necessary optimality conditions for optimization problems with variational inequality constraints. SIAM J Control Optim 10(4):943–962
138. Ye JJ (2005) Necessary and sufficient optimality conditions for mathematical programs with equilibrium constraints. J Math Anal Appl 307(1):350–369
139. Yun Y, Gen M, Seo S Various hybrid methods based on genetic algorithm with fuzzy logic controller. J Intell Manuf 14(3–4):401–419
140. Zimmermann H (1987) Fuzzy sets, decision making, and expert systems. Springer, Berlin

Chapter 3
Bi-Level Decision Making in Ra-Ra Phenomenon

Abstract The Ra-Ra phenomenon has been studied in many problems, such as flow shop scheduling problems (Zhou and Xu, Int J Uncertain Fuzziness Knowl-Based Syst 17(6):807–831, 2009), portfolio selection (Yan, Mod Appl Sci 3(6):126–131, 2009), and vendor selection (Xu and Ding, Int J Prod Econ 131(2):709–720, 2011). These studies have demonstrated that imprecision and complexity cannot always be dealt using simple random variables, so it is necessary to be mindful of the Ra-Ra when seeking to solve practical problems. To clearly describe the Ra-Ra phenomenon in bi-level decision making models, this chapter examines a practical construction example, namely, a transport flow distribution problem (TFDP) in a large-scale construction project. First, the problem background is introduced and the bi-level problem described to elucidate the motivations for considering the Ra-Ra phenomenon in the TFDP. Then, the form, transformation methods, and the properties of the general bi-level decision making model with Ra-Ra coefficients are discussed, after which a heuristic algorithm to solve the bi-level model is designed and a numerical example used to exemplify the effectiveness of the algorithm. Finally, a real-world TFDP case at a construction site is given.

Keywords Ra-Ra phenomenon • Ra-Ra EEDE model • Ra-Ra ECEC model • Ra-Ra DCCC model • Transport flow distribution

3.1 Transport Flow Distribution Problem

Network flow optimization has a significant role in combinatorial optimization. The transport flow distribution problem (TFDP) is a branch of the minimal cost network flow problem (MCNFP), which encompasses a wide category of problems. The MCNFP is important for many real-world applications such as communications [24, 33], informatics [5], and transportation [21]. Other well-known problems such as shortest path problems and assignment problems are considered special MCNFP cases [31, 42].

J. Xu et al., *Random-Like Bi-level Decision Making*, Lecture Notes in Economics and Mathematical Systems 688, DOI 10.1007/978-981-10-1768-1_3

In recent decades, the TFDP has been widely researched, with many models and algorithms being developed, such as [13, 14, 53]. These studies, however, have not often considered carrier type selection and transportation time. Yet, both costs and time control are vital considerations in construction projects, and especially in large-scale construction projects in which transportation costs are largely based on the carrier rates, which in turn, have a significant influence on transportation time. Under real conditions, there is increasing pressure to shorten transportation times to reduce or eliminate additional project expenses, as the early arrival of materials shortens construction project completion time and improves construction efficiency. In these cases, it is necessary to include both the carrier type selection and transportation time when conducting the transportation network analysis. In this chapter, a multi-objective bi-level TFDP is studied. On the upper level, the construction contractor determines the material flow on each transportation network path, with the criteria being a minimization of both direct costs and total transportation time costs. On the lower level, the transportation manager controls each carrier's flow so that total transportation costs are minimized.

All research presented above was based on deterministic transportation networks. However, transportation systems are often complex, so decision-makers inevitably encounter uncertain parameters when seeking to make decisions. Within the last two decades, studies into network flow problems have increasingly exploited the use of multilevel-models with uncertain parameters. For example, Watling [46] studied a user equilibrium traffic network assignment problem with stochastic travel times and a late arrival penalty. Chen and Zhou [10] developed a α-reliable mean-excess traffic equilibrium model with stochastic travel times. Lin [22] constructed a revised stochastic flow network to model a realistic computer network in which each arc had a lead time and a stochastic capacity. Sumalee et al. [43] dealt with a reliable network design problem which examined uncertain demand and total travel time reliability. In actual analyses, randomness is considered an important source of uncertainty. Yet within the TFDP, randomness is seen to be increasingly complex because of often incomplete or uncertain information. To date, there has been little research which has consideredmulti-level two-fold uncertainty coefficients for a TFDP.

The TFDP proposed in this chapter is a multi-objective bi-level programming problem. First introduced by Geoffrion and Hogan [19], and consequently developed by researchers such as Tarvainen and Haimes [44], Osman et al. [30], and Calvete and Galéb [8], multi-objective bi-level programming has greatly improved in both theory and practice. While previous studies have significantly contributed to many applications, to the best of our knowledge, there is no known research that has considered TFDP modeling. As bi-level programming problems are intrinsically difficult, it is not surprising that most exact algorithms to date have focused on the simplest bi-level programming cases; that is, problems which have had relatively easy to determine properties such as linear, quadratic or convex objectives and/or constraint functions [12]. Since the proposed bi-level TFDP model is nonlinear, non-convex and non-differentiable, it follows that a search for exact formally efficient algorithms is futile, so instead, effective heuristic algorithms need to be found to solve the TFDP. Determining a global optimal solution to the TFDP is very important. Specifically, the multiple objectives discussed in this chapter are dealt with by employing non-dominated solutions rather than applying weighted sum scalarization. Further, a multi-objective bi-level particle swarm optimization (MOBLPSO) is developed to solve a real world TFDP at the SBY Hydropower Project.

In construction projects, and especially in large scale construction projects, TFDPs are important. Here, we discuss a multi-objective bi-level TFDP in a large-scale construction project. To establish the model, a full description is given.

3.1.1 Ra-Ra Phenomenon

There is a strong motivation for taking the Ra-Ra environment into account for the TFDP.

In real world situations, transportation plans are usually made before any transportation activity commences; however, as a determined value for some parameters cannot be estimated in advance, uncertainty needs to be accounted for in such transportation problems. The large scale construction project TFDP considered in this chapter is subject to twofold randomness and has incomplete and/or uncertain information. As an example of this uncertainty, transportation time on an arc cannot be fixed because of possible transportation environmental effects, such as traffic accidents, traffic congestion, vehicle breakdowns, bad weather, natural disasters, and special events [17]. Therefore, transportation time is often subject to a stochastic distribution. Generally, transportation time approximately follows a normal distribution, which is expressed as $\mathcal{N}(\tilde{\mu}, \sigma^2)$ [25, 34], and which has to be truncated to avoid negative values. However, the expected transportation time value $\tilde{\mu}$ is also uncertain because this is dependent on the carrier's speed, which in turn is influenced by such uncertainties as vehicle condition and weather. Despite this, it is possible to specify a realistic distribution (e.g., normal distribution) for the parameter $\tilde{\mu}$ using statistical methods, related expertise or other knowledge. When the value of $\tilde{\mu}$ is a random variable which approximately follows a normal distribution, this overlapping randomness pattern is said to be Ra-Ra. A flowchart which shows transportation time as a Ra-Ra variable is shown in Fig. 3.1.

The situation is similar with transportation costs. However, due to the fluctuation in gasoline prices over time, the mean transportation costs are considered to approximately follow a normal distribution, which results in a Ra-Ra randomness in the transportation costs. From the description above, the Ra-Ra variable is employed to take account of hybrid uncertainty to establish a more feasible network flow scheme.

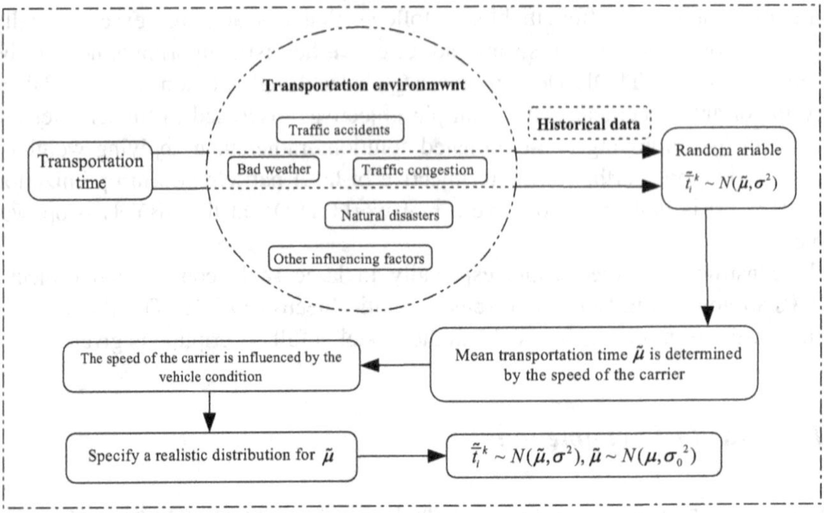

Fig. 3.1 Flowchart of transportation time as a Ra-Ra variable

3.1.2 Bi-Level Description

The TFDP considers the construction contractor and the transportation manager as the two participants. In a large-scale construction project, materials often have supply (origin) and receipt (destination) nodes, with the construction contractor generally assigning the work to a specialized transportation company. The bi-level model concurrently considers the construction contractor and the transportation manager in the specialized transportation company, but gives priority to the contractors benefit by considering the influence of the their decision-making on the flow distribution of the transportation carriers. As both cost and time control are important in construction projects, overall effectiveness also needs to be considered. The construction contractor assigns the material flows to each transportation path to minimize both the direct costs and the transportation time costs, while the transportation manager aims to minimize the transportation costs by making decisions about each carriers material flows on the transportation path based on the construction contractor's decision making, which in turn influences the contractor's decision-making through adjustments to each carrier material flow along the transportation path.

Therefore, the TFDP in this chapter can be abstracted as a bi-level programming problem. To conveniently model the problem, the transportation network is considered a bipartite network represented by a graph with sets of nodes and arcs. In the network, a node represents the network facilities such as a station or a yard, and an arc represents the line between two adjacent facilities. The TFDP model structure is shown in Fig. 3.2.

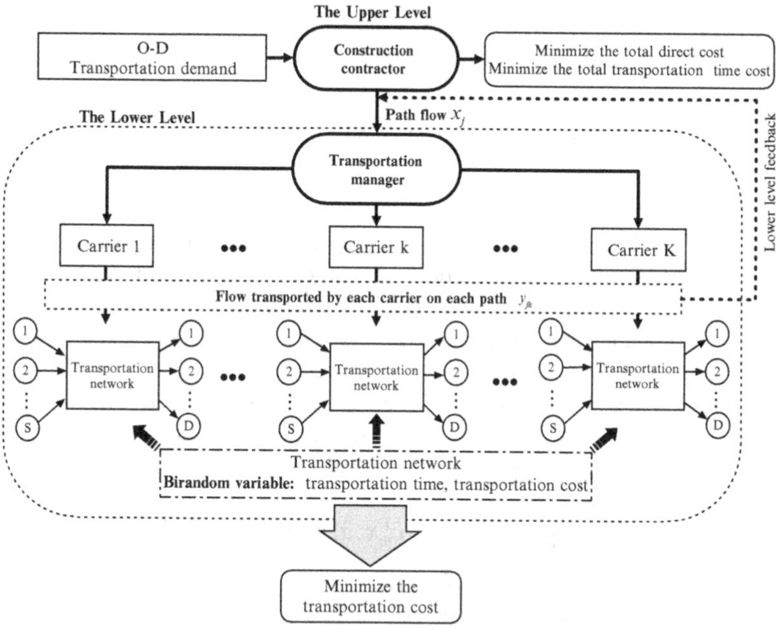

Fig. 3.2 Model structure for TFDP

3.2 Bi-Level Decision Making Model with Ra-Ra Coefficients

The general form for the single-objective bi-level decision making model with Ra-Ra coefficients is as follows:

$$\begin{cases} \max\limits_{x \in R^{n_1}} F(\boldsymbol{x}, \boldsymbol{y}, \boldsymbol{\xi}) \\ \text{s.t.} \begin{cases} G_i(\boldsymbol{x}, \boldsymbol{y}, \boldsymbol{\xi}) \le 0, i = 1, 2, \cdots, q \\ \text{where } y \text{ solves:} \\ \begin{cases} \max\limits_{y \in R^{n_2}} f(\boldsymbol{x}, \boldsymbol{y}, \boldsymbol{\xi}) \\ \text{s.t. } g_j(\boldsymbol{x}, \boldsymbol{y}, \boldsymbol{\xi}) \le 0, j = 1, 2, \cdots, p, \end{cases} \end{cases} \end{cases} \tag{3.1}$$

where, $x \in R^{n_1}$ is decision variable for the upper-level decision maker and $y \in R^{n_2}$ is decision variable for the lower-level decision maker; $F(\boldsymbol{x}, \boldsymbol{y}, \boldsymbol{\xi})$ is objective function of the upper-level programming and $f(\boldsymbol{x}, \boldsymbol{y}, \boldsymbol{\xi})$ is objective function of the lower-level programming; $G_i(\boldsymbol{x}, \boldsymbol{y}, \boldsymbol{\xi})$ is the constraint of the upper-level programming and $g_j(\boldsymbol{x}, \boldsymbol{y}, \boldsymbol{\xi})$ is the constraint of the lower-level programming. $\boldsymbol{\xi}$ is a Ra-Ra vector on the probability space $(\Omega, \mathscr{A}, Pr)$.

We give the general form of the bi-level multi-objective decision making model with Ra-Ra coefficients as follows, and the discussion of this section is based on this model,

$$\begin{cases} \min\limits_{x \in R^{n_1}} [F_1(\boldsymbol{x},\boldsymbol{y},\boldsymbol{\xi}), F_2(\boldsymbol{x},\boldsymbol{y},\boldsymbol{\xi}), \cdots, F_m(\boldsymbol{x},\boldsymbol{y},\boldsymbol{\xi})] \\ \text{s.t.} \begin{cases} G_i(\boldsymbol{x},\boldsymbol{y},\boldsymbol{\xi}) \leq 0, i = 1, 2, \cdots, q \\ \text{where } y \text{ solves:} \\ \begin{cases} \min\limits_{y \in R^{n_2}} [f_1(\boldsymbol{x},\boldsymbol{y},\boldsymbol{\xi}), f_2(\boldsymbol{x},\boldsymbol{y},\boldsymbol{\xi}), \cdots, f_m(\boldsymbol{x},\boldsymbol{y},\boldsymbol{\xi})] \\ \text{s.t. } g_j(x,y,\xi) \leq 0, j = 1, 2, \cdots, p, \end{cases} \end{cases} \end{cases} \tag{3.2}$$

where, $x \in R^{n_1}$ is decision variable for the upper-level decision maker and $y \in R^{n_2}$ is decision variable for the lower-level decision maker; $F(\boldsymbol{x},\boldsymbol{y},\boldsymbol{\xi})$ is objective function of the upper-level programming and $f(\boldsymbol{x},\boldsymbol{y},\boldsymbol{\xi})$ is objective function of the lower-level programming; $G_i(\boldsymbol{x},\boldsymbol{y},\boldsymbol{\xi})$ is the constraint of the upper-level programming and $g_j(\boldsymbol{x},\boldsymbol{y},\boldsymbol{\xi})$ is the constraint of the lower-level programming.

We also give the general form of the linear bi-level multi-objective decision making model with Ra-Ra coefficients as follows, and the discussion of transformation technology in this section is based on this model,

$$\begin{cases} \max\limits_{x \in R^{n_1}} \left[\bar{\bar{c}}_1^T \boldsymbol{x}, \bar{\bar{c}}_2^T \boldsymbol{x}, \cdots, \bar{\bar{c}}_m^T \boldsymbol{x}, \bar{\bar{a}}_1^T \boldsymbol{y}, \bar{\bar{a}}_2^T \boldsymbol{y}, \cdots, \bar{\bar{a}}_m^T \boldsymbol{y}\right] \\ \text{s.t.} \begin{cases} \bar{\bar{e}}_i^T \boldsymbol{x} \leq \bar{\bar{b}}_i, i = 1, 2, \cdots, q \\ \text{where } \boldsymbol{y} \text{ solves:} \\ \begin{cases} \min\limits_{y \in R^{n_2}} \left[\bar{\bar{a}}_1^T \boldsymbol{y}, \bar{\bar{a}}_2^T \boldsymbol{y}, \cdots, \bar{\bar{a}}_m^T \boldsymbol{y}\right] \\ \text{s.t. } \bar{\bar{k}}_j^T \boldsymbol{x} \leq \bar{\bar{d}}_j, j = 1, 2, \cdots, p \\ \boldsymbol{x} > 0, \boldsymbol{y} > 0, \end{cases} \end{cases} \end{cases} \tag{3.3}$$

where $\bar{\bar{c}}_i = (\bar{\bar{c}}_{i1}, \bar{\bar{c}}_{i2}, \cdots, \bar{\bar{c}}_{im})^T$, $\bar{\bar{a}}_j = (\bar{\bar{a}}_{j1}, \bar{\bar{a}}_{j2}, \cdots, \bar{\bar{a}}_{jm})^T$, $\bar{\bar{e}} = (\bar{\bar{e}}_1, \bar{\bar{e}}_2, \cdots, \bar{\bar{e}}_q)^T$, $\bar{\bar{k}} = (\bar{\bar{k}}_1, \bar{\bar{k}}_2, \cdots, \bar{\bar{k}}_p)^T$ are Ra-Ra vectors; $\bar{\bar{b}}_i$ and $\bar{\bar{d}}_j$ are Ra-Ra variables, $i = 1, 2, \cdots, q$, $j = 1, 2, \cdots, p$. This is a typical linear bi-level Ra-Ra multi-objective problem.

In this chapter, three kinds of bi-level multi-objective models "EEDE", "ECCE" and "DCCC" are introduced. Those three kinds of models contains all six philosophies for a single level programming, namely EE, EC, DE, DC, CE and CC. Before that, some elements of Ra-Ra operators are provided as follows.

1. Ra-Ra Expected Value Operator

   The expected value of a Ra-Ra variable is somewhat similar to random expected value in form. The expected value operator and variance of the Ra-Ra variable are defined as follows [47].

Definition 3.1. Let ξ be a Ra-Ra variable defined on the probability space $(\Omega, \mathscr{A}, Pr)$. Then the expected value of Ra-Ra variable ξ is defined as

$$E[\xi] = \int_0^{\infty} Pr\{\omega \in \Omega | E[\xi(\omega)] \geq t\} dt - \int_{-\infty}^{0} Pr\{\omega \in \Omega | E[\xi(\omega)] \leq t\} dt,$$

provided that at least one of the above two integrals is finite.

Similarly to the variance of random variables, the variance of the Ra-Ra variable is given as follows.

Definition 3.2. Let ξ be a Ra-Ra variable with finite expected value $E[\xi]$. Then the variance of Ra-Ra variable ξ is defined as

$$V[\xi] = E[(\xi - E[\xi])^2]. \tag{3.4}$$

By the definition of expected value operator of Ra-Ra variables, we can compute the expected value of some Ra-Ra variables with special distribution. Next, let's define the expected value of discrete Ra-Ra variables in another way.

Definition 3.3 (Expected value of discrete Ra-Ra variables). Let ξ be a discrete Ra-Ra variable on $(\Omega, \mathscr{A}, Pr)$, its expected value can be defined as follows,

$$E[\xi] = \sum_{\omega_i \in \Omega} Pr\{\omega = \omega_i\} E[\xi_i], \tag{3.5}$$

where $\xi_i (i = 1, 2, \cdots)$ are random variables on the probability space $(\Omega_i, \mathscr{A}_i, Pr_i)$.

By the above definition, we know that if ξ_i degenerates to be a certain number or a certain function, then ξ degenerates to be a random variable on $(\Omega, \mathscr{A}, Pr)$. Next, let's restrict our attention to three kinds of special discrete Ra-Ra variables and discuss their expected value.

Theorem 3.1. *Let ξ be a 0–1 distributed Ra-Ra variable and $\bar{p} \sim \mathscr{U}(a, b)$, where $0 < a < b < 1$. Then we have*

$$E[\xi] = \frac{a + b}{2}. \tag{3.6}$$

Theorem 3.2. *Let ξ be a 0–1 distributed Ra-Ra variable and $\bar{p}$ holds that*

$$\begin{pmatrix} a_1 & a_2 & \cdots & a_m \\ p_1 & p_2 & \cdots & p_m \end{pmatrix} \tag{3.7}$$

where $0 < a_i < 1$ and $\sum_{i=1}^{m} p_i = 1$. Then we have

$$E[\xi] = \sum_{i=1}^{m} a_i p_i. \tag{3.8}$$

From Theorems 3.1 and 3.2, we know that the probability $\bar{p}$ of $Pr\{\omega = 1\}$ and only varies between 0 and 1. Then $0 < E[\xi] < 1$ must hold. The following theorem will present the variance of a 0–1 distributed Ra-Ra variable.

Theorem 3.3. *Let ξ be a 0–1 distributed Ra-Ra variable defined in Theorem 3.1. Then we have*

$$V[\xi] = \frac{a+b}{2} - \left(\frac{a+b}{2}\right)^2. \tag{3.9}$$

Theorem 3.4. *Let ξ be a 0–1 distributed Ra-Ra variable defined in Theorem 3.2. Then we have*

$$V[\xi] = \sum_{i=1}^{m} a_i^2 p_i - \left(\sum_{i=1}^{m} a_i p_i\right)^2. \tag{3.10}$$

The expected value of binomially distributed Ra-Ra variables is given as follows.

Theorem 3.5. *Let $\xi \sim B(n, k, \bar{p})$ be a binomially distributed Ra-Ra variable and $\bar{p} \sim \mathscr{U}(a, b)$, where $0 < a < b < 1$. Then we have*

$$E[\xi] = \frac{n(a+b)}{2}, \tag{3.11}$$

$$V[\xi] = \frac{(a+b)n}{2} + \frac{(b-a)^2 n^2}{12} - \frac{(b^2+ab+a^2)n}{3}. \tag{3.12}$$

Theorem 3.6. *Let ξ be a binomially distributed Ra-Ra variable and $\bar{p}$ holds that*

$$\begin{pmatrix} a_1 \ a_2 \cdots a_m \\ p_1 \ p_2 \cdots p_m \end{pmatrix} \tag{3.13}$$

where $0 < a_i < 1$ and $\sum_{i=1}^{m} p_i = 1$. Then we have

$$E[\xi] = n \sum_{i=1}^{m} a_i p_i, \tag{3.14}$$

$$V[\xi] = n \sum_{i=1}^{m} a_i p_i + (n^2 - n) \sum_{i=1}^{m} a_i^2 p_i - n^2 \left(\sum_{i=1}^{m} a_i p_i\right)^2. \tag{3.15}$$

For the poisson distributed Ra-Ra variables, we also obtain the following results.

Theorem 3.7. *Let $\xi \sim B(k, \bar{\lambda})$ be a Ra-Ra variable with $\bar{\lambda} \sim \mathscr{U}(a, b)$, where $b > a > 0$. Then we have*

$$E[\xi] = \frac{a+b}{2}, \tag{3.16}$$

$$V[\xi] = E[\xi^2] - (E[\xi])^2 = \frac{a+b}{2} + \frac{(b-a)^2}{12}. \tag{3.17}$$

Theorem 3.8. *Let ξ be a binomially distributed Ra-Ra variable and $\bar{\lambda}$ holds that*

$$\begin{pmatrix} a_1 \; a_2 \cdots a_m \\ p_1 \; p_2 \cdots p_m \end{pmatrix} \tag{3.18}$$

where $0 < a_i < 1$ and $\sum_{i=1}^{m} p_i = 1$. Then we have

$$E[\xi] = \sum_{i=1}^{m} a_i p_i, \tag{3.19}$$

$$V[\xi^2] = \sum_{i=1}^{m} a_i^2 p_i + \sum_{i=1}^{m} a_i p_i - \left(\sum_{i=1}^{m} a_i p_i\right)^2. \tag{3.20}$$

For the continuous Ra-Ra variable, we have the following definition which is different from Definition 3.1 in form.

Definition 3.4 (Expected value of continuous Ra-Ra variables). Let ξ be a continuous Ra-Ra variable on $(\Omega, \mathscr{A}, Pr)$ with the density function $\bar{p}(x)$, then its expected value can be defined as follows,

$$E[\xi] = \int_0^{\infty} Pr\left\{\int_{x\in\Omega^*} x\bar{p}(x)dx \geq r\right\} dr - \int_{-\infty}^{0} Pr\left\{\int_{x\in\Omega^*} x\bar{p}(x)dx \leq r\right\} dr, \tag{3.21}$$

where $\bar{p}(x)$ is a random function defined on $(\Omega^*, \mathscr{A}^*, Pr^*)$.

The proofs for above theorems can be found in [47].

2. Ra-Ra Chance-Constrained Operator

   Let's recall the basic definition and property of the chance of Ra-Ra events.

Definition 3.5 (Peng and Liu [32]). Let ξ be a Ra-Ra variable, and B a Borel set of **R**. Then the chance of Ra-Ra event $\xi \in B$ is a function from (0, 1] to [0, 1], defined as

$$Ch\{\xi \in B\}(\alpha) = \sup_{Pr\{A\}\geq\alpha} \inf_{\omega\in A} Pr\{\xi(\omega) \in B\}, \tag{3.22}$$

where α is a prescribed probability level. The value of primitive chance at α is called α-chance.

We all note that the symbol Pr appears twice in the right side of equality (3.22). In fact, they represent different meanings. In other words, the overloading allows us to use the same symbol Pr for different probability measures, because we can deduce the exact meaning in the context.

Based on Definition 3.5, a natural idea is to provide a confidence level α at which it is desired that the stochastic constrains hold. Let's still consider the following model,

$$\begin{cases} \max f(\boldsymbol{x}, \boldsymbol{\xi}) \\ \text{s.t.} \begin{cases} g_r(\boldsymbol{x}, \boldsymbol{\xi}) \leq 0, r = 1, 2, \cdots, p \\ \boldsymbol{x} \in X, \end{cases} \end{cases} \tag{3.23}$$

where $f(\boldsymbol{x}, \boldsymbol{\xi})$ and $g_r(\boldsymbol{x}, \boldsymbol{\xi})$, $r = 1, 2 \cdots, p$ are continuous functions in X and $\boldsymbol{\xi} = (\xi_1, \xi_2, \cdots, \xi_n)$ is a Ra-Ra vector on the probability space $(\Omega, \mathscr{A}, Pr)$. Based on the chance-constraint operator, the maximax Ra-Ra chance-constrained programming model (CCM) was proposed:

$$\begin{cases} \max \bar{f} \\ \text{s.t.} \begin{cases} Ch\{f(\boldsymbol{x}, \boldsymbol{\xi}) \geq \bar{f}\}(\alpha) \geq \beta \\ Ch\{g_r(\boldsymbol{x}, \boldsymbol{\xi}) \leq 0\}(\eta_r) \geq \theta_r, r = 1, 2, \cdots, p \\ x \in X, \end{cases} \end{cases} \tag{3.24}$$

where $\alpha, \beta, \eta_r, \theta_r$ are the predetermined confidence levels.

Definition 3.6. A solution $\mathbf{x} \in X$ is said to be the feasible solution of problem (3.24) if and only if $Ch\{f(\boldsymbol{x}, \boldsymbol{\xi}) \geq \bar{f}\} \geq \beta$ and $Ch\{g_r(\boldsymbol{x}, \boldsymbol{\xi}) \leq 0\} \geq \alpha_r$ for all r. For any feasible $\boldsymbol{x}$, if there is a solution $\boldsymbol{x}^*$ such $\bar{f}^* > \bar{f}$, then $\boldsymbol{x}^*$ is called the optimal solution.

If the objective is to be minimized (for example, the objective is a cost function), the CCP model should be as follows,

$$\begin{cases} \min \bar{f} \\ \text{s.t.} \begin{cases} Ch\{f(\boldsymbol{x}, \boldsymbol{\xi}) \leq \bar{f}\}(\alpha) \geq \beta \\ Ch\{g_r(\boldsymbol{x}, \boldsymbol{\xi}) \leq 0\}(\eta_r) \geq \theta_r, r = 1, 2, \cdots, p \\ \boldsymbol{x} \in X, \end{cases} \end{cases} \tag{3.25}$$

where $\alpha, \beta, \eta_r, \theta_r$ are the predetermined confidence levels. Similarly, we have the following definition,

Definition 3.7. A solution $\boldsymbol{x} \in X$ is said to be the feasible solution of problem (3.25) if and only if $Ch\{f(\boldsymbol{x}, \boldsymbol{\xi}) \leq \bar{f}\} \geq \beta$ and $Ch\{g_r(\boldsymbol{x}, \boldsymbol{\xi}) \leq 0\} \geq \alpha_r$ for all r. For any feasible $\mathbf{x}$, if there is a solution $\boldsymbol{x}^*$ such $\bar{f}^* < \bar{f}$, then $\boldsymbol{x}^*$ is called the optimal solution.

According to the above definition, we know that

$$Ch\{g_r(\boldsymbol{x}, \boldsymbol{\xi}) \leq 0\}(\eta_r) \geq \theta_r \Leftrightarrow Pr\{\omega | Pr\{g_r(\boldsymbol{x}, \boldsymbol{\xi}) \leq 0\} \geq \theta_r\} \geq \eta_r, r = 1, 2, \cdots, p.$$

For given confidence levels η_r, θ_r, using the primitive chance measure we have the chance constraints as follows,

$$Pr\{\omega | Pr\{g_r(\boldsymbol{x}, \boldsymbol{\xi}) \leq 0\} \geq \theta_r\} \geq \eta_r, r = 1, 2, \cdots, p.$$

Thus a point $\boldsymbol{x}(\geq 0)$ is called feasible for problem (3.25) if and only if the probability measures of the random events $\{\omega|Pr\{g_r(\boldsymbol{x},\boldsymbol{\xi}) \leq 0\} \geq \theta_r\}$ are at least $\eta_r, r = 1, 2, \cdots, p$. Since

$$Ch\{f(\boldsymbol{x},\boldsymbol{\xi}) \leq \bar{f}\}(\alpha) \geq \beta \Leftrightarrow Pr\{\omega|Pr\{f(\boldsymbol{x},\boldsymbol{\xi}) \leq \bar{f}\} \geq \beta\} \geq \alpha,$$

then problem (3.25) can be rewritten as

$$\begin{cases} \min \bar{f} \\ \text{s.t.} \begin{cases} Pr\{\omega|Pr\{f(\boldsymbol{x},\boldsymbol{\xi}) \leq \bar{f}\} \geq \beta\} \geq \alpha \\ Pr\{\omega|Pr\{g_r(\boldsymbol{x},\boldsymbol{\xi}) \leq 0\} \geq \theta_r\} \geq \eta_r, r = 1, 2, \cdots, p \\ \boldsymbol{x} \in X, \end{cases} \end{cases} \tag{3.26}$$

where $\alpha, \beta, \eta_r, \theta_r$ are the predetermined confidence levels, $r = 1, 2, \cdots, p$. Similarly, if decision makers want to maximize the objective value, then problem (3.24) can be rewritten as

$$\begin{cases} \max \bar{f} \\ \text{s.t.} \begin{cases} Pr\{\omega|Pr\{f(\boldsymbol{x},\boldsymbol{\xi}) \geq \bar{f}\} \geq \beta\} \geq \alpha \\ Pr\{\omega|Pr\{g_r(\boldsymbol{x},\boldsymbol{\xi}) \leq 0\} \geq \theta_r\} \geq \eta_r, r = 1, 2, \cdots, p \\ \boldsymbol{x} \in X, \end{cases} \end{cases} \tag{3.27}$$

where $\alpha, \beta, \eta_r, \theta_r$ are the predetermined confidence levels, $r = 1, 2, \cdots, p$.

3. Ra-Ra dependent constraint operator

   Uncertain environment, event, and the chance function are key elements in DCM. Let us redefine them in Ra-Ra decision systems, and introduce the principle of uncertainty. By uncertain environment (in this case the Ra-Ra environment) we mean the Ra-Ra constraints represented by

$$g_j(\boldsymbol{x},\boldsymbol{\xi}) \leq 0, j = 1, 2, \cdots, p, \tag{3.28}$$

   where $\boldsymbol{x}$ is a decision vector, and $\boldsymbol{\xi}$ is a Ra-Ra vector. By the event we mean the system of inequalities

$$h_k(\boldsymbol{x},\boldsymbol{\xi}) \leq 0, k = 1, 2, \cdots, q. \tag{3.29}$$

   The chance function of an event ε characterized by (3.29) is defined as the chance measure of the event ε, i.e.,

$$f(\boldsymbol{x}) = Ch\{h_k(\boldsymbol{x},\boldsymbol{\xi}) \leq 0, k = 1, 2, \cdots, q\}, \tag{3.30}$$

   subject to the uncertain environment (3.28).

For each decision $\boldsymbol{x}$ and realization $\boldsymbol{\xi}$, an event ε is said to be consistent in the uncertain environment if the following two conditions hold: (i) $h_k(\boldsymbol{x}, \boldsymbol{\xi}) \leq 0, k = 1, 2, \cdots, q$; and (ii) $g_j(\boldsymbol{x}, \boldsymbol{\xi}) \leq 0, j \in J$, where J is the index set of all dependent constraints.

Assume that there are m events ε_i characterized by $h_{ik}(\boldsymbol{x}, \boldsymbol{\xi}) \leq 0, k = 1, 2, \cdots, q_i$ for $i = 1, 2, \cdots, m$ in the uncertain environment $g_j(\boldsymbol{x}, \boldsymbol{\xi}) \leq 0, j = 1, 2, \cdots, p$. The principle of uncertainty implies that the chance function of the ith event ε_i in the uncertain environment is

$$f_i(\boldsymbol{x}) = Ch \left\{ \begin{array}{l} h_{ik}(\boldsymbol{x}, \boldsymbol{\xi}) \leq 0, k = 1, 2, \cdots, q \\ g_j(\boldsymbol{x}, \boldsymbol{\xi}) \leq 0, j \in J_i, \end{array} \right\} \tag{3.31}$$

where J_i are defined by

$$J_i = \{j \in \{1, 2, \cdots, p\} | g_j(\boldsymbol{x}, \boldsymbol{\xi}) \leq 0 \textit{ is a dependent constraint of } \varepsilon_i\},$$

for $i = 1, 2, \cdots, m$.

3.3 Ra-Ra EEDE Model

Because of the existence of twofold random parameters, we usually cannot find the precise decision for a complicated real-life problem. Hence, an efficient tool should be provided to convert the Ra-Ra parameter into a crisp one. EEDE denotes that expected value operator is used to deal with the uncertain variable in upper-level objectives and constraints as well as the lower-level constraints. At the same time, dependent-chance operator is used to deal with the lower-level objectives.

3.3.1 General Form of Ra-Ra EEDE Model

The expected value of uncertain variable serves as a powerful tool for a wide variety of applications. It is naturally desirable to introduce the concept of the expected value of a Ra-Ra variable. The expected value operator will be adopted when decisions are made from the view point the expectation. The relative definitions, theorems, lemmas and proofs can be found in [47]. When the decision-maker wishes to maximize the chance functions of these events (i.e. the probabilities of satisfying the events), dependent-chance operator can be used [38]. When adopting EEDE operator, we optimize the expected values of the objective functions subject to the expected constraints in the upper-level programming, and we optimize the dependent-chance of the objective functions subject to the expected constraints in the lower-level programming.

Let's consider the typical single objective bi-level model (3.1) with Ra-Ra parameters. It follows from the expected operator and dependent-chance operator (with α-chance measure) that

$$\begin{cases} \max\limits_{x\in R^{n_1}} E[F(\boldsymbol{x},\boldsymbol{y},\boldsymbol{\xi})] \\ \text{s.t.} \begin{cases} E[G_i(\boldsymbol{x},\boldsymbol{y},\boldsymbol{\xi})] \le 0, i = 1,2,\cdots,q \\ \text{where } \boldsymbol{y} \text{ solves:} \\ \begin{cases} \max\limits_{y\in R^{n_2}} Ch\{f(\boldsymbol{x},\boldsymbol{y},\boldsymbol{\xi}) \ge \bar{f}\}(\alpha) \\ \text{s.t. } E[g_j(\boldsymbol{x},\boldsymbol{y},\boldsymbol{\xi})] \le 0, j = 1,2,\cdots,p. \end{cases} \end{cases} \end{cases} \tag{3.32}$$

After being dealt with by expected value operator and dependent-chance operator, the problem (3.1) has been converted into a certain programming and then decision makers can easily obtain the optimal solution.

Remark 3.1. If the Ra-Ra vector ξ degenerates to a random vector, then for any given $\alpha > 0$,

$$Ch\{f(x,\xi) \le 0\}(\alpha) \equiv Pr\{h(x,\xi) \le 0\}.$$

Thus, the model (3.32) becomes

$$\begin{cases} \max\limits_{x\in R^{n_1}} E[F(\boldsymbol{x},\boldsymbol{y},\boldsymbol{\xi})] \\ \text{s.t.} \begin{cases} E[G_i(\boldsymbol{x},\boldsymbol{y},\boldsymbol{\xi})] \le 0, i = 1,2,\cdots,q \\ \text{where } \boldsymbol{y} \text{ solves:} \\ \begin{cases} \max\limits_{y\in R^{n_2}} Pr\{f(\boldsymbol{x},\boldsymbol{y},\boldsymbol{\xi}) \ge \bar{f}\} \\ \text{s.t. } E[g_j(\boldsymbol{x},\boldsymbol{y},\boldsymbol{\xi})] \le 0, j = 1,2,\cdots,p, \end{cases} \end{cases} \end{cases} \tag{3.33}$$

which is a standard single objective bi-level EEDE model.

Definition 3.8. The feasible set of problem (3.33) is defined as

$$\Omega = \{(x,y) : \boldsymbol{x} \in R^{n_1}, \boldsymbol{y} \in R^{n_2}, E[G_i(\boldsymbol{x},\boldsymbol{y},\boldsymbol{\xi})] \le 0, E[g_j(\boldsymbol{x},\boldsymbol{y},\boldsymbol{\xi})] \le 0\}.$$

Definition 3.9. $\boldsymbol{x}$ is said to be a *upper-level feasible solution* of problem (3.32) if and only if, for any given y, $E[g_j(\boldsymbol{x},\boldsymbol{y},\boldsymbol{\xi})] \le 0$ $(j = 1,2,\cdots,p)$. For any feasible solution $\boldsymbol{x}$, if $E[F(\boldsymbol{x}^*,\boldsymbol{y},\boldsymbol{\xi})] \ge E[F(\boldsymbol{x},\boldsymbol{y},\boldsymbol{\xi})]$, then $\boldsymbol{x}^*$ is the optimal solution of problem (3.32).

Definition 3.10. $\boldsymbol{y}$ is said to be a *lower-level feasible solution* of problem (3.32) if and only if, for any given x, $E[G_i(\boldsymbol{x},\boldsymbol{y},\boldsymbol{\xi})] \le 0$ $(i = 1,2,\cdots,q)$. For any feasible solution $\boldsymbol{y}$, if $E[f(\boldsymbol{x},\boldsymbol{y}^*,\boldsymbol{\xi})] \ge E[f(\boldsymbol{x},\boldsymbol{y},\boldsymbol{\xi})]$, then $\boldsymbol{y}^*$ is the optimal solution of problem (3.32).

In this section, we focus on the multi-objective bi-level EEDE model, which can be developed as follows,

$$\begin{cases} \max\limits_{x\in R^{n_1}} \{E[F_1(x,y,\xi)], E[F_2(x,y,\xi)], \cdots, E[F_m(x,y,\xi)]\} \\ \text{s.t.} \begin{cases} E[G_i(x,y,\xi)] \le 0, i = 1,2,\cdots,m_1 \\ \text{where } y \text{ solves:} \\ \begin{cases} \max\limits_{y\in R^{n_2}} \begin{bmatrix} Ch\{f_1(x,y,\xi) \ge \bar{f}_1\}(\alpha_1) \\ Ch\{f_2(x,y,\xi) \ge \bar{f}_2\}(\alpha_2) \\ \cdots \\ Ch\{f_r(x,y,\xi) \ge \bar{f}_{m_2}\}(\alpha_{m_2}) \end{bmatrix} \\ \text{s.t. } E[g_i(x,y,\xi)] \le 0, i = 1,2,\cdots,p_2, \end{cases} \end{cases} \end{cases} \tag{3.34}$$

which has the equivalent form as follows:

$$\begin{cases} \max\limits_{x\in R^{n_1}} \{E[F_1(x,y,\xi)], E[F_2(x,y,\xi)], \cdots, E[F_{m_1}(x,y,\xi)]\} \\ \text{s.t.} \begin{cases} E[G_i(x,y,\xi)] \le 0, i = 1,2,\cdots,p_1 \\ \text{where } y \text{ solves:} \\ \begin{cases} \max\limits_{y\in R^{n_2}} \begin{bmatrix} Pr\{\omega|Pr\{f_1(x,y,\xi(\omega)) \ge \bar{f}_1\} \ge \alpha_1\} \\ Pr\{\omega|Pr\{f_1(x,y,\xi(\omega)) \ge \bar{f}_2\} \ge \alpha_2\} \\ \cdots \\ Pr\{\omega|Pr\{f_1(x,y,\xi(\omega)) \ge \bar{f}_{m_2}\} \ge \alpha_{m_2}\} \\ \end{bmatrix} \\ \text{s.t. } E[g_i(x,y,\xi)] \le 0, i = 1,2,\cdots,p_2. \end{cases} \end{cases} \end{cases} \tag{3.35}$$

For problem (3.34), some concepts of solutions are defined.

Definition 3.11. x^* is said a feasible solution to (3.34) if and only if it is a Pareto optimal solution of the lower model, as well as it satisfies $E[G_i(x,y,\xi)] \le 0, i = 1,2,\cdots,m_1$.

Definition 3.12. A feasible solution x^* is said to be Pareto optimal solution of problem (3.34), if and only if there doesn't exist feasible solutions x such that

$$E[F_u(x,y,\xi)] \ge E[F_u(x^*,y,\xi)], u = 1,2,\cdots,m,$$

and there is at least one $u(u = 1,2,\cdots,m)$ such that $E[F_u(x,y,\xi)] > E[F_u(x^*,y,\xi)]$.

3.3.2 The Steepest Descent Direction Method for Linear Model

In this section, we focus on the linear version of EEDE model as follows:

$$\begin{cases} \max\limits_{x\in R^{n_1}} [E[\widetilde{\widetilde{C}}_1^T x + \widetilde{\widetilde{D}}_1^T y], E[\widetilde{\widetilde{C}}_2^T x + \widetilde{\widetilde{D}}_2^T y], \cdots, E[\widetilde{\widetilde{C}}_{m_1}^T x + \widetilde{\widetilde{D}}_{m_1}^T y]] \\ \text{s.t.} \begin{cases} E[\widetilde{\widetilde{A}}_i^T x + \widetilde{\widetilde{B}}_i^T y] \le 0, i = 1, 2, \cdots, p_1 \\ \text{where } y \text{ solves:} \\ \begin{cases} \max\limits_{y\in R^{n_2}} [Ch\{\tilde{\bar{c}}_1^T x + \tilde{\bar{d}}_1^T y \ge \bar{f}_1\}(\alpha_1^L), Ch\{\tilde{\bar{c}}_2^T x + \tilde{\bar{d}}_2^T y \ge \bar{f}_2\}(\alpha_2^L), \\ \cdots, Ch\{\tilde{\bar{c}}_{m_2}^T x + \tilde{\bar{d}}_{m_2}^T y \ge \bar{f}_{m_2}\}(\alpha_{m_2}^L)] \\ \text{s.t. } E[\tilde{\bar{a}}_i^T x + \tilde{\bar{b}}_i^T y \le 0] \le 0, i = 1, 2, \cdots, p_2, \end{cases} \end{cases} \end{cases} \quad (3.36)$$

where α_i^L are given confidence levels, $\bar{f}_i$ are predetermined objective values. Model (3.36) also can be written as

$$\begin{cases} \max\limits_{x\in R^{n_1}} [E[\widetilde{\widetilde{C}}_1^T x + \widetilde{\widetilde{D}}_1^T y], E[\widetilde{\widetilde{C}}_2^T x + \widetilde{\widetilde{D}}_2^T y], \cdots, E[\widetilde{\widetilde{C}}_{m_1}^T x + \widetilde{\widetilde{D}}_{m_1}^T y]] \\ \text{s.t.} \begin{cases} E[\widetilde{A}_i^T x + \widetilde{B}_i^T y] \le 0, i = 1, 2, \cdots, p_1 \\ \text{where } y \text{ solves:} \\ \begin{cases} \max\limits_{y\in R^{n_2}} [\beta_1^L, \beta_2^L, \cdots, \beta_{m_2}^L] \\ \text{s.t.} \begin{cases} Ch\{\tilde{c}_1^T x + \tilde{d}_1^T y \ge \bar{f}_1\}(\alpha_1^L) \ge \beta_i^L, i = 1, 2, \cdots, m_2 \\ E[\tilde{a}_i^T x + \tilde{b}_i^T y] \le 0, i = 1, 2, \cdots, p_2. \end{cases} \end{cases} \end{cases} \end{cases} \quad (3.37)$$

Crisp Equivalent Model

In some cases, (3.36) can be converted into its crisp equivalent model.

Theorem 3.9. *Assume that Ra-Ra vector* $\widetilde{\widetilde{C}}_i = (\widetilde{\widetilde{C}}_{i1}, \widetilde{\widetilde{C}}_{i2}, \cdots, \widetilde{\widetilde{C}}_{in})^T$ *is normally distributed with mean vector* $\widetilde{C}_i = (\widetilde{C}_{i1}, \widetilde{C}_{i2} \cdots, \widetilde{C}_{in})^T$ *and positive definite covariance matrix* V_i^C*, written as* $\widetilde{\widetilde{C}}_i \sim \mathcal{N}(\widetilde{C}_i, V_i^C)$*, where* $\widetilde{C}_i$ *are normally distributed random vectors, written as* $\widetilde{C}_i \sim \mathcal{N}(\mu_i^C, V_i'^C)$*. Similarly,* $\widetilde{\widetilde{D}}_i \sim \mathcal{N}(\widetilde{D}_i, V_i^D)$*, where* $\widetilde{D}_i \sim \mathcal{N}(\mu_i^D, V_i'^D)$*. All the Ra-Ra variables are assumed to be independent to each other. Then* $E[\widetilde{\widetilde{C}}_1^T x + \widetilde{\widetilde{D}}_1^T y]$ *is equivalent to*

$$\sum_{j=1}^{n} (\mu_i^{CT} x + \mu_i^{DT} y).$$

Proof. It follows from the normally distribution and independence that

$$\widetilde{\widetilde{C}}_1^T x + \widetilde{\widetilde{D}}_1^T y \sim \mathcal{N}\left(\sum_{j=1}^{n} \widetilde{C}_i^T x + \widetilde{D}_i^T y, x^T V_i^C x + y^T V_i^D D\right)$$

(linearity of Ra-Ra expected operator). Thus,

$$E\left[\widetilde{\widetilde{C}}_1^T x + \widetilde{\widetilde{D}}_1^T y\right] = E\left[\sum_{j=1}^{n} \widetilde{C}_i^T x + \widetilde{D}_i^T y\right] = \sum_{j=1}^{n} \left(\mu_i^{CT} x + \mu_i^{DT} y\right).$$

This completes he proof. □

Theorem 3.9 implies that expected values depend on mean values of the random variables but not the variances. For $E[\widetilde{A}_i^T x + \widetilde{B}_i^T y] \le 0$ and $E[\widetilde{a}_i^T x + \widetilde{b}_i^T y] \le 0$, there are the similar conclusions.

For $Ch\{\tilde{\tilde{c}}_{m_2}^T x + \tilde{\tilde{d}}_{m_2}^T y$, we have the following theorem.

Theorem 3.10. *Assume that* $\tilde{\tilde{c}}_i \sim \mathcal{N}(\bar{c}_i, V_i^c)$ *with* $\bar{c}_i \sim \mathcal{N}(\mu_i^c, V_i'^c)$ *and* $\tilde{\tilde{d}}_i \sim \mathcal{N}(\bar{d}_i, V_i^d)$ *with* $\bar{d}_i \sim \mathcal{N}(\mu_i^d, V_i'^d)$. *Then* $Ch\{\tilde{\tilde{c}}_i^T x + \tilde{\tilde{d}}_i^T y \ge \bar{f}_i\}(\alpha_i^L) \ge \beta_i^L$ *is equivalent to*

$$\bar{f}_i \le (\mu_i^{cT} x + \mu_i^{dT} y) + \Phi^{-1}(1 - \beta_i^L)\sqrt{x^T V_i^c x + y^T V_i^d y}$$
$$+ \Phi^{-1}(1 - \alpha_i^L)\sqrt{x^T V_i'^c x + y^T V_i'^d y}.$$

Proof. It follows from the definition of Ra-Ra chance that

$$Ch\{\tilde{\tilde{c}}_i^T x + \tilde{\tilde{d}}_i^T y \ge \bar{f}_1\}(\alpha_1^L) \ge \beta_i^L$$

can be written as

$$Pr\{\omega | Pr\{\tilde{\tilde{c}}_i^T x + \tilde{\tilde{d}}_i^T y \ge \bar{f}_i\} \ge \beta_i^L\} \ge \alpha_i^L.$$

From the assumption,

$$\tilde{\tilde{c}}_i^T x + \tilde{\tilde{d}}_1^T y \sim \mathcal{N}(\bar{c}_i^T x + \bar{d}_i^T y, x^T V_i^c x + y^T V_i^d y).$$

From Theorem 2.5,

$$Pr\{\tilde{\tilde{c}}_i^T x + \tilde{\tilde{d}}_i^T y \ge \bar{f}_i\} \ge \beta_i^L$$

is equivalent to

$$\bar{f}_i \le \Phi^{-1}(1 - \beta_i^L)\sqrt{x^T V_i^c x + y^T V_i^d y} + \bar{c}_i^T x + \bar{d}_i^T y.$$

So

$$
\begin{aligned}
&Ch\{\tilde{\bar{c}}_1^T \boldsymbol{x} + \tilde{\bar{d}}_1^T \boldsymbol{y} \geq \bar{f}_1\}(\alpha_1^L) \geq \beta_i^L \\
&\Leftrightarrow Pr\{\omega | Pr\{\tilde{\bar{c}}_i^T \boldsymbol{x} + \tilde{\bar{d}}_i^T \boldsymbol{y} \geq \bar{f}_i\} \geq \beta_i^L\} \geq \alpha_i^L \\
&\Leftrightarrow Pr\left\{\tilde{\boldsymbol{c}}_i^T \boldsymbol{x} + \tilde{\boldsymbol{d}}_i^T \boldsymbol{y} \geq \bar{f}_i - \Phi^{-1}(1-\beta_i^L)\sqrt{\boldsymbol{x}^T V_i^c \boldsymbol{x} + \boldsymbol{y}^T V_i^d \boldsymbol{y}}\right\} \geq \alpha_i^L \\
&\Leftrightarrow Pr\left\{\frac{(\tilde{\boldsymbol{c}}_i^T \boldsymbol{x} + \tilde{\boldsymbol{d}}_i^T \boldsymbol{y}) - (\mu_i^{cT} \boldsymbol{x} + \mu_i^{dT} \boldsymbol{y})}{\sqrt{\boldsymbol{x}^T V_i'^c \boldsymbol{x} + \boldsymbol{y}^T V_i'^d \boldsymbol{y}}} \geq \right. \\
&\qquad \left. \frac{(\bar{f}_i - \Phi^{-1}(1-\beta_i^L)\sqrt{\boldsymbol{x}^T V_i^c \boldsymbol{x} + \boldsymbol{y}^T V_i^d \boldsymbol{y}}) - (\mu_i^{cT} \boldsymbol{x} + \mu_i^{dT} \boldsymbol{y})}{\sqrt{\boldsymbol{x}^T V_i'^c \boldsymbol{x} + \boldsymbol{y}^T V_i'^d \boldsymbol{y}}}\right\} \geq \alpha_i^L \\
&\Leftrightarrow Pr\left\{\frac{(\tilde{\boldsymbol{c}}_i^T \boldsymbol{x} + \tilde{\boldsymbol{d}}_i^T \boldsymbol{y}) - (\mu_i^{cT} \boldsymbol{x} + \mu_i^{dT} \boldsymbol{y})}{\sqrt{\boldsymbol{x}^T V_i'^c \boldsymbol{x} + \boldsymbol{y}^T V_i'^d \boldsymbol{y}}} \leq \right. \\
&\qquad \left. \frac{(\bar{f}_i - \Phi^{-1}(1-\beta_i^L)\sqrt{\boldsymbol{x}^T V_i^c \boldsymbol{x} + \boldsymbol{y}^T V_i^d \boldsymbol{y}}) - (\mu_i^{cT} \boldsymbol{x} + \mu_i^{dT} \boldsymbol{y})}{\sqrt{\boldsymbol{x}^T V_i'^c \boldsymbol{x} + \boldsymbol{y}^T V_i'^d \boldsymbol{y}}}\right\} \leq 1 - \alpha_i^L \\
&\Leftrightarrow \Phi\left(\frac{(\bar{f}_i - \Phi^{-1}(1-\beta_i^L)\sqrt{\boldsymbol{x}^T V_i^c \boldsymbol{x} + \boldsymbol{y}^T V_i^d \boldsymbol{y}}) - (\mu_i^{cT} \boldsymbol{x} + \mu_i^{dT} \boldsymbol{y})}{\sqrt{\boldsymbol{x}^T V_i'^c \boldsymbol{x} + \boldsymbol{y}^T V_i'^d \boldsymbol{y}}}\right) \leq 1 - \alpha_i^L \\
&\Leftrightarrow \left(\bar{f}_i - \Phi^{-1}(1-\beta_i^L)\sqrt{\boldsymbol{x}^T V_i^c \boldsymbol{x} + \boldsymbol{y}^T V_i^d \boldsymbol{y}}\right) - (\mu_i^{cT} \boldsymbol{x} + \mu_i^{dT} \boldsymbol{y}) \\
&-\Phi^{-1}(1-\alpha_i^L)\sqrt{\boldsymbol{x}^T V_i'^c \boldsymbol{x} + \boldsymbol{y}^T V_i'^d \boldsymbol{y}} \leq 0 \\
&\Leftrightarrow \bar{f}_i \leq (\mu_i^{cT} \boldsymbol{x} + \mu_i^{dT} \boldsymbol{y}) + \Phi^{-1}(1-\beta_i^L)\sqrt{\boldsymbol{x}^T V_i^c \boldsymbol{x} + \boldsymbol{y}^T V_i^d \boldsymbol{y}} \\
&+\Phi^{-1}(1-\alpha_i^L)\sqrt{\boldsymbol{x}^T V_i'^c \boldsymbol{x} + \boldsymbol{y}^T V_i'^d \boldsymbol{y}}.
\end{aligned}
$$

This completes the proof. □

The Steepest Descent Direction Method

By Theorems 3.9 and 3.10, model (3.36) can be converted into its crisp equivalent model. We can also use the classical multiple objective technique to deal with the objectives. Next, we use the steepest descent direction method to handle bi-level structure. This method is proposed by Savard and Gauvin [36].

For convenience, We consider the nonlinear bi-level programming problem (BLP):

$$\min_x F(x, y),$$

where y, slightly abusing notation, is solution of the lower level program

$$\begin{cases} \min_y f_0(x, y) \\ s.t. \begin{cases} f_i(x, y) \leq 0, i \in I \\ f_j(x, y) \leq 0, j \in J, \end{cases} \end{cases}$$

and I, J are finite sets of indices. For a fixed x, we denote $y(x)$ an optimal solution of the lower level program and $\Omega(x) = \{y : f_i(x, y) \leq 0, i \in I, f_i(x, y) = 0, j \in J\}$, the lower level feasible region. A point $(x, y(x))$ is called a rational point.

Assumption 1. The problem is well-posed, i.e. that for any x, the optimal solution $y(x)$ of the lower level problem is unique.

Assumption 2. If $I(x) = \{i \in I | f_i(x, y(x)) = 0\}$ denotes that the set of indices for the binding inequality constraints at $(x, y(x))$, then we have that the vectors

$$\nabla_y f_i(x, y(x)), i \in I(x) \cup J,$$

are linearly independent. This implies the existence of a unique Kuhn-Tucker vector $\lambda(x)$ corresponding to the optimal solution $y(x)$. We denote

$$E(x) = \{v | \nabla_y f_i(x, y(x))v = 0, i \in I(x) \cup J\},$$

the tangent subspace at $y(x)$.

Assumption 3. We assume a second-order sufficient optimality condition:

$$v^T \nabla_y^2 L(x, y(x); \lambda(x))v > 0, \forall v \in E(x), v \neq 0,$$

where $L(x, y, \lambda) = f_0(x, y) + \sum_{i \in I(x) \cup J} \lambda_i f_i(x, y)$ is the Lagrangian corresponding to the lower level program.

The first assumption ensures that the bi-level problem has a solution. Otherwise, the first level cannot impose his own choice within the set $y(x)$ which can lead to an ill-posed problem. The second assumption assumes a first-order necessary condition for the lower level problem. The second and third assumptions guarantee at least Lipschitzian behavior of the function $y(x)$. The second assumption can be replaced by the Mangasarian Fromowitz constraint qualification, in which case the Kuhn Tucker vectors are not necessarily unique, at the cost of a lightly more complicated formulation of the quadratic program defined below. The third assumption cannot be significantly relaxed.

Under the previous assumptions, we have the following first-order necessary optimality conditions.

Lemma 3.1. *Let $(x^*, y(x^*))$ be an optimal solution for BLP. Then for any first level direction z at x^*, the directional derivative of the objective function of the first level problem satisfies:*

$$F'(x^*, y(x^*); z) = \nabla_x F(x^*, y(x^*))z + \nabla_y F(x^*, y(x^*))w(x^*, z) \geq 0,$$

where $w(x, z)$ is the optimal solution for $x = x*$ the quadratic program:*

$$(QP(x;z)) : \begin{cases} \min\limits_{w} (z^T, w^T)\nabla^2_{(x,y)} L(x, y(x); \lambda(x))(z, w) \\ s.t. \begin{cases} \nabla_y f_i(x, y(x))w \leq -\nabla_x f_i(x, y(x))w, i \in I(x) \\ \nabla_y f_i(x, y(x))w = -\nabla_x f_i(x, y(x))w, i \in J \\ \nabla_y f_0(x, y(x))w = -\nabla_x f_0(x, y(x))w, i \in J + \nabla_x L(x, y(x); \lambda(x))z. \end{cases} \end{cases}$$

Now, we show in some general manner how the previous result can be exploited to develop a descent method for BLP. Let $(x_k, y(x_k))$ denote the solution at iteration k.

Step 1. Steepest descent direction. Solve the quadratic bi-level programming problem $QBLP(x_k)$ to obtain the steepest descent direction z_k with the corresponding direction $w(x_k, z_k)$ for the lower level program. If it happens that the optimal value of the quadratic $QBLP(x_k)$ is nonnegative then stop; $(x_k, y(x_k))$ satisfies the necessary optimality condition. Otherwise go to step 2.

Step 2. Step length computation. Compute a step length t_k such that

$$F(x_k + t_k z_k, y(x_k + t_k z_k)) < F(x_k, y(x_k)).$$

Return to Step 1 with the feasible point $(x_{k+l}, y(x_{k+l})) = (x_k + t_k z_k, y(x_k + t_k z_k))$ and with $k = k + 1$.

If convergence is obtained, the steepest descent algorithm will usually yield a locally optimal solution for BLP. Some global optimization scheme is then needed to explore for a global solution.

Remark 3.2. At Step 2 of the algorithm, it should be noticed that whenever $F(x_k + tz_k, y(x_k + tz_k))$ is required, the lower level program LLP$(x_k + tz_k)$ must be solved. The assumptions made on the problem assure that there exists $t^* > 0$ such that for all t with $0 \leq t < t^*$, the feasible set for the lower level $\Omega(x_k + tz_k)$ is nonempty and the lower level's optimal solution $y_k = y(x_k + tz_k)$ exists. Hence, the use of an approximate line search technique, adapted for nondifferential function, is justified

Remark 3.3. In the implementation of the algorithm, it will not be necessary to solve QBLP(x_k) exactly, as any rational solution with nonpositive objective yields a descent direction; this direction has less chance to be extremal. In any way, an efficient implementation of the algorithm must exploit the structure of the feasible set.

Numerical Example

In what follows, a numerical example is presented to illustrate the effectiveness of the proposed solution method to the linear Ra-Ra EEDE model.

Example 3.1. We consider the following Ra-Ra bi-level multiobjective linear programming problem, which has two objective functions of the leader and two objective functions of the follower:

$$\begin{cases} \max\limits_{x_1,x_2} F_1 = \tilde{\tilde{\xi}}_{11}x_1 + \tilde{\tilde{\xi}}_{12}x_2 + \tilde{\tilde{\xi}}_{13}y_1 + \tilde{\tilde{\xi}}_{14}y_2 \\ \max\limits_{x_1,x_2} F_2 = \tilde{\tilde{\xi}}_{21}x_1 + \tilde{\tilde{\xi}}_{22}x_2 + \tilde{\tilde{\xi}}_{23}y_1 + \tilde{\tilde{\xi}}_{24}y_2 \\ \text{s.t.} \begin{cases} \text{where } (y_1, y_2) \text{ solves} \\ \max\limits_{y_1,y_2} f_1 = \tilde{\tilde{\eta}}_{11}x_1 + \tilde{\tilde{\eta}}_{12}x_2 + \tilde{\tilde{\eta}}_{13}y_1 + \tilde{\tilde{\eta}}_{14}y_2 \\ \max\limits_{y_1,y_2} f_2 = \tilde{\tilde{\eta}}_{21}x_1 + \tilde{\tilde{\eta}}_{22}x_2 + \tilde{\tilde{\eta}}_{23}y_1 + \tilde{\tilde{\eta}}_{24}y_2 \\ \text{s.t.} \begin{cases} \tilde{\tilde{\zeta}}_1 x_1 + \tilde{\tilde{\zeta}}_2 x_2 + \tilde{\tilde{\zeta}}_3 y_1 + \tilde{\tilde{\zeta}}_4 y_2 \le 200 \\ 2x_1 - x_2 + 3y_1 + y_2 \le 100 \\ x_1, x_2, x_3, y_1, y_2 \ge 0, \end{cases} \end{cases} \end{cases} \tag{3.38}$$

where $\tilde{\tilde{\xi}}_{ij}, \tilde{\tilde{\eta}}_{ij}, \tilde{\tilde{\zeta}}_j (i = 1,2; j = 1,2,4)$ are Ra-Ra variables characterized as:

$\tilde{\tilde{\xi}}_{11} \sim \mathcal{N}(\tilde{\xi}_{11}, 2)$, with $\tilde{\xi}_{11} \sim \mathcal{N}(6,2)$; $\tilde{\tilde{\xi}}_{12} \sim \mathcal{N}(\tilde{\xi}, 2)$, with $\tilde{\xi}_{12} \sim \mathcal{N}(10,2)$;

$\tilde{\tilde{\xi}}_{13} \sim \mathcal{N}(\tilde{\tilde{\xi}}_{13}, 2)$, with $\tilde{\xi}_{13} \sim \mathcal{N}(8,2)$; $\tilde{\tilde{\xi}}_{14} \sim \mathcal{N}(\tilde{\xi}_{14}, 2)$, with $\tilde{\xi}_{14} \sim \mathcal{N}(12,2)$;

$\tilde{\tilde{\xi}}_{21} \sim \mathcal{N}(\tilde{\xi}_{21}, 2)$, with $\tilde{\xi}_{21} \sim \mathcal{N}(3,2)$; $\tilde{\tilde{\xi}}_{22} \sim \mathcal{N}(\tilde{\xi}_{22}, 2)$, with $\tilde{\xi}_{22} \sim \mathcal{N}(-5,2)$;

$\tilde{\tilde{\xi}}_{23} \sim \mathcal{N}(\tilde{\xi}_{23}, 2)$, with $\tilde{\xi}_{23} \sim \mathcal{N}(7,2)$; $\tilde{\tilde{\xi}}_{24} \sim \mathcal{N}(\tilde{\xi}_{24}, 2)$, with $\tilde{\xi}_{24} \sim \mathcal{N}(-9,2)$;

$\tilde{\tilde{\eta}}_{11} \sim \mathcal{N}(\tilde{\eta}_{11}, 1)$, with $\tilde{\eta}_{11} \sim \mathcal{N}(12,1)$; $\tilde{\tilde{\eta}}_{12} \sim \mathcal{N}(\tilde{\eta}_{12}, 1)$, with $\tilde{\eta}_{12} \sim \mathcal{N}(10,1)$;

$\tilde{\tilde{\eta}}_{13} \sim \mathcal{N}(\tilde{\eta}_{13}, 1)$, with $\tilde{\eta}_{13} \sim \mathcal{N}(8,1)$; $\tilde{\tilde{\eta}}_{14} \sim \mathcal{N}(\tilde{\eta}_{14}, 1)$, with $\tilde{\eta}_{14} \sim \mathcal{N}(6,1)$;

$\tilde{\tilde{\eta}}_{21} \sim \mathcal{N}(\tilde{\eta}_{21}, 1)$, with $\tilde{\eta}_{21} \sim \mathcal{N}(12,1)$; $\tilde{\tilde{\eta}}_{22} \sim \mathcal{N}(\tilde{\eta}, 1)$, with $\tilde{\eta}_{22} \sim \mathcal{N}(5,1)$;

$\tilde{\tilde{\eta}}_{23} \sim \mathcal{N}(\tilde{\eta}_{23}, 1)$, with $\tilde{\eta}_{23} \sim \mathcal{N}(18,1)$; $\tilde{\tilde{\eta}}_{24} \sim \mathcal{N}(\tilde{\eta}_{24}, 1)$, with $\tilde{\eta}_{24} \sim \mathcal{N}(20,1)$;

$\tilde{\tilde{\zeta}}_1 \sim \mathcal{N}(\tilde{\zeta}_1, 1)$, with $\tilde{\zeta}_1 \sim \mathcal{N}(6,1)$; $\tilde{\tilde{\zeta}}_2 \sim \mathcal{N}(\tilde{\zeta}_2, 1)$, with $\tilde{\zeta}_2 \sim \mathcal{N}(12,1)$;

$\tilde{\tilde{\zeta}}_3 \sim \mathcal{N}(\tilde{\zeta}_3, 1)$, with $\tilde{\zeta}_3 \sim \mathcal{N}(18,1)$; $\tilde{\tilde{\zeta}}_4 \sim \mathcal{N}(\tilde{\zeta}_4, 1)$, with $\tilde{\zeta}_4 \sim \mathcal{N}(24,1)$.

If the confidence levels are given by the decision makers, the EEDE model of (3.38) is:

$$\begin{cases} \max\limits_{x_1,x_2} E[\tilde{\bar{\xi}}_{11}x_1 + \tilde{\bar{\xi}}_{12}x_2 + \tilde{\bar{\xi}}_{13}y_1 + \tilde{\bar{\xi}}_{14}y_2] \\ \max\limits_{x_1,x_2} E[\tilde{\bar{\xi}}_{21}x_1 + \tilde{\bar{\xi}}_{22}x_2 + \tilde{\bar{\xi}}_{23}y_1 + \tilde{\bar{\xi}}_{24}y_2] \\ \text{s.t.} \begin{cases} \text{where } (y_1, y_2) \text{ solves} \\ \max\limits_{y_1,y_2}\{\beta_1^L, \beta_2^L\} \\ \text{s.t.} \begin{cases} Pr\{\omega | Pr\{\tilde{\bar{\eta}}_{11}(\omega)x_1 + \tilde{\bar{\eta}}_{12}(\omega)x_2 + \tilde{\bar{\eta}}_{13}(\omega)y_1 + \tilde{\bar{\eta}}_{14}(\omega)y_2 \le 300\} \ge \beta_1^L\} \ge 0.8 \\ Pr\{\omega | Pr\{\tilde{\bar{\eta}}_{21}(\omega)x_1 + \tilde{\bar{\eta}}_{22}(\omega)x_2 + \tilde{\bar{\eta}}_{23}(\omega)y_1 + \tilde{\bar{\eta}}_{24}(\omega)y_2 \le 200\} \ge \beta_2^L\} \ge 0.8 \\ E[\tilde{\bar{\zeta}}_1 x_1 + \tilde{\bar{\zeta}}_2 x_2 + \tilde{\bar{\zeta}}_3 y_1 + \tilde{\bar{\zeta}}_4 y_2] \le 200 \\ 2x_1 - x_2 + 3y_1 + y_2 \le 100 \\ x_1, x_2, y_1, y_2 \ge 0. \end{cases} \end{cases} \end{cases} \tag{3.39}$$

It follows from Theorems 3.9 and 3.10 that (3.39) is equivalent to

$$\begin{cases} \max\limits_{x_1,x_2} E[F_1] = 6x_1 + 10x_2 + 8y_1 + 12y_2 \\ \max\limits_{x_1,x_2} E[F_1] = 3x_1 - 5x_2 + 7y_1 - 9y_2 \\ \text{s.t.} \begin{cases} \text{where } (y_1, y_2) \text{ solves} \\ \max\limits_{y_1,y_2}\{\beta_1^L, \beta_2^L\} \\ \text{s.t.} \begin{cases} 300 \le 12x_1 + 10x_2 + 8y_1 + 6y_2 - 0.84\left(\sqrt{x_1^2 + x_2^2 + y_1^2 + y_2^2}\right) \\ +\Phi^{-1}(1 - \beta_1^L)\left(\sqrt{x_1^2 + x_2^2 + y_1^2 + y_2^2}\right) \\ 200 \le 12x_1 + 5x_2 + 18y_1 + 20y_2 - 1.68\left(\sqrt{x_1^2 + x_2^2 + y_1^2 + y_2^2}\right) \\ +\Phi^{-1}(1 - \beta_2^L)\left(\sqrt{x_1^2 + x_2^2 + y_1^2 + y_2^2}\right) \\ 6x_1 + 12x_2 + 18y_1 + 24y_2 \le 200 \\ 2x_1 - x_2 + 3y_1 + y_2 \le 100 \\ x_1, x_2, y_1, y_2 \ge 0. \end{cases} \end{cases} \end{cases} \tag{3.40}$$

By using the steepest descent direction method mentioned above, the solutions is $(x_1^*, x_2^*, y_1^*, y_2^*) = (22.6, 36.9, 0.7, 42.8)$, the corresponding objective function values are $(E[F_1^*], E[F_1^*], \beta_1^{*L}, \beta_2^{*L}) = (320.9, 114, 7, 0.7, 0.6)$.

3.3.3 Ra-Ra Simulation-Based PSO for Nonlinear Models

If (3.34) is a nonlinear model, i.e., some functions in model (3.34) is nonlinear, the random features make it impossible to be converted into its crisp equivalent model. In order solve nonlinear (3.34), three issues are needed to solve: the trade-off of the multiple objectives, the twofold randomness and the bi-level structure. In this section, we design a hybrid method consisting of simulation technique, lexicographic method and particle swarm optimization algorithm, where they are deal with twofold randomness, multiple objectives and bi-level structure, respectively.

Ra-Ra Simulation for Expected Value and Chance

Assume that $\boldsymbol{\xi}$ is an n-dimensional Ra-Ra vector on the probability space $(\Omega, \mathscr{A}, Pr)$, and $f : \mathbf{R}^n \rightarrow \mathbf{R}$ is a measurable function. One problem is to calculate the expected value $E[f(\boldsymbol{x}, \boldsymbol{\xi})]$ for the given $\boldsymbol{x}$. Note that, for each $\omega \in \Omega$, we may calculate the expected value $E[f(\boldsymbol{x}, \boldsymbol{\xi}(\omega))]$ by random simulation. Since $E[f(\boldsymbol{x}, \boldsymbol{\xi})]$ is essentially the expected value of random variable $E[f(\boldsymbol{x}, \boldsymbol{\xi}(\omega))]$, we may produce a Ra-Ra simulation by the following process.

Firstly, we sample $\omega_1, \omega_2, \cdots, \omega_N$ from Ω according to Pr, where ω_k is an n-dimensional vector. For each $\omega_i (i = 1, 2, \cdots, N)$, $\boldsymbol{\xi}(\omega_i)$ are all random variables. Then we can apply stochastic simulation to get their expected values, respectively. Randomly generate $\omega_i^1, \omega_i^2, \cdots, \omega_i^M$ from Ω according to the probability measure Pr for each $\boldsymbol{\xi}(\omega_i) (i = 1, 2, \cdots, N)$. Compute $f(\boldsymbol{x}, \boldsymbol{\xi}(\omega_i^1)), f(\boldsymbol{x}, \boldsymbol{\xi}(\omega_i^2)), \cdots, f(\boldsymbol{x}, \boldsymbol{\xi}(\omega_i^M))$. Then

$$E[f(\boldsymbol{x}, \boldsymbol{\xi}(\omega_i))] = \frac{\sum_{j=1}^{M} f(\boldsymbol{x}, \boldsymbol{\xi}(\omega_i^j))}{M}.$$

Next, we can get the expected value of $f(\boldsymbol{x}, \boldsymbol{\xi})$ as follows,

$$E[f(\boldsymbol{x}, \boldsymbol{\xi})] = \frac{\sum_{i=1}^{N} E[f(\boldsymbol{x}, \boldsymbol{\xi}(\omega_i))]}{N}.$$

Then the procedure simulating the expected value of the function $f(\boldsymbol{x}, \boldsymbol{\xi})$ can be summarized as follows:

Step 1. Set $e = l = 0$;
Step 2. Generate ω from Ω according to the probability measure Pr;
Step 3. For the ω, generate ω^i according to the probability measure Pr;
Step 4. Repeat the third step M times;
Step 5. $l \leftarrow l + f(\boldsymbol{x}, \boldsymbol{\xi}(\omega^i))$;
Step 6. Repeat the second to fifth step N times;
Step 7. $e \leftarrow e + l$;
Step 8. Return e/MN.

Example 3.2. We employ the Ra-Ra simulation to calculate the expected value of $\sqrt{\bar{\bar{\xi}}_1^2 + \bar{\bar{\xi}}_2^2 + \bar{\bar{\xi}}_3^2}$, where $\bar{\bar{\xi}}_1$, $\bar{\bar{\xi}}_2$ and $\bar{\bar{\xi}}_3$ are Ra-Ra variables defined as

$$\begin{aligned}
\bar{\bar{\xi}}_1 &= \mathscr{U}(\tilde{\rho}_1, \tilde{\rho}_1 + 2), &&\text{with } \tilde{\rho}_1 \sim \mathscr{N}(4, 1),\\
\bar{\bar{\xi}}_2 &= \mathscr{N}(\tilde{\rho}_2, 1), &&\text{with } \tilde{\rho}_2 \sim \mathscr{U}(3, 5),\\
\bar{\bar{\xi}}_3 &= exp(\tilde{\rho}_3), &&\text{with } \tilde{\rho}_3 \sim \mathscr{U}(1, 2).
\end{aligned}$$

A run of Ra-Ra simulation with 10,000 cycles shows that

$$E\left[\sqrt{\bar{\bar{\xi}}_1^2 + \bar{\bar{\xi}}_2^2 + \bar{\bar{\xi}}_3^2}\right] = 6.8409.$$

Since the existence of nonlinear functions $f_k(\boldsymbol{x}, \boldsymbol{\xi})$, it is difficult to convert the objective functions into the crisp ones. Then the Ra-Ra simulation technique is used to obtain the approximative chance measure. Suppose that $\boldsymbol{\xi}$ is an n-dimensional Ra-Ra vector defined on the probability space $(\Omega, \mathscr{A}, Pr)$, and $f : \mathbf{R}^n \to \mathbf{R}$ is a measurable function. For any real number $\alpha \in (0, 1]$, we design a Ra-Ra simulation to compute the α-chance $Ch\{f(\boldsymbol{x}, \boldsymbol{\xi}) \geq f\}(\alpha)$. That is, we should find the supremum $\bar{\beta}$ such that

$$Pr\{\omega \in \Omega | Pr\{f(\boldsymbol{x}, \boldsymbol{\xi}(\omega)) \geq f\} \geq \bar{\beta}\} \geq \alpha.$$

First, we sample $\omega_1, \omega_2, \cdots, \omega_N$ from Ω according to the probability measure Pr, and estimate $\beta_k = Pr\{f(\boldsymbol{x}, \boldsymbol{\xi}(\omega_k)) \geq f\}$ for $k = 1, 2, \cdots, N$ by random simulation. Let N' be the integer part of αN. Then the value $\bar{\beta}$ can be taken as the N'th largest element in the sequence $\{\beta_1, \beta_2, \cdots, \beta_N\}$.

Then the procedure simulating the α-chance $Ch\{f(\boldsymbol{x}, \boldsymbol{\xi}) \geq f\}(\alpha)$ can be summarized as follows:

Step 1. Generate $\omega_1, \omega_2, \omega_N$ from Ω according to the probability measure Pr;
Step 2. Compute the probability $\beta_k = Pr\{f(\boldsymbol{x}, \boldsymbol{\xi}(\omega_k)) \geq f\}$ for $k = 1, 2, \cdots, N$ by random simulation;
Step 3. Set N' as the integer part of αN
Step 4. Return the N'th largest element in $\{\beta_1, \beta_2, \cdots, \beta_N\}$.

Example 3.3. Consider the following Ra-Ra variables,

$$\begin{aligned}
\bar{\bar{\xi}}_1 &= \mathscr{U}(\tilde{\rho}_1, \tilde{\rho}_1 + 2), &&\text{with } \tilde{\rho}_1 \sim \mathscr{N}(4, 1),\\
\bar{\bar{\xi}}_2 &= \mathscr{N}(\tilde{\rho}_2, 1), &&\text{with } \tilde{\rho}_2 \sim \mathscr{U}(3, 5),\\
\bar{\bar{\xi}}_3 &= exp(\tilde{\rho}_3), &&\text{with } \tilde{\rho}_3 \sim \mathscr{U}(1, 2).
\end{aligned}$$

A run of Ra-Ra simulation with 10,000 cycles shows that

$$Ch\left\{\sqrt{\bar{\bar{\xi}}_1^2 + \bar{\bar{\xi}}_2^2 + \bar{\bar{\xi}}_3^2} \geq 6\right\}(0.8) = 0.5348.$$

Lexicographic Method

The basic idea of lexicographic method is to rank the objective function by its importance to decision makers and then resolve the next objective function after resolving the above one. We take the solution of the last programming problem as the final solution.

Consider the following multi-objective programming problem,

$$\begin{cases} \min[f_1(\boldsymbol{x}), f_2(\boldsymbol{x}), \cdots, f_m(\boldsymbol{x})] \\ \text{s.t. } \boldsymbol{x} \in X. \end{cases} \tag{3.41}$$

Without loss of generality, assume the rank as $f_1(\boldsymbol{x}), f_2(\boldsymbol{x}), \cdots, f_m(\boldsymbol{x})$ according to different importance. Solve the following single objective problem in turn,

$$\begin{cases} \min_{\boldsymbol{x}\in X} f_1(\boldsymbol{x}) \\ \text{s.t.} \begin{cases} \min f_i(\boldsymbol{x}) \\ \text{s.t.} \begin{cases} f_k(\boldsymbol{x}) = f_k(\boldsymbol{x}^k),\ k = 1, 2, \cdots, i-1, \\ \boldsymbol{x} \in X, \end{cases} \end{cases} \end{cases} \tag{3.42}$$

where $i = 1, 2, \cdots, m$, X is the feasible area and denote the feasible area of problem (3.42) as X^i.

Lemma 3.2. *Let $X \subset \mathbf{R}^n$, $f : X \to \mathbf{R}^m$. If $\boldsymbol{x}^m$ be the optimal solution by the lexicographic method, then $\boldsymbol{x}^m$ is an efficient solution of problem (3.41).*

Proof. If $\boldsymbol{x}^m$ is not an efficient solution of problem (3.41), there exists $\bar{\boldsymbol{x}} \in X$ such that $\boldsymbol{f}(\bar{\boldsymbol{x}}) \le \boldsymbol{f}(\boldsymbol{x}^m)$. Since $f_1(\boldsymbol{x}^m) = f_1^* = f_1(\boldsymbol{x}^1)$, $f_1(\bar{\boldsymbol{x}}) < f_1(\boldsymbol{x}^m)$ cannot hold. It necessarily follows that $f_1(\bar{\boldsymbol{x}}) = f_1(\boldsymbol{x}^m)$.

If we have proved $f_k(\bar{\boldsymbol{x}}) = f_k(\boldsymbol{x}^m)(k = 1, 2, \cdots, i-1)$, but $f_i(\bar{\boldsymbol{x}}) < f_i(\boldsymbol{x}^m)$. It follows that $\bar{\boldsymbol{x}}$ is a feasible solution of problem (3.42). Since $f_i(\bar{\boldsymbol{x}}) < f_i(\boldsymbol{x}^m) = f_i(\boldsymbol{x}^i)$, this results in the conflict with that $\boldsymbol{x}^i$ the optimal solution of problem (3.42). Thus, $f_k(\bar{\boldsymbol{x}}) = f_k(\boldsymbol{x}^m)(k = 1, 2, \cdots, i)$ necessarily holds. Then we can prove $f_k(\bar{\boldsymbol{x}}) = f_k(\boldsymbol{x}^m)(k = 1, 2, \cdots, m)$ by the mathematical induction. This conflicts with $\boldsymbol{f}(\bar{\boldsymbol{x}}) \le \boldsymbol{f}(\boldsymbol{x}^m)$. This completes the proof. □

C-PSO

In this section, we recall the C-PSO algorithm proposed by Zhang et al. [51].

The PSO algorithm is a member of the wide category of swarm intelligence methods used to solve global optimization problems. PSO is easily implemented and is computationally inexpensive and has been found to be quite successful in a wide variety of optimization problems [4, 7, 35, 50]. In these previous studies, almost all the algorithms used were improved PSO versions. The main reason for the development of these improved PSO versions is that premature PSO convergence is inevitable, particularly in problems with high dimensionality. Usually, the original PSO creates a new solution from the past solution, the previous best solution and the global best solution. The only information the original PSO borrows from other particles is the global best solution. In the C-PSO, the crossover operator in the real-coded genetic algorithm is implemented on the swarm members by borrowing the linear combination concept for two vectors, in which one vector is as warm member and the other vector is randomly selected from an elite set (see Algorithm: step 3). Note that the crossover probability is set at 100 % in this study. The crossover operator can improve the information exchange between the particles, thereby preventing a premature swarm convergence. For the ith particle at iteration k, the C-PSO algorithm can be described as in the following:

Suppose that the search space is D-dimensional, with the velocity and position represented as $v_i^k = (v_{i1}^k, v_{i2}^k, \cdots, v_{iD}^k)$ and $x_i^k = (x_{i1}^k, x_{i2}^k, \cdots, x_{iD}^k)$ respectively. The personal best position is presented as $p_i^k = (p_{i1}^k, p_{i2}^k, \cdots, p_{iD}^k)$ and the global best position is $p_g^k = (p_{g1}^k, p_{g2}^k, \cdots, p_{gD}^k)$. At each step, the particle is updated using the following equations:

$$\overline{x_i^k} = p_i * (x_i^k) + (1 - p_i) * (a_i^k), \tag{3.43}$$

$$v_i^{k+1} = w * (v_i^k) + c_1 * \xi * (p_i^k - \overline{x_i^k}) + c_2 * \xi\eta(p_i^k - \overline{x_i^k}), \tag{3.44}$$

$$x_i^{k+1} = x_i^k + v_i^{k+1}, \tag{3.45}$$

where x_i denotes the ith particle's position, ξ, η and p_i are random numbers distributed uniformly on $(0, 1)$ respectively. w, c_1 and c_2 are constants and a_i is selected randomly from A_i (A_i is the elite set which is introduced in the following part, see Algorithm: step 3).

The proposed algorithmic process is an interactive co-evolutionary process. We first initialize the population, and then interactively solve the high dimensional multi-objective optimization problems in the upper level and the lower levels using the C-PSO algorithm. From the first iteration, a set of approximate Pareto optimal solutions is obtained using the elite strategy. This interactive procedure is then repeated until accurate Pareto optimal solutions are found. The details of the proposed algorithm are given as follows:

Numerical Example

In this section, a numerical example is given to illustrate the effectiveness of the application of the models and algorithms.

Algorithm

Step 1. Initializing. (1) Initialize the population P_0 with N_u particles, each of which are composed of $n_s = N_u/N_l$ sub-swarms of size N_l. For the kth ($k = 1, 2, \cdots, n_s$) sub-swarm, the jth particle's position is presented as: $z_j = (x_j, y_j)(j = 1, 2, \cdots, n_l)$; the corresponding velocity is presented as: $v_j = (v_{x_j}, v_{y_j}(j = 1, 2, \cdots, n_l)$; and z_j and v_j are randomly sampled in the feasible space. (2) Initialize the external loop counter $t := 0$.
Step 2. For the kth sub-swarm, ($k = 1, 2, \cdots, n_s$), each particle is assigned a non-domination rank ND_l and a crowding value CD_l in F space. Then, all resulting sub-swarms are combined into one population named P_t, after which, each particle is assigned a non-domination rank ND_u and a crowding value CD_u in F space.
Step 3. The non-domination particles assigned to both $ND_u = 1$ and $ND_l = 1$ from P_t are saved in the elite set A_t.
Step 4. For the k-th sub-swarm, ($k = 1, 2, \cdots, n_s$), the lower level decision variables are updated.
(1) Initialize the lower level loop counter $tl := 0$. (2) Update the jth ($j = 1, 2, \cdots, N_l$) particle's position with the fixed x_j and the fixed v_{y_j} according to Eqs. (3.43), (3.44) and (3.45). (3) $t_l := t_l + 1$. (4) If $t_l \geq T_l$, go to step 4 (5). Otherwise, go to step 4(2). (5) Each particle of the ith sub-swarm is reassigned a non-domination rank ND_l and a crowding value CD_l in F space. Then, all resulting sub-swarms are combined into one population renamed Q_t, after which each particle is reassigned a non-domination rank ND_u and a crowding value CD_u in F space.
Step 5. Combine population P_t and Q_t to form R_t. The combined population R_t is reassigned a non-domination rank ND_u, and the particles within an identical nondomination rank are assigned a crowding distance value CD_u in the F space.
Step 6. Choose half the particles from Rt. The particles with the rank $ND_u = 1$ are considered first. From the particles with the rank $ND_u = 1$, the particles with $ND_l = 1$ are aligned one by one in the order of the reduced crowded distance CD_u, and for each such particle the corresponding sub-swarm from its source population (either P_t or Q_t) is copied to an intermediate population S_t. If a sub-swarm has already been copied to S_t and a future particle from the same sub-swarm is found to have $ND_u = NDl = 1$, the sub-swarm is not copied again. When all particles from $ND_u = 1$ have been considered, a similar consideration is then given to those particles from $ND_u = 2$, and so on until exactly n_s subswarms have been copied to S_t.
Step 7. Update the elite set At. The non-domination particles assigned both $ND_u = 1$ and $ND_l = 1$ from S_t are saved in the elite set A_t.
Step 8. Update the upper level decision variables in S_t. (1) Initiate the upper level loop counter $t_u := 0$. (2) Update the ith ($i = 1, 2, \cdots, N_u$) particle's position with the fixed y_i and the fixed v_{x_i} according to Eqs. (3.43), (3.44) and (3.45). (3) $t_u := t_u + 1$. (4) If $t_u \geq Tu$, go to step 8(5). Otherwise, go to step 8(2). (5) Every member is then assigned a non-domination rank ND_u and a crowding distance value CD_u in F space.
Step 9. $t := t + 1$.
Step 10. If $t \geq T$, output the elite setA_t. Otherwise, go to step 2.
Step 11. Select randomly from the elite set A_t. The criterion for the personal best position choice is; if the current position is dominated by the previous position, then the previous position is retained; otherwise, the previous position is replaced; however, if neither dominates the other, then one is selected randomly. A relatively simple scheme is used to handle the constraints. Whenever two individuals are compared, the respective constraints are checked. If both are feasible, non-domination sorting technology is directly applied to decide which one is to be selected. If one is feasible and the other is infeasible, the feasible constraint dominates. If both are infeasible, then the constraint with the lowest violation is said to dominate the other. The notations used in the proposed algorithm are detailed in Table 3.1.

Table 3.1 The notations of the algorithm

x_i	The ith particle's position for the upper level problem
y_j	The jth particle's position for the lower level problem
z_j	The jth particle's position in the BLMPP
v_{x_i}	The velocity of x_i
v_{y_i}	The velocity of y_i
N_u	The population size of the upper level problem
N_l	The sub-swarm size of the lower level problem
t_u	Current iteration number for the upper level problem
t_l	Current iteration number for the lower level problem
T_u	The pre-defined max iteration number for t_u
T_l	The pre-defined max iteration number for t_l
ND_u	Non-domination sorting rank for the upper level problem
CD_u	Crowding distance value of the upper level problem
ND_l	Non-domination sorting rank for the lower level problem
CD_l	Crowding distance value of the lower level problem
P_t	The tth iteration population
Q_t	The offspring of P_t
S_t	Intermediate population

Example 3.4. Consider the problem

$$
\begin{cases}
\max\limits_{x_1,x_2,x_3} F_1 = 2\tilde{\tilde{\xi}}_1^2 x_1 + \sqrt{\tilde{\tilde{\xi}}_2 x_2 + \tilde{\tilde{\xi}}_3 x_3} + \tilde{\tilde{\xi}}_4 y_1 + \sqrt{\tilde{\tilde{\xi}}_5 y_2} + \sqrt{\tilde{\tilde{\xi}}_6 y_3} \\
\max\limits_{x_1,x_2,x_3} F_2 = \tilde{\tilde{\xi}}_1^2 x_1 + \tilde{\tilde{\xi}}_2^2 x_2 + x_3 + y_1 + y_2 + y_3 \\
\text{s.t.} \begin{cases}
\text{where } (y_1, y_2, y_3) \text{ solves} \\
\max\limits_{y_1,y_2,y_3} f_1 = \sqrt{7\tilde{\tilde{\xi}}_1 x_1 + 3\tilde{\tilde{\xi}}_2 x_2 + 8\tilde{\tilde{\xi}}_3 x_3} + (12\tilde{\tilde{\xi}}_4 y_1 + 11\tilde{\tilde{\xi}}_5 y_2 + 5\tilde{\tilde{\xi}}_6 y_3)^2 \\
\max\limits_{y_1,y_2,y_3} f_2 = x_1 + 5x_2 + 7x_3 + y_1 + 11y_2 - 8\tilde{\tilde{\xi}}_6^3 y_3 \\
\text{s.t.} \begin{cases}
\sqrt{x_1\tilde{\tilde{\xi}}_1 + x_2\tilde{\tilde{\xi}}_2} + x_3^2\tilde{\tilde{\xi}}_3 + \tilde{\tilde{\xi}}_4 y_2 + 17y_3\tilde{\tilde{\xi}}_6 \leq 1540 \\
3x_1 + 6x_2 + 12x_3 + 14y_1 + 19y_2 \leq 2400 \\
x_1 + x_2 + x_3 + y_1 + y_2 - y_3 \geq 150 \\
x_1 + x_2 + x_3 + y_1 + y_2 + y_3 \leq 200 \\
x_1, x_2, x_3, y_1, y_2, y_3 \geq 0,
\end{cases}
\end{cases}
\end{cases}
\tag{3.46}
$$

where $\xi_j, j = 1, 2, \cdots, 6$ are Ra-Ra variables characterized as

$$
\begin{array}{llll}
\tilde{\tilde{\xi}}_1 \sim \mathcal{N}(\tilde{\xi}_1, 1), & \text{with } \tilde{\xi}_1 \sim \mathcal{N}(8, 1); & \tilde{\tilde{\xi}}_2 \sim \mathcal{N}(\tilde{\xi}_2, 1), & \text{with } \tilde{\xi}_2 \sim \mathcal{N}(10, 1); \\
\tilde{\tilde{\xi}}_3 \sim \mathcal{N}(\tilde{\xi}_3, 1), & \text{with } \tilde{\xi}_3 \sim \mathcal{N}(12, 1); & \tilde{\tilde{\xi}}_4 \sim \mathcal{N}(\tilde{\xi}, 1) & \text{with } \tilde{\xi}_4 \sim \mathcal{N}(14, 1); \\
\tilde{\tilde{\xi}}_5 \sim \mathcal{N}(\tilde{\xi}, 1) & \text{with } \tilde{\xi}_5 \sim \mathcal{N}(16, 1); & \tilde{\tilde{\xi}}_6 \sim \mathcal{N}(\tilde{\xi}_6, 1) & \text{with } \tilde{\xi}_6 \sim \mathcal{N}(18, 1).
\end{array}
$$

We can obtain the EEDE model as follows:

$$
\left\{
\begin{array}{l}
\max\limits_{x_1,x_2,x_3} E\left[2\bar{\bar{\xi}}_1^2 x_1 + \sqrt{\bar{\bar{\xi}}_2 x_2 + \bar{\bar{\xi}}_3 x_3 + \bar{\bar{\xi}}_4 y_1 + \sqrt{\bar{\bar{\xi}}_5 y_2} + \sqrt{\bar{\bar{\xi}}_6 y_3}}\right] \\
\max\limits_{x_1,x_2,x_3} E\left[\bar{\bar{\xi}}_1^2 x_1 + \bar{\bar{\xi}}_2^2 x_2 + x_3 + y_1 + y_2 + y_3\right] \\
\text{s.t.} \left\{
\begin{array}{l}
\text{where } (y_1, y_2, y_3) \text{ solves} \\
\max\limits_{y_1,y_2,y_3} \{\beta_1^L, \beta_2^L\} \\
\text{s.t.} \left\{
\begin{array}{l}
Pr\left\{\omega | Pr\left\{\sqrt{7\bar{\bar{\xi}}_1 x_1 + 3\bar{\bar{\xi}}_2 x_2 + 8\bar{\bar{\xi}}_3 x_3 + (12\bar{\bar{\xi}}_4 y_1 + 11\bar{\bar{\xi}}_5 y_2} \right.\right. \\
\qquad \left.\left. + 5\bar{\bar{\xi}}_6 y_3)^2 \geq 200\right\} \geq \beta_1^L\right\} \geq 0.8 \\
Pr\{\omega | Pr\{= x_1 + 5x_2 + 7x_3 + y_1 + 11y_2 - 8\bar{\bar{\xi}}_6^3 y_3\} \geq \beta_2^L\} \geq 0.8 \\
E\left[\sqrt{x_1\bar{\bar{\xi}}_1 + x_2\bar{\bar{\xi}}_2 + x_3^2\bar{\bar{\xi}}_3 + \bar{\bar{\xi}}_4 y_2 + 17 y_3 \bar{\bar{\xi}}_6}\right] \leq 1540 \\
3x_1 + 6x_2 + 12x_3 + 14y_1 + 19y_2 \leq 2400 \\
x_1 + x_2 + x_3 + y_1 + y_2 - y_3 \geq 150 \\
x_1 + x_2 + x_3 + y_1 + y_2 + y_3 \leq 200 \\
x_1, x_2, x_3, y_1, y_2, y_3 \geq 0.
\end{array}
\right.
\end{array}
\right.
\end{array}
\right.
\tag{3.47}
$$

After running the algorithm above, we obtain $(x_1^*, x_2^*, x_3^*, y_1^*, y_2^*, y_3^*) = (11.32, 26.81, 12.63, 20.15, 10.72, 0.31)$ and $(E[F_1^*], E[F_1^*], \beta_1^{*L}, \beta_2^{*L}) = (69.80, 11.72, 0.68, 0.92)$.

3.4 Ra-Ra ECEC Model

Because of the existence of Ra-Ra parameters, we usually cannot find the precise decision for a complicated real-life problem. Hence, an efficient tool should be provided to convert the Ra-Ra parameter into a crisp one. ECEC denotes that expected value operator is used to deal with the uncertain variable in upper-level objectives and lower-level constraints, at the same time, chance-constrained operator is used to deal with the upper-level constraints and lower-level objectives.

3.4.1 General Form of Ra-Ra ECEC Model

Let's consider the typical bi-level single objective model with Ra-Ra parameters. Then it follows from the ECEC operator that:

$$\begin{cases} \min_{x \in R^{n_1}} E[F(\boldsymbol{x}, \boldsymbol{y}, \boldsymbol{\xi})] \\ \text{s.t.} \begin{cases} Ch\{G_i(\boldsymbol{x}, \boldsymbol{y}, \boldsymbol{\xi}) \geq \overline{G}_i\}(\alpha_i^U) \geq \beta_i^U, i = 1, 2, \cdots, p_1 \\ \text{where } \boldsymbol{y} \text{ solves:} \\ \begin{cases} \min_{y \in R^{n_2}} E[f(\boldsymbol{x}, \boldsymbol{y}, \boldsymbol{\xi})] \\ \text{s.t. } Ch\{g_j(\boldsymbol{x}, \boldsymbol{y}, \boldsymbol{\xi}) \geq \bar{g}_i\}(\alpha_i^L) \geq \beta_i^L, i = 1, 2, \cdots, p_2, \end{cases} \end{cases} \end{cases} \tag{3.48}$$

where $\alpha_i^U, \beta_i^U, \alpha_i^L, \beta_i^L$ are the predetermined confidence levels in the lower-level model $\overline{G}_i, \bar{g}_i$ are predetermined objective values. $\boldsymbol{\xi}$ is a Ra-Ra vector. Similarly, if decision makers want to maximize the objective value, then,

$$\begin{cases} \max_{x \in R^{n_1}} E[F(\boldsymbol{x}, \boldsymbol{y}, \boldsymbol{\xi})] \\ \text{s.t.} \begin{cases} Ch\{G_i(\boldsymbol{x}, \boldsymbol{y}, \boldsymbol{\xi}) \geq \overline{F}_i\}(\alpha_i^U) \geq \beta_i^U, i = 1, 2, \cdots, p_1 \\ \text{where } \boldsymbol{y} \text{ solves:} \\ \begin{cases} \max_{y \in R^{n_2}} E[f(\boldsymbol{x}, \boldsymbol{y}, \boldsymbol{\xi})] \\ s.t. \;\; Ch\{g_j(\boldsymbol{x}, \boldsymbol{y}, \boldsymbol{\xi}) \geq \bar{f}_i\}(\alpha_i^L) \geq \beta_i^L, i = 1, 2, \cdots, p_2. \end{cases} \end{cases} \end{cases} \tag{3.49}$$

In what follows, we focus on the cases. of multiple objective If all the objectives are benefit-type, the corresponding ECEC model is developed as

$$\begin{cases} \max_{x \in R^{n_1}} \begin{bmatrix} E[F_1(\boldsymbol{x}, \boldsymbol{y}, \boldsymbol{\xi})] \\ E[F_2(\boldsymbol{x}, \boldsymbol{y}, \boldsymbol{\xi})] \\ \cdots \\ E[F_{m_1}(\boldsymbol{x}, \boldsymbol{y}, \boldsymbol{\xi})] \end{bmatrix} \\ \text{s.t.} \begin{cases} Ch\{G_i(\boldsymbol{x}, \boldsymbol{y}, \boldsymbol{\xi}) \geq \overline{G}_i\}(\alpha_i^U) \geq \beta_i^U, i = 1, 2, \cdots, p_1 \\ \text{where } \boldsymbol{y} \text{ solves:} \\ \max_{y \in R^{n_2}} \begin{bmatrix} E[f_1(\boldsymbol{x}, \boldsymbol{y}, \boldsymbol{\xi})] \\ E[f_2(\boldsymbol{x}, \boldsymbol{y}, \boldsymbol{\xi})] \\ \cdots \\ E[f_{m_1}(\boldsymbol{x}, \boldsymbol{y}, \boldsymbol{\xi})] \end{bmatrix} \\ s.t. \;\; Ch\{g_i(\boldsymbol{x}, \boldsymbol{y}, \boldsymbol{\xi}) \geq \bar{g}_i\}(\alpha_i^U) \geq \beta_i^U, i = 1, 2, \cdots, p_2, \end{cases} \end{cases} \tag{3.50}$$

where $\alpha_i^U, \beta_i^U, \alpha_i^L, \beta_i^L$ are the predetermined confidence levels in the lower-level model $\overline{G}_i, \bar{g}_i$ are predetermined objective values.

If all the objectives are cost-type, then we obtain

$$\begin{cases} \min_{x \in R^{n_1}} \begin{bmatrix} E[F_1(x,y,\xi)] \\ E[F_2(x,y,\xi)] \\ \cdots \\ E[F_{m_1}(x,y,\xi)] \end{bmatrix} \\ \text{s.t.} \begin{cases} Ch\{G_i(x,y,\xi) \geq \overline{G}_i\}(\alpha_i^U) \geq \beta_i^U, i = 1,2,\cdots,p_1 \\ \text{where } y \text{ solves:} \\ \begin{cases} \min_{y \in R^{n_2}} \begin{bmatrix} E[f_1(x,y,\xi)] \\ E[f_2(x,y,\xi)] \\ \cdots \\ E[f_{m_1}(x,y,\xi)] \end{bmatrix} \\ s.t.\ Ch\{g_i(x,y,\xi) \geq \bar{g}_i\}(\alpha_i^U) \geq \beta_i^U, i = 1,2,\cdots,p_2. \end{cases} \end{cases} \end{cases} \tag{3.51}$$

For model (3.50), some concepts are defined as follows:

Definition 3.13. x^* is said to be a feasible solution for the leader at $(\alpha_i^U, \beta_i^U$-levels, if and only if y is a Pareto optimal solution for the follower at (α_i^L, β_i^L)-levels, when $x = x^*$, and satisfies $Ch\{G_i(x,y,\xi) \geq \overline{G}_i\}(\alpha_i^U) \geq \beta_i^U, i = 1,2,\cdots,p_1$.

Definition 3.14. x^* is said to be a Pareto optimal solution of the leader at $(\alpha_i^U, \beta_i^L$-levels if and only if that it is a feasible solution for the leader at $(\alpha_i^U, \beta_i^U$-levels, and there exists no other feasible solution x or the leader at $(\alpha_i^U, \beta_i^U$-levels such that $E[F_i(x^*,y,\xi)] \geq E[F_i(x,y,\xi)]$ for all i and strict inequity holds for at least one $i_0 \in \{1,2,\cdots,m_1\}$.

3.4.2 A Homotopy Method for Linear Model

In this section, we concentrate on the ECEC version of the bi-level multi-objective linear programming problem with Ra-Ra coefficients as follows.

$$\begin{cases} \max_{x \in R^{n_1}} \left[E\left[\tilde{\tilde{C}}_1^T x + \tilde{\tilde{D}}_1^T y\right], E\left[\tilde{\tilde{C}}_2^T x + \tilde{\tilde{D}}_2^T y\right], \cdots, E\left[\tilde{\tilde{C}}_{m_1}^T x + \tilde{\tilde{D}}_{m_1}^T y\right]\right] \\ \text{s.t.} \begin{cases} Ch\left\{\tilde{\tilde{A}}_i^T x + \tilde{\tilde{B}}_i^T y \geq \overline{F}_i\right\}(\alpha_i^U) \geq \beta_i^U, i = 1,2,\cdots,p_1 \\ \text{where } y \text{ solves:} \\ \begin{cases} \max_{y \in R^{n_2}} \left[E[\tilde{\bar{c}}_1^T x + \tilde{\bar{d}}_1^T y], E[\tilde{\bar{c}}_2^T x + \tilde{\bar{d}}_2^T y \geq \bar{f}_2], \cdots, E[\tilde{\bar{c}}_{m_2}^T x + \tilde{\bar{d}}_{m_2}^T y \geq \bar{f}_{m_2}]\right] \\ \text{s.t. } Ch\left\{\tilde{\bar{a}}_i^T x + \tilde{\bar{b}}_i^T y \geq \bar{f}_i\right\}(\alpha_i^L) \geq \beta_i^L, i = 1,2,\cdots,p_2. \end{cases} \end{cases} \end{cases} \tag{3.52}$$

If the Ra-Ra variables in (3.52) are normally distributed, then (3.52) can be converted in its crisp equivalent model by Theorems 3.9 and 3.10. Thus, we can solve it by classical bi-level programming techniques.

Homotopy Method

1. The homotopy equation for follower's optimization problem
   Consider the basic bi-level programming problem:

$$\begin{cases} \min\limits_{x,y} f(x,y) \\ \text{s.t.} \begin{cases} g(x,y) \leq 0 \\ \text{where } y \text{ solves:} \\ \begin{cases} \min\limits_{y} F(x,y) \\ \text{s.t. } G(x,y) \leq 0, \end{cases} \end{cases} \end{cases} \tag{3.53}$$

where f, F; g and G are sufficient smooth and convex. First, we consider the follower's optimization problem,

$$\begin{cases} \min_y \ F(x,y) \\ \text{s.t. } G(x,y) \leq 0. \end{cases} \tag{3.54}$$

Assume that for given point $x \in R^n$, the set $\{y \in R^n : G(x,y) \leq 0\}$ is nonempty. Then optimization problem (3.54) is equivalent to the following KKT system.

$$\begin{aligned} &\nabla_y F(x,y) + \nabla_y G(x,y)u = 0, \\ &UG(x,y) = 0, \ \ u \geq 0, \ \ G(x,y) \leq 0, \end{aligned}$$

where $u \in R_+^{m_2}$ and $U = diag(u)$. Therefore, bi-level problem (3.53) can be expressed as:

$$\begin{cases} \min_{x,y} \ f(x,y) \\ \text{s.t.} \begin{cases} g(x,y) \leq 0, \\ \nabla_y F(x,y) + \nabla_y G(x,y)u = 0, \\ UG(x,y) = 0, \ \ u \geq 0, \ \ G(x,y) \leq 0. \end{cases} \end{cases} \tag{3.55}$$

The homotopy equation for the follower's optimization can be represented as [52]:

$$h(w,t) \triangleq \begin{pmatrix} \nabla_y F(x,y) + \nabla_y G(x,y)u \\ UG(x,y) + te \end{pmatrix} = 0, \tag{3.56}$$

where $w = (x,y,u)^T$, $e = (1,\cdots,1)^T \in R^{m_2}$, and $t \in (0,1]$.

Set $\tilde{f}(w) = f(x,y)$, $\tilde{g}(w) = g(x,y)$. Some feasible sets and index set are denoted as follows:

$$\Omega_1(t) \triangleq \{w \in R^{n+m} \times R^{m_2}_{++} : \tilde{g}(w) \le 0, h(w,t) = 0\},$$

$$\Omega(t) \triangleq \Omega_1(t) \times R^{m_1}_{++} \times R^{m+m_2},$$

$$\partial\Omega_1(t) \triangleq \left\{ w \in \bar{\Omega_1}(t) : \prod_{i=1}^{m_1} \tilde{g}_i(w) = 0 \right\},$$

$$I(w) \triangleq \{i \in \{1, \cdots, m_1\} : \tilde{g}_i(w) = 0\},$$

$$I_G(x,y) = \{i \in \{1, \cdots, m_2\} : \tilde{G}_i(x,y) = 0\}.$$

Condition A [52].

(A.1) $\forall t \in [0,1]$, $\Omega_1(t)$ is nonempty and bounded;
(A.2) $\{\nabla_y G_i(x,y),\ i \in I_G(x,y)\}$ is a matrix of full column rank;
(A.3) for $t \in [0,1]$, $w \in \Omega_1(t)$, $\{\nabla \tilde{g}_i(w),\ i \in I(w),\ \nabla_w h(w,t)\}$ is full column rank;
(A.4) $\forall w \in \bar{\Omega}_1(1)$, $\left\{w + \sum_{i \in I(w)} \upsilon_i \nabla \tilde{g}_i(w) + \nabla_w h(w,1)\lambda\right\} \cap \bar{\Omega_1(1)} = \{w\}$;
(A.5) for any $(x,y) \in \{(x,y) : g(x,y) \le 0, G(x,y) \le 0,\}$,

$$\nabla^2_{yy} F(x,y) + \sum_{i=1}^{m_2} \left(\nabla^2_{yy} G_i(x,y) + \nabla_y G_i(x,y) \nabla_y G_i(x,y)^T\right)$$

is positive definite.

Remark 3.4. Condition (A.5) can guarantee the uniqueness of solution for the follower problem of bi-level programming problem. Geometrical interpretation of condition (A.4) is that $\bar{\Omega}_1(1)$ satisfies the normal cone condition.

Lemma 3.3 (McCormick [26]). *Suppose that Condition (A.1), (A.2) and (A.5) hold, and the set $\Omega = \{(x,y) \in R^{n+m} : g(x,y) \le 0, G(x,y) \le 0\}$ is nonempty. Then for any given point $x \in \bar{\Omega}$ the solution curve $\{y(t), u(t), t \in (0,1]\}$ for follower's homotopy equation (3.56) is continuous, bounded and unique, and as $t \to 0, y(t)$ tend to the solution of the follower problem.*

2. The homotopy equation for bi-level programming problem and convergence of the homotopy pathway [52]

Consider the following parameterized programming problem:

$$\begin{cases} \min_{x,y} f(x,y) \\ \text{s.t.} \begin{cases} g(x,y) \le 0 \\ h(w,t) = 0. \end{cases} \end{cases} \tag{3.57}$$

When t is descending to 0, by Lemma 3.3, parameterized programming problem (3.57) becomes (3.55). Problem (3.57) can be state as:

$$\begin{cases} \min_w \tilde{f}(x,y) \\ \text{s.t.} \begin{cases} \tilde{g}(w) \le 0 \\ h(w,t) = 0. \end{cases} \end{cases} \tag{3.58}$$

Its KKT system is

$$\begin{aligned} &\nabla\tilde{f}(w) + \nabla\tilde{g}(w)\upsilon + \nabla_w h(w,t)\lambda = 0, \\ &h(w,t) = 0, \\ &V\tilde{g}(w) = 0, \ \ \upsilon \ge 0, \ \tilde{g}(w) \le 0, \end{aligned} \tag{3.59}$$

where $\lambda \in R^{m+m_2}$, $\upsilon \in R_+^{m_1}$ are multipliers, and $V = diag(\upsilon)$. When t is descending to 0, by Lemma (3.3), (3.59) is transformed as

$$\tilde{H}(w,\upsilon,\lambda) \triangleq \begin{pmatrix} \nabla\tilde{f}(w) + \nabla\tilde{g}(w)\upsilon + \nabla_w h(w,0)\lambda \\ h(w,0) \\ V\tilde{g}(w) \end{pmatrix} = 0, \ \ \upsilon \ge 0, \ \tilde{g}(w) \le 0. \tag{3.60}$$

The mapping $\tilde{H} : R^{n+2m+M-1+2m_2} \to R^{n+2m+M-1+2m_2}$ is sufficient smooth with respect to (w,υ,λ). Equation (3.60) is the KKT system for problem (3.55).

Now, a homotopy equation for solving the bi-level programming problem needs to be constructed. The following homotopy equation for bi-level programming problem can be defined [52]:

$$\tilde{H}(\theta,\theta_0^0,t) \triangleq \begin{pmatrix} (1-t)(\nabla\tilde{f}(w) + \nabla\tilde{g}(w)\upsilon) + \nabla_w h(w,t)\lambda + t(w-w^0) \\ h(w,0) \\ V\tilde{g}(w) - tV^0\tilde{g}(w^0) \end{pmatrix} = 0, \tag{3.61}$$

where $\theta = (w,\upsilon,\lambda)^T$, $\theta_0^0 = (w^0,\upsilon^0)^T$, $t \in (0,1]$, $w = (x,y,u) \in R^{n+m} \times R_+^{m_2}$, $\upsilon \in R_+^{m_1}$, $\lambda \in R^{m+M-2}$, and $U = diag(u)$, $V = diag(\upsilon)$.

Lemma 3.4 ([52]). *For any given $\theta_0^0 = (w^0,\upsilon^0) \in \Omega_1(1) \times R^{m_1} + +$, then the equation $H(\theta,\theta_0^0,1) = 0$ has a unique solution $\theta = (\theta_0^0,0)^T$.*

Proof. When $t = 1$, (3.61) becomes

$$\begin{aligned} &\nabla_w h(w,1)\lambda + w - w^0 = 0, \\ &h(w,1) = 0, \\ &V\tilde{g}(w) = V^0\tilde{g}(w^0). \end{aligned}$$

By $\tilde{g}(w) \leq 0$ and $h(w,1) = 0$, we have $I(w) = \varnothing$ and $w \in \Omega_1(1)$, then by Condition (A.4) and $\nabla_w h(w,1)\lambda + w - w^0 = 0$, we have $\lambda = 0$, $w = w^0$, hence $\upsilon = \upsilon^0$ and $\theta = (\theta_0^0, 0)^T \in \Omega_1(1)$.

In some cases, we rewrite (3.61) as follows form:

$$H_{\theta_0^0}(\theta, t) = H(\theta, \theta_0^0, t),$$

and the solution set of H as follows:

$$H_{\theta_0^0}^{-1} = \{(\theta, t) \in \Omega(t) \times (0,1] : H_{\theta_0^0}(\theta, t) = 0\}.$$

The following parameterized Sard theorem on the smooth manifold will be used.

Lemma 3.5 (Parameterized Sard theorem [3]). *Let Q, N and P be smooth manifolds of dimensions q, m and p, respectively. And let $\Phi : Q \times N \to P$ be a C^r map, where $r \geq \max\{0, m-p\}$. If $0 \in P$ is a regular value of Φ, then for almost all $a \in Q$, 0 is a regular value of $\Phi_a \equiv \Phi(a, dot)$.*

Lemma 3.6 ([52]). *If $H(\theta, \theta_0^0, t)$ is defined as (3.61) and Condition (A.1)–(A.5) is true, then for almost all $\theta^0 = (\theta_0^0, 0)^T \in \Omega(1)$, 0 is the regular value of map H, and $H_{\theta_0^0}^{-1}$ consists of some smooth curves, which starts from $(\theta^0, 1)$, denoted by $\Gamma_{\theta_0^0}$.*

Proof. $\forall (\theta, t) \in \Omega(t) \times (0,1]$

$$\frac{\partial H(\theta, \theta_0^0, t)}{\partial(w, w^0, \upsilon^0)} = \begin{pmatrix} Q & -tI & 0 \\ \nabla_w h(w,t)^T & 0 & 0 \\ V\nabla\tilde{g}(w)^T & -tV^0\nabla_w\tilde{g}(w^0) & -t\mathrm{diag}(\tilde{g}(w^0)) \end{pmatrix}$$

where

$$Q = (1-t)\left(\nabla^2\tilde{f}(w) + \sum_{i=1}^{m_1} \upsilon_i \nabla^2\tilde{g}(w)\right) + \sum_{i=1}^{m+m_2} \lambda_i \nabla^2 h_i(w,t) + tI.$$

By $w^0 \in \Omega_1(1)$, $\tilde{g}(w^0) \leq 0$, and since $\nabla_w h(w,t)$ is full column rank, then we have $\partial H(\theta, \theta_0^0, t)/\partial(w, w^0, \upsilon^0)$ is full row rank. Hence, $\partial H(\theta, \theta_0^0, t)/\partial(\theta, \theta_0^0, t)$ is also full row rank, and 0 is a regular value.

By parameterized Sard theorem on smooth manifold, for almost all $(\theta^0, 0)^T \in \Omega(t)$, 0 is a regular value of $H(\theta, \theta_0^0, t)$. By the inverse image theorem, $H_{\theta_0^0}^{-1}(0)$ consists of some smooth curves. And, because $H_{\theta_0^0}^{-1}((\theta, \theta_0^0, t), 1) = 0$, there must be a smooth $\Gamma_{\theta_0^0}$ starting from $((\theta, \theta_0^0, t), 1)$.

Lemma 3.7 ([52]). *Suppose that Condition A hold. Given $\theta^0 \in \Omega(t)$, if 0 is a regular value of H, then $\Gamma_{\theta_0^0}$ is a bounded curve in $\Omega(t) \times (0,1]$.*

Proof. Suppose $\Gamma_{\theta_0^0} \subset \Omega(t) \times (0, 1]$ is unbounded, then there exists $\{(\theta, t_k)\} \subset \Gamma_{\theta_0^0}$, and $\| (\theta, t_k) \| \to \infty$. Noting that (x, y) and $t \in (0, 1]$ is bounded, hence, by Lemma 3.3, (x^k, y^k, u^k) is bounded, then there exists a subsequence of $\{(\upsilon_i^k, \lambda_i^k)\}$ (denoted also by $\{(\upsilon_i^k, \lambda_i^k)\}$), which goes to infinity, we set $(\upsilon_i^k, \lambda_i^k)$ is unbounded. Let

$$I_1 = \{\imath \in \{1 \cdots, m_1\} : \upsilon_i^k \to \infty\}, \quad I_2 = \{\imath \in \{1 \cdots, m + m_2\} : \lambda_i^k \to \infty\}.$$

By $H(\theta^k, \theta_0^0, t_k) = 0$, i.e., (θ^k, t_k) satisfies

$$H(\theta, \theta_0^0, t) = \begin{pmatrix} (1-t)(\nabla \tilde{f}(w) + \nabla \tilde{g}(w)\upsilon) + \nabla_w h(w,t)\lambda + t(w - w^0) \\ h(w,t) \\ V\tilde{g}(w) - tV^0\tilde{g}(w^0) \end{pmatrix} = 0,$$

we have [52]:

1. when $t^* \in (0, l]$, by $\{w^k\}$ is bounded, and $\tilde{f}$, $\tilde{g}$, h is sufficient smooth, and if $k \to \infty$, then $I_1 \subset I(w)$, and

$$\left(\sum_{i \in I_1} \upsilon_i^k \nabla \tilde{g}_i(w^k) + \sum_{i \in I_2} \lambda_i^k \nabla_w h_i(w^k, t_k)\right) \Big/ \min_{i \in I_1, j \in I_2} \{|\upsilon_i^k|, |\lambda_i^k|\} \to 0,$$

   this contradicts to Condition (A.3);
2. when $t^* = 1$, by the third equation of (3.61), we get $I_1 = I(w)$, hence, we have

$$\sum_{i \in I} \upsilon_i^k \nabla \tilde{g}_i(w^k) + \lambda^k \nabla_w h_i(w^k, t_k) + w - w^0 = 0,$$

   and this contradicts to Condition (A.4). Hence, $\{(\upsilon^k, \lambda^k)\}$ is bounded, and the result is true.

By Lemmas 3.6 and 3.7, the main result can be obtained.

Theorem 3.11. *Suppose that Condition A hold. Then (3.60) has solution as t is descending to 0, for almost all $\theta^0 \in \Omega(t)$, $H_{\theta_0^0}^{-1}(0)$ contains a smooth curve, which stars from $(\theta^0, 1)$. As $t \to 0$, the limit set $\Theta \times \{0\} \subset \Omega(1) \times \{0\}$ of $\Gamma_{\theta_0^0}$ is nonempty, and every point of Θ is a solution of (3.60), i.e., the KKT solution. Especially, if $\Gamma_{\theta_0^0}$ is finite, and $(\theta^*, 0)$ is the end point of $\Gamma_{\theta_0^0}$, then θ^* is a the KKT solution [52].*

Proof ([52]). By Lemma 3.6, for almost all $\theta^0 \in \Omega(t)$, 0 is a regular value of $H_{\theta_0^0}$, and $\Gamma_{\theta_0^0}$ consists of some smooth curves, among them, a smooth curve is starting from $(\theta^0, 1)$.

By the classification theorem of one-dimensional smooth manifold, $\Gamma_{\theta_0^0}$ is diffeomorphic to a unit circle or the unit interval (0, 1]. Because

$$\frac{\partial H_{\theta_0^0}(\theta,1)}{\partial \theta}|_{\theta=(w^0,\upsilon^0,0)} \tag{3.62}$$

$$= \begin{pmatrix} \sum_{i=1}^{m+m_2} \lambda_i \nabla_w^2 h(w^0,1) + I \; 0 & & \nabla_w h(w^0,1) \\ \nabla_w^2 h(w^0,1)^T & 0 & 0 \\ V^0 \nabla \tilde{g}(w^0)^T & \mathrm{diag}(\tilde{g}(w^0)) \; 0 & \end{pmatrix}. \tag{3.63}$$

By $\tilde{g}(w^0) \leq 0$, we have $\partial H_{\theta_0^0}(\theta,1)/\partial\theta$ is nonsingular, hence $\Gamma_{\theta_0^0}$ is diffeomorphic to unit interval. Let (w^*, t^*) be a limit point of $\Gamma_{\theta_0^0}$, then three cases are possible.

1. $(\theta, t^*) \in \Omega(t^*) \times \{1\}$;
2. $(\theta, t^*) \in \partial\Omega(t^*) \times (0, 1]$;
3. $(\theta, t^*) \in \Omega(t^*) \times \{0\}$.

By Lemma 3.4 and the equation $H(\theta, \theta_0^0, 1)$ has only one solution $((\theta_0^0, 0)^T, 1)$ in $\Omega(1) \times \{1\}$, the case (1) is impossible. In case (2), there must exist a sequence of $(\theta^k, t_k) \in \Gamma_{\theta_0^0}$ such that $1 \leq j \leq m_1$, $g_j(x^k, y^k) \to 0$. From the third Eq. (3.61), it follows that $\| \upsilon_j^k \| \to \infty$, which contradicts Lemma 3.7. Thus, case (2) is impossible. As a conclusion, (3) is the only possible case.

3. The predictor-corrector algorithm for homotopy equation

   This part discusses how to trace numerically the homotopy equation (3.61). By Theorem 3.11, for almost all $\theta^0 \in \Omega(t)$, the homotopy (3.61) generates a smooth curve $\Gamma_{\theta_0^0}$. We call $\Gamma_{\theta_0^0}$ the homotopy path. Tracing numerically from $(\theta^0, 1)$ until $\mu \to 0$, one can find a solution of (3.60). The tangent-vector at a point on $\Gamma_{\theta_0^0}$ has two opposite directions, one (the positive direction) makes s increase, and another (the negative direction) makes s decrease. The negative direction will lead us back to the initial point, so we must go along the positive direction. The criterion in step 1 (b) of Algorithm 1 that determines the positive direction is based on a basic theory of homotopy method (see [2, 3, 16]), that is, the positive direction – at any point (w, t) on $\Gamma_{\theta_0^0}$ keeps the sign of the determinant

$$\begin{vmatrix} H'_{\theta_0^0}(\theta^0,1) \\ \eta^T \end{vmatrix}$$

invariant. Then we can determine the direction by following result.

Proposition 3.1 (Garcia and Zangwill [18], Allgower and Georg [2]). *If Γ is smooth, then the positive direction $\eta^{(0)}$ at the initial point $(\theta_0^0, 0)^T$ satisfies*

$$sign \begin{vmatrix} H'_{\theta_0^0}(\theta^0, 1) \\ \eta^T \end{vmatrix} = (-1)^{m+m_1+m_2+1}.$$

The algorithm can be given as follows [52].

(Euler.Newton method).

1. Give $t_0 = 1$, an initial point $\theta^0 \in \Omega(1)$, step length $h_0 \geq 0$ and two small positive numbers $\varepsilon_1, t_\varepsilon \geq 0$, $k := 0$.
2. Compute a predictor point $(\theta^{(k+1,0)}, t_{k+1}, 0)$:
   (a) Compute a unit tangent vector $\xi^{(k)} \in R^{n+2m+m_1+m_2+1}$;
   (b) Determine the direction $\eta^{(k)}$ of predictor step:
   If the sign of the determinant

$$\begin{vmatrix} H'_{\theta_0^0}(\theta^0, 1) \\ \eta^T \end{vmatrix} = (-1)^{m+m_1+m_2+1},$$

   then $\eta^{(k)} = \xi^{(k)}$, else if the sign of the determinant

$$\begin{vmatrix} H'_{\theta_0^0}(\theta^0, 1) \\ \eta^T \end{vmatrix} = (-1)^{m+m_1+m_2},$$

   then $\eta^{(k)} = -\xi^{(k)}$;
   (c) $(\theta^{(k,0)}, t_k, 0) = (\theta^{(k)}, t_k) + h_k \eta^{(k)}$.
3. Compute a corrector point $(\theta^{(k+1,0)}, t_{k+1})$:

$$(\theta^{(k+1,j)}, t_{k+1}, j) = (\theta^{(k,j-1)}, t_{k,j-1}) - M^+_{k,j-1} H_{\theta_0^0}(\theta^{(k,j)}, t_{k,j}), \ j = 1, 2, \cdots$$

   until $\| H_{\theta_0^{(0)}}(\theta^{(k+1,j)}, t_{k+1}, j) \| \leq \varepsilon_1$. Set $(\theta^{(k+1)}, t_{k+1}) = (\theta^{(k+1,j)}, t_{k+1}, j)$.
4. If $t_{k+1} \leq t_\varepsilon$, then stop, else choose a new step-length $h_{k+1} \geq 0$. $k := k+1$, and go to (1).

   In the above algorithm,

$$M_{k,j-1} = H'_{\theta_0^0}(\theta^{(k,j-1)}, t_k, j-1), \text{ and } M^+ = M^T(MM^T)^{-1} \tag{3.64}$$

is the Moore-Penrose inverse of M [52].

Numerical Example

In what follows, a numerical example is presented to illustrate the effectiveness of the proposed solution method to the linear Ra-Ra EEDE model.

Example 3.5. We consider the following Ra-Ra bi-level multiobjective linear programming problem, which has two objective functions of the leader and two objective functions of the follower:

$$\begin{cases} \max\limits_{x_1,x_2} F_1 = \tilde{\bar{\xi}}_{11}x_1 + \tilde{\bar{\xi}}_{12}x_2 + \tilde{\bar{\xi}}_{13}y_1 + \tilde{\bar{\xi}}_{14}y_2 \\ \max\limits_{x_1,x_2} F_2 = \tilde{\bar{\xi}}_{21}x_1 + \tilde{\bar{\xi}}_{22}x_2 + \tilde{\bar{\xi}}_{23}y_1 + \tilde{\bar{\xi}}_{24}y_2 \\ \text{s.t.} \begin{cases} \text{where } (y_1, y_2) \text{ solves} \\ \max\limits_{y_1,y_2} f_1 = \tilde{\bar{\eta}}_{11}x_1 + \tilde{\bar{\eta}}_{12}x_2 + \tilde{\bar{\eta}}_{13}y_1 + \tilde{\bar{\eta}}_{14}y_2 \\ \max\limits_{y_1,y_2} f_2 = \tilde{\bar{\eta}}_{21}x_1 + \tilde{\bar{\eta}}_{22}x_2 + \tilde{\bar{\eta}}_{23}y_1 + \tilde{\bar{\eta}}_{24}y_2 \\ \text{s.t.} \begin{cases} \tilde{\bar{\zeta}}_1 x_1 + \tilde{\bar{\zeta}}_2 x_2 + \tilde{\bar{\zeta}}_3 y_1 + \tilde{\bar{\zeta}}_4 y_2 \le 200 \\ 2x_1 - x_2 + 3y_1 + y_2 \le 100 \\ x_1, x_2, y_1, y_2 \ge 0, \end{cases} \end{cases} \end{cases} \tag{3.65}$$

where $\tilde{\bar{\xi}}_{ij}, \tilde{\bar{\eta}}_{ij}, \tilde{\bar{\zeta}}_j (i = 1,2; j = 1,2,4)$ are Ra-Ra variables characterized as:

$\tilde{\bar{\xi}}_{11} \sim \mathcal{N}(\bar{\xi}_{11}, 2)$, with $\bar{\xi}_{11} \sim \mathcal{N}(6, 2)$; $\tilde{\bar{\xi}}_{12} \sim \mathcal{N}(\bar{\xi}, 2)$, with $\bar{\xi}_{12} \sim \mathcal{N}(10, 2)$;

$\tilde{\bar{\xi}}_{13} \sim \mathcal{N}(\tilde{\bar{\xi}}_{13}, 2)$, with $\bar{\xi}_{13} \sim \mathcal{N}(8, 2)$; $\tilde{\bar{\xi}}_{14} \sim \mathcal{N}(\bar{\xi}_{14}, 2)$, with $\bar{\xi}_{14} \sim \mathcal{N}(12, 2)$;

$\tilde{\bar{\xi}}_{21} \sim \mathcal{N}(\bar{\xi}_{21}, 2)$, with $\bar{\xi}_{21} \sim \mathcal{N}(3, 2)$; $\tilde{\bar{\xi}}_{22} \sim \mathcal{N}(\bar{\xi}_{22}, 2)$, with $\bar{\xi}_{22} \sim \mathcal{N}(-5, 2)$;

$\tilde{\bar{\xi}}_{23} \sim \mathcal{N}(\bar{\xi}_{23}, 2)$, with $\bar{\xi}_{23} \sim \mathcal{N}(7, 2)$; $\tilde{\bar{\xi}}_{24} \sim \mathcal{N}(\bar{\xi}_{24}, 2)$, with $\bar{\xi}_{24} \sim \mathcal{N}(-9, 2)$;

$\tilde{\bar{\eta}}_{11} \sim \mathcal{N}(\bar{\eta}_{11}, 1)$, with $\bar{\eta}_{11} \sim \mathcal{N}(12, 1)$; $\tilde{\bar{\eta}}_{12} \sim \mathcal{N}(\bar{\eta}_{12}, 1)$, with $\bar{\eta}_{12} \sim \mathcal{N}(10, 1)$;

$\tilde{\bar{\eta}}_{13} \sim \mathcal{N}(\bar{\eta}_{13}, 1)$, with $\bar{\eta}_{13} \sim \mathcal{N}(8, 1)$; $\tilde{\bar{\eta}}_{14} \sim \mathcal{N}(\bar{\eta}_{14}, 1)$, with $\bar{\eta}_{14} \sim \mathcal{N}(6, 1)$;

$\tilde{\bar{\eta}}_{21} \sim \mathcal{N}(\bar{\eta}_{21}, 1)$, with $\bar{\eta}_{21} \sim \mathcal{N}(12, 1)$; $\tilde{\bar{\eta}}_{22} \sim \mathcal{N}(\bar{\eta}, 1)$, with $\bar{\eta}_{22} \sim \mathcal{N}(5, 1)$;

$\tilde{\bar{\eta}}_{23} \sim \mathcal{N}(\bar{\eta}_{23}, 1)$, with $\bar{\eta}_{23} \sim \mathcal{N}(18, 1)$; $\tilde{\bar{\eta}}_{24} \sim \mathcal{N}(\bar{\eta}_{24}, 1)$, with $\bar{\eta}_{24} \sim \mathcal{N}(20, 1)$;

$\tilde{\bar{\zeta}}_1 \sim \mathcal{N}(\bar{\zeta}_1, 1)$, with $\bar{\zeta}_1 \sim \mathcal{N}(6, 1)$; $\tilde{\bar{\zeta}}_2 \sim \mathcal{N}(\bar{\zeta}_2, 1)$, with $\bar{\zeta}_2 \sim \mathcal{N}(12, 1)$;

$\tilde{\bar{\zeta}}_3 \sim \mathcal{N}(\bar{\zeta}_3, 1)$, with $\bar{\zeta}_3 \sim \mathcal{N}(18, 1)$; $\tilde{\bar{\zeta}}_4 \sim \mathcal{N}(\bar{\zeta}_4, 1)$, with $\bar{\zeta}_4 \sim \mathcal{N}(24, 1)$.

If the confidence levels are given by the decision makers, the ECEC model of (3.65) is:

$$\begin{cases} \max\limits_{x_1,x_2} E[\tilde{\bar{\xi}}_{11}x_1 + \tilde{\bar{\xi}}_{12}x_2 + \tilde{\bar{\xi}}_{13}y_1 + \tilde{\bar{\xi}}_{14}y_2] \\ \max\limits_{x_1,x_2} E[\tilde{\bar{\xi}}_{21}x_1 + \tilde{\bar{\xi}}_{22}x_2 + \tilde{\bar{\xi}}_{23}y_1 + \tilde{\bar{\xi}}_{24}y_2] \\ \text{where } (y_1, y_2) \text{ solves} \\ \begin{cases} \max\limits_{y_1,y_2} E[\tilde{\bar{\eta}}_{11}x_1 + \tilde{\bar{\eta}}_{12}x_2 + \tilde{\bar{\eta}}_{13}y_1 + \tilde{\bar{\eta}}_{14}y_2] \\ \max\limits_{y_1,y_2} E[\tilde{\bar{\eta}}_{21}x_1 + \tilde{\bar{\eta}}_{22}x_2 + \tilde{\bar{\eta}}_{23}y_1 + \tilde{\bar{\eta}}_{24}y_2] \\ \text{s.t.} \begin{cases} Pr\{\omega | Pr\{\tilde{\bar{\xi}}_1(\omega)x_1 + \tilde{\bar{\xi}}_2(\omega)x_2 + \tilde{\bar{\xi}}_3(\omega)y_1 + \tilde{\bar{\xi}}_4(\omega)y_2] \le 200\} \ge 0.8\} \\ \qquad \ge 0.8 \\ 2x_1 - x_2 + 3y_1 + y_2 \le 100 \\ x_1, x_2, y_1, y_2 \ge 0. \end{cases} \end{cases} \end{cases} \tag{3.66}$$

It follows from Theorems 3.9 and 3.10 that (3.39) is equivalent to

$$\begin{cases} \max\limits_{x_1,x_2} E[F_1] = 6x_1 + 10x_2 + 8y_1 + 12y_2 \\ \max\limits_{x_1,x_2} E[F_1] = 3x_1 - 5x_2 + 7y_1 - 9y_2 \\ \text{s.t.} \begin{cases} \text{where } (y_1, y_2) \text{ solves} \\ \max\limits_{y_1,y_2} E[f_1] = 12x_1 + 10x_2 + 8y_1 + 6y_2 \\ \max\limits_{y_1,y_2} E[f_2] = 12x_1 + 5x_2 + 18y_1 + 20y_2 \\ \text{s.t.} \begin{cases} 6x_1 + 12x_2 + 18y_1 + 24y_2 + 1.68\sqrt{x_1^2 + x_2^2 + y_1^2 + y_2^2} \le 200 \\ 2x_1 - x_2 + 3y_1 + y_2 \le 100 \\ x_1, x_2, y_1, y_2 \ge 0. \end{cases} \end{cases} \end{cases} \tag{3.67}$$

By using the homotopy method mentioned above, the solutions is $(x_1^*, x_2^*, y_1^*, y_2^*) = (23.6, 35.2, 17.3, 42.8)$, the corresponding objective function values are $(E[F_1^*], E[F_2^*], E[f_1^*], E[f_2^*]) = (321.9, 114.8, 34.8, 16.5)$.

3.4.3 Ra-Ra Simulation-Based CHK-PSO for Nonlinear Models

If (3.50) or (3.51) is a nonlinear model, i.e., some functions in model (3.50) or (3.51) are nonlinear, the Ra-Ra features make it impossible to be converted into its crisp equivalent model. In order solve nonlinear (3.50) or (3.51), three issues are needed to solve: the trade-off of the multiple objectives, the twofold randomness and the bi-level structure. In this section, we design a hybrid method consisting of simulation technique, step method and particle swarm optimization algorithm, where they are deal with twofold randomness, multiple objectives and bi-level structure, respectively.

Ra-Ra Simulation for Critical Values

Let $\boldsymbol{\xi}$ be an n-dimensional Ra-Ra vector on the probability space $(\Omega, \mathscr{A}, Pr)$, and $f : \mathbf{R}^n \to \mathbf{R}$ be a measurable function. For any given confidence levels α and β, we find the maximal value $\bar{f}$ such that

$$Ch\{f(\boldsymbol{x}, \boldsymbol{\xi}) \geq \bar{f}\}(\alpha) \geq \beta$$

holds. That is, we should compute the maximal value $\bar{f}$ such that

$$Pr\{\omega \in \Omega | Pr\{f(\boldsymbol{x}, \boldsymbol{\xi}(\omega)) \geq \bar{f}\} \geq \beta\} \geq \alpha$$

holds. We sample $\omega_1, \omega_2, \cdots, \omega_N$ from Ω according to the probability measure Pr, where ω_k is an n-dimensional vector, and estimate $\bar{f}_k = \sup\{f_k | Pr\{f(\boldsymbol{x}, \boldsymbol{\xi}(\omega_k))\} \geq \beta\}$ for $k = 1, 2, \cdots, N$ by the stochastic simulation. Let N' be the integer part of αN. Then $\bar{f}$ can be taken as the N'th largest element in the sequence $\{\bar{f}_1, \bar{f}_2, \cdots, \bar{f}_N\}$.

Then the procedure simulating the critical value $\bar{f}$ of the function $Pr\{\omega \in \Omega | Pr\{f(\boldsymbol{x}, \boldsymbol{\xi}(\omega)) \geq \bar{f}\} \geq \beta\} \geq \alpha$ can be summarized as follows:

Step 1. Generate $\omega_1, \omega_2, \cdots, \omega_N$ from Ω according to the probability measure Pr;
Step 2. Find $\bar{f}_k = \sup\{f_k | Pr\{f(\boldsymbol{x}, \boldsymbol{\xi}(\omega_k)) \geq f_k\} \geq \beta\}$ for $k = 1, 2, \cdots, N$ by random simulation
Step 3. Set N' as the integer part of αN;
Step 4. Return the N'th largest element in $\{\bar{f}_1, \bar{f}_2, \cdots, \bar{f}_N\}$.

Example 3.6. Find the maximal value $\bar{f}$ such that

$$Ch\left\{\sqrt{\bar{\bar{\xi}}_1^2 + \bar{\bar{\xi}}_2^2 + \bar{\bar{\xi}}_3^2} \geq \bar{f}\right\}(0.9) \geq 0.9,$$

where $\bar{\bar{\xi}}_1$, $\bar{\bar{\xi}}_2$ and $\bar{\bar{\xi}}_3$ are Ra-Ra variables defined as

$$\begin{aligned}
\bar{\bar{\xi}}_1 &= \mathscr{U}(\tilde{\rho}_1, \tilde{\rho}_1 + 2), && \text{with } \tilde{\rho}_1 \sim \mathscr{N}(0, 1),\\
\bar{\bar{\xi}}_2 &= \mathscr{N}(\tilde{\rho}_2, 1), && \text{with } \tilde{\rho}_2 \sim \mathscr{U}(3, 5),\\
\bar{\bar{\xi}}_3 &= exp(\tilde{\rho}_3), && \text{with } \tilde{\rho}_3 \sim \mathscr{U}(1, 2).
\end{aligned}$$

A run of Ra-Ra simulation with 10,000 cycles shows that $\bar{f} = 3.2188$.

The Step Method

In this section, we use the step method, which is also called STEM method and is the interactive programming method to deal with the multi-objective structure [6], to resolve the problem.

The STEM method is based on the norm ideal point method and its resolving process includes the analysis and decision stage. In the analysis stage, analyzer resolves the problem by the norm ideal point method and provides the decision makers with the solutions and the related objective values and the ideal objective values. In the decision stage, decision maker gives the tolerance level of the satisfied object to the dissatisfied object to make its objective value better after comparing the objective values obtained in the analysis stage with the ideal point, then provides the analyzer with the information to go on resolving. Do it repeatedly and decision making will get the final satisfied solution.

Shimizu once extent the STEM method to deal with the general nonlinear multi-objective programming problem. Interested readers can refer to literatures [37, 40] and others [11, 29, 41] about its further development.

Consider the following multi-objective programming problem,

$$\begin{cases} \min \boldsymbol{f}(\boldsymbol{x}) = (f_1(\boldsymbol{x}), f_2(\boldsymbol{x}), \cdots, f_m(\boldsymbol{x})) \\ \text{s.t.} \quad \boldsymbol{x} \in X, \end{cases} \tag{3.68}$$

where $\boldsymbol{x} = (x_1, x_2, \cdots, x_n)$ and $X = \{\boldsymbol{x} \in \mathbf{R}^n | A\boldsymbol{x} = \boldsymbol{b}, \boldsymbol{x} \geq 0\}$. Let $\boldsymbol{x}^i$ be the optimal solution of the problem $\min_{\boldsymbol{x} \in X} f_i(\boldsymbol{x})$ and compute each objective function $f_i(\boldsymbol{x})$ at $\boldsymbol{x}^k$, then we get m^2 objective function value,

$$f_{ik} = f_i(\boldsymbol{x}^k), \ i, k = 1, 2, \cdots, m.$$

Denote $f_i^* = f_{ii} = f_i(\boldsymbol{x}^i), \boldsymbol{f}^* = (f_1^*, f_2^*, \cdots, f_m^*)^T$ and f_i^* is a ideal point of the problem (3.68). Compute the maximum value of the objective function $f_i(\boldsymbol{x})$ at every minimum point $\boldsymbol{x}^k$

$$f_i^{\max} = \max_{1 \leq k \leq m} f_{ik}, i = 1, 2, \cdots, m.$$

To make it more clearly, we list it in Table 3.2.

According to Table 3.2, we only look for the solution $\boldsymbol{x}$ such that the distance between $\boldsymbol{f}(\boldsymbol{x})$ and $\boldsymbol{f}^*$ is minimum, that is, the solution such that each objective is close to the ideal point. Consider the following problem,

Table 3.2 Payoff table

$\boldsymbol{f}$	$\boldsymbol{x}^1$	$\cdots$	$\boldsymbol{x}^i$	$\cdots$	$\boldsymbol{x}^m$	max
f_1	$f_{11} = f_1^*$	$\cdots$	f_{1i}	$\cdots$	f_{1m}	$f_1^{\max}$
$\vdots$	$\vdots$	$\vdots$		$\vdots$	$\vdots$	
f_i	f_{i1}	$\cdots$	$f_{ii} = f_i^*$	$\cdots$	f_{im}	$f_i^{\max}$
$\vdots$	$\vdots$	$\vdots$		$\vdots$	$\vdots$	
f_m	f_{m1}	$\cdots$	f_{mi}	$\cdots$	$f_{mm} = f_m^*$	$f_m^{\max}$

$$\min_{x\in X}\max_{1\le i\le m} w_i\left|f_i(\boldsymbol{x})-f_i^*\right| = \min_{x\in X}\max_{1\le i\le m} w_i\left|\sum_{j=1}^{n} c_{ij}x_j - f_i^*\right|, \tag{3.69}$$

where $\boldsymbol{w} = (w_1, w_2, \cdots, w_m)^T$ is the weight vector and w_i is the ith weight which can be decided as follows,

$$\alpha_i = \begin{cases} \frac{f_i^{\max}-f_i^*}{f_i^{\max}}\frac{1}{||\boldsymbol{c}_i||}, f_i^{\max} > 0, \\ \frac{f_i^*-f_i^{\max}}{f_i^{\max}}\frac{1}{||\boldsymbol{c}_i||}, f_i^{\max} \le 0, \end{cases}, i = 1, 2, \cdots, m, \tag{3.70}$$

$$w_i = \alpha_i / \sum_{i=1}^{m} \alpha_i, \ i = 1, 2, \cdots, m, \tag{3.71}$$

where $||\boldsymbol{c}_i|| = \sqrt{\sum_{j=1}^{n} c_{ij}^2}$. Then the problem (3.68) is equivalent to

$$\begin{cases} \min \lambda \\ \text{s.t.} \begin{cases} w_i\left(\sum\limits_{j=1}^{n} c_{ij}x_j - f_i^*\right) \le \lambda, \ i = 1, 2, \cdots, m \\ \lambda \ge 0, \boldsymbol{x} \in X. \end{cases} \end{cases} \tag{3.72}$$

Assume that the optimal solution of the problem (3.72) is $(\tilde{\boldsymbol{x}}, \tilde{\lambda})^T$. It is obvious that $(\tilde{\boldsymbol{x}}, \tilde{\lambda})^T$ is a weak efficient solution of the problem (3.2). In order to check if $\tilde{\boldsymbol{x}}$ is satisfied, the decision maker needs to compare $f_i(\tilde{\boldsymbol{x}})$ with the ideal objective value f_i^*, $i = 1, 2, \cdots, m$. If the decision maker has been satisfied with $f_s(\tilde{\boldsymbol{x}})$, but dissatisfied with $f_t(\tilde{\boldsymbol{x}})$, we add the following constraint in the next step in order to improve the objective value f_t,

$$f_t(\boldsymbol{x}) \le f_t(\tilde{\boldsymbol{x}}).$$

For the satisfied object f_s, we add one tolerance level δ_s,

$$f_s(\boldsymbol{x}) \le f_s(\tilde{\boldsymbol{x}}) + \delta_s.$$

Thus, in the problem (3.72), we replace X with the following constraint set,

$$X^1 = \{\boldsymbol{x} \in X | f_s(\boldsymbol{x}) \le f_s(\tilde{\boldsymbol{x}}) + \delta_s, f_t(\boldsymbol{x}) \le f_t(\tilde{\boldsymbol{x}})\},$$

and delete the objective f_s (do it by letting $w_s = 0$), then resolve the new problem to get better solutions.

In a word, the STEM method can be summarized as follows:

Step 1. Compute every single objective programming problem,

$$f_i(\boldsymbol{x}^i) = \min_{\boldsymbol{x}\in X} f_i(\boldsymbol{x}),\ i = 1, 2, \cdots, m.$$

If $\boldsymbol{x}^1 = \cdots = \boldsymbol{x}^m$, we obtain the optimal solution $\boldsymbol{x}^* = \boldsymbol{x}^1 = \cdots = \boldsymbol{x}^m$ and stop.

Step 2. Compute the objective value of $f_i(\boldsymbol{x})$ at every minimum point $\boldsymbol{x}^k$, then get m^2 objective values $f_{ik} = f_i(\boldsymbol{x}^k)(i, k = 1, 2, \cdots, m)$. List Table 3.2 and we have

$$f_i^* = f_{ii},\ f_i^{\max} = \max_{1\le k\le m} f_{ik},\ i = 1, 2, \cdots, m.$$

Step 3. Give the initial constraint set and let $X^1 = X$.

Step 4. Compute the weight coefficients $w_1, w_2, \cdots, w_m$ by Eqs. (3.70) and (3.71).

Step 5. Solve the auxiliary problem,

$$\begin{cases} \min \lambda \\ \text{s.t.} \begin{cases} w_i\left(\sum\limits_{j=1}^{n} c_{ij}x_j - f_i^*\right) \le \lambda,\ i = 1, 2, \cdots, m \\ \lambda \ge 0, \boldsymbol{x} \in X^k. \end{cases} \end{cases} \tag{3.73}$$

Let the optimal of problem (3.73) be $(\boldsymbol{x}^k, \lambda^k)^T$.

Step 6. Decision maker compares the reference value $f_i(\boldsymbol{x}^k)(i = 1, 2, \cdots, m)$ with the ideal objective value f_i^*. (1) If decision maker is satisfied with all objective values, output $\tilde{\boldsymbol{x}} = \boldsymbol{x}^k$. (2) If the decision maker is dissatisfied with all objective values, there doesn't exists any satisfied solutions and stop the process. (3) If the decision maker is satisfied with the object $f_{s_k}(1 \le s_k \le m, k < m)$, turn to **Step 7**.

Step 7. Decision maker gives the tolerance level $\delta_{s_k} > 0$ to the object f_{s_k} and construct the new constraint set as follows,

$$X^{k+1} = \{\boldsymbol{x} \in X^k | f_{s_k}(\boldsymbol{x}) \le f_{s_k}(\boldsymbol{x}^k) + \delta_{s_k}, f_i(\boldsymbol{x}) \le f_i(\boldsymbol{x}^k), i \ne s_k\}.$$

Let $\delta_{s_k} = 0$, $k = k + 1$ and turn to **Step 4**.

CHKS-PSO

After dealing with twofold randomness and multiple objective, we considered how to handle bi-level structure. In this section, particle swarm optimization based on Chen-Harker-Kanzow-Smale smoothing function (CHKS-PSO), proposed by Jiang et al. [20].

We consider the nonlinear bi-level programming formulated as follows

$$\begin{cases} \min_{x} F(\boldsymbol{x},\boldsymbol{y}) \\ \text{s.t.} \begin{cases} h(\boldsymbol{x},\boldsymbol{y}) \leq 0 \\ \text{where } \boldsymbol{y} \text{ solves} \\ \begin{cases} \min_{y} F(\boldsymbol{x},\boldsymbol{y}) \\ \text{s.t. } g(\boldsymbol{x},\boldsymbol{y}) \leq 0. \end{cases} \end{cases} \end{cases} \tag{3.74}$$

Then we can reduce the nonlinear bi-level programming problem to the one-level programming problem:

$$\begin{cases} \min_{x,y,\lambda} F(\boldsymbol{x},\boldsymbol{y}) \\ \text{s.t.} \begin{cases} h(\boldsymbol{x},\boldsymbol{y}) \leq 0 \\ \nabla_y L(\boldsymbol{x},\boldsymbol{y},\boldsymbol{\lambda}) = 0 \\ \boldsymbol{\lambda}^T g(\boldsymbol{x},\boldsymbol{y}) = 0 \\ g(\boldsymbol{x},\boldsymbol{y}) = 0 \\ \boldsymbol{\lambda} \geq \boldsymbol{0}, \end{cases} \end{cases} \tag{3.75}$$

where $L(\boldsymbol{x},\boldsymbol{y},\boldsymbol{\lambda}) = f(\boldsymbol{x},\boldsymbol{y}) + \boldsymbol{\lambda}^T g(\boldsymbol{x},\boldsymbol{y})$ are Lagrange multipliers.

Consider the nonlinear complementarity problem (NCP for short) in the problem (3.75):

$$g(\boldsymbol{x},\boldsymbol{y}) \leq 0, \boldsymbol{\lambda} \geq 0, \boldsymbol{\lambda}^T g(\boldsymbol{x},\boldsymbol{y}) = 0. \tag{3.76}$$

We note that the problem (3.76) is a non-smooth problem, hence standard methods are not guaranteed to solve such problem. In the last few decades, various methods have been developed for solving NCP, where the smoothing-type algorithm is one of the most effective methods for NCP. Fukushima and Pang [16] proposed a smooth method to solve the mathematical programming problem with complementarity constraints. The problem (3.76) with nonlinear complementarity condition is non-convex and non-differential, and even not satisfies the regularity assumptions. For this reason, we apply smoothing method to solve this problem.

Definition 3.15. The Chen-Mangasarian smoothing function is $\phi : \Re^2 \rightarrow \Re$ defined by $\phi(a,b) = a + b - \sqrt{(a-b)^2}$ By introducing a smoothing parameter $\epsilon \in \Re$ into ϕ, we obtain the Chen-Harker-Kanzow-Smale (CHKS) smoothing function

$$\phi_\epsilon(a,b) = a + b - \sqrt{(a-b)^2 + 4\epsilon^2}.$$

The Chen-Mangasarian smoothing function has the property $\phi(a,b) = 0$ if and only if $a \geq 0, b \geq 0, ab = 0$, but it is nondifferentiable at $a = b = 0$. But, the CHKS smoothing function has the property $\phi(a,b,\epsilon) = 0$ if and only if

$a \geq 0, b \geq 0$, $ab = \epsilon/2$ for $\epsilon \geq 0$, and the function is smooth with respect to a; b for $\epsilon \geq 0$. Hence, by applying the CHKS smoothing function

$$\phi(a, b, \epsilon) = a + b - \sqrt{(a-b)^2 + 4\epsilon^2},$$

the problem (3.76) can be approximated by:

$$\begin{cases} \nabla_y L(\boldsymbol{x}, \boldsymbol{y}, \boldsymbol{\lambda}) = 0 \\ \lambda_j - g_j(x, y) - \sqrt{(\lambda_j + g_j(x, y))^2 + 4\epsilon^2} = 0. \end{cases} \tag{3.77}$$

Hence, the problem (3.75) can be transformed as follows:

$$\begin{cases} \min\limits_{x,y,\lambda} F(\boldsymbol{x}, \boldsymbol{y}) \\ \text{s.t.} \begin{cases} h(\boldsymbol{x}, \boldsymbol{y}) \leq 0 \\ \nabla_y L(\boldsymbol{x}, \boldsymbol{y}, \boldsymbol{\lambda}) = 0 \\ \boldsymbol{\lambda}^T g(\boldsymbol{x}, \boldsymbol{y}) = 0 \\ \lambda_j - g_j(x, y) - \sqrt{(\lambda_j + g_j(x, y))^2 + 4\epsilon^2} = 0. \end{cases} \end{cases} \tag{3.78}$$

The smoothing factor ϵ is treated as a variable rather than a parameter in the problem (3.77) and (3.78), which avoids the difficulty of not satisfying the regularity assumptions induced by complementarity condition in the problem (3.75). For a given $\epsilon \geq 0$ and x fixed, if one wants to get the optimal solution y of the lower level problem, what he needs to do is to solve the problem (3.77). If $f(x, y)$ satisfies all constraints of the lower level problem, then $f(x, y)$ gets its optimal value at y. Thus what we need to do is to solve the following single programming problem to get the optimal solution x and (x, y) is then the approximate optimal solution of $F(x, y)$:

$$\begin{cases} \min\limits_{x,y,\lambda} F(\boldsymbol{x}, \boldsymbol{y}) \\ \text{s.t. } h(\boldsymbol{x}, \boldsymbol{y}) \leq 0. \end{cases} \tag{3.79}$$

To solve the above problems, PSO is used.

Overview of Particle Swarm Optimization

PSO is a population-based heuristic algorithm that simulates the social behavior of birds flocking to achieve precise objectives in a multidimensional space. In the PSO, the population is referred as a swarm and individuals are called particles. Similar to other evolutionary algorithms, PSO performs searches using a population of individuals that are updated from iteration to iteration. To find the optimal or approximately optimal solution, each particle changes its search direction based its own best previous experience and the best past experience of the entire swarm.

Each particle is a potential solution to the optimization problem. A particle represents a point in a D-dimension space, and its status is characterized from its position and velocity. The position the particle i at iteration k can be represented by a D-dimensional vector $X_i^k = \{x_{i1}^k, x_{i2}^k, \cdots, x_{iD}^k\}$ and the velocity is represented by another D-dimensional vector $V_i^k = \{v_{i1}^k, v_{i2}^k, \cdots, v_{iD}^k\}$. The fitness of each particle is evaluated using the optimization problem objective function. The best previous position of the ith particle is remembered until iteration k, which is represented as $PB_i^k = \{p_{i1}^k, p_{i2}^k, \cdots, p_{iD}^k\}$. The best position in the entire swarm is denoted $GB_i^k = \{g_{i1}^k, g_{i2}^k, \cdots, g_{iD}^k\}$.

To search for the optimal solution, each particle changes velocity and position based on the following two formulas:

$$v_{id}^k = w \cdot v_{id}^{k-1} + c_1 \cdot r_1 \cdot (p_{id}^{k-1} - x_{id}^{k-1}) + c_2 \cdot r_2 \cdot (g_{id}^{k-1} - x_{id}^{k-1}), \tag{3.80}$$

$$x_{id}^k = x_{id}^{k-1} + v_{id}^k, d = 1, 2, \cdots, D, \tag{3.81}$$

where w is the inertia weight that controls the impact of the previous velocity of the particle on the current velocity. r_1, r_2 are random numbers between zero and one; and c_1, c_2 are the positive constant parameters or acceleration factors that control the maximum step size.

For the PSO, Eq. (3.80) is used to calculate the new velocity according to the previous velocity and the distance of its current position from both its own best historical position and the best position of the entire population. Generally, the value for each component in V in the range $[-V_{\max}, V_{\max}]$ to control the excessive roaming of the particles outside the search space. Then, using Eq. (3.81), the particle flies toward a new position. This process is repeated until a user-defined stopping criterion is reached.

Fitness Function

In the problem (3.79), let $\varepsilon = (\varepsilon_1, \varepsilon_2, \cdots, \varepsilon_{m_1})$, where ε_i are small positive numbers and tend to zero with the increasing of the generations, $\overline{S} = \{(x, y) \in S | h(x, y) \leq \varepsilon\}$, $b(x, y) = \max\{\max\{h(x, y) - \varepsilon\}, 0\}$ The fitness function $\overline{F}(x, y)$ is defined as following

$$\overline{F}(x, y) = \begin{cases} F(x, y), & b(x, y) = 0, \\ F_{\max} + b(x, y), & b(x, y) \neq 0, \end{cases} \tag{3.82}$$

where the parameter $F_{\max}$ is the objective function value of the worst feasible solution in the population. If there is no feasible solution in the population, $F_{\max}$ returns a value zero. In the process of evolution, infeasible solutions gradually close to feasible region under the pressure of $F_{\max} + b(x, y)$, and after they enter into feasible region, they close to optimal solution under the pressure of $F(x, y)$.

Steps

Step 1. In this step, the population size M and the maximal iterated generation T are given. Randomly generate a set of initial population in X. Set $t = 1$.

Step 2. For a given $\epsilon \geq 0$ and each $x \in X$ fixed, we solve the problem (3.77) for y.

Step 3. For each $y \in Y$, we solve the problem (3.79) using PSO for optimal solution x and optimal value F. Set $x^* = x, F^* = F, t = t + 1$.

Step 4. The algorithm terminates when the generation t is greater than the maximal iterated generation T. Otherwise, set $t = t + 1$ and goto Step 2, then we can get the another solution (x', y') and value $F(x', y')$ If $F' < F^*$, then set $x^* = x', y^* = y', F^* = F'$.

Step 5. After T iterations, we will get the approximate optimal solution (x^*, y^*) and approximate optimal value $F(x^*, y^*)$ of nonlinear bi-level programming.

Numerical Example

In this section, a numerical example is given to illustrate the effectiveness of the application of the models and algorithms proposed earlier.

Example 3.7. Consider the problem

$$\left\{\begin{array}{l}
\max\limits_{x_1,x_2,x_3} F_1 = 2\tilde{\tilde{\xi}}_1^2 x_1 + \sqrt{\tilde{\tilde{\xi}}_2 x_2 + \tilde{\tilde{\xi}}_3 x_3} + \tilde{\tilde{\xi}}_4 y_1 + \sqrt{\tilde{\tilde{\xi}}_5 y_2} + \sqrt{\tilde{\tilde{\xi}}_6 y_3} \\
\max\limits_{x_1,x_2,x_3} F_2 = \tilde{\tilde{\xi}}_1^2 x_1 + \tilde{\tilde{\xi}}_2^2 x_2 + x_3 + y_1 + y_2 + y_3 \\
\text{s.t.} \left\{\begin{array}{l}
\text{where } (y_1, y_2, y_3) \text{ solves} \\
\max\limits_{y_1,y_2,y_3} f_1 = \sqrt{7\tilde{\tilde{\xi}}_1 x_1 + 3\tilde{\tilde{\xi}}_2 x_2 + 8\tilde{\tilde{\xi}}_3 x_3} + (12\tilde{\tilde{\xi}}_4 y_1 + 11\tilde{\tilde{\xi}}_5 y_2 + 5\tilde{\tilde{\xi}}_6 y_3)^2 \\
\max\limits_{y_1,y_2,y_3} f_2 = x_1 + 5x_2 + 7x_3 + y_1 + 11y_2 - 8\tilde{\tilde{\xi}}_6^3 y_3 \\
\text{s.t.} \left\{\begin{array}{l}
\sqrt{x_1\tilde{\tilde{\xi}}_1 + x_2\tilde{\tilde{\xi}}_2} + x_3^2\tilde{\tilde{\xi}}_3 + \tilde{\tilde{\xi}}_4 y_2 + 17y_3\tilde{\tilde{\xi}}_6 \leq 1540 \\
3x_1 + 6x_2 + 12x_3 + 14y_1 + 19y_2 \leq 2400 \\
x_1 + x_2 + x_3 + y_1 + y_2 - y_3 \geq 150 \\
x_1 + x_2 + x_3 + y_1 + y_2 + y_3 \leq 200 \\
x_1, x_2, x_3, y_1, y_2, y_3 \geq 0,
\end{array}\right.
\end{array}\right.
\end{array}\right. \tag{3.83}$$

where $\xi_j, j = 1, 2, \cdots, 6$ are Ra-Ra variables characterized as

$$\begin{array}{llll}
\tilde{\tilde{\xi}}_1 \sim \mathcal{N}(\tilde{\xi}_1, 1), & \text{with } \tilde{\xi}_1 \sim \mathcal{N}(8, 1); & \tilde{\tilde{\xi}}_2 \sim \mathcal{N}(\tilde{\xi}_2, 1), & \text{with } \tilde{\xi}_2 \sim \mathcal{N}(10, 1); \\
\tilde{\tilde{\xi}}_3 \sim \mathcal{N}(\tilde{\xi}_3, 1), & \text{with } \tilde{\xi}_3 \sim \mathcal{N}(12, 1); & \tilde{\tilde{\xi}}_4 \sim \mathcal{N}(\tilde{\xi}, 1) & \text{with } \tilde{\xi}_4 \sim \mathcal{N}(14, 1); \\
\tilde{\tilde{\xi}}_5 \sim \mathcal{N}(\tilde{\xi}, 1) & \text{with } \tilde{\xi}_5 \sim \mathcal{N}(16, 1); & \tilde{\tilde{\xi}}_6 \sim \mathcal{N}(\tilde{\xi}_6, 1) & \text{with } \tilde{\xi}_6 \sim \mathcal{N}(18, 1).
\end{array}$$

We can obtain the ECEC model as follows.

$$\begin{cases} \max_{x_1,x_2,x_3} E\left[2\bar{\bar{\xi}}_1^2 x_1 + \sqrt{\bar{\bar{\xi}}_2 x_2 + \bar{\bar{\xi}}_3 x_3} + \bar{\bar{\xi}}_4 y_1 + \sqrt{\bar{\bar{\xi}}_5 y_2} + \sqrt{\bar{\bar{\xi}}_6 y_3}\right] \\ \max_{x_1,x_2,x_3} E\left[\bar{\bar{\xi}}_1^2 x_1 + \bar{\bar{\xi}}_2^2 x_2 + x_3 + y_1 + y_2 + y_3\right] \\ \text{s.t.} \begin{cases} \text{where } (y_1, y_2, y_3) \text{ solves} \\ \max_{y_1,y_2,y_3} E\left[\sqrt{7\bar{\bar{\xi}}_1 x_1 + 3\bar{\bar{\xi}}_2 x_2 + 8\bar{\bar{\xi}}_3 x_3} + (12\bar{\bar{\xi}}_4 y_1 + 11\bar{\bar{\xi}}_5 y_2 + 5\bar{\bar{\xi}}_6 y_3)^2 \geq 200\right] \\ \max_{y_1,y_2,y_3} E\left[x_1 + 5x_2 + 7x_3 + y_1 + 11y_2 - 8\bar{\bar{\xi}}_6^3 y_3\right] \\ \text{s.t.} \begin{cases} Pr\{\omega | Pr\{\sqrt{x_1\bar{\bar{\xi}}_1 + x_2\bar{\bar{\xi}}_2 + x_3^2\bar{\bar{\xi}}_3} + \bar{\bar{\xi}}_4 y_2 + 17y_3\bar{\bar{\xi}}_6 \leq 1540\} \geq 0.8\} \geq 0.8 \\ 3x_1 + 6x_2 + 12x_3 + 14y_1 + 19y_2 \leq 2400 \\ x_1 + x_2 + x_3 + y_1 + y_2 - y_3 \geq 150 \\ x_1 + x_2 + x_3 + y_1 + y_2 + y_3 \leq 200 \\ x_1, x_2, x_3, y_1, y_2, y_3 \geq 0. \end{cases} \end{cases} \end{cases} \tag{3.84}$$

After running the algorithm above, we obtain $(x_1^*, x_2^*, x_3^*, y_1^*, y_2^*, y_3^*) = (12.32, 26.51, 13.63, 20.12, 10.22, 0.45)$ and $(E[F_1]^*, E[F_2]^*, E[f_1]^*, E[f_2]^*) = (69.80, 11.72, 1.68, 20.92)$.

3.5 Ra-Ra DCCC Model

The existence of Ra-Ra parameters leads the difficulty in finding the precise decision for a complicated real-life problem. Hence, an efficient tool should be provided to convert the random parameter into a crisp one. DCCC denotes that dependent chance operator is used to deal with the uncertain variable in upper-level objectives, at the same time, chance constrained operator is used to deal with the lower-level objectives.

3.5.1 *General Form of Ra-Ra DCCC Model*

In this situation, people will optimize some ideal values of the objective functions subject to some chance constraints under some predetermined confidence levels.

Based on dependent chance and chance constraint definition, a natural idea is to provide a confidence level α at which it is desired that the stochastic constrains hold. Let's still consider the typical single objective bi-level decision making model. Based on the dependent-chance operator and chance-constraint operator, the maximax type of Ra-Ra DCCC model is proposed as follows:

$$\begin{cases} \max\limits_{x \in R^{n_1}} Ch\{F(\boldsymbol{x},\boldsymbol{y},\boldsymbol{\xi}) \geq \overline{F}\}(\eta) \\ \text{s.t.} \begin{cases} Ch\{G_i(\boldsymbol{x},\boldsymbol{y},\boldsymbol{\xi}) \leq 0\}(\alpha_i^U) \geq \beta_i^U, i = 1,2,\cdots,p_1 \\ \text{where } y \text{ solves:} \\ \max\limits_{y \in R^{n_2}} \bar{f} \\ \text{s.t.} \begin{cases} Ch\{f(\boldsymbol{x},\boldsymbol{y},\boldsymbol{\xi}) \geq \bar{f}\}(\alpha^L) \geq \beta^L \\ Ch\{g_i(\boldsymbol{x},\boldsymbol{y},\boldsymbol{\xi}) \leq 0\}(\eta_i^L) \geq \theta_i^L, i = 1,2,\cdots,p_2, \end{cases} \end{cases} \end{cases} \tag{3.85}$$

where $\eta, \alpha_i^U, \beta_i^U$ are the predetermined confidence levels in the upper-level programming, $\alpha^L, \beta^L, \eta_i^L, \theta_i^L$ are the predetermined confidence levels in the lower-level programming; $\overline{F}$ is the aspired objective value determined by the upper level decision maker.

If the objective is to be minimized (for example, the objective is a cost function), the DCCC model should be as follows,

$$\begin{cases} \min\limits_{x \in R^{n_1}} Ch\{F(\boldsymbol{x},\boldsymbol{y},\boldsymbol{\xi}) \leq \overline{F}\}(\eta) \\ \text{s.t.} \begin{cases} Ch\{G_i(\boldsymbol{x},\boldsymbol{y},\boldsymbol{\xi}) \leq 0\}(\alpha_i^U) \geq \beta_i^U, i = 1,2,\cdots,p_1 \\ \text{where } y \text{ solves:} \\ \min\limits_{y \in R^{n_2}} \bar{f} \\ \text{s.t.} \begin{cases} Ch\{f(\boldsymbol{x},\boldsymbol{y},\boldsymbol{\xi}) \leq \bar{f}\}(\alpha^L) \geq \beta^L \\ Ch\{g_i(\boldsymbol{x},\boldsymbol{y},\boldsymbol{\xi}) \leq 0\}(\eta_i^L) \geq \theta_i^L, i = 1,2,\cdots,p_2. \end{cases} \end{cases} \end{cases} \tag{3.86}$$

From the definition of the Ra-Ra primitive chance measure, model (3.85) and (3.86) can be rewritten as

$$\begin{cases} \max\limits_{x \in R^{n_1}} \theta \\ \text{s.t.} \begin{cases} Pr\{\omega | Pr\{F(\boldsymbol{x},\boldsymbol{y},\boldsymbol{\xi}(\boldsymbol{\omega})) \geq \overline{F}\} \geq \theta\} \geq \eta \\ Pr\{\omega | Pr\{G_i(\boldsymbol{x},\boldsymbol{y},\boldsymbol{\xi}(\boldsymbol{\omega})) \leq 0\} \geq \beta_i^U\} \geq \alpha_i^U, i = 1,2,\cdots,p_1 \\ \text{where } \boldsymbol{y} \text{ solves:} \\ \max\limits_{y \in R^{n_2}} \bar{f} \\ \text{s.t.} \begin{cases} Pr\{\omega | Pr\{f(\boldsymbol{x},\boldsymbol{y},\boldsymbol{\xi}(\boldsymbol{\omega})) \geq \bar{f}\} \geq \beta^L\} \geq \alpha^L \\ Pr\{\omega | Pr\{g_i(\boldsymbol{x},\boldsymbol{y},\boldsymbol{\xi}(\boldsymbol{\omega})) \leq 0\} \geq \theta_i^L \geq \eta_i^L, i = 1,2,\cdots,p_2 \end{cases} \end{cases} \end{cases} \tag{3.87}$$

and

$$\begin{cases} \min\limits_{x \in R^{n_1}} \theta \\ \text{s.t.} \begin{cases} Pr\{\omega | Pr\{F(\boldsymbol{x},\boldsymbol{y},\boldsymbol{\xi}(\boldsymbol{\omega})) \leq \overline{F}\} \geq \theta\} \geq \eta \\ Pr\{\omega | Pr\{G_i(\boldsymbol{x},\boldsymbol{y},\boldsymbol{\xi}(\boldsymbol{\omega})) \leq 0\} \geq \beta_i^U\} \geq \alpha_i^U, i = 1,2,\cdots,p_1 \\ \text{where } \boldsymbol{y} \text{ solves:} \\ \min\limits_{y \in R^{n_2}} \bar{f} \\ \text{s.t.} \begin{cases} Pr\{\omega | Pr\{f(\boldsymbol{x},\boldsymbol{y},\boldsymbol{\xi}(\boldsymbol{\omega})) \leq \bar{f}\} \geq \beta^L\} \geq \alpha^L \\ Pr\{\omega | Pr\{g_i(\boldsymbol{x},\boldsymbol{y},\boldsymbol{\xi}(\boldsymbol{\omega})) \leq 0\} \geq \theta_i^L \geq \eta_i^L, i = 1,2,\cdots,p_2 \end{cases} \end{cases} \end{cases} \tag{3.88}$$

respectively.

As $\boldsymbol{\xi}$ degenerate to a random vector, (3.87) and (3.88) degenerate to

$$\left\{\begin{array}{l} \max\limits_{x \in R^{n_1}} \theta \\ \text{s.t.} \left\{\begin{array}{l} Pr\{F(\boldsymbol{x},\boldsymbol{y},\boldsymbol{\xi}) \geq \overline{F}\} \geq \theta \\ Pr\{G_i(\boldsymbol{x},\boldsymbol{y},\boldsymbol{\xi}) \leq 0\} \geq \beta_i^U\}, i = 1,2,\cdots,p_1 \\ \text{where } \boldsymbol{y} \text{ solves:} \\ \max\limits_{y \in R^{n_2}} \bar{f} \\ \text{s.t.} \left\{\begin{array}{l} Pr\{f(\boldsymbol{x},\boldsymbol{y},\boldsymbol{\xi}) \geq \bar{f}\} \geq \beta^L \\ Pr\{g_i(\boldsymbol{x},\boldsymbol{y},\boldsymbol{\xi}) \leq 0\} \geq \theta_i^L, i = 1,2,\cdots,p_2 \end{array}\right. \end{array}\right. \end{array}\right. \tag{3.89}$$

and

$$\left\{\begin{array}{l} \min\limits_{x \in R^{n_1}} \theta \\ \text{s.t.} \left\{\begin{array}{l} Pr\{F(\boldsymbol{x},\boldsymbol{y},\boldsymbol{\xi}) \leq \overline{F}\} \leq \theta \\ Pr\{G_i(\boldsymbol{x},\boldsymbol{y},\boldsymbol{\xi}) \leq 0\} \geq \beta_i^U, i = 1,2,\cdots,p_1 \\ \text{where } \boldsymbol{y} \text{ solves:} \\ \max\limits_{y \in R^{n_2}} \bar{f} \\ \text{s.t.} \left\{\begin{array}{l} Pr\{f(\boldsymbol{x},\boldsymbol{y},\boldsymbol{\xi}) \leq \bar{f}\} \leq \beta^L \\ Pr\{g_i(\boldsymbol{x},\boldsymbol{y},\boldsymbol{\xi}) \leq 0\} \geq \theta_i^L, i = 1,2,\cdots,p_2 \end{array}\right. \end{array}\right. \end{array}\right. \tag{3.90}$$

respectively.

Consider the following bi-level multi-objective programming problem (3.2) with Ra-Ra coefficients, for a fixed decision vector $\boldsymbol{x}$, it is meaningless to maximize the objectives $F(\boldsymbol{x},\boldsymbol{y},\boldsymbol{\xi})$, before we know the exact value of the Ra-Ra vector $\boldsymbol{\xi}$, just as we can not maximize a random function in stochastic programming. Also, we can not judge weather or not a decision $\boldsymbol{x}$ is feasible before we know the value of $\boldsymbol{\xi}$. Hence, both the objectives and constraints in problem (3.2) are ill-defined. For presenting a mathematically meaningful Ra-Ra programming, we build a new class of Ra-Ra programming to model Ra-Ra decision problems via chance measure which was proposed above. We present the DCCC bi-level multi-objective programming as follows:

$$\left\{\begin{array}{l} \max\limits_{x \in R^{n_1}} \begin{bmatrix} Ch\{F_1(\boldsymbol{x},\boldsymbol{y},\boldsymbol{\xi}) \geq \overline{F}_1\}(\eta_1^U) \\ Ch\{F_2(\boldsymbol{x},\boldsymbol{y},\boldsymbol{\xi}) \geq \overline{F}_2\}(\eta_2^U) \\ \cdots \\ Ch\{F_m(\boldsymbol{x},\boldsymbol{y},\boldsymbol{\xi}) \geq \overline{F}_m\}(\eta_{m_1}^U) \end{bmatrix} \\ \text{s.t.} \left\{\begin{array}{l} Ch\{G_i(\boldsymbol{x},\boldsymbol{y},\boldsymbol{\xi}) \leq 0\}(\alpha_i^U) \geq \beta_i^U, i = 1,2,\cdots,q \\ \text{where } \boldsymbol{y} \text{ solves:} \\ \max\limits_{y \in R^{n_2}} [f_1, f_2, \cdots, f_{m_2}] \\ \text{s.t.} \left\{\begin{array}{l} Ch\{f_i(\boldsymbol{x},\boldsymbol{y},\boldsymbol{\xi}) \geq \bar{f}_i\}(\alpha_i^L) \geq \beta_i^L, i = 1,2,\cdots,m_2 \\ Ch\{g_j(\boldsymbol{x},\boldsymbol{y},\boldsymbol{\xi}) \leq 0\}(\eta_i^L) \geq \theta_i^L, i = 1,2,\cdots,p_2, \end{array}\right. \end{array}\right. \end{array}\right. \tag{3.91}$$

where η_i^U, α_i^U and β_i^U are predetermined confidence levels in upper-level programming; α_i^L, β_i^L, η_i^L and θ_i^L; $\overline{F}_i$ are aspired objectives' values determined by the upper level decision maker.

Model (3.91) also can be written as

$$\begin{cases} \max\limits_{x\in R^{n_1}} \{\theta_1^U, \theta_2^U, \cdots, \theta_{m_1}^U\} \\ \text{s.t.} \begin{cases} Pr\{\omega|Pr\{F_i(\boldsymbol{x},\boldsymbol{y},\boldsymbol{\xi}(\omega)) \geq \overline{F}_i\} \geq \theta_i^U\} \geq \eta_i^U, i=1,2,\cdots,m_1 \\ Pr\{\omega|Pr\{G_i(\boldsymbol{x},\boldsymbol{y},\boldsymbol{\xi}(\omega)) \leq 0\} \geq \beta_i^U\} \geq \alpha_i^U, i=1,2,\cdots,p_1 \\ \text{where } \boldsymbol{y} \text{ solves:} \\ \max\limits_{y\in R^{n_2}} [f_1, f_2, \cdots, f_{m_2}] \\ \text{s.t.} \begin{cases} Pr\{\omega|Pr\{f_l(\boldsymbol{x},\boldsymbol{y},\boldsymbol{\xi}(\omega)) \geq \bar{f}_l\} \geq \beta_i^U\} \geq \alpha_i^L, l=1,2,\cdots,m_2 \\ Pr\{\omega|Pr\{g_j(\boldsymbol{x},\boldsymbol{y},\boldsymbol{\xi}(\omega)) \leq 0\} \geq \theta_i^L\} \geq \eta_i^L i=1,2,\cdots,p_2. \end{cases} \end{cases} \end{cases} \tag{3.92}$$

We can also develop the min-version of (3.91) and (3.92). As $\boldsymbol{\xi}$ degenerates to random vector, (3.91) and (3.92) degenerates to

$$\begin{cases} \max\limits_{x\in R^{n_1}} \{\gamma_1^U, \gamma_2^U, \cdots, \gamma_{m_1}^U\} \\ \text{s.t.} \begin{cases} Pr\{F_i(\boldsymbol{x},\boldsymbol{y},\boldsymbol{\xi}) \geq \overline{F}_1\} \geq \gamma_i^U, i=1,2,\cdots,m_1 \\ Pr\{G_i(\boldsymbol{x},\boldsymbol{y},\boldsymbol{\xi}) \leq 0\} \geq \beta_i^U, i=1,2,\cdots,p_1 \\ \text{where } \boldsymbol{y} \text{ solves:} \\ \max\limits_{y\in R^{n_2}} [f_1, f_2, \cdots, f_{m_2}] \\ \text{s.t.} \begin{cases} Pr\{f_l(\boldsymbol{x},\boldsymbol{y},\boldsymbol{\xi}) \geq \bar{f}_l\} \geq \beta_i^U, l=1,2,\cdots,m_2 \\ Pr\{g_j(\boldsymbol{x},\boldsymbol{y},\boldsymbol{\xi}) \leq 0\} \geq \theta_i^L i=1,2,\cdots,p_2. \end{cases} \end{cases} \end{cases} \tag{3.93}$$

3.5.2 Interactive Balance Space Approach for Linear Model

In this section, we concentrate on the DCCC version of the bi-level multi-objective linear programming problem with Ra-Ra coefficients as follows.

$$\begin{cases} \max\limits_{x\in R^{n_1}} \Big\{Ch\Big\{\tilde{\bar{\boldsymbol{C}}}_1^T\boldsymbol{x}+\tilde{\bar{\boldsymbol{D}}}_1^T\boldsymbol{y} \geq \overline{F}_1\Big\}(\eta_1^U), Ch\Big\{\tilde{\bar{\boldsymbol{C}}}_2^T\boldsymbol{x}+\tilde{\bar{\boldsymbol{D}}}_2^T\boldsymbol{y} \geq \overline{F}_2\Big\}(\eta_2^U), \\ \cdots, Ch\Big\{\tilde{\bar{\boldsymbol{C}}}_{m_1}^T\boldsymbol{x}+\tilde{\bar{\boldsymbol{D}}}_{m_1}^T\boldsymbol{y} \geq \overline{F}_{m_1}\Big\}(\eta_{m_1}^U)\Big\} \\ \text{s.t.} \begin{cases} Ch\Big\{\tilde{\bar{\boldsymbol{A}}}_i^T\boldsymbol{x}+\tilde{\bar{\boldsymbol{B}}}_i^T\boldsymbol{y} \geq \overline{F}_i\Big\}(\alpha_i^U) \geq \beta_i^U, i=1,2,\cdots,p_1 \\ \text{where } \boldsymbol{y} \text{ solves:} \\ \max\limits_{y\in R^{n_2}} [\bar{f}_1, \bar{f}_2, \cdots, \bar{f}_{m_2}] \\ \text{s.t.} \begin{cases} Ch\Big\{\tilde{\bar{\boldsymbol{c}}}_i^T\boldsymbol{x}+\tilde{\bar{\boldsymbol{d}}}_i^T\boldsymbol{y} \geq \bar{f}_i\Big\}(\alpha_i^L) \geq \beta_i^L, i=1,2,\cdots,p_2 \\ Ch\Big\{\tilde{\bar{\boldsymbol{a}}}_i^T\boldsymbol{x}+\tilde{\bar{\boldsymbol{b}}}_i^T\boldsymbol{y} \geq \bar{f}_i\Big\}(\eta_i^L) \geq \theta_i^L, i=1,2,\cdots,m_2 \end{cases} \end{cases} \end{cases} \tag{3.94}$$

or

$$\begin{cases} \max_{x \in R^{n_1}} [\theta_1^U, \theta_2^U, \cdots, \theta_{m_1}^U] \\ \text{s.t.} \begin{cases} Pr\left\{\omega | Pr\left\{\tilde{\bar{A}}_i^T x + \tilde{\bar{B}}_i^T y \geq \bar{F}_i\right\} \geq \beta_i^U\right\} \geq \alpha_i^U, i = 1, 2, \cdots, p_1 \\ Pr\left\{\omega | Pr\left\{\tilde{\bar{C}}_1^T x + \tilde{\bar{D}}_1^T y \geq \bar{F}_1\right\} \geq \theta_1^U\right\} \geq \eta_i^U, i = 1, 2, \cdots, m_1 \\ \text{where } y \text{ solves:} \\ \max_{y \in R^{n_2}} [\bar{f}_1, \bar{f}_2, \cdots, \bar{f}_{m_2}] \\ \text{s.t.} \begin{cases} Pr\left\{\omega | Pr\left\{\tilde{\bar{c}}_i^T x + \tilde{\bar{d}}_i^T y \geq \bar{f}_i\right\} \geq \beta_i^L\right\} \geq \alpha_i^L, i = 1, 2, \cdots, p_2 \\ Pr\left\{\omega | Pr\left\{\tilde{\bar{a}}_i^T x + \tilde{\bar{b}}_i^T y \geq \bar{f}_i\right\} \geq \theta_i^L\right\} \geq \eta_i^L, i = 1, 2, \cdots, m_2. \end{cases} \end{cases} \end{cases} \tag{3.95}$$

Interactive Balance Space Approach

If the Ra-Ra variables in (3.94) or (3.95) are normally distributed, then (3.94) or (3.95) can be converted in its crisp equivalent model by Theorems 3.9 and 3.10. Thus, we can solve it by classical bi-level programming techniques. In this section, we adopt the revised version of interactive balance space approach proposed by Abo-Sinna and Baky [1].

Assume that there are three levels in a hierarchy structure. Let a vector of decision variable $\boldsymbol{x} = (x_1, x_2) \in \Re^n$ be partitioned among the three planners. The first-level decision maker (FLDM) has control over the vector $x_1 \in \Re^{n_1}$, and the second-level decision maker (SLDM) has control over the vector $x_2 \in \Re^{n_2}$, where $n = n_1 + n_2$. Furthermore, assume that $F_i(x_{1,2}) : \Re^n \to \Re^{m_i}, i = 1, 2.$ are the first-level and second-level vector objective functions, respectively. So the problem may be formulated as follows:

$$\min_{x_1} F_1(x_1, x_2) = \{f_{11}(x_1, x_2), f_{12}(x_1, x_2), \cdots, f_{1m_1}(x_1, x_2)\}, \tag{3.96}$$

where x_2 solves

$$\min_{x_2} F_2(x_1, x_2) = \{f_{21}(x_1, x_2), f_{22}(x_1, x_2), \cdots, f_{2m_1}(x_1, x_2)\}, \tag{3.97}$$

subject to

$$G = \{(x_1 x_2) | g_i(x_1, x_2) \leq 0, i = 1, 2, \ldots, q\},$$

where G is the bi-level convex constraint feasible choice set, m_1 is the number of first-level objective functions, m_2 is the number of second-level objective functions.

The decision mechanism of the problem is that the FLDM and SLDM, adopt the two-planer Stackelberg game. The solution of the problem is obtained by solving the FLDM and SLDM problems separately. In this way, we can present satisfactoriness and the preferred solution from the perspective of singular-level multi-objective decision-making problems. We now introduce several theorems with the help of the quality of balance space approach to provide a theoretical basis for the first-level and second-level multi-objective decision-making problems. Consider a multi-objective optimization (MOO) problem:

$$\min_{x \in X \subseteq \Re^n} f(\boldsymbol{x}), \tag{3.98}$$

where $\boldsymbol{x}$ denotes the decision making variable, $f = (f_1 f_2, \cdots, f_m) : \Re^n \to \Re^m$ are continuous real-valued functions over X, and X is a compact robust set in $\Re^n$ defined by $X = \{x | h_j(x) \leq 0, j = 1, 2, \cdots, q\}$. Let $a_i = \min_{x \in X} f_i(\boldsymbol{x})$ and $b_i = \max_{x \in X} f_i(\boldsymbol{x})$. On $[a_i, b_i]$ define $A_i \in f_i(u_i)$ whose membership function $\mu_{A_i}(f_i(\boldsymbol{x}))$ meets (i) and (ii) as below:

1. When the objective value $f_i(\boldsymbol{x})$ approaches or equals the decision maker's ideal value, $\mu_{A_i}(f_i(\boldsymbol{x}))$ approaches or equal 1. Otherwise, 0.
2. If $f_i(\boldsymbol{x}_1) \leq f_i(\boldsymbol{x}_2)$, then $\mu_{A_i}(f_i(\boldsymbol{x}_1)) \geq \mu_{A_i}(f_i(\boldsymbol{x}_2)), i = 1, 2, \cdots, m$.
   Let $X_i^0 = \{\boldsymbol{x} \in X | f_i(\boldsymbol{x}) = a_i\}$ be the set of all global optimizers, and $\boldsymbol{a} = (a_1, a_2, \cdots, a_m)$ be the vector of partial global solutions, and let $X_i^0(\eta) = \{\boldsymbol{x} \in X | f_i(\boldsymbol{x}) - a_i \leq \eta\}$ be the g-suboptimal solution sets, then we have the following definitions.

Definition 3.16. A MOO problem (3.98) is said to be balanced if $X^0 = \bigcap_{i=1}^m X_i^0 \neq \emptyset$. Otherwise, it is unbalanced. X^0 is called the global optimal solution of the MOO problem (3.98).

Definition 3.17. A MOO problem (3.98) is said to be η-balanced if $X^0 = \emptyset$, but there is a nonempty intersection, $X_\eta^0 = \bigcap_{i=1}^m X_i^0(\eta) \neq \emptyset$. Otherwise, it is η-unbalanced. X_η^0 is called the g-optimal solution set of the MOO problem (3.98), and each $x \in X_\eta^0$ is called a η-optimal point.

If we allow a different bound η_i of deviations from partial global minimum values $\boldsymbol{a} = a_1, a_2, \cdots, a_m$ for all objective functions, then we have

$$X_i^0(\eta_i) = \{\boldsymbol{x} \in X | f_i(\boldsymbol{x}) - a_i \leq \eta_i\}.$$

Definition 3.18. The point $\eta = (\eta_1, \eta_2, \cdots, \eta_m) \in \Re^m, \eta_i \geq 0$ is said to be a balance point if $X_\eta^0 = \bigcap_{i=1}^m X_i^0(\eta) \neq \emptyset$ and for every $\eta' \in \Re^m$ such that $0 \leq \eta_i' \leq \eta_i, i = 1, 2, \cdots, m$ and $\eta_i' < \eta_j$ for at least one $j \in \{1, 2, \cdots, m\}$ the set $X_{\eta'}^0 = \emptyset$.

Definition 3.19. The balance set is the set of all the balance points and is denoted by $\boldsymbol{B}$.

If we want to reflect the relative importance of each objective function one can apportion $\eta_i = \alpha_i\eta, \alpha_i \geq 0, i = 1, 2, \cdots, m, \sum_{i=1}^{m} \alpha_i = 1, \eta > 0, (\alpha_i, i = 1, 2, \cdots, m$: the apportion numbers or weighting numbers or proportion of deviations), and we have the following definition

Definition 3.20. The apportioned balance number $\eta_0(\alpha) \in R, \alpha = (\alpha_1, \alpha_2, \cdots, \alpha_m)$ is the smallest number $\eta \in R$ for which the following set is non-empty:

$$X_\eta^0(\alpha) = \bigcap_{i=1}^{m} X_i^0(\eta\alpha_i) = \bigcap_{i=1}^{m} \{x \in X | f_i(x) - a_i \leq \alpha_i\eta\} \neq \emptyset \tag{3.99}$$

i.e.,

$$\eta_0(\alpha) = \min \left\{ \eta : X_\eta^0(\alpha) = \bigcap_{i=1}^{m} X_i^0(\eta\alpha_i) \neq \emptyset \right\} . \tag{3.100}$$

The following Theorem and Proposition show how the apportioned balance number can be calculated.

Theorem 3.12. *Let $a_i > 0, i = 1, 2, \cdots, m$. Then the apportioned balance number $\eta_0(\alpha)$ for the MOO problem is determined by following min-max problem:*

$$\eta_0(\alpha) = \min_{x \in X} \max_{1 \leq i \leq m} \left[\frac{f_i(x) - \alpha_i}{\alpha_i} \right] . \tag{3.101}$$

Proposition 3.2. *Let $\alpha_i > 0$ for all $i = 1, 2, \cdots Q, Q \leq m$, the apportioned balance number $\eta_0(\alpha)$ with respect to α is:*

$$\eta_0(\alpha) = \min_{\eta \in \mathbf{B}} \max_{1 \leq i \leq Q} \left[\frac{\eta_i}{\alpha_i} \right] . \tag{3.102}$$

If the equation of the balance set is known, and $\alpha_i > 0$ for all $i = 1, 2, \cdots Q, Q \leq m$, the apportioned balance number $\eta_0(\alpha)$ can be found by solving the following problem on the balance set

$$\begin{cases} \min Z \\ \text{s.t.} \begin{cases} Z \geq \frac{\eta_i}{\alpha_i}, i = 1, 2, \cdots Q, Q \leq m \\ \eta \in \mathbf{B}. \end{cases} \end{cases} \tag{3.103}$$

The advantages of the balanced space approach over other methods and techniques for solving MOO problems are presented in the following paragraph.

For a MOO problem (3.98) with several conflicting objectives, scalarized approaches are generally used. In other words, most known methods involve the construction of a substitute single objective problem with a scalar cost function, such as the utility function method, the global criterion method, the bounded objective function method, and the e-constraint method. The deficiencies in such approaches have long been recognized and different schemes with less straightforward scalarization have been developed, such as the lexicographic method, the goal-programming method, and the cone construction and perturbation methods. For non-reconcilable objectives, scalarization may serve only to distort the original problem, and replacing the objectives with a new single-objective problem means that there is a possibility that certain objectives may totally disappear from the results.

Recently developed set-to-set full global optimization methods, which are capable of providing an entire set of global optimizers, allow for a different approach, under which the vector optimization problems appear as natural as the scalar optimization problems. The balanced space approach uses set-to-set full global optimization methods, provides a g-optimal solution for (3.98), and is able to treat the scalarization method defects. The methods concerned with computing the g-optimal solution offer solutions to non-scalarized vector optimization problems. Solutions using non-scalarized global optimization methods, such as the balanced space approach, give a measure of satisfaction for each conflicting objective. This measure of satisfaction is determined using the balanced point together with the balanced set. In other words, the components g_i from a balanced point g represent the measure of achievement of a certain objective within the equilibrium defined by the balanced set, which corresponds to the best possible solution for a vector optimization problem with conflicting objectives. This demonstrates the importance of the balanced set which, when presented to decision makers, gives a clear picture of what can be achieved when there are several conflicting objectives.

According to the η-optimal solution for the balanced space approach, we present new definitions for the satisfactory solution and the preferred solution for single-level multi-objective decision-making problems. New definitions for the feasible solution and the preferred solution (η-optimal point) are also given.

Definition 3.21. If $\boldsymbol{x}*$ is an η-optimal point (3.98), then $\mu_{A_i}(f_i(\boldsymbol{x}^*))$ is defined as the satisfactory solution $\boldsymbol{x}*$ to objective $f_i(\boldsymbol{x})$.

Definition 3.22. With certain values $s_1, s_2, \cdots, s_m$ given in advance by the decision maker for the objective functions $f_1(\boldsymbol{x}), f_2(\boldsymbol{x}), \cdots, f_m(\boldsymbol{x})$, if η-optimal point $\boldsymbol{x}^*$ satisfies $\mu_{A_i} f_i(\boldsymbol{x}^*) > s_i, i = 1, 2, \cdots, m$, then $\boldsymbol{x}^*$ is the preferred solution that corresponds to the satisfaction level $s_i, i = 1, 2, \cdots, m$.

Definition 3.23. If x and x^* are two η-optimal points, then x^* is more preferred than x if $\mu_{A_i} f_i(\boldsymbol{x}^*) \geq \mu_{A_i} f_i(\boldsymbol{x}), i = 1, 2, \cdots, m$ and $\mu_{A_i} f_i(\boldsymbol{x}^*) > \mu_{A_i} f_i(\boldsymbol{x})$ for at least one index i.

The membership function $\mu_{A_i} f_i(\boldsymbol{x})$ is given as below

$$\mu_{A_i} f_i(\boldsymbol{x}) = \frac{b_i - f_i(\boldsymbol{x})}{b_i - a_i}, \tag{3.104}$$

which is decided according to the decision-maker's requirements. Obviously, Eq. (3.104) meets the two requirements (i) and (ii) for $\mu_{A_i} f_i(\boldsymbol{x})$.

The following new formula is introduced to interconnect the satisfaction level given by the decision-maker for each objective function with the apportioned numbers (deviation proportions) that are needed to reflect the relative importance of each objective function and to calculate the apportioned balance number.

Given a satisfaction level $s_i, i = 1, 2, \cdots, m$, by normalizing s_i we obtain:

$$\alpha_i = \frac{s_i}{\sum_{i=1}^{m} s_i}, i = 1, 2, \cdots, m. \tag{3.105}$$

Therefore, we have apportioned the numbers $\alpha_i, i = 1, 2, \cdots, m$ to include the satisfaction level, which can then be used to derive the interactive algorithm for the problem solution.

According to the Stackelberg game and mathematical programming, the following new definitions for the feasible problem solution and the preferred problem solution are given.

Definition 3.24. For any x_1 given by the leader, if the decision making variable x_2 on the lower level is an η-optimal point of the follower, then (x_1, x_2) is a feasible solution to the bi-level problem.

Definition 3.25. If (x_1, x_2) is a feasible solution of the bi-level problem and if x_1 is the η-point of the leader, then (x_1, x_2) is the preferred solution (η-optimal point) for the bi-level problem.

To solve a bi-level problem using the three-planner Stackelberg game, the leader lists the preferred or satisfactory solutions in rank order to the follower. Then, the attained solution is sent to the follower who seeks a solution using the balanced space approach, finally arriving at a solution that approaches the preferred solution of the leader. Finally, the leader decides on the preferred solution (η-optimal point) to the bi-level problem according to its satisfaction level.

This solution method simplifies the bi-level problem by transforming it into three separate multi-objective decision making problems, thus, the avoiding the difficulties associated with non-convexity.

The upper-level decision making problem for this bi-level problem is as follows:

$$\min_{x_1 \in G_1} (f_{11}(x_1, x_2), f_{12}(x_1, x_2), \cdots, f_{1m_1}(x_1, x_2)). \tag{3.106}$$

To obtain the preferred solution to the FLDM problem, (3.106) can be rewritten as the following multi-objective decision making problem:

$$\min_{\boldsymbol{x}} (f_{11}(\boldsymbol{x}), f_{12}(\boldsymbol{x}), \cdots, f_{1m_1}(\boldsymbol{x})). \tag{3.107}$$

As per the definitions and theorems listed above, the algorithmic steps to solving the FLDM problem (3.107) using the balanced space approach and satisfaction levels, are as follows (where $s_{ij}^F, i = 1, 2, \cdots, m_1, j = 0, 1, 2, \cdots$ means a j satisfaction level for objective i in the FLDM problem):

Algorithm I

Step 1. Set the satisfaction level at $s_{ij}^F, i = 1, 2, \cdots, m_1, j = 0, 1, 2, \cdots$. Let $s_i = s_{i0}^F, i = 1, 2, \cdots, m_1$ at the beginning, and let $s_i = s_{i1}^F, s_{i2}^F, \cdots, s_{i3}^F, i = 1, 2, \cdots, m_1$ respectively.
Step 2. Calculate the apportioned numbers

$$\alpha_i = \frac{s_i}{\sum_{i=1}^{m_1} s_i}, i = 1, 2, \cdots, m_1.$$

Calculate the apportioned balanced number $\eta_0^F(\alpha)$ and the η-optimal solution set X_η^{0F} for (3.107).
Step 3. If the FLDM is satisfied with any $x \in X_\eta^{0F}(\alpha)$, then $\boldsymbol{x}$ is the preferred solution to the FLDM, so go to Step 5. Otherwise go to Step 4.
Step 4. Adjust the satisfaction level. Let $s_i = s_{ij+1}^F \geq s_{ij}^F, i = 1, 2, \cdots, m_1$ and go to step 2.
Step 5. Stop.
As per the aforementioned definitions and theorems, and similar to Algorithm I, the algorithmic steps for solving an SLDM problem using the balanced space approach and satisfaction level are as follows: (where $s_{ij}^S, i = 1, 2, \cdots, m_2, j = 0, 1, 2, \cdots$. means a j satisfaction level for objective i in the SLDM problem):

Algorithm II

Step 1. Set the satisfaction level at $s_{ij}^S, i = 1, 2, \cdots, m_2, j = 0, 1, 2, \cdots$ Let $s_i = s_{i0}^S, i = 1, 2, \cdots, m_2$ at the beginning, and $s_i = s_{i1}^S, s_{i2}^S, \cdots, s_{i3}^S, i = 1, 2, \cdots, m_2$ respectively.
Step 2. Calculate the apportioned numbers

$$\alpha_i = \frac{s_i}{\sum_{i=1}^{m_2} s_i}, i = 1, 2, \cdots, m_2.$$

Calculate the apportioned balanced number $\eta_0^F(\alpha)$ and the η-optimal solution set X_η^{0F}.
Step 3. If the SLDM is satisfied with any $x \in X_\eta^{0F}(\alpha)$, then $\boldsymbol{x}$ is the preferred solution to the SLDM, so go to Step 5. Otherwise go to Step 4.
Step 4. Adjust the satisfaction level. Let $s_i = s_{ij+1}^S \geq s_{ij}^S, i = 1, 2, \cdots, m_2$ and go to Step 2.
Step 5. Stop.

The following interactive algorithm controls the solution steps between the previous algorithms, Algorithm I and Algorithm II, for the first and second levels of the bi-level multi-objective decision making problem. The steps are as follows:

Step 1. Set $k = 0$, solve the FLDM problem following Algorithm I to obtain a set of preferred solutions that are acceptable to the FLDM. The FLDM then puts the solutions in order, according to the following format:

Preferred solution:

$$(x_1^k, x_2^k), (x_1^{k+1}, x_2^{k+1}), \cdots, (x_1^{k+p}, x_2^{k+p}).$$

Preferred ranking (satisfaction ranking):

$$(x_1^k, x_2^k) \succ (x_1^{k+1}, x_2^{k+1}), \succ \cdots \succ (x_1^{k+p}, x_2^{k+p}).$$

Step 2. Given $x_1^F = x_1^k$ in the SLDM problem, solve the SLDM problem following Algorithm II and obtain x_2^S. Let $x_2^F = x_2^k$

Step 3. If

$$\frac{||F_1(x_1^F, x_2^F) - F_1(x_1^F, x_2^S)||}{||F_1(x_1^F, x_2^F)||} \leq \delta^F, \tag{3.108}$$

(where δ^F is a fairly small positive number given by the FLDM) then go to Step 5. Otherwise go to Step 4.

Step 4. Let $k = k + 1$, and go to Step 2.

Step 5. Stop.

Numerical Example

In what follows, a numerical example is presented to illustrate the effectiveness of the proposed solution method for the linear Ra-Ra DCCC model.

Example 3.8. We consider the following Ra-Ra bi-level multiobjective linear programming problem , which has two objective functions of the leader and two objective functions of the follower:

$$\begin{cases} \max\limits_{x_1,x_2} F_1 = \bar{\bar{\xi}}_{11}x_1 + \bar{\bar{\xi}}_{12}x_2 + \bar{\bar{\xi}}_{13}y_1 + \bar{\bar{\xi}}_{14}y_2 \\ \max\limits_{x_1,x_2} F_2 = \bar{\bar{\xi}}_{21}x_1 + \bar{\bar{\xi}}_{22}x_2 + \bar{\bar{\xi}}_{23}y_1 + \bar{\bar{\xi}}_{24}y_2 \\ \text{s.t.} \begin{cases} \text{where } (y_1, y_2) \text{ solves} \\ \max\limits_{y_1,y_2} f_1 = \bar{\bar{\eta}}_{11}x_1 + \bar{\bar{\eta}}_{12}x_2 + \bar{\bar{\eta}}_{13}y_1 + \bar{\bar{\eta}}_{14}y_2 \\ \max\limits_{y_1,y_2} f_2 = \bar{\bar{\eta}}_{21}x_1 + \bar{\bar{\eta}}_{22}x_2 + \bar{\bar{\eta}}_{23}y_1 + \bar{\bar{\eta}}_{24}y_2 \\ \text{s.t.} \begin{cases} \bar{\bar{\zeta}}_1x_1 + \bar{\bar{\zeta}}_2x_2 + \bar{\bar{\zeta}}_3y_1 + \bar{\bar{\zeta}}_4y_2 \leq 200 \\ 2x_1 - x_2 + 3y_1 + y_2 \leq 100 \\ x_1, x_2, y_1, y_2 \geq 0, \end{cases} \end{cases} \end{cases} \tag{3.109}$$

where $\tilde{\tilde{\xi}}_{ij}, \tilde{\tilde{\eta}}_{ij}, \tilde{\tilde{\zeta}}_j (i = 1,2; j = 1,2,4)$ are Ra-Ra variables characterized as:

$$
\begin{array}{l}
\tilde{\tilde{\xi}}_{11} \sim \mathcal{N}(\tilde{\xi}_{11},2),\ \text{with}\ \tilde{\xi}_{11} \sim \mathcal{N}(6,2);\ \tilde{\tilde{\xi}}_{12} \sim \mathcal{N}(\tilde{\xi},2),\ \text{with}\ \tilde{\xi}_{12} \sim \mathcal{N}(10,2);\\
\tilde{\tilde{\xi}}_{13} \sim \mathcal{N}(\tilde{\tilde{\xi}}_{13},2),\ \text{with}\ \tilde{\xi}_{13} \sim \mathcal{N}(8,2);\ \tilde{\tilde{\xi}}_{14} \sim \mathcal{N}(\tilde{\xi}_{14},2),\ \text{with}\ \tilde{\xi}_{14} \sim \mathcal{N}(12,2);\\
\tilde{\tilde{\xi}}_{21} \sim \mathcal{N}(\tilde{\xi}_{21},2),\ \text{with}\ \tilde{\xi}_{21} \sim \mathcal{N}(3,2);\ \tilde{\tilde{\xi}}_{22} \sim \mathcal{N}(\tilde{\xi}_{22},2),\ \text{with}\ \tilde{\xi}_{22} \sim \mathcal{N}(-5,2);\\
\tilde{\tilde{\xi}}_{23} \sim \mathcal{N}(\tilde{\xi}_{23},2),\ \text{with}\ \tilde{\xi}_{23} \sim \mathcal{N}(7,2);\ \tilde{\tilde{\xi}}_{24} \sim \mathcal{N}(\tilde{\xi}_{24},2),\ \text{with}\ \tilde{\xi}_{24} \sim \mathcal{N}(-9,2);\\
\tilde{\tilde{\eta}}_{11} \sim \mathcal{N}(\tilde{\eta}_{11},1),\ \text{with}\ \tilde{\eta}_{11} \sim \mathcal{N}(12,1);\ \tilde{\tilde{\eta}}_{12} \sim \mathcal{N}(\tilde{\eta}_{12},1),\ \text{with}\ \tilde{\eta}_{12} \sim \mathcal{N}(10,1);\\
\tilde{\tilde{\eta}}_{13} \sim \mathcal{N}(\tilde{\eta}_{13},1),\ \text{with}\ \tilde{\eta}_{13} \sim \mathcal{N}(8,1);\ \tilde{\tilde{\eta}}_{14} \sim \mathcal{N}(\tilde{\eta}_{14},1),\ \text{with}\ \tilde{\eta}_{14} \sim \mathcal{N}(6,1);\\
\tilde{\tilde{\eta}}_{21} \sim \mathcal{N}(\tilde{\eta}_{21},1),\ \text{with}\ \tilde{\eta}_{21} \sim \mathcal{N}(12,1);\ \tilde{\tilde{\eta}}_{22} \sim \mathcal{N}(\tilde{\eta},1),\ \text{with}\ \tilde{\eta}_{22} \sim \mathcal{N}(5,1);\\
\tilde{\tilde{\eta}}_{23} \sim \mathcal{N}(\tilde{\eta}_{23},1),\ \text{with}\ \tilde{\eta}_{23} \sim \mathcal{N}(18,1);\ \tilde{\tilde{\eta}}_{24} \sim \mathcal{N}(\tilde{\eta}_{24},1),\ \text{with}\ \tilde{\eta}_{24} \sim \mathcal{N}(20,1);\\
\tilde{\tilde{\zeta}}_{1} \sim \mathcal{N}(\tilde{\zeta}_{1},1),\ \text{with}\ \tilde{\zeta}_{1} \sim \mathcal{N}(6,1);\ \tilde{\tilde{\zeta}}_{2} \sim \mathcal{N}(\tilde{\zeta}_{2},1),\ \text{with}\ \tilde{\zeta}_{2} \sim \mathcal{N}(12,1);\\
\tilde{\tilde{\zeta}}_{3} \sim \mathcal{N}(\tilde{\zeta}_{3},1),\ \text{with}\ \tilde{\zeta}_{3} \sim \mathcal{N}(18,1);\ \tilde{\tilde{\zeta}}_{4} \sim \mathcal{N}(\tilde{\zeta}_{4},1),\ \text{with}\ \tilde{\zeta}_{4} \sim \mathcal{N}(24,1).
\end{array}
$$

If the confidence levels are given by the decision makers, the DCCC model of (3.109) is:

$$
\begin{cases}
\max\limits_{x_1,x_2}[\beta_1,\beta_2]\\
\text{s.t.}\begin{cases}
Pr\{\omega|Pr\{\tilde{\tilde{\xi}}_{11}(\omega)x_1 + \tilde{\tilde{\xi}}_{12}(\omega)x_2 + \tilde{\tilde{\xi}}_{13}(\omega)y_1 + \tilde{\tilde{\xi}}(\omega)_{14}y_2 \ge 200\} \ge \beta_1\} \ge 0.8\\
Pr\{\omega|Pr\{\tilde{\tilde{\xi}}_{21}x_1(\omega) + \tilde{\tilde{\xi}}_{22}(\omega)x_2 + \tilde{\tilde{\xi}}_{23}(\omega)y_1 + \tilde{\tilde{\xi}}_{24}(\omega)y_2 \ge 120\} \ge \beta_2\} \ge 0.8\\
\text{where } (y_1,y_2) \text{ solves}\\
\max\limits_{y_1,y_2}[\bar{f}_1,\bar{f}_2]\\
\text{s.t.}\begin{cases}
Pr\{\omega|Pr\{\tilde{\tilde{\eta}}_{11}x_1 + \tilde{\tilde{\eta}}_{12}x_2 + \tilde{\tilde{\eta}}_{13}y_1 + \tilde{\tilde{\eta}}_{14}y_2 \ge \bar{f}_1\} \ge 0.8\} \ge 0.8\\
Pr\{\omega|Pr\{\tilde{\tilde{\eta}}_{21}x_1 + \tilde{\tilde{\eta}}_{22}x_2 + \tilde{\tilde{\eta}}_{23}y_1 + \tilde{\tilde{\eta}}_{24}y_2 \ge \bar{f}_2\} \ge 0.8\} \ge 0.8\\
Pr\{\omega|Pr\{\tilde{\tilde{\zeta}}_1(\omega)x_1 + \tilde{\tilde{\zeta}}_2(\omega)x_2 + \tilde{\tilde{\zeta}}_3(\omega)y_1 + \tilde{\tilde{\zeta}}_4(\omega)y_2] \le 200\} \ge 0.8\} \ge 0.8\\
2x_1 - x_2 + 3y_1 + y_2 \le 100\\
x_1, x_2, y_1, y_2 \ge 0.
\end{cases}
\end{cases}
\end{cases}
\tag{3.110}
$$

It follows from Theorems 3.9 and 3.10 that (3.110) is equivalent to

$$\begin{cases} \max_{x_1,x_2}[\beta_1,\beta_2] \\ \text{s.t.} \begin{cases} 200 \le 6x_1 + 10x_2 + 8y_1 + 12y_2 - 0.84\left(\sqrt{2(x_1^2 + x_2^2 + y_1^2 + y_2^2)}\right) \\ +\Phi^{-1}(1-\beta_1^L)\left(\sqrt{2(x_1^2 + x_2^2 + y_1^2 + y_2^2)}\right) \\ 120 \le 3x_1 - 5x_2 + 7y_1 - 9y_2 - 0.84\left(\sqrt{2(x_1^2 + x_2^2 + y_1^2 + y_2^2)}\right) \\ +\Phi^{-1}(1-\beta_1^L)\left(\sqrt{2(x_1^2 + x_2^2 + y_1^2 + y_2^2)}\right) \\ \text{where } (y_1,y_2) \text{ solves} \\ \max_{y_1,y_2}[\bar{f}_1,\bar{f}_2] \\ \text{s.t.} \begin{cases} \bar{f}_1 \le 12x_1 + 10x_2 + 8y_1 + 6y_2 - 1.68\left(\sqrt{x_1^2 + x_2^2 + y_1^2 + y_2^2}\right) \\ \bar{f}_2 \le 12x_1 + 5x_2 + 18y_1 + 20y_2 - 1.68\left(\sqrt{x_1^2 + x_2^2 + y_1^2 + y_2^2}\right) \\ 200 \le 6x_1 + 12x_2 + 18y_1 + 24y_2 - 1.68\left(\sqrt{x_1^2 + x_2^2 + y_1^2 + y_2^2}\right) \\ x_1, x_2, y_1, y_2 \ge 0. \end{cases} \end{cases} \end{cases} \tag{3.111}$$

By using the interactive balance space method mentioned above, the solutions is $(x_1^*, x_2^*, y_1^*, y_2^*) = (22.6, 35.1, 18.3, 41.8)$, the corresponding objective function values are $(\beta_1^* \beta_2^*, \bar{f}_1^*, \bar{f}_2^*) = (0.91, 0.76, 34.61, 16.52)$.

3.5.3 Ra-Ra Simulation-Based CST-PSO for Nonlinear Models

If (3.91) or (3.92) is a nonlinear model, i.e., some functions in model (3.91) or (3.92) are nonlinear, the Ra-Ra features make it impossible to be converted into its crisp equivalent model. In order solve nonlinear (3.91) or (3.92), three issues are needed to solve: the trade-off of the multiple objectives, the twofold randomness and the bi-level structure. In this section, we design a hybrid method consisting of simulation technique, the satisfying trade-off method and CST-PSO algorithm, where they are deal with twofold randomness, multiple objectives and bi-level structure, respectively. Since the Ra-Ra simulation techniques have been proposed in Sects. 3.3.3 and 3.4.3. Next, we just present the satisfying trade-off method for multiple objectives and CST-PSO for bi-level structure.

The Satisfying Trade-Off Method

The satisfying trade-off method for multi-objective programming problems was proposed by Nakayama [27] and Sawaragi et al. [37]. It is an interactive method combining the satisfying level method with the ideal point method. This method can be applied to not only the linear multi-objective but also the nonlinear multi-objective programming.

Consider

$$\begin{cases} \max[H_1(\boldsymbol{x}), H_2(\boldsymbol{x}), \cdots, H_m(\boldsymbol{x})] \\ \text{s.t. } \boldsymbol{x} \in X. \end{cases} \tag{3.112}$$

In the begin, let's briefly introduce the simple satisfying level method which in mainly referred in [28]. In some real decision making problems, the decision maker usually provides a reference objective values $\bar{\boldsymbol{H}} = (\bar{H}_1, \bar{H}_2, \cdots, \bar{H}_m)^T$. If the solution satisfies the reference value, take it. The simple satisfying level method can be summarized as follows:

Step 1. Decision maker gives the reference objective values $\bar{\boldsymbol{H}}$.
Step 2. Solve the following programming problem,

$$\begin{cases} \max \sum_{i=1}^{m} H_i(\boldsymbol{x}) \\ \text{s.t.} \begin{cases} H_i(\boldsymbol{x}) \geq \bar{H}_i, \ i = 1, 2, \cdots, m \\ \boldsymbol{x} \in X. \end{cases} \end{cases} \tag{3.113}$$

Step 3. If the problem (3.113) doesn't have the feasible solution, turn to Step 4. If the problem (3.113) has the optimal solution $\bar{\boldsymbol{x}}$, output $\bar{\boldsymbol{x}}$.
Step 4. Decision maker re-gives the reference objective values $\bar{\boldsymbol{H}}$ and turn to Step 2.

The satisfying trade-off method can be summarized as follows:

Step 1. Take the ideal point $\boldsymbol{H}^* = (H_1^*, H_2^*, \cdots, H_m^*)^T$ such that $H_i^* > \max_{\boldsymbol{x} \in X} f_i(\boldsymbol{x})$ $(i = 1, 2, \cdots, m)$.
Step 2. Decision maker gives the objective level $\bar{\boldsymbol{H}}^k = (\bar{H}_1^k, \bar{H}_2^k, \cdots, \bar{H}_m^k)^T$ and $\bar{H}_i^k < \bar{H}_i^* (i = 1, 2, \cdots, m)$. Let $k = 1$.
Step 3. Compute the weight and solve the following problem to get the efficient solution.

$$w_i^k = \frac{1}{H_i^* - \bar{H}_i^k}, \ i = 1, 2, \cdots, m. \tag{3.114}$$

$$\min_{\boldsymbol{x} \in X} \max_{1 \leq i \leq m} w_i^k |H_i^* - H_i(\boldsymbol{x})|, \tag{3.115}$$

or the equivalent problem,

$$\begin{cases} \min \lambda \\ \text{s.t.} \begin{cases} w_i^k (H_i^* - H_i(\boldsymbol{x})) \leq \lambda, \ i = 1, 2, \cdots, m \\ \boldsymbol{x} \in X. \end{cases} \end{cases} \tag{3.116}$$

Suppose that the optimal solution is $\boldsymbol{x}^k$.

Step 4. According to the objective value $\boldsymbol{H}(\boldsymbol{x}^k) = (H_1(\boldsymbol{x}^k), H_2(\boldsymbol{x}^k), \cdots, H_m(\boldsymbol{x}^k))^T$, decision maker divides them into three classes: (1) which needs to improve, denote the related subscript set I_I^k, (2) which is permitted to release, denote the related subscript set I_R^k, (3) which is accepted, denote the related subscript set I_A^k. If $I_I^k = \Phi$, stop the iteration and output $\boldsymbol{x}^k$. Otherwise, decision maker gives the new reference objective values $\tilde{H}_i^k, i \in I_I^k \cup I_R^k$ and let $\tilde{H}_i^k = H_i(\boldsymbol{x}^k), i \in I_A^k$.

Step 5. Let $u_i(i = 1, 2, \cdots, m)$ be the optimal Kuhn-Tucker operator of the first constraints. If there exists a minimal nonnegative number ε such that

$$\sum_{i=1}^{m} u_i w_i^k (\tilde{H}_i^k - H_i(\boldsymbol{x}^k)) \geq -\varepsilon,$$

then we deem that $\tilde{H}_i^k$ passes the check for feasibility. Let $\tilde{H}_{i+1} = \tilde{H}_i^k$ $(i = 1, 2, \cdots, m)$, turn to Step 3. Otherwise, $\tilde{H}_i^k$ isn't feasible. The detail can be referred in [37]. Decision maker should re-give $\tilde{H}_i^k, i \in I_I^k \cup I_R^k$ and recheck it.

CST-PSO

After dealing with twofold randomness and multiple objective, we considered how to handle bi-level structure. In this section, particle swarm optimization based on a hybrid intelligent algorithm by combining particle swarm optimization with chaos searching technique (CST-PSO), proposed by Wan et al. [45].

Consider

$$\begin{cases} \min_x F(x, y) \\ \text{s.t.} \begin{cases} g(x, y) \leq 0 \\ \text{where } y \text{ solves} \\ \begin{cases} \min_y f(x, y) \\ \text{s.t. } h(x, y) \leq 0. \end{cases} \end{cases} \end{cases} \tag{3.117}$$

If $(\bar{x}, \bar{y})$ is a feasible solution to (3.117), then

$$\begin{cases} \min_\lambda ||\nabla_y f(\bar{x}, \bar{y}) + \lambda^T \nabla_y h(\bar{x}, \bar{y})||^2 + ||\lambda^T h(\bar{x}, \bar{y})||^2 \\ \text{s.t. } \lambda \geq 0. \end{cases} \tag{3.118}$$

Particle swarm optimization (PSO), a population-based algorithm, was inspired by the social behavior of animals such as fish schooling and birds flocking. Similar to other population-based algorithms, such as evolutionary algorithms, PSO can solve a variety of difficult optimization problems, and has shown a faster convergence rate than other evolutionary algorithms on some problems. Another advantage of PSO is that it has very few parameters to adjust, which makes it particularly easy to implement.

Suppose that the search space is D-dimensional, so then the ith particle of the swarm can be represented by a D-dimensional vector, $X_i = (x_{i1}, x_{i2}, \cdots, x_{iD})$. The velocity of this particle can be represented by another D-dimensional vector $V_i = (v_{i1}, v_{i2}, \cdots, v_{iD}$. The best previously visited position of the ith particle is denoted as $p_i^{best} = (p_{i1}, p_{i2}, \cdots, p_{iD})$ The best previously visited position of the swarm is denoted as$g_i^{best} = (g_{i1}, g_{i2}, \cdots, g_{iD})$. Change the velocity and position of the ith particle according to the following equation (see notes below):

$$v_i^{k+1} = v_i^k + c_1 r_1 (p_{best} - x_i^k) + c_2 r_2 (g_{best} - x_i^k), \tag{3.119}$$

$$x_i^{k+1} = v_i^k + v_i^{k+1}, \tag{3.120}$$

where c_1 and c_2 are positive constants, called acceleration, and r_1 and r_2 are two random numbers, uniformly distributed in $[0, 1]$. To prevent the particle from moving far from the search space, the constant $V_{\max}$ is implemented to limit the velocity.

Chaotic variables, which are found in nonlinear systems, are non-periodic, dynamic and unique. They can appear stochastic but can be generated using deterministic means. Chaotic variables are in an unshaped out-of-order state, and can blend with other specific forms which are relative to "immobile points", or "periodic points". Chaotic variables have subtle internal structures with "strange attractors" that attract system movement and confine it within the specified range.

The chaos searching technique (CST) is a new searching method. The basic idea of the algorithm is to transform the problem variables from the solution space to the chaotic space and then perform a search to determine the solution using the randomness, orderliness and ergodicity of the chaotic variable.

The chaos searching technique has two steps: Firstly, all points are searched in turn within the changing variable range, and, taking the better point as the current optimum point, regard this point as the center. A tiny chaotic disturbance is imposed and a more careful search performed to determine the optimum point. The chaos search technique has many advantages. It is not sensitive to the initial value, easily moves out of the local minimum value, and has a rapid search velocity and gradual global convergence.

The following Logistics map is used to generate the chaotic sequence as it is more convenient:

$$z_{i+1} = \mu z_i (1 - z_i), \tag{3.121}$$

where $z_i \in [0, 1] (i = 1, 2, \cdots)$ is the chaotic variable, $i (= 1, 2, \cdots)$ is the number of iterations; and μ is the control parameter. The system is entirely in a chaotic situation when $\mu = 4$ and the chaos space belongs to [0, 1].

The main idea of this algorithm is to embed the CST into the PSO to solve nonlinear bi-level programming problems, thereby combining the advantages and avoiding the disadvantages. The algorithm is described in detail as follows: firstly, the swarm particles are randomly initialed. Then, problem (3.118) is solved to judge whether the particle is feasible. If there is a solution to problem (3.118) and the objective function equals zero, then the particle is feasible. The particle is then added to the feasible list and the upper level's objective function is set as the fitness value for the particle. Conversely, if the particle is infeasible, the particle is added to the infeasible list and the objective function for problem (3.118) is set as the particles fitness value. Secondly, the particles in the feasible list are ranked in ascending order, and then the particles in the infeasible list are also ranked in ascending order. The velocity of the particle nearest the top and its new position are assigned according to Eqs. (3.119) and (3.120), and the particles at the bottom of the list are updated using the CST. After one iteration, the fitness values of the particles are again computed. The above steps are repeated until the stopping criterion is met.

The steps for the proposed algorithm are detailed in the following:

Step 1. Initialize the parameters. Population size (the number of particles) is set at $M = m + n$, where the m particles are updated using PSO and the n particles are updated using CST. Maximal velocity, $V_{\max}$, two learning factors, c_1 and c_2, and two random variables, $r_1, r_2 \in [0, 1]$, are initially set. The maximum number of iterations (T) is set at the algorithmic termination conditions and the iteration counter is set at $t = 0$.

Step 2. Initialize the particles. Initialize the ith ($i = 1, 2, \cdots, M$) particle randomly with an initial position, X_i, within the pre-specified range and velocity, V_i, in the maximal speed range, $V_{\max}$. The best previously visited position of the ith ($i = 1, 2, \cdots, M$) particle, P_i^{best}, is initialized as X_i.

Step 3. Compute the fitness values of the particles. Equation (3.118) is solved using CST. If there is a solution to problem (3.118) and the objective function value equals zero, then the particle is feasible, so is added to the feasible list and the upper level's objective function is set as the particles fitness value. Conversely, if the particle is infeasible, it is added to the infeasible list and the objective function for problem (3.118) is set as the particles fitness value.

Step 4. Rank the particles. The particles in the feasible list are ranked in ascending order; Then the particles in the infeasible list are also ranked in ascending order.

Step 5. Update the local best position, P_i^{best}, and the global best position, g^{best}. For the ith ($i = 1, 2, \cdots, M$) particle, compare the particle's fitness evaluation with its P_i^{best}. If the current value is better than P_i^{best}, then the current value is set as the P_i^{best}. Compare the first particle's fitness evaluation with the global best position, g^{best}. If the current value is better than g^{best}, then the current value is set as g^{best}.

Step 6. Update the particles. For the first m particles, the particle velocity and its new position are assigned according to Eqs. (3.119) and (3.120); Then the n particles at the bottom of the list are updated using the CST with an initial chaos variable X_i.

Step 7. Terminal conditions. $t = t + 1$, If the number of iterations is larger than the maximum number of iterations (T), goto Step 8, otherwise go to Step3.

Step 8. Out put the results. Out put the optimal particle, compute and out put the upper level and lower level's objective function values.

Notes: Firstly, the CST is not only used to solve problem (3.118) but also embedded in the PSO to improve the worse particles, which not only enhances the particles but also improves the diversity of the particle swarm to avoid the PSO becoming trapped in the local optima. Secondly, the upper level and lower level decision variables are all randomly generated and updated by the PSO or CST in our algorithm. This is quite different from when only the upper level decision variable is encoded and the lower level decision variable is computed according to the upper level decision variable. The feasible weighting value is introduced to replace the particle fitness value if it is infeasible, which can force the infeasible particle to become feasible. Thirdly, in the early stages of the algorithm, a feasible weighting value is used to rank the infeasible particles, which can avoid situations in which the infeasible particle is denoted as the best particle, even though it is only the best of all when the fitness value is used by all particles.

Numerical Example

In this section, a numerical example is given to illustrate the effectiveness of the application of the models and algorithms proposed earlier.

Example 3.9. Consider the problem

$$
\begin{cases}
\max\limits_{x_1,x_2,x_3} F_1 = 2\tilde{\tilde{\xi}}_1^2 x_1 + \sqrt{\tilde{\tilde{\xi}}_2 x_2 + \tilde{\tilde{\xi}}_3 x_3} + \tilde{\tilde{\xi}}_4 y_1 + \sqrt{\tilde{\tilde{\xi}}_5 y_2} + \sqrt{\tilde{\tilde{\xi}}_6 y_3} \\
\max\limits_{x_1,x_2,x_3} F_2 = \tilde{\tilde{\xi}}_1^2 x_1 + \tilde{\tilde{\xi}}_2^2 x_2 + x_3 + y_1 + y_2 + y_3 \\
\text{s.t.} \begin{cases}
\text{where } (y_1, y_2, y_3) \text{ solves} \\
\max\limits_{y_1,y_2,y_3} f_1 = \sqrt{7\tilde{\tilde{\xi}}_1 x_1 + 3\tilde{\tilde{\xi}}_2 x_2 + 8\tilde{\tilde{\xi}}_3 x_3} + (12\tilde{\tilde{\xi}}_4 y_1 + 11\tilde{\tilde{\xi}}_5 y_2 + 5\tilde{\tilde{\xi}}_6 y_3)^2 \\
\max\limits_{y_1,y_2,y_3} f_2 = x_1 + 5x_2 + 7x_3 + y_1 + 11y_2 - 8\tilde{\tilde{\xi}}_6^3 y_3 \\
\text{s.t.} \begin{cases}
\sqrt{x_1\tilde{\tilde{\xi}}_1 + x_2\tilde{\tilde{\xi}}_2 + x_3^2\tilde{\tilde{\xi}}_3 + \tilde{\tilde{\xi}}_4 y_2 + 17y_3\tilde{\tilde{\xi}}_6} \le 1540 \\
3x_1 + 6x_2 + 12x_3 + 14y_1 + 19y_2 \le 2400 \\
x_1 + x_2 + x_3 + y_1 + y_2 - y_3 \ge 150 \\
x_1 + x_2 + x_3 + y_1 + y_2 + y_3 \le 200 \\
x_1, x_2, x_3, y_1, y_2, y_3 \ge 0,
\end{cases}
\end{cases}
\end{cases}
\tag{3.122}
$$

where $\xi_j, j = 1, 2, \cdots, 6$ are Ra-Ra variables characterized as

$$
\begin{array}{lll}
\tilde{\tilde{\xi}}_1 \sim \mathcal{N}(\tilde{\xi}_1, 1), \text{ with } \tilde{\xi}_1 \sim \mathcal{N}(8, 1); & \tilde{\tilde{\xi}}_2 \sim \mathcal{N}(\tilde{\xi}_2, 1), & \text{with } \tilde{\xi}_2 \sim \mathcal{N}(10, 1); \\
\tilde{\tilde{\xi}}_3 \sim \mathcal{N}(\tilde{\xi}_3, 1), \text{ with } \tilde{\xi}_3 \sim \mathcal{N}(12, 1); & \tilde{\tilde{\xi}}_4 \sim \mathcal{N}(\tilde{\xi}, 1) & \text{with } \tilde{\xi}_4 \sim \mathcal{N}(14, 1); \\
\tilde{\tilde{\xi}}_5 \sim \mathcal{N}(\tilde{\xi}, 1) \quad \text{with } \tilde{\xi}_5 \sim \mathcal{N}(16, 1); & \tilde{\tilde{\xi}}_6 \sim \mathcal{N}(\tilde{\xi}_6, 1) & \text{with } \tilde{\xi}_6 \sim \mathcal{N}(18, 1).
\end{array}
$$

We can obtain the DCCC model as follows.

$$\begin{cases} \max\limits_{x_1,x_2,x_3} [\beta_1, \beta_2] \\ \text{s.t.} \begin{cases} Pr\{\omega|Pr\{2\bar{\bar{\xi}}_1^2(\omega)x_1 + \sqrt{\bar{\bar{\xi}}_2(\omega)x_2 + \bar{\bar{\xi}}_3(\omega)x_3} + \bar{\bar{\xi}}_4(\omega)y_1 + \sqrt{\bar{\bar{\xi}}_5(\omega)y_2} \\ \quad + \sqrt{\bar{\bar{\xi}}_6(\omega)y_3}\} \geq \beta_1\} \geq 0.8 \\ Pr\{\omega|Pr\{\bar{\bar{\xi}}_1^2(\omega)x_1 + \bar{\bar{\xi}}_2^2(\omega)x_2 + x_3 + y_1 + y_2 + y_3\} \geq \beta_2\} \geq 0.8 \\ \text{where } (y_1, y_2, y_3) \text{ solves} \\ \max\limits_{y_1,y_2,y_3} [\bar{f}_1, \bar{f}_2] \\ \text{s.t.} \begin{cases} Pr\{\omega|Pr\{\sqrt{7\bar{\bar{\xi}}_1(\omega)x_1 + 3\bar{\bar{\xi}}_2(\omega)x_2 + 8\bar{\bar{\xi}}_3(\omega)x_3} + (12\bar{\bar{\xi}}_4(\omega)y_1 \\ \quad + 11\bar{\bar{\xi}}_5(\omega)y_2 + 5\bar{\bar{\xi}}_6(\omega)y_3)^2 \geq \bar{f}_1\} \geq 0.8\} \geq 0.8 \\ Pr\{\omega|Pr\{x_1 + 5x_2 + 7x_3 + y_1 \\ +11y_2 - 8\bar{\bar{\xi}}_6^3(\omega)y_3 \geq \bar{f}_2\} \geq 0.8\} \geq 0.8 \\ Pr\{\omega|Pr\{\sqrt{x_1\bar{\bar{\xi}}_1(\omega) + x_2\bar{\bar{\xi}}_2(\omega) + x_3^2\bar{\bar{\xi}}_3(\omega)a} + \bar{\bar{\xi}}_4(\omega)y_2 + 17y_3\bar{\bar{\xi}}_6(\omega) \\ \leq 1540\} \geq 0.8\} \geq 0.8 \\ 3x_1 + 6x_2 + 12x_3 + 14y_1 + 19y_2 \leq 2400 \\ x_1 + x_2 + x_3 + y_1 + y_2 - y_3 \geq 150 \\ x_1 + x_2 + x_3 + y_1 + y_2 + y_3 \leq 200 \\ x_1, x_2, x_3, y_1, y_2, y_3 \geq 0. \end{cases} \end{cases} \end{cases} \tag{3.123}$$

After running the algorithm above, we obtain $(x_1^*, x_2^*, x_3^*, y_1^*, y_2^*, y_3^*) = (11.56, 26.52, 13.73, 20.15, 10.24, 0.25)$ and $(\beta_1^*, \beta_2^*, \bar{f}_1^*, \bar{f}_2^*) = (0.88, 0.74, 14.68, 20.92)$.

3.6 Transport Flow Distribution of the SBY Hydropower Project

In this section, the expectation multi-objective bi-level programming model with chance constraints under a Ra-Ra environment based on the philosophy is proposed: decisions are selected by optimizing the expected values of the objective functions subject to chance constraints with some predetermined confidence levels given by the actual decision-makers.

3.6.1 Modelling

The mathematical model described has the following assumptions.

1. The proposed transportation network is a single material transportation network composed of nodes and road sections.

2. The capacities of the different arcs are satisfactorily independent, and the total flow of carriers on each arc cannot exceed its capacity.
3. Flows on all transportation paths between OD pairs satisfy the feasible flow conservation [15].
4. All transportation paths in the network are known.
5. The transportation cost of every road section, and transportation time are considered Ra-Ra variables, with the attributes determined from available statistics and historical data as well as forecast transportation environments. They are considered to be independent.
6. The demand at every reception node must be met on schedule. The material has a given transportation duration. If the transportation time exceeds the given duration, then a delay cost is added.

The following mathematical notations are used to describe the TFDP:

o : index of origin node;
d : index of destination node;
Ψ : set of carriers, $k \in \Psi$ is an index;
Φ : set of arcs in the transportation network, $i \in \Phi$ is an index;
Ω : set of paths in the transportation network, $j \in \Omega$ in an index;
E : set of origin-destination (OD) pairs, $(o, d) \in E$;
A_j : set of arcs in transportation path j; and
P_{od} : set of paths from origin node o to destination node d.

Certain parameters

r_i : maximal passing capacity of arc i;
w_k : weight for carrier k in the transportation network;
v_k : volume capacity of carrier k;
T_j : transportation time constraint, represents the time of transportation path j is T_j units in which the material
demand between all OD pairs have to be transported;
γ_i^j : a binary variable equal to 1 if and only if arc i is a segment of transportation path j for carrier k;
Q_{od} : transportation demand of the material from origin node o to destination node d;
c_j : direct cost of unit volume material using transportation j; and
$\lceil\,\rceil$: ceiling operator rounding upward to integer.

Uncertain parameters

$\bar{\tilde{e}}_i^k$: unit transportation cost of material flow on arc a_i for carrier k;
$\bar{\tilde{t}}_{i0}^k$: free transportation time of material flow on arc i for carrier k; and
$\bar{\tilde{t}}_i^k$: transportation time of material flow on arc a_i for carrier k.

Decision variables

x_j : volume of material flow on transportation path j, which is the decision variable of the upper level; and

y_{kj} : volume of material flow transported by carrier k through path j, which is the decision variable of the lower level.

Lower Level Model for the Bi-Level TFDP

The problem posed on the lower level is how to make decisions on material flow by each type of carrier on transportation path y_{kj} while satisfying all capacity constraints, with the main objective being to minimize expected total transportation cost.

1. Objective functions of the lower level

   The total transportation cost of the material is calculated by taking the sum of the carriers of each arc's transportation cost and the number of carriers needed to transport the material across the network. In real conditions, it is desirable that each service carrier be fully loaded, so the numbers of carrier k through path j can be denoted as $\lceil y_{kj}/v_k \rceil$. Since $\tilde{\bar{e}}_i^k$ is considered a Ra-Ra variable, the total transportation cost of material is considered under a Ra-Ra environment. Generally, it is difficult to completely minimize total transportation costs because of the Ra-Ra variables. Because decision-makers expect minimal cost, the expected value of the total transportation cost is the objective of the lower level. Denote the expected total transportation cost of material as $C(y_{kj})$, then the objective function of the lower level model can be formulated as:

$$\min \ C(y_{kj}) = E\left[\sum_{k\in\Psi}\sum_{j\in\Omega}\sum_{i\in A_j} \gamma_i^j \tilde{\bar{e}}_i^k \lceil y_{kj}/v_k \rceil\right]. \tag{3.124}$$

   Generally, a path can be represented by a sequence of adjacent arcs. A binary variable γ_i^j is introduced to determine whether an arc i is a segment of path j for carrier k:

$$\gamma_i^j = \begin{cases} 1, & \text{if } i \in A_j, \quad i \in \Phi, j \in \Omega, \\ 0, & \text{otherwise.} \end{cases} \tag{3.125}$$

2. Constraints of the lower level

   For transportation time, each carrier requires the transport of material from the source to the destination on schedule T_j. If not, a delay cost is applied. $\sum_{i\in A_j} \gamma_i^j \tilde{\bar{t}}_i^k$ represents the total travel time of carrier k on transportation path j, in which $\tilde{\bar{t}}_i^k$ is usually represented by a non-decreasing function (i.e., Bureau of Public Roads (BPR) function) [39] as follows:

$$\tilde{\bar{t}}_i^k = \tilde{\bar{t}}_{i0}^k \left[1 + \alpha \left(\frac{\lceil y_{kj}/v_k \rceil}{r_i} \right)^\beta \right], \quad i \in \Phi, k \in \Psi, \tag{3.126}$$

where α and β are user-defined parameters, and in this problem are set to 0.15 and 2.0 respectively.

Technically, it is not possible to strictly ensure that the random event $\sum_{i \in A_j} \gamma_i^j \tilde{\bar{t}}_i^k$ doesn't exceed T_j because of the Ra-Ra variable $\tilde{\bar{t}}_{i0}^k$. In practical problem, the decision makers often provide an appropriate budget T_j in advance, to ensure the restriction is, to a certain extent, satisfied, that is to maximize the probability of the random event:

$$Pr \left\{ \sum_{i \in A_j} \gamma_i^j \tilde{\bar{t}}_{i0}^k(\omega) \left[1 + \alpha \left(\frac{\lceil y_{kj}/v_k \rceil}{r_i} \right)^\beta \right] \le T_j \right\},$$

under a given confidence level, which can be written as follows:

$$Pr \left\{ \omega \middle| Pr \left\{ \sum_{i \in A_j} \gamma_i^j \tilde{\bar{t}}_{i0}^k(\omega) \left[1 + \alpha \left(\frac{\lceil y_{kj}/v_k \rceil}{r_i} \right)^\beta \right] \le T_j \right\} \ge \theta \right\} \ge \delta, \ j \in \Omega, k \in \Psi. \tag{3.127}$$

Here the decision makers' aspiration level is indicated as θ, so we use a "Pr" to ensure the constraint holds at the predetermined confidence level. Additionally, based on probability theory, a further "Pr" is needed to describe the random elements, which guarantee the establishment of a certain confidence level δ, resembling the P-model (probability maximization model) presented in [9].

The transportation flow may exceed some arcs' capacity because of uncertainties such as the condition of the construction project road. Such conditions may require the manager to select another path. Thus, the total amount of capacity on arc i cannot exceed the maximal capacity of the arc i, which produces the following constraint:

$$\sum_{j \in \Omega} \sum_{k \in \Psi} \gamma_i^j \lceil y_{kj}/v_k \rceil \le r_i, \quad i \in \Phi. \tag{3.128}$$

Actually, it is difficult to ensure each service carrier is fully loaded, so the sum of all flows transported by all kinds of carriers through each path cannot be less than the material flow assigned to it, thus the following constraint is obtained:

$$\sum_{k \in \Psi} y_{kj} \ge x_j, \quad j \in \Omega. \tag{3.129}$$

The path flow in path j used by carrier k should not be negative, such that

$$y_{kj} \ge 0, \quad j \in \Omega, k \in \Psi. \tag{3.130}$$

Upper Level Model for the Bi-Level TFDP

The problem the construction contractor on the upper level faces is how to assign material flow among the transportation paths across the complete transportation network, i.e., how to decide the material flow x_j through transportation path j. Thus, the decision variable on the upper level is x_j.

1. Objective functions of the upper level

   For large-scale construction projects, cost and time control are both important, so minimizing total direct costs and total transportation time costs are the two objectives of the upper level model. The two objectives of the upper level can be described as follows:

   Firstly, the upper level decision maker attempts to minimize the direct costs of the complete network by assigning the flow of the material to each transportation path to achieve a system optimized flow pattern, thus, the total direct cost is the sum of all transportation costs from different transportation paths. The first objective function of the upper level model can be formulated as:

$$C(x_j, y_{kj}) = \sum_{(o,d)\in E} \sum_{j\in P_{od}} c_j x_j. \tag{3.131}$$

   In real conditions, there is increasing pressure to shorter transportation time to reduce or eliminate extra project expenses, with the early arrival of materials shortening the completion time of the construction project and improving construction efficiency. Thus, the total transportation time for carrier k in each path can be described as $\sum_{i\in A_j} \gamma_i^j \tilde{\bar{t}}_i^k$, and, since $\tilde{\bar{t}}_i^k$ is a Ra-Ra variable, $\sum_{i\in A_j} \gamma_i^j \tilde{\bar{t}}_i^k$ can be regarded as a special Ra-Ra variable. Similarly, the expected value of the total transportation time cost is one of the objectives on the upper level. Different carriers are given different weights, and

$$\tilde{\bar{t}}_i^k = \tilde{\bar{t}}_{i0}^k \left[1 + \alpha \left(\frac{\lceil y_{kj}/v_k \rceil}{r_i} \right)^{\beta} \right],$$

   so the second objective function of the upper level can be described as follows:

$$T(x_j, y_{kj}) = E\left[\sum_{k\in\Psi} w_k \sum_{j\in\Omega} \sum_{i\in A_j} \gamma_i^j \tilde{\bar{t}}_{i0}^k \left[1 + \alpha \left(\frac{\lceil y_{kj}/v_k \rceil}{r_i} \right)^{\beta} \right] \right]. \tag{3.132}$$

2. Constraints of the upper level

   According to the basic assumptions, it is stipulated that the demands between all OD pairs should be satisfied. Thus:

$$\sum_{j\in P_{od}} x_j = Q_{od},\ (o,d) \in E. \tag{3.133}$$

The following constraint ensures that the sum of the weights is equal to 1:

$$\sum_{k\in\Psi} w_k = 1. \tag{3.134}$$

In order to describe the non-negative variables, the constraints in (3.135) are presented:

$$x_j \geq 0, \quad j \in \Omega. \tag{3.135}$$

Global Model for the Bi-Level TFDP

Based on the above discussion, by integrating of Eqs. (3.124)~(3.135), the following global model for the nonlinear multi-objective bi-level programming with Ra-Ra variables is formulated for the TFDP in a large-scale construction project:

$$\left\{\begin{array}{l}
\min\ C(x_j, y_{kj}) = \sum_{(o,d)\in E}\sum_{j\in P_{od}} c_j x_j \\
\min\ T(x_j, y_{kj}) = E\left[\sum_{k\in\Psi} w_k \sum_{j\in\Omega}\sum_{i\in A_j} \gamma_i^j \tilde{\bar{t}}_{i0}^k\left[1+\alpha\left(\frac{\lceil y_{kj}/v_k\rceil}{r_i}\right)^\beta\right]\right] \\
\text{s.t.} \left\{\begin{array}{l}
\sum_{j\in P_{od}} x_j = Q_{od}, \quad (o,d)\in E \\
\sum_{k\in\Psi} w_k = 1 \\
x_j \geq 0, \quad j\in\Omega \\
\min\ C(y_{kj}) = E\left[\sum_{k\in\Psi}\sum_{j\in\Omega}\sum_{i\in A_j} \gamma_i^j \tilde{\bar{e}}_i^k \lceil y_{kj}/v_k\rceil\right] \\
\text{s.t.} \left\{\begin{array}{l}
Pr\left\{\omega \middle| Pr\left\{\sum_{i\in A_j} \gamma_i^j \tilde{\bar{t}}_{i0}^k(\omega)\left[1+\alpha\left(\frac{\lceil y_{kj}/v_k\rceil}{r_i}\right)^\beta\right] \leq T_j\right\} \geq \theta\right\} \geq \delta \\
\sum_{j\in\Omega}\sum_{k\in\Psi} \gamma_i^j \lceil y_{kj}/v_k\rceil \leq r_i, i\in\Phi, j\in\Omega, k\in\Psi \\
\gamma_i^j = \left\{\begin{array}{ll} 1, \text{ if } i\in A_j, & j\in\Omega \\ 0 & \text{otherwise} \end{array}\right. \\
\sum_{k\in\Psi} y_{kj} \geq x_j, \quad j\in\Omega \\
y_{kj} \geq 0, \quad j\in\Omega, k\in\Psi.
\end{array}\right.
\end{array}\right.
\end{array}\right. \tag{3.136}$$

3.6.2 Presentation of the SBY Hydropower Project

In this section, an earth-rock work transportation project in a large-scale water conservancy and hydropower construction project is taken as an example for our optimization method. The SBY Hydropower Project is located in the middle reaches of Qingjiang River in Badong County, Sichuan province, China. The project is the first cascading dam project on the Qingjiang main stream and the third most important project after the Geheyan and Gaobazhou dam projects in China. Once completed, it will provide a major power source to meet peak load demand in the Central China Power Grid. The installed capacity and annual output of the SBY Power Plant are 1,600 MW and 3.92 GWh. The project has a powerful regulating ability with a normal pool level of 400 m and reservoir capacity of $4.58 \times 10^9\ m^3$. The project consists of a concrete-faced rock fill dam (CFRD), an underground power house, a chute spillway on the left bank, and a sluice tunnel on the right bank. The dam is 233 m high and is the tallest of its kind in the world at present with a total volume of $15.64 \times 10^6\ m^3$.

Detailed data for the SBY Hydropower Project were obtained from the Hubei Qingjiang Shuibuya Project Construction Company. In a large scale construction project, and especially in a large water conservancy and hydropower construction project, the earth-rock work is usually the primary material, and everyday earth-rock work transportation is required for excavation projects, borrow areas, filling projects, dump sites and stockpile areas, as these areas are turned over frequently and therefore need to be replaced frequently (see Fig. 3.3). The SBY Hydropower Project has 4 excavation projects, 2 borrow areas, 3 stockpile areas and 2 dump sites. The location and detailed information regarding the borrow areas, dump sites and stockpile areas at the SBY Hydropower Project are illustrated in Fig. 3.4.

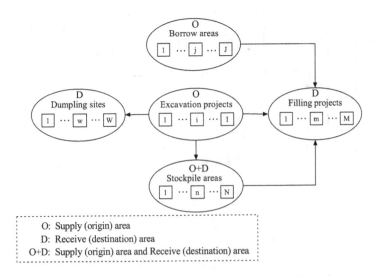

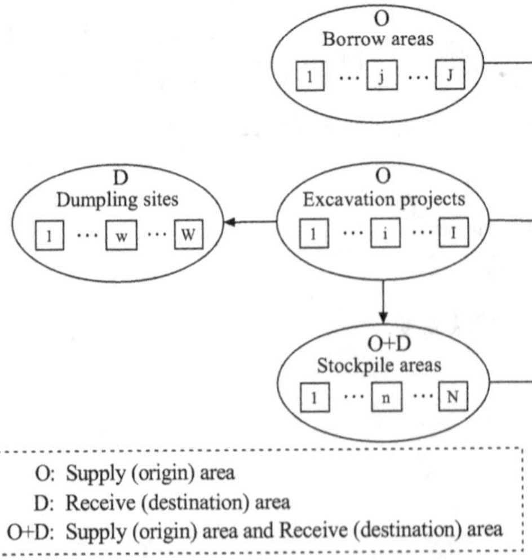

Fig. 3.3 Earth-rock work transportation relations in large scale construction project

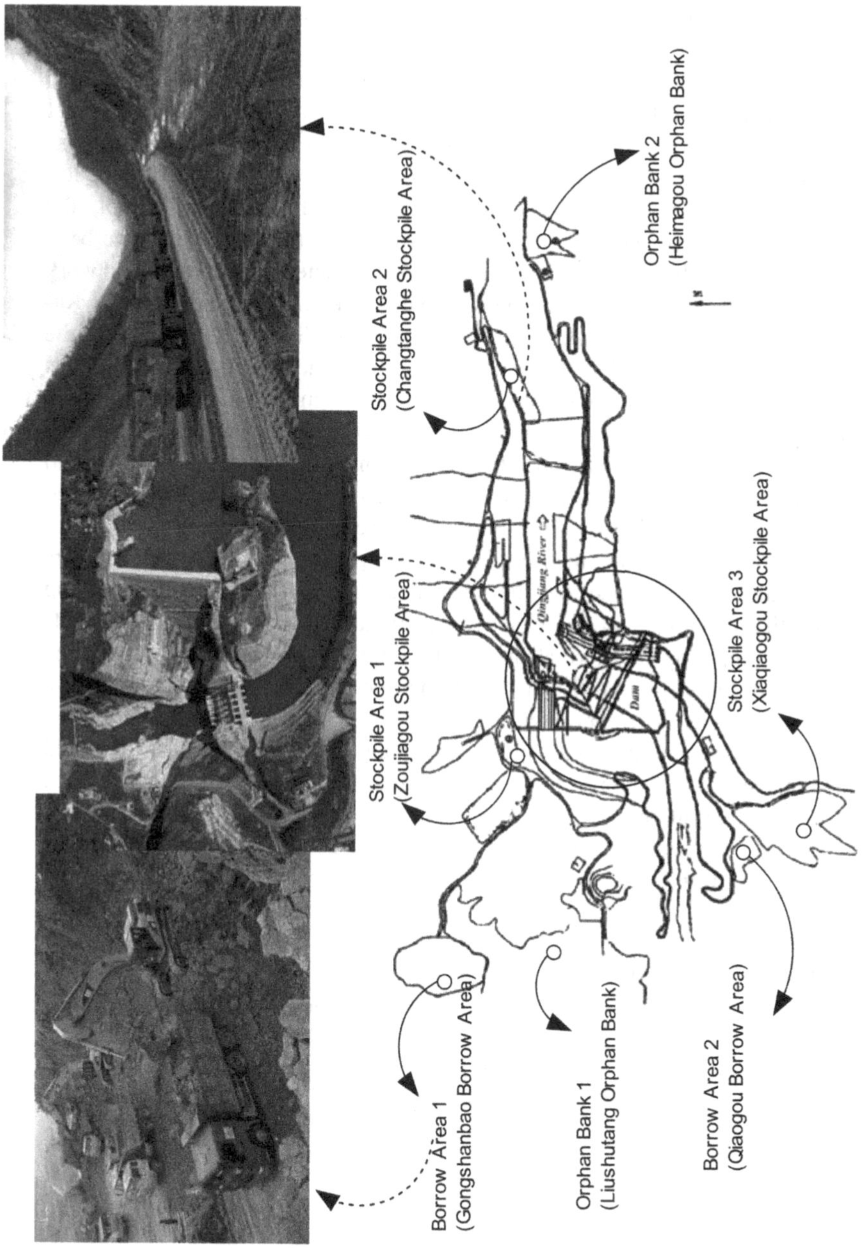

Fig. 3.4 Layout of borrow areas, dumpling sites and stockpile areas of SBY Hydropower Project

Three types of dump trucks (carriers) are considered for the construction project to transport the earth-rock work along the different paths connecting the OD pairs, with the destination nodes having a practical timeliness demand. All necessary data for each kind of carrier was calculated as shown in Table 3.3. Table 3.4 gives details of the paths across the whole road network.

The transportation network in the project includes an internal road network and an external road network. In this case, only the internal road network is considered. The internal road network has 20 trunk roads located on the left and right banks, with 11 on the left and 9 on the right, with a cross river bridge connecting the left and right banks. Therefore, 21 links are considered in this section. To apply the proposed methods more conveniently, adjacent roads of the same type have been combined and road shapes have been ignored. An abstracted transportation network is illustrated in Fig. 3.5. For each transportation network link, there is a free flow Ra-Ra travel time $\tilde{\bar{t}}^k_{i0}$, and Ra-Ra transportation cost $\tilde{\bar{e}}^k_i$. (See Tables 3.5 and 3.6 for the corresponding data). To collect the transportation time and cost data, investigations and surveys were conducted to obtain historical data from the financial department and the opinions of the experienced engineers in construction team at the Hubei Qingjiang SBY Project Construction Company. Since the transportation time and costs on each arc of the path change over time, the data were classified based on different periods. Both the transportation time and costs were assumed to approximately follow normal distributions in each period, and the two parameters (expected value and variance) for the normal distributions were estimated using maximum likelihood estimation, which was justified by a chi-square goodness-of-fit test. By comparing the normal distributions for the same transportation time and costs in different periods, it was found that the expected values for the above normal distributions also approximately followed a similar random distribution pattern, which was also justified using a chi-square goodness-of-fit test. It should be noted that since the variance fluctuations were insignificant, the median variance values in the different periods were selected as the variances for the above normal distributions. The predicted confidence levels given by the decision maker were respectively $\theta = 0.9, \delta = 0.85$.

Table 3.3 Information of carriers in SBY Hydropower Project

Carrier	Kind index k	Type (heaped capacity; maximum payload)	Weight w_k
Dump truck	1	K30N-8.4 (16 m^3; 26 t)	0.3
	2	Terex TA28 (17 m^3; 28 t)	0.3
	3	Perlini DP366 (20 m^3; 36 t)	0.4

Table 3.4 Related data about information of paths in earth-rock work transportation network

OD pairs	Demand Q_{od} (m^3)	Transportation path j	Direct cost (CNY/m^3)	Constraint time (h)
Excavation project 1,2→ Dumpling site 1	950	1 ⟺ #5→#7→#9	1.80	1.45
Excavation project 1,2→ Filling project 1,2	2650	2 ⟺ #5→#3→#1→#6	2.50	2.15
		3 ⟺ #5→#1→#6	1.70	1.35
		4 ⟺ #5→#7→#8→#6	2.40	2.20
		5 ⟺ #5→#7→#6	1.90	1.40
Excavation project 1,2→ Stockpile area 1	1700	6 ⟺ #5→#7	1.20	1.25
Excavation project 3,4→ Filling project 3,4	2300	7 ⟺ #11→#10→#14→#16	2.30	2.25
		8 ⟺ #11→#13→#16	1.90	1.45
		9 ⟺ #11→#10→#13→#16	2.70	2.30
		10 ⟺ #11 → #10 → #14 → #15 →#16	3.00	3.50
Excavation project 3,4→ Dumpling site 2	1650	11 ⟺ #11→#13→#12	1.60	1.50
		12 ⟺ #11→#10→#13→#12	2.33	2.50
Excavation project 3,4→ Stockpile area 2	1480	13 ⟺ #11→#10→#13	2.60	1.75
		14 ⟺ #11→#13	1.10	1.25
Excavation project 3,4→ Stockpile area 3	1380	15 ⟺ #11→#10→#14→#15	2.10	2.33
Borrow area 1→ Filling project 1,2	1700	16 ⟺ #2→#4→#6	1.50	1.50
Borrow area 2→ Filling project 3,4	1250	17 ⟺ #14→#15→#16	1.80	1.55
		18 ⟺ #14→#16	1.20	1.20
Stockpile area 1→ Filling project 1,2	1150	19 ⟺ #7→#8→#6	1.90	1.45
Stockpile area 2→ Filling project 3,4	1300	20 ⟺ #13→#16	1.40	1.17
Stockpile area 3→ Filling project 1,2	1260	21 ⟺ #15→#7→#8→#6	2.40	2.17
		22 ⟺ #15→#7→#6	2.20	1.60

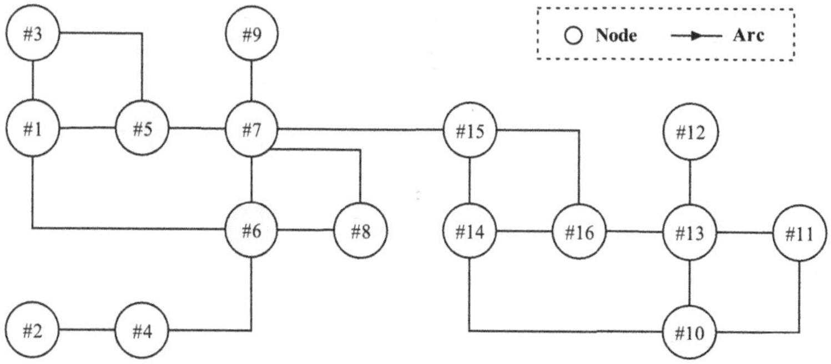

Fig. 3.5 Illustrations of road network of SBY Hydropower Project

Table 3.5 Ra-Ra free flow travel time $\tilde{\bar{t}}_{i0}^k$ and Ra-Ra transportation cost $\tilde{\bar{e}}_i^k$ 1

Arc i	Corresponding nodes	Arc capacity r_i (n/h)	Transportation cost e_i^k (CNY/unit)			Free transportation time t_{i0}^k (h)		
			Carrier 1	Carrier 2	Carrier 3	Carrier 1	Carrier 2	Carrier 3
1	#1, #3	105	$\mathcal{N}(\mu, 0.64)$,with $\mu \sim \mathcal{N}(5.2, 0.1)$	$\mathcal{N}(\mu, 1.00)$,with $\mu \sim \mathcal{N}(6.0, 0.09)$	$\mathcal{N}(\mu, 0.01)$,with $\mu \sim \mathcal{N}(6.8, 1.00)$	$\mathcal{N}(\mu, 0.25)$,with $\mu \sim \mathcal{N}(0.2, 0.01)$	$\mathcal{N}(\mu, 0.81)$,with $\mu \sim \mathcal{N}(0.24, 0.09)$	$\mathcal{N}(\mu, 0.16)$,with $\mu \sim \mathcal{N}(0.28, 0.04)$
2	#1, #5	110	$\mathcal{N}(\mu, 1.00)$,with $\mu \sim \mathcal{N}(3.2, 0.64)$	$\mathcal{N}(\mu, 0.49)$,with $\mu \sim \mathcal{N}(3.5, 0.16)$	$\mathcal{N}(\mu, 0.81)$,with $\mu \sim \mathcal{N}(3.8, 0.25)$	$\mathcal{N}(\mu, 0.36)$,with $\mu \sim \mathcal{N}(0.15, 0.04)$	$\mathcal{N}(\mu, 0.25)$,with $\mu \sim \mathcal{N}(0.16, 0.04)$	$\mathcal{N}(\mu, 0.09)$,with $\mu \sim \mathcal{N}(0.18, 0.01)$
3	#1, #6	135	$\mathcal{N}(\mu, 4.00)$,with $\mu \sim \mathcal{N}(7.2, 1.00)$	$\mathcal{N}(\mu, 0.81)$,with $\mu \sim \mathcal{N}(7.5, 0.64)$	$\mathcal{N}(\mu, 0.64)$,with $\mu \sim \mathcal{N}(7.8, 0.49)$	$\mathcal{N}(\mu, 0.25)$,with $\mu \sim \mathcal{N}(0.30, 0.16)$	$\mathcal{N}(\mu, 0.36)$,with $\mu \sim \mathcal{N}(0.35, 0.09)$	$\mathcal{N}(\mu, 0.16)$,with $\mu \sim \mathcal{N}(0.38, 0.01)$
4	#2, #4	115	$\mathcal{N}(\mu, 1.00)$,with $\mu \sim \mathcal{N}(3.7, 0.49)$	$\mathcal{N}(\mu, 4.00)$,with $\mu \sim \mathcal{N}(4.0, 0.81)$	$\mathcal{N}(\mu, 0.01)$,with $\mu \sim \mathcal{N}(4.3, 0.01)$	$\mathcal{N}(\mu, 0.81)$,with $\mu \sim \mathcal{N}(0.14, 0.16)$	$\mathcal{N}(\mu, 0.36)$,with $\mu \sim \mathcal{N}(0.18, 0.09)$	$\mathcal{N}(\mu, 0.64)$,with $\mu \sim \mathcal{N}(0.19, 0.01)$
5	#3, #5	108	$\mathcal{N}(\mu, 0.36)$,with $\mu \sim \mathcal{N}(2.2, 0.01)$	$\mathcal{N}(\mu, 0.81)$,with $\mu \sim \mathcal{N}(2.4, 0.09)$	$\mathcal{N}(\mu, 0.64)$,with $\mu \sim \mathcal{N}(2.6, 0.16)$	$\mathcal{N}(\mu, 0.01)$,with $\mu \sim \mathcal{N}(0.14, 0.01)$	$\mathcal{N}(\mu, 0.25)$,with $\mu \sim \mathcal{N}(0.18, 0.01)$	$\mathcal{N}(\mu, 0.09)$,with $\mu \sim \mathcal{N}(0.22, 0.01)$
6	#4, #6	112	$\mathcal{N}(\mu, 0.81)$,with $\mu \sim \mathcal{N}(3.2, 0.16)$	$\mathcal{N}(\mu, 1.00)$,with $\mu \sim \mathcal{N}(4.0, 0.09)$	$\mathcal{N}(\mu, 0.49)$,with $\mu \sim \mathcal{N}(4.8, 0.09)$	$\mathcal{N}(\mu, 0.64)$,with $\mu \sim \mathcal{N}(0.12, 0.16)$	$\mathcal{N}(\mu, 0.09)$,with $\mu \sim \mathcal{N}(0.15, 0.01)$	$\mathcal{N}(\mu, 0.25)$,with $\mu \sim \mathcal{N}(0.18, 0.01)$
7	#5, #7	180	$\mathcal{N}(\mu, 1.00)$,with $\mu \sim \mathcal{N}(5.0, 0.01)$	$\mathcal{N}(\mu, 0.64)$,with $\mu \sim \mathcal{N}(5.2, 0.25)$	$\mathcal{N}(\mu, 0.36)$,with $\mu \sim \mathcal{N}(5.4, 0.16)$	$\mathcal{N}(\mu, 0.01)$,with $\mu \sim \mathcal{N}(0.25, 0.01)$	$\mathcal{N}(\mu, 0.04)$,with $\mu \sim \mathcal{N}(0.28, 0.01)$	$\mathcal{N}(\mu, 0.01)$,with $\mu \sim \mathcal{N}(0.32, 0.01)$
8	#6, #7	145	$\mathcal{N}(\mu, 0.49)$,with $\mu \sim \mathcal{N}(4.5, 0.01)$	$\mathcal{N}(\mu, 0.36)$,with $\mu \sim \mathcal{N}(4.2, 0.16)$	$\mathcal{N}(\mu, 0.49)$,with $\mu \sim \mathcal{N}(3.4, 0.16)$	$\mathcal{N}(\mu, 0.09)$,with $\mu \sim \mathcal{N}(0.16, 0.01)$	$\mathcal{N}(\mu, 0.04)$,with $\mu \sim \mathcal{N}(0.19, 0.01)$	$\mathcal{N}(\mu, 0.09)$,with $\mu \sim \mathcal{N}(0.23, 0.01)$
9	#6, #8	160	$\mathcal{N}(\mu, 0.16)$,with $\mu \sim \mathcal{N}(3.6, 0.25)$	$\mathcal{N}(\mu, 0.81)$,with $\mu \sim \mathcal{N}(4.0, 1.00)$	$\mathcal{N}(\mu, 1.00)$,with $\mu \sim \mathcal{N}(4.4, 0.09)$	$\mathcal{N}(\mu, 0.04)$,with $\mu \sim \mathcal{N}(0.13, 0.01)$	$\mathcal{N}(\mu, 0.01)$,with $\mu \sim \mathcal{N}(0.15, 0.04)$	$\mathcal{N}(\mu, 0.09)$,with $\mu \sim \mathcal{N}(0.17, 0.01)$
10	#7, #8	165	$\mathcal{N}(\mu, 1.00)$,with $\mu \sim \mathcal{N}(4.0, 0.81)$	$\mathcal{N}(\mu, 1.00)$,with $\mu \sim \mathcal{N}(4.4, 0.49)$	$\mathcal{N}(\mu, 0.64)$,with $\mu \sim \mathcal{N}(4.8, 0.25)$	$\mathcal{N}(\mu, 0.01)$,with $\mu \sim \mathcal{N}(0.16, 0.01)$	$\mathcal{N}(\mu, 0.04)$,with $\mu \sim \mathcal{N}(0.19, 0.01)$	$\mathcal{N}(\mu, 0.09)$,with $\mu \sim \mathcal{N}(0.23, 0.04)$

Table 3.6 Ra-Ra free flow travel time $\tilde{\tilde{t}}_{i0}^k$ and Ra-Ra transportation cost $\tilde{\tilde{e}}_i^k$ 2

Arc i	Corresponding nodes	Arc capacity r_i (n/h)	Transportation cost e_i^k (CNY/unit)			Free transportation time t_{i0}^k (h)		
			Carrier 1	Carrier 2	Carrier 3	Carrier 1	Carrier 2	Carrier 3
11	#7, #9	100	$\mathcal{N}(\mu, 4.00)$,with $\mu \sim \mathcal{N}(7.2, 0.09)$	$\mathcal{N}(\mu, 0.16)$,with $\mu \sim \mathcal{N}(7.6, 0.81)$	$\mathcal{N}(\mu, 0.25)$,with $\mu \sim \mathcal{N}(8.0, 0.16)$	$\mathcal{N}(\mu, 0.01)$,with $\mu \sim \mathcal{N}(0.36, 0.01)$	$\mathcal{N}(\mu, 0.04)$,with $\mu \sim \mathcal{N}(0.39, 0.01)$	$\mathcal{N}(\mu, 0.09)$,with $\mu \sim \mathcal{N}(0.42, 0.01)$
12	#7, #15	130	$\mathcal{N}(\mu, 0.81)$,with $\mu \sim \mathcal{N}(8.0, 0.81)$	$\mathcal{N}(\mu, 0.64)$,with $\mu \sim \mathcal{N}(8.5, 0.5)$	$\mathcal{N}(\mu, 1.00)$,with $\mu \sim \mathcal{N}(8.7, 0.01)$	$\mathcal{N}(\mu, 0.01)$,with $\mu \sim \mathcal{N}(0.33, 0.01)$	$\mathcal{N}(\mu, 0.01)$,with $\mu \sim \mathcal{N}(0.36, 0.01)$	$\mathcal{N}(\mu, 0.01)$,with $\mu \sim \mathcal{N}(0.38, 0.01)$
13	#10, #11	225	$\mathcal{N}(\mu, 4.00)$,with $\mu \sim \mathcal{N}(5.0, 0.81)$	$\mathcal{N}(\mu, 0.64)$,with $\mu \sim \mathcal{N}(6.0, 0.16)$	$\mathcal{N}(\mu, 0.25)$,with $\mu \sim \mathcal{N}(6.5, 0.09)$	$\mathcal{N}(\mu, 0.01)$,with $\mu \sim \mathcal{N}(0.30, 0.01)$	$\mathcal{N}(\mu, 0.01)$,with $\mu \sim \mathcal{N}(0.34, 0.01)$	$\mathcal{N}(\mu, 0.01)$,with $\mu \sim \mathcal{N}(0.36, 0.01)$
14	#10, #13	155	$\mathcal{N}(\mu, 0.81)$,with $\mu \sim \mathcal{N}(5.5, 0.36)$	$\mathcal{N}(\mu, 1.00)$,with $\mu \sim \mathcal{N}(5.6, 0.49)$	$\mathcal{N}(\mu, 0.04)$,with $\mu \sim \mathcal{N}(5.8, 0.16)$	$\mathcal{N}(\mu, 0.01)$,with $\mu \sim \mathcal{N}(0.22, 0.09)$	$\mathcal{N}(\mu, 0.04)$,with $\mu \sim \mathcal{N}(0.26, 0.16)$	$\mathcal{N}(\mu, 0.04)$,with $\mu \sim \mathcal{N}(0.30, 0.09)$
15	#10, #14	150	$\mathcal{N}(\mu, 0.64)$,with $\mu \sim \mathcal{N}(4.5, 0.49)$	$\mathcal{N}(\mu, 1.00)$,with $\mu \sim \mathcal{N}(4.6, 0.25)$	$\mathcal{N}(\mu, 0.09)$,with $\mu \sim \mathcal{N}(4.8, 0.09)$	$\mathcal{N}(\mu, 0.04)$,with $\mu \sim \mathcal{N}(0.22, 0.01)$	$\mathcal{N}(\mu, 0.01)$,with $\mu \sim \mathcal{N}(0.25, 0.04)$	$\mathcal{N}(\mu, 0.01)$,with $\mu \sim \mathcal{N}(0.27, 0.01)$
16	#11, #13	170	$\mathcal{N}(\mu, 0.64)$,with $\mu \sim \mathcal{N}(3.6, 0.16)$	$\mathcal{N}(\mu, 0.09)$,with $\mu \sim \mathcal{N}(4.0, 0.01)$	$\mathcal{N}(\mu, 0.16)$,with $\mu \sim \mathcal{N}(4.4, 0.25)$	$\mathcal{N}(\mu, 0.04)$,with $\mu \sim \mathcal{N}(0.16, 0.01)$	$\mathcal{N}(\mu, 0.09)$,with $\mu \sim \mathcal{N}(0.18, 0.04)$	$\mathcal{N}(\mu, 0.01)$,with $\mu \sim \mathcal{N}(0.20, 0.01)$
17	#12, #13	125	$\mathcal{N}(\mu, 0.25)$,with $\mu \sim \mathcal{N}(4.2, 0.16)$	$\mathcal{N}(\mu, 0.49)$,with $\mu \sim \mathcal{N}(4.3, 0.09)$	$\mathcal{N}(\mu, 1.00)$,with $\mu \sim \mathcal{N}(4.4, 0.16)$	$\mathcal{N}(\mu, 0.01)$,with $\mu \sim \mathcal{N}(0.22, 0.01)$	$\mathcal{N}(\mu, 0.04)$,with $\mu \sim \mathcal{N}(0.24, 0.09)$	$\mathcal{N}(\mu, 0.09)$,with $\mu \sim \mathcal{N}(0.26, 0.01)$
18	#13, #16	145	$\mathcal{N}(\mu, 1.00)$,with $\mu \sim \mathcal{N}(4.5, 0.16)$	$\mathcal{N}(\mu, 0.04)$,with $\mu \sim \mathcal{N}(4.7, 0.81)$	$\mathcal{N}(\mu, 0.49)$,with $\mu \sim \mathcal{N}(4.9, 1.00)$	$\mathcal{N}(\mu, 0.01)$,with $\mu \sim \mathcal{N}(0.21, 0.64)$	$\mathcal{N}(\mu, 1.00)$,with $\mu \sim \mathcal{N}(0.23, 0.01)$	$\mathcal{N}(\mu, 0.09)$,with $\mu \sim \mathcal{N}(0.25, 0.16)$
19	#14, #15	140	$\mathcal{N}(\mu, 0.25)$,with $\mu \sim \mathcal{N}(5.6, 0.49)$	$\mathcal{N}(\mu, 0.09)$,with $\mu \sim \mathcal{N}(6.0, 0.01)$	$\mathcal{N}(\mu, 0.04)$,with $\mu \sim \mathcal{N}(6.5, 0.04)$	$\mathcal{N}(\mu, 0.01)$,with $\mu \sim \mathcal{N}(0.30, 0.01)$	$\mathcal{N}(\mu, 0.16)$,with $\mu \sim \mathcal{N}(0.33, 0.01)$	$\mathcal{N}(\mu, 0.09)$,with $\mu \sim \mathcal{N}(0.35, 0.04)$
20	#14, #16	135	$\mathcal{N}(\mu, 0.36)$,with $\mu \sim \mathcal{N}(4.6, 0.25)$	$\mathcal{N}(\mu, 0.04)$,with $\mu \sim \mathcal{N}(4.8, 0.16)$	$\mathcal{N}(\mu, 0.04)$,with $\mu \sim \mathcal{N}(5.0, 0.01)$	$\mathcal{N}(\mu, 0.09)$,with $\mu \sim \mathcal{N}(0.18, 0.04)$	$\mathcal{N}(\mu, 0.16)$,with $\mu \sim \mathcal{N}(0.22, 0.09)$	$\mathcal{N}(\mu, 0.25)$,with $\mu \sim \mathcal{N}(0.25, 0.49)$
21	#15, #16	140	$\mathcal{N}(\mu, 1.00)$,with $\mu \sim \mathcal{N}(5.8, 0.09)$	$\mathcal{N}(\mu, 1.00)$,with $\mu \sim \mathcal{N}(6.2, 0.16)$	$\mathcal{N}(\mu, 0.81)$,with $\mu \sim \mathcal{N}(6.5, 0.04)$	$\mathcal{N}(\mu, 0.01)$,with $\mu \sim \mathcal{N}(0.30, 0.09)$	$\mathcal{N}(\mu, 0.16)$,with $\mu \sim \mathcal{N}(0.32, 0.25)$	$\mathcal{N}(\mu, 0.04)$,with $\mu \sim \mathcal{N}(0.35, 0.01)$

3.6.3 Solutions and Discussion

To verify the practicality and efficiency of the TFDP optimization model under a Ra-Ra environment presented previously, the proposed MOBLPSO was implemented to determine the flow assignment amongst the transportation paths and amongst the carriers over a certain period using actual data from the Hubei Qingjiang Shuibuya Project Construction Company. After running the proposed MOBLPSO using MATLAB 7.0, the computational results were obtained and the efficiency of the proposed algorithm was proven.

The computer running environment was an Intel core 2 Duo 2.26 GHz clock pulse with 2048 MB memory. The problem was solved using the proposed algorithm with satisfactory solutions being obtained within 21 min on average, and the optimal solutions for the lower level programming and the Pareto optimal solution set for the upper level programming were determined.

The red dots in Fig. 3.6 show the pareto-optimal solutions, while the blue dots show the best position of the particles in this iteration. Depending on their preference, the decision makers can choose a plan from these pareto-optimal solutions. For example, if the decision makers felt that the total direct cost objective

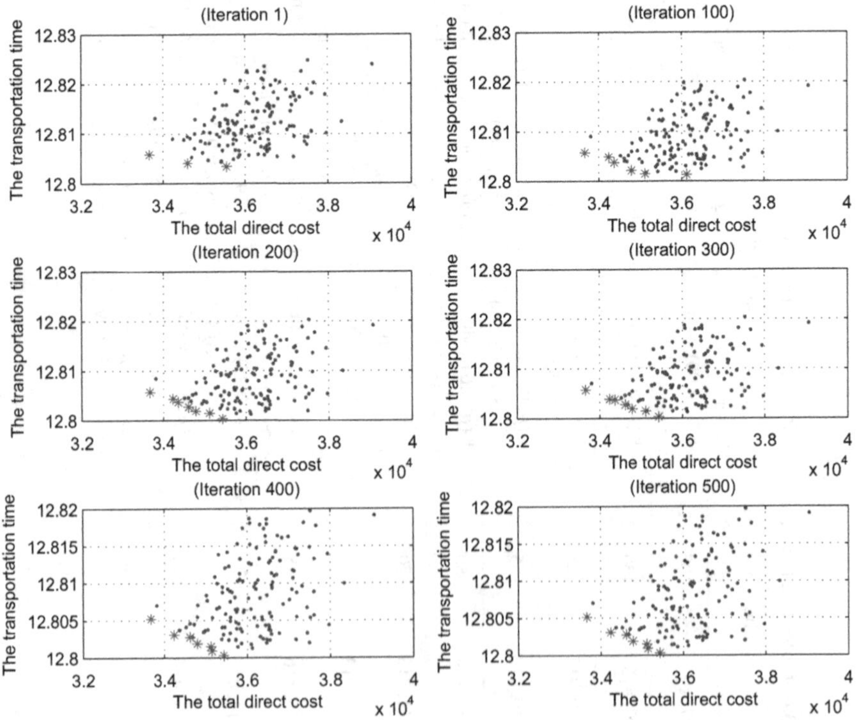

Fig. 3.6 Pareto optimal solutions of upper level programming for SBY Hydropower Project

Table 3.7 The most cost-effective plan for TFDP in SBY case

Transportation path j	1	2	3	4	5	6	7	8
Material volume x_j	950.000	543.763	1672.488	256.317	177.432	1700.000	766.646	1411.958
Material volume y_{1j}	426.432	309.723	576.206	71.020	93.874	505.161	371.885	449.943
Material volume y_{2j}	413.345	107.514	587.051	92.521	44.360	620.743	267.585	521.378
Material volume y_{3j}	369.912	159.467	509.230	52.741	127.025	574.096	285.792	440.637
Transportation path j	9	10	11	12	13	14	15	16
Material volume x_j	57.966	63.430	1279.675	370.325	161.357	1318.643	1380.000	1700.000
Material volume y_{1j}	144.947	43.373	487.119	398.457	650.263			
Material volume y_{2j}	36.224	46.715	520.477	195.901	42.686	421.532	476.339	598.634
Material volume y_{3j}	42.172	48.536	364.225	155.662	46.672	409.992	505.204	451.103
Transportation path j	17	18	19	20	21	22		
Material volume x_j	346.384	903.616	1150.000	1300.000	747.349	512.651		
Material volume y_{1j}	158.874	427.931	381.462	382.515	363.827	63.158		
Material volume y_{2j}	111.618	200.388	430.262	446.545	264.172	80.990		
Material volume y_{3j}	171.539	325.558	338.276	470.940	373.284	484.426		

Table 3.8 The most time-saving plan for TFDP in SBY case

Transportation path j	1	2	3	4	5	6	7	8
Material volume x_j	950.000	1013.517	758.507	820.2407	57.735	1700.000	948.762	1056.185
Material volume y_{1j}	265.783	354.500	308.790	329.115	45.111	573.575	492.092	422.106
Material volume y_{2j}	427.952	324.088	400.510	476.403	39.031	523.953	354.168	405.054
Material volume y_{3j}	315.831	293.735	334.929	390.005	432.165	38.571	602.472	398.627
Transportation path j	9	10	11	12	13	14	15	16
Material volume x_j	253.392	41.661	647.444	1002.556	200.435	1279.565	1380.000	1700.000
Material volume y_{1j}	157.707	21.322	363.277	489.996	100.614	337.177	417.547	552.807
Material volume y_{2j}	48.491	28.744	222.921	402.297	109.078	535.589	454.315	539.657
Material volume y_{3j}	56.696	39.224	199.717	265.074	31.314	406.799	508.137	607.536
Transportation path j	17	18	19	20	21	22		
Material volume x_j	1057.049	192.951	1150.000	1300.000	500.306	759.694		
Material volume y_{1j}	274.888	127.423	337.068	377.977	274.467	369.143		
Material volume y_{2j}	493.531	131.399	451.603	525.488	255.822	299.626		
Material volume y_{3j}	439.660	47.452	361.329	396.535	65.138	331.525		

was more important, they may sacrifice transportation time for a more economical scheme, and would choose the absolute left of the pareto-optimal solution in the transport flow distribution plan shown in Table 3.7. On the contrary, if decision makers felt that transportation time objective was more important, they would choose minimum total transportation time or the lowest of the pareto-optimal solutions and incur greater costs for the TFDP. The minimum transportation time plan is shown in Table 3.8.

To highlight the advantages of our mathematical model (3.136), additional computational work was done using the proposed MOBLPSO to solve a similar TFDP under a different uncertain environment, the random environment. To conduct these comparisons under similar circumstances, the analyses were conducted based running the test problem 10 times. A detailed analysis follows.

Table 3.9 Comparisons of two types of model

	The direct cost objective ($\times 10^3$)		The transportation time cost objective		The transportation cost objective ($\times 10^3$)	
Results	Ra-Ra	Random	Ra-Ra	Random	Ra-Ra	Random
Best result	33.662	33.832	12.8003	12.7997	18.036	18.067
Average result	34.546	35.022	12.8027	12.8045	18.239	18.285
Worst result	35.429	36.211	12.8051	12.8092	18.441	18.503

To guarantee a fair comparison between the TFDP model with birandomness (denoted by TFDP-birm) and a model that only considers single randomness (denoted by TFDP-rm), the random distribution for each related uncertain parameter in the TFDP-rm was selected in the following way. Take the transportation cost $\bar{\tilde{e}}_1^1$ (i.e., $\bar{\tilde{e}}_1^1 \sim \mathcal{N}(\mu, 0.64)$, with $\mu \sim \mathcal{N}(5.2, 0.1)$) for example. The stochastic nature of the expected value μ in the normal distribution $\mathcal{N}(\mu, 0.64)$ was ignored by using its expectation 5.2 as a representation while the variance, 0.64 was retained. Thus, for the TFDP-rm, the birandomness of the transportation cost $\bar{\tilde{e}}_1^1$ was degenerated to a single randomness, in which the distribution of the transportation cost could be expressed as $\mathcal{N}(5.2, 0.64)$. Since the variance in the random variable μ was sufficiently small (i.e., $0.1 \leqslant 1$), the expectation of the random variable μ essentially reflected the most possible value over time. Thus, it was reasonable to select $\mathcal{N}(5.2, 0.64)$ as the normal distribution for the transportation cost in the TFDP-rm to compare with that in the TFDP-birm. The transformation of the other related uncertain parameters followed a similar pattern. Thus the model for the TFDP-rm was formulated and solved using the proposed MOBLPSO ater running 10 times.

As shown in Table 3.9, the best, the worst, and average results for the random type were higher than their counterparts for the TFDP-birm. It is worth noting that the gaps between the best and the worst and between the best and the average solutions for TFDP-rm were wider than the gaps of the Ra-Ra type counterparts. This indicates that randomness creates a much larger solution space when uncertainty is introduced. Fortunately, the widened solution space with the further stochastic nature in the TFDP-birm provided better solutions, which were successfully located by the MOBLPSO, and can be seen in the narrower gaps between the best and the worst and between the best and the average solutions found for the TFDP-birm. This suggests that MOBLPSO is an effective and relatively efficient approach for solving a TFDP under a Ra-Ra environment.

Since the PSOPC for the lower level is nested in the MOPSO for the upper level, and the MOPSO is the main body of the proposed MOBLPSO, the MOPSO evaluation was the main focus. In the MOPSO, a multi-objective method was introduced to derive the Pareto optimal solution set for the upper level programming. This provided effective and non-dominated alternate schemes for the construction contractor. Compared to the weight-sum method which deals with the multi-objectives in [49], the solutions here were confirmed to be more practical.

Table 3.10 Algorithm evaluation by metrics of performance for Pareto optimal sets

Iteration	The average distance metric	The distribution metric	The extent metric
1	0.0781	0.1474	48.6227
100	0.0738	0.3460	58.1423
200	0.0878	0.4031	61.1722
300	0.0526	0.6967	69.2970
400	0.0554	0.8082	64.7773
500	0.0435	0.8938	68.8648

Comparing different optimization techniques experimentally always involves a performance analysis. In the case of multi-objective optimization, the definition of quality is substantially more complex than for single-objective optimization problems. There are many performance metrics to measure the distance of the resulting non-dominated set from the Pareto-optimal front, the distribution of the solution found, and the extent of the obtained non-dominated front [54].

To gain further insight into the performance of the multi-objective method in the proposed algorithm, the procedure was run different times and the results are summarized in Fig. 3.6 and Table 3.10. As shown in Fig. 3.6, the amount and the distribution of Pareto optimal solutions in each iteration were satisfactory. For a further expression of the efficiency of the convergence, the three performance metrics proposed by Zitzler et al. [54] were introduced: (1) The average distance that the resulting non-dominated set is from the Pareto-optimal front, the value of which should decrease with an increase in the iterations, thus indicating that the program results are closer to the Pareto-optimal front, and demonstrating the convergence of the algorithm; (2) The distribution in combination with the number of non-dominated solutions found: the higher the value, the better the distribution for an appropriate neighboring parameter; (3) The extent of the obtained non-dominated fronts: the maximum extent in each dimension that is used to estimate the spread of the range, which, in this section, is equal to the distance between the two outer solutions.

The performance metrics for the Pareto optimal set shown in Table 3.10 provide a satisfactory result for convergence efficiency. Although there are some fluctuations in the three metrics over the 500 iterations, these did not affect the final results.

To assess the efficiency and effectiveness of the MOBLPSO for the proposed TFDP, the MOBLPSO results for the TFDP in the SBY Project were compared with two other state-of-the-art heuristic algorithms; the genetic algorithm for a multi-objective bi-level model (denoted MOBLGA) [48], and the simulated annealing algorithm for a multi-objective bi-level model (denoted MOBLSA) [23].

To conduct the comparisons under similar circumstances, the parameter selections for the MOBLGA and MOBLSA were the same as those for the MOBLPSO, and non-dominated alternate schemes were also employed for both. To measure the quality of the results obtained using the three algorithms, a weight sum method was introduced to determine one minimal weight sum for the objectives from the non-

Table 3.11 Computation time and memory used by MOBLPSO, MOBLGA and MOBLSA

	Combination of weights		Minimal weight sum value of the two objectives			Average computation time (min)		
Type	ω_1	ω_2	MOBLPSO	MOBLGA	MOBLSA	MOBLPSO	MOBLGA	MOBLSA
1	$\omega_1 = 0.1$	$\omega_2 = 0.9$	1152.027	1160.989	1152.531	20.856	25.899	34.458
2	$\omega_1 = 0.2$	$\omega_2 = 0.8$	1024.272	1032.419	1024.273	20.926	26.032	33.921
3	$\omega_1 = 0.3$	$\omega_2 = 0.7$	896.238	903.848	897.214	21.052	25.679	34.068
4	$\omega_1 = 0.4$	$\omega_2 = 0.6$	769.472	775.277	769.555	21.158	25.329	35.568
5	$\omega_1 = 0.5$	$\omega_2 = 0.5$	641.786	646.706	641.897	20.963	26.169	35.012
6	$\omega_1 = 0.6$	$\omega_2 = 0.4$	514.148	518.136	514.216	20.678	26.969	35.142
7	$\omega_1 = 0.7$	$\omega_2 = 0.3$	386.486	388.529	386.521	21.203	26.135	34.791
8	$\omega_1 = 0.8$	$\omega_2 = 0.2$	258.792	260.836	258.792	20.734	25.475	35.291
9	$\omega_1 = 0.9$	$\omega_2 = 0.1$	131.081	132.143	131.081	21.134	26.189	34.815

dominated solutions. Thus, the comparison was implemented based on a unique measuring criterion (i.e., the minimal weight sum of the objectives). To ensure conformity validity for the multi-objectives, the dimension divisions and a unifying of the order of magnitude needed to be performed before the weight-sum procedure.

Table 3.11 shows the comparison results; i.e., the minimal weight sum value of the two objectives, and the average computation times obtained using the preceding approaches for different weight combinations (i.e., ω_1 and ω_2 represent the weights of the two objectives respectively). It was demonstrated that the MOBLPSO for the TFDP performed the optimization better than the MOBLGA, as MOBLGA could lead to a local search and require more computation time. On the other hand, the MOBLSA had similar results to the MOBLPSO, but the computation time was much slower.

References

1. Abo-Sinna M, Baky I (2007) Interactive balance space approach for solving multi-level multi-objective programming problems. Inf Sci 177(16):3397–3410
2. Allgower EL, Georg K (1980) Simplicial and continuation methods for approximating fixed points and solutions to systems of equations. SIAM Rev 22:28–85
3. Allgower EL, Georg K (1990) Numerical continuation methods: an introduction. Springer, Berlin/New York
4. Alves M, Costa J (2014) An algorithm based on particle swarm optimization for multiobjective bilevel linear problems. Appl Math Comput 247:547–561
5. Avestimehr AS, Diggavi SN, Tse DNC (2011) Wireless network information flow: a deterministic approach. IEEE Trans Inf Theory 57(4):1872–1905
6. Benayoun R, De Montgolfier J, Tergny J, Laritchev O (1971) Linear programming with multiple objective functions: step method (STEM). Math Program 1(1):366–375
7. Blackwell T, Branke J (2006) Multiswarms, exclusion, and anti-convergence in dynamic environments. IEEE Trans Evol Comput 10(4):459–472
8. Calvete HI, Galé C (2010) Linear bilevel programs with multiple objectives at the upper level. J Comput Appl Math 234(4):950–959

9. Charnes A, Cooper WW (1963) Deterministic equivalents for optimizing and satisficing under chance constraints. Oper Res 11(1):18–39
10. Chen A, Zhou Z (2010) The α-reliable mean-excess traffic equilibrium model with stochastic travel times. Transp Res B 44(4):493–513
11. Choo EV, Atkins D (1980) An interactive algorithm for multiobjective programming. Comput Oper Res 7:81–88
12. Colson B, Marcotte P, Savard G (2005) Bilevel programming: a survey. 4OR: Q J Oper Res 3(2):87–107
13. Dang C, Sun Y, Wang Y, Yang Y (2011) A deterministic annealing algorithm for the minimum concave cost network flow problem. Neural Netw 24(7):699–708
14. Fontes DBMM, Hadjiconstantinou E, Christofides N (2006) A dynamic programming approach for solving single-source uncapacitated concave minimum cost network flow problems. Eur J Oper Res 174(2):1205–1219
15. Ford LR, Fulkerson DR (1962) Flows in networks. Princeton University Press, New Jersey
16. Fortuny-Amat J, McCarl B (1981) A representation and economic interpretation of a two-level programming problem. J Oper Res Soc 32:783–792
17. Gao S, Chabini I (2006) Optimal routing policy problems in stochastic time-dependent networks. Transp Res B 40(2):93–122
18. Garcia CB, Zangwill WI (1981) Pathways to solutions, fixed points and equilibria. Prentice-Hall, Englewood Cliffs
19. Geoffrion AM, Hogan WW (1972) Coordination of two-level organizations with multiple objectives. Techniques of optimization. Academic, New York, pp 455–466
20. Jiang Y, Li X, Huang C, Wu X (2013) Application of particle swarm optimization based on CHKS smoothing function for solving nonlinear bilevel programming problem. Appl Math Comput 219(9):4332–4339
21. Lee WS, Lim WI, Koo PH (2009) Transporter scheduling based on a network flow model under a dynamic block transportation environment. In: Proceedings of international conference on computers & industrial engineering, Troyes, pp 311–316
22. Lin Y (2010) Reliability evaluation of a revised stochastic flow network with uncertain minimum time. Physica A 389(6):1253–1258
23. Liu H, Wang L, Song H (2009) Bi-Level model of railway transportation existing network and simulated annealing algorithm. In: Proceedings of the second international conference on transportation engineering, Chengdu
24. Lopes Y, Fernandes NC, Bastos CAM et al (2015) SMARTFlow: a solution for autonomic management and control of communication networks for smart grids. In: ACM symposium on applied computing, Salamanca, pp 2212–2217
25. Mazloumi E, Currie G, Rose G (2010) Using GPS data to gain insight into public transport travel time variability. J Transp Eng 136(7):623–631
26. McCormick PG (1989) The projective SUMT method for convex programming. Math Oper Res 14:203–223
27. Nakayama H (1984) Proposal of satisfying trade-off method for multiobjective programming. Trans Soc Inst Control Eng (in Japanese) 20:29–53
28. Nakayama H (1999) Basic handbook of operation
29. Nijkamp P, Spronk J (1978) Interactive multiple goal programming. International series in management science/operations research. Martinus Nijhoff Publishing
30. Osman MS, Abo-Sinna MA, Amer AH, Emam OE (2004) A multi-level non-linear multi-objective decision-making under fuzziness. Appl Math Comput 153(1):239–252
31. Paparrizos K, Samaras N, Sifaleras A (2009) An exterior simplex type algorithm for the minimum cost network flow problem. Comput Oper Res 36(4):1176–1190
32. Peng J, Liu B (2007) Birandom variables and birandom programming. Comput Ind Eng 53(3):433–453
33. Pióro M, Medhi D (2004) Routing, flow, and capacity design in communication and computer networks. Morgan Kaufmann, Amsterdam/Boston

34. Rakha H, El-Shawarby I, Arafeh M (2010) Trip travel-time reliability: issues and proposed solutions. J Intell Transp Syst 14(4):232–250
35. Reyes-Sierra M, Coello C (2006) Multi-objective particle swarm optimizers: a survey of the state-of-the-art. Int J Comput Intell Res 2(3):287–308
36. Savard G, Gauvin J (1994) The steepest descent direction for the nonlinear bilevel programming problem. Oper Res Lett 15(5):265–272
37. Sawaragi Y, Nakayama H, Tanino T (1997) Theory of multiobjective optimization. Academic, Orlando
38. Sha DY, Hsu CY (2006) A hybrid particle swarm optimization for job shop scheduling problem. Comput Ind Eng 51(4):791–808
39. Sheffi Y (1984) Urban transportation network: equilibrium analysis with mathematical programming methods. Prentice-Hall, Englewood Cliffs
40. Shimizu K (1982) Theory of multiobjective and conflict. Kyoritsu Syuppan (in Japanese)
41. Spronk J, Telgen J (1981) An ellipsoidal interactive multiple goal programming method. Lecture notes in economics and mathematical systems, vol 190, pp 380–387
42. Stetsyuk PI (2016) Problem statements for *k*-node shortest path and *k*-node shortest cycle in a complete graph. Cybern Syst Anal 52(1):1–5
43. Sumalee A, Watling DP, Nakayama S (2006) Reliable network design problem: case with uncertain demand and total travel time reliability. J Transp Res Board 1964:81–90
44. Tarvainen K, Haimes YY (1982) Coordination of hierarchical multiobjective systems: theory and methodology. IEEE Trans Syst Man Cybern 12(6):751–764
45. Wan Z, Wang G, Sun B (2013) A hybrid intelligent algorithm by combining particle swarm optimization with chaos searching technique for solving nonlinear bilevel programming problems. Swarm Evol Comput 8:26–32
46. Watling D (2006) User equilibrium traffic network assignment with stochastic travel times and late arrival penalty. Eur J Oper Res 175(3):1539–1556
47. Xu JP, Yao LM (2011) Random-like multiple objective decision making. Springer, Berlin/Heidelberg
48. Yin Y (2002) Multiobjective bilevel optimization for transportation planning and management problems. J Adv Transp 36(1):93–105
49. Zhang G, Lu J, Dillon T (2007) Decentralized multi-objective bilevel decision making with fuzzy demands. Knowl-Based Syst 20(5):495–507
50. Zhang T, Hu T, Zheng Y, Guo X (2012) An improved particle swarm optimization for solving bilevel multiobjective programming problem. J Appl Math. Hindawi Publishing Corporation, doi:10.1155/2012/626717
51. Zhang T, Hu T, Guo X, Chen Z, Zheng Y (2013) Solving high dimensional bilevel multiobjective programming problem using a hybrid particle swarm optimization algorithm with crossover operator. Knowl-Based Syst 53:13–19
52. Zhu DL, Xu Q, Lin ZH (2004) A homotopy method for solving bilevel programming problem. Nonlinear Anal 57:917–928
53. Zhu X, Yuan Q, Garcia-Diaz A, Dong L (2011) Minimal-cost network flow problems with variable lower bounds on arc flows. Comput Oper Res 38(8):1210–1218
54. Zitzler E, Deb K, Thiele L (2000) Comparison of multiobjective evolutionary algorithms: empirical results. Evol Comput 8(2):173–195

Chapter 4
Bi-Level Decision Making in Ra-Fu Phenomenon

Abstract In the real-life world, there are two kinds of common uncertainties, i.e., randomness and fuzziness. Accordingly, two efficient theories, probability theory and possibility theory are developed to handle them. For the possibility theory, readers may refer to Dubois and Prade (Possibility theory: an approach to computerized processing of uncertainty. Plenum Press, New York, 1988), Dubois and Prade (Possibility theory: qualitative and quantitative aspects. In: Quantified representation of uncertainty and imprecision. Springer, Berlin/New York, pp 169–226, 1998), and Zadeh (Fuzzy Sets Syst 1(1):3–28, 1978). In some decision problems such as supply chain network problems (Xu et al., Inf Sci 178(8):2022–2043, 2008), inventory problems (Xu and Liu, Inf Sci 178(14):2899–2914, 2008), portfolio selection problems (Li and Xu, Omega 37(2):439–449, 2009), the mixture of randomness and fuzziness is required to be considered simultaneously. From a viewpoint of ambiguity and randomness different from fuzzy random variables (Krätschmer V, Fuzzy Sets Syst 123(1):1–9, 2001; Kruse and Meyer, Statistics with vague data. Springer, Dordrecht, 1987; Kwakernaak, Inf Sci 15(1):1–29, 1978; Puri and Ralescu, J Math Anal Appl 114(2):409–422, 1986), by considering the experts' ambiguous understanding of means and variances of random variables, a concept of random fuzzy variables is proposed. In this chapter, we consider the construction site security planning (CSSP) problems with Ra-Fu phenomenon. Then three classes of bi-level mathematical models with Ra-Fu parameters are developed. Some theoretical results and algorithms are proposed for the models. At the end of the chapter, a practical case study demonstrates the feasibility and efficiency of the proposed methods.

Keywords Ra-Fu phenomenon • Ra-Fu EECC model • Ra-Fu CCDD model • Ra-Fu DDEE model • Construction site security planning

4.1 Construction Site Security Planning Problem

Vulnerability of construction facilities due to man-made threats and accidents is one of the major threats to the construction industry. At the same time, thefts and vandalism at construction sites are a growing problem. Large equipments such as graders, backhoes, and other large pieces of equipment that cost thousands of dollars

J. Xu et al., *Random-Like Bi-level Decision Making*, Lecture Notes in Economics and Mathematical Systems 688, DOI 10.1007/978-981-10-1768-1_4

are prime targets. These dangerous accidents, thefts and vandalism are not only monetary, but put the construction behind schedule [40]. Therefore, construction site security planning is becoming more and more important. Federal regulations have been produced to establish security requirements and arrangements during the construction of critical infrastructure projects, such as the construction security certification program [11–13], the recommended security guidelines issued by the Federal Aviation Administration [9] and the National Industrial Security Program [25].

The National Institute of Standards and Technology (NIST) and the Construction Industry Institute (CII) recognized that the effectiveness and efficiency of security level are influenced by decisions made during the planning and construction phases [8, 23]. However, few research studies investigated the implementation of security measures during the construction projects phase [2, 35]. Khalafallah and El-Rayes [15] developed a multi-objective optimization model of static site layout planning that is capable of maximizing construction-related airport security while minimizing site layout costs which include the costs of security systems and the travel cost of construction resources on site. Said and El-Rayes [27] presented the development of an automated multi-objective optimization framework for the planning of construction site layout and security systems of critical infrastructure projects that provides the capability of minimizing overall security risks and minimizing overall site costs.

Although previous studies addressed security needs and considerations during the construction phase, all of them focused on the implementation of physical security measures without considering bi-level relationship between defender and attacker, namely, they all stood in the planner's position ignoring the attacker's strategies. In fact, CSSP is in nature bi-level because it can be described as a leader-follower or a Stackelberg game [34]. The leader is the project security officer (the defender), who must first decide which facilities should be secured with limited funds. The follower is the potential attacker who then has the possibility to destroy or attack a subset of the facilities to inflict effectiveness loss of the construction facilities system. Attacker's strategies will in turn affect defender's strategies.

Bi-level fortification/interdiction median model has already been adopted as a strong tool for assessing network vulnerabilities to linkage or node disruptions in supply chain, critical infrastructure protection planning, homeland security improvements etc. [7, 22, 24, 30]. Church and Scaparra [7] demonstrated that protecting the most vulnerable facilities or predictable targets is not necessarily the most cost-effective way of confronting threats. Fortification patterns which take into account the interdependency among the system components and the effect of multiple, simultaneous losses can produce better and more resilient protection plans. This theory also can be applied in construction site. This chapter presents a bi-level multi-objective decision-making model for CSSP. In practice, there is often a complex uncertain environment in construction projects, especially in large scale construction projects. The imprecision and complexity of large scale construction projects cannot be dealt with simple random variables [42, 44, 46]. In this chapter, a type of Ra-Fu variable is defined to deal with the complex random phenomena in CSSP.

Bi-level programming problem falls into the class of NP-hard problems [4, 21, 45]. Many scholars have tried to find an effective way to solve this type of problem. Uno and Katagiri [39] applied random search algorithm, genetic algorithm and tabu algorithm to solve bi-level defensive location problems. Scaparra and Church [30] proposed a specialized tree search algorithm to solve the bi-level mixed-integer program for critical infrastructure protection planning. Liberatore et al. [20] presented heuristic approaches based on heuristic concentration-type rules to solve the stochastic R-interdiction median problem with fortification. All these algorithms need some external parameters and suffer from the parameters bias problem. To avoid the bias generated by parameters, this chapter applies the Plant Growth Simulation Algorithm (PGSA) to solve CSSP.

The development of analytical approaches, mathematical models with uncertainty as well as algorithms which are able to solve bi-level CSSP are still largely unexplored. This chapter will describe the problem clearly, provide a bi-level multi-objective model for CSSP and design an effective algorithm to solve it.

The main purpose of developing CSSP model is to deploy security countermeasures in a way that collectively achieve the following four main security functions: deterrence, detection, delay, and detainment [37]. All security countermeasures in construction site can be grouped in three main layers [8]:

(1) site fence or the outer layer;
(2) site grounds or the intermediate layer;
(3) target fence or the inner layer.

Considering the large area and so many facilities in the large-scale construction project, the intrusion detection systems in the target fences, the response forces, area lighting and natural surveillance in the site grounds cannot be applied in all the facilities. The project security officer must first decide which facilities to be secured under limited securing fund. In feedback, attacker will make his decision with given securing information.

CSSP has distinct characteristics which make traditional fortification/interdiction median models inapplicable. The characteristics are listed as follows:

(1) In normal supply chain systems, facilities can be divided into "suppliers" and "customers", but there is no such distinct demarcation in construction site. One facility can be both a "supplier" and a "customer" simultaneously and sometimes neither.
(2) The costs of securing different facilities in construction site are different.
(3) Almost all traditional fortification models are based on a fundamental assumption that once a facility is attacked, it is interdicted and impossible to offer services or products [7, 22, 24, 30], which is only the "worst-case" scenario at construction site. On most occasions, intentional disruption would cause economic loss and reduce efficiency. Moreover, the degree of loss is decided by the severity of the attack, the vulnerability of facilities, the complexity of the construction environment and many other random factors.

To state the problem clearly, the bi-level problem and motivations for considering Ra-Fu phenomena are described in the following.

4.1.1 Ra-Fu Phenomena

It is difficult to describe the practical situation of security problems at construction site with simple random parameters or crisp ones. Facility losses are influenced by attack severity, facilities vulnerability, the complexity of construction environment, and many other random factors [14, 38]. Therefore the facility losses are random variables themselves. Denote the facility loss by $\tilde{\bar{C}}$. Assume that $\tilde{\bar{C}}$ follows from normal distribution, i.e.,

$$\tilde{\bar{C}} \sim \mathcal{N}(\tilde{\mu}, \delta^2).$$

In addition to historical data, the experts' knowledge and experiences can be also used to increase the reliability of the estimation. If $\tilde{\mu}$ is regarded as a triangular fuzzy number, i.e.,

$$\tilde{\mu} = (a, b, c),$$

where a, b, c are constant numbers. Then the facility loss $\tilde{\bar{C}}$ can be described by Ra-Fu variables.

4.1.2 Bi-Level Description

When adopting a security framework, CSSP can be described as a leader-follower or a Stackelberg game [34]. The leader is the project security officer (the defender) who must first decide which facilities should be secured with limited funds. The objectives are to maximize the effectiveness of the construction facilities system and minimize the economic loss in potential attacks and the security cost. The follower is the potential attackers which include individuals or groups that aim to destroy onsite critical assets and inflict project efficiency [8, 18, 23]. They have the possibility to destroy or attack a subset R of the facilities to inflict maximum loss of effectiveness of the construction facilities system. The site targets include the critical facility under construction, site office trailers that contain sensitive information, and/or storage areas of classified equipment or materials [27].

The Ra-Fu bi-level problem description and Ra-Fu phenomena are shown by Fig. 4.1 below.

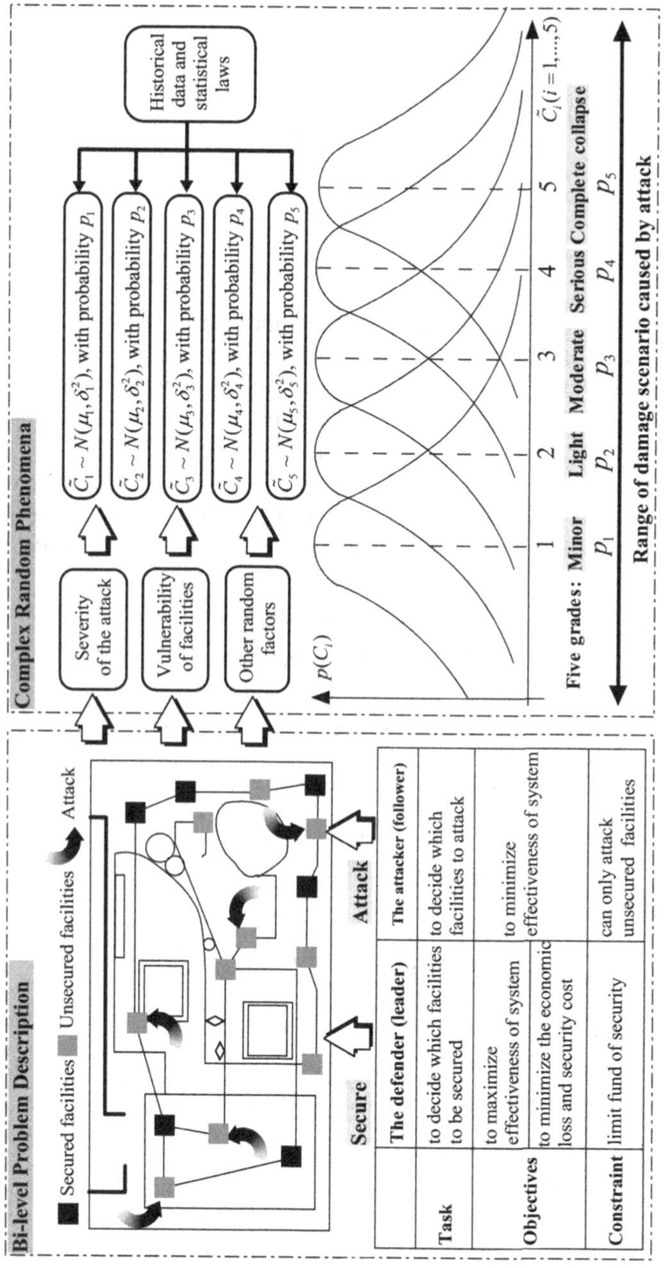

Fig. 4.1 Bi-level problem description and Ra-Fu phenomena consideration

4.2 Bi-Level Decision Making Model with Ra-Fu Coefficients

We give the general form of the single-objective bi-level decision making model with Ra-Fu coefficients as follows,

$$
\begin{cases}
\max\limits_{x\in R^{n_1}} F(\boldsymbol{x},\boldsymbol{y},\boldsymbol{\xi}) \\
\text{s.t.} \begin{cases}
G_i(\boldsymbol{x},\boldsymbol{y},\boldsymbol{\xi}) \leq 0, i=1,2,\cdots,p_1 \\
\text{where } \boldsymbol{y} \text{ solves:} \\
\begin{cases}
\max\limits_{y\in R^{n_2}} f(\boldsymbol{x},\boldsymbol{y},\boldsymbol{\xi}) \\
\text{s.t. } g_j(\boldsymbol{x},\boldsymbol{y},\boldsymbol{\xi}) \leq 0, j=1,2,\cdots,p_2,
\end{cases}
\end{cases}
\end{cases}
\tag{4.1}
$$

where, $\boldsymbol{x} \in R^{n_1}$ is decision variable for the upper-level decision maker and $y \in R^{n_2}$ is decision variable for the lower-level decision maker; $F(\boldsymbol{x},\boldsymbol{y},\boldsymbol{\xi})$ is the objective function of the upper-level model and $f(\boldsymbol{x},\boldsymbol{y},\boldsymbol{\xi})$ is the objective function of the lower-level model $G_i(\boldsymbol{x},\boldsymbol{y},\boldsymbol{\xi})$ is the constraint of the upper-level model and $g_j(\boldsymbol{x},\boldsymbol{y},\boldsymbol{\xi})$ is the constraint of the lower-level model; $\boldsymbol{\xi}$ is a Ra-Fu vector.

If $F(\boldsymbol{x},\boldsymbol{y},\boldsymbol{\xi})$ and $f(\boldsymbol{x},\boldsymbol{y},\boldsymbol{\xi})$ are vectors, i.e., $F(\boldsymbol{x},\boldsymbol{y},\boldsymbol{\xi}) = [F_1(\boldsymbol{x},\boldsymbol{y},\boldsymbol{\xi}), F_2(\boldsymbol{x},\boldsymbol{y},\boldsymbol{\xi}), \cdots, F_{m_1}(\boldsymbol{x},\boldsymbol{y},\boldsymbol{\xi})]$ and $f(\boldsymbol{x},\boldsymbol{y},\boldsymbol{\xi}) = [f_1(\boldsymbol{x},\boldsymbol{y},\boldsymbol{\xi}), f_2(\boldsymbol{x},\boldsymbol{y},\boldsymbol{\xi}), \cdots, f_{m_2}(\boldsymbol{x},\boldsymbol{y},\boldsymbol{\xi})]$, then model (4.1) become a Ra-Fu bi-level multi-objective decision-making model below.

$$
\begin{cases}
\min\limits_{x\in R^{n_1}} [F_1(\boldsymbol{x},\boldsymbol{y},\boldsymbol{\xi}), F_2(\boldsymbol{x},\boldsymbol{y},\boldsymbol{\xi}), \cdots, F_{m1}(\boldsymbol{x},\boldsymbol{y},\boldsymbol{\xi})] \\
\text{s.t.} \begin{cases}
G_i(\boldsymbol{x},\boldsymbol{y},\boldsymbol{\xi}) \leq 0, i=1,2,\cdots,p_1 \\
\text{where } y \text{ solves:} \\
\begin{cases}
\min\limits_{y\in R^{n_2}} [f_1(\boldsymbol{x},\boldsymbol{y},\boldsymbol{\xi}), f_2(\boldsymbol{x},\boldsymbol{y},\boldsymbol{\xi}), \cdots, f_{m2}(\boldsymbol{x},\boldsymbol{y},\boldsymbol{\xi})] \\
\text{s.t. } g_j(x,y,\xi) \leq 0, j=1,2,\cdots,p_2.
\end{cases}
\end{cases}
\end{cases}
\tag{4.2}
$$

In this chapter, the models with the form of (4.2) are mainly discussed. As $F_i(\boldsymbol{x},\boldsymbol{y},\boldsymbol{\xi})(i = 1,2,\cdots,m_1)$ and $f_i(\boldsymbol{x},\boldsymbol{y},\boldsymbol{\xi})(i = 1,2,\cdots,m_2)$, the linear version of (4.2) is obtained below.

$$
\begin{cases}
\max \left[\widetilde{\overline{\boldsymbol{C}}}{}_1^T\boldsymbol{x} + \widetilde{\overline{\boldsymbol{D}}}{}_1^T\boldsymbol{y}, \widetilde{\overline{\boldsymbol{C}}}{}_2^T\boldsymbol{x} + \widetilde{\overline{\boldsymbol{D}}}{}_2^T\boldsymbol{y}, \cdots, \widetilde{\overline{\boldsymbol{C}}}{}_{m_1}^T\boldsymbol{x} + \widetilde{\overline{\boldsymbol{D}}}{}_{m_1}^T\boldsymbol{y}\right] \\
\text{s.t.} \begin{cases}
\widetilde{\overline{\boldsymbol{A}}}{}_i^T\boldsymbol{x} + \widetilde{\overline{\boldsymbol{B}}}{}_i^T\boldsymbol{y} \leq \widetilde{\overline{E}}_i, i=1,2,\cdots,p_1 \\
\text{where } \boldsymbol{y} \text{ solves} \\
\begin{cases}
\max \left[\bar{\tilde{\boldsymbol{c}}}{}_1^T\boldsymbol{x} + \bar{\tilde{\boldsymbol{d}}}{}_1^T\boldsymbol{y}, \bar{\tilde{\boldsymbol{c}}}{}_2^T\boldsymbol{x} + \bar{\tilde{\boldsymbol{d}}}{}_2^T\boldsymbol{y}, \cdots, \bar{\tilde{\boldsymbol{c}}}{}_{m_1}^T\boldsymbol{x} + \bar{\tilde{\boldsymbol{d}}}{}_{m_2}^T\boldsymbol{y}\right] \\
\text{s.t.} \bar{\tilde{\boldsymbol{a}}}{}_i^T\boldsymbol{x} + \bar{\tilde{\boldsymbol{b}}}{}_i^T\boldsymbol{y} \leq \bar{\tilde{e}}_i, i=1,2,\cdots,p_2 \\
\boldsymbol{x} \in \Re^{n_1}, \boldsymbol{x} \in \Re^{n_2},
\end{cases}
\end{cases}
\end{cases}
\tag{4.3}
$$

where $\widetilde{\overline{\boldsymbol{C}}}_1 = (\widetilde{\overline{C}}_{11}, \widetilde{\overline{C}}_{12}, \cdots, \widetilde{\overline{C}}_{1n_1})^T, \widetilde{\overline{\boldsymbol{C}}}_2 = (\widetilde{\overline{C}}_{21}, \widetilde{\overline{C}}_{22}, \cdots, \widetilde{\overline{\boldsymbol{C}}}_{2m_1})^T, \cdots, \widetilde{\overline{\boldsymbol{C}}}_{m_1} = (\widetilde{\overline{C}}_{m_{11}}, \widetilde{\overline{C}}_{m_{12}}, \cdots, \widetilde{\overline{C}}_{m_1n_1})^T, \widetilde{\overline{\boldsymbol{D}}}_1 = (\widetilde{\overline{D}}_{11}, \widetilde{\overline{D}}_{12}, \cdots, \widetilde{\overline{D}}_{1n_1})^T, \widetilde{\overline{\boldsymbol{D}}}_2 = (\widetilde{\overline{D}}_{21}, \widetilde{\overline{D}}_{22}, \cdots, \widetilde{\overline{D}}_{2n_1})^T, \cdots,$ $\widetilde{\overline{\boldsymbol{D}}}_{m_1} = (\widetilde{\overline{D}}_{m_11}, \widetilde{\overline{D}}_{m_12}, \cdots, \widetilde{\overline{D}}_{m_1n_1})^T, \widetilde{\overline{\boldsymbol{A}}}_1 = (\widetilde{\overline{A}}_{11}, \widetilde{\overline{A}}_{12}, \cdots, \widetilde{\overline{A}}_{1n_1})^T, \widetilde{\overline{\boldsymbol{A}}}_2 = (\widetilde{\overline{A}}_{21}, \widetilde{\overline{A}}_{22}, \cdots, \widetilde{\overline{A}}_{2n_1})^T,$

$\cdots, \tilde{\bar{\boldsymbol{A}}}_{p_1} = (\tilde{\bar{A}}_{p_1 1}, \tilde{\bar{A}}_{p_1 2}, \cdots, \tilde{\bar{A}}_{p_1 n_1})^T, \tilde{\bar{\boldsymbol{B}}}_1 = (\tilde{\bar{B}}_{11}, \tilde{\bar{B}}_{12}, \cdots, \tilde{\bar{B}}_{1n_1})^T, \tilde{\bar{\boldsymbol{B}}}_2 = (\tilde{\bar{B}}_{21}, \tilde{\bar{B}}_{22}, \cdots, \tilde{\bar{B}}_{2n_1})^T,$ $\cdots, \tilde{\bar{\boldsymbol{B}}}_{p_1} = (\tilde{\bar{B}}_{p_1 1}, \tilde{\bar{B}}_{p_1 2}, \cdots, \tilde{\bar{B}}_{p_1 n_1})^T, \tilde{\bar{\boldsymbol{c}}}_1 = (\tilde{\bar{c}}_{11}, \tilde{\bar{c}}_{12}, \cdots, \tilde{\bar{c}}_{1n_2})^T, \tilde{\bar{\boldsymbol{c}}}_2 = (\tilde{\bar{c}}_{21}, \tilde{\bar{c}}_{22}, \cdots, \tilde{\bar{\boldsymbol{c}}}_{2n_2})^T,$ $\cdots, \tilde{\bar{\boldsymbol{c}}}_{m_2} = (\tilde{\bar{c}}_{m_2 1}, \tilde{\bar{C}}_{m_2 2}, \cdots, \tilde{\bar{c}}_{m_2 n_2})^T, \tilde{\bar{\boldsymbol{d}}}_1 = (\tilde{\bar{d}}_{11}, \tilde{\bar{d}}_{12}, \cdots, \tilde{\bar{d}}_{1n_2})^T, \tilde{\bar{\boldsymbol{d}}}_2 = (\tilde{\bar{d}}_{21}, \tilde{\bar{d}}_{22}, \cdots, \tilde{\bar{d}}_{2n_2})^T,$ $\cdots, \tilde{\bar{\boldsymbol{d}}}_{m_2} = (\tilde{\bar{d}}_{m_2 1}, \tilde{\bar{d}}_{m_2 2}, \cdots, \tilde{\bar{d}}_{m_2 n_2}), \tilde{\bar{\boldsymbol{a}}}_1 = (\tilde{\bar{a}}_{11}, \tilde{\bar{a}}_{12}, \cdots, \tilde{\bar{a}}_{1n_2})^T, \tilde{\bar{\boldsymbol{a}}}_2 = (\tilde{\bar{a}}_{21}, \tilde{\bar{a}}_{22}, \cdots, \tilde{\bar{a}}_{2n_2})^T,$ $\cdots, \tilde{\bar{\boldsymbol{a}}}_{p_2} = (\tilde{\bar{a}}_{p_2 1}, \tilde{\bar{a}}_{p_2 2}, \cdots, \tilde{\bar{a}}_{p_2 n_2})^T, \tilde{\bar{\boldsymbol{b}}}_1 = (\tilde{\bar{b}}_{11}, \tilde{\bar{b}}_{12}, \cdots, \tilde{\bar{b}}_{1n_2})^T, \tilde{\bar{\boldsymbol{b}}}_2 = (\tilde{\bar{b}}_{21}, \tilde{\bar{b}}_{22}, \cdots, \tilde{\bar{b}}_{2n_2})^T,$ $\cdots, \tilde{\bar{\boldsymbol{b}}}_{p_2} = (\tilde{\bar{b}}_{p_2 1}, \tilde{\bar{b}}_{p_2 2}, \cdots, \tilde{\bar{b}}_{p_2 n_2})^T$ are Ra-Fu vectors, and $\tilde{\bar{E}}_1, \tilde{\bar{E}}_2, \cdots, \tilde{\bar{E}}_{p_1}, \tilde{\bar{e}}_1, \tilde{\bar{e}}_2, \cdots, \tilde{\bar{e}}_{p_2}$ are Ra-Fu variables.

Since there exist the Ra-Fu variables, for each given decision vector $\boldsymbol{x}, \boldsymbol{y}$, it is meaningless to maximize the objective functions before we know the realization of Ra-Fu variables. Also, we cannot judge whether or not a decision vector $\boldsymbol{x}, \boldsymbol{y}$ is feasible before we know the precise value of Ra-Fu variables. Hence both the objectives and the constraints in Problem (4.2) and (4.3) are not well defined mathematically.

In order to present mathematically meaningful Ra-Fu models, the expected operator and chance operators for Ra-Fu variables are able to proposed to deal with the Ra-Fu objectives and constrains. There are 36 different models by the law of combination. We discuss three representative models in this Chapter while others can be handled with the techniques by the revelation from the proposed models.

4.3 Ra-Fu EECC Model

Since the title of the model follows from the order of leader's objectives, leader's constraints, follower's objectives, follower's constraints. Thus, EECC stands for the model with expected operator to the leader's model and chance operator to the follower's model.

4.3.1 General Form of Ra-Fu EECC Model

The EECC model of (4.1) is formulated as follows.

$$\begin{cases} \max\limits_{\boldsymbol{x} \in R^{n_1}} E[F(\boldsymbol{x}, \boldsymbol{y}, \boldsymbol{\xi})] \\ \text{s.t.} \begin{cases} E[G_i(\boldsymbol{x}, \boldsymbol{y}, \boldsymbol{\xi})] \leq 0, i = 1, 2, \cdots, p_1 \\ \text{where } \boldsymbol{y} \text{ solves:} \\ \max\limits_{\boldsymbol{y} \in R^{n_2}} \bar{f}, \\ \text{s.t.} \begin{cases} Ch\{f(\boldsymbol{x}, \boldsymbol{y}, \boldsymbol{\xi}) \geq \bar{f}\}(\alpha) \geq \beta \\ Ch\{g_j(\boldsymbol{x}, \boldsymbol{y}, \boldsymbol{\xi}) \leq 0\}(\gamma_i) \geq \delta_i, i = 1, 2, \cdots, p_2, \end{cases} \end{cases} \end{cases} \tag{4.4}$$

where α and β are predetermined confidence levels. Hence the follower need to obtain the (α, β) optimistic values to the return function $f(\boldsymbol{x}, \boldsymbol{y}, \boldsymbol{\xi})$.

Similarly, the EECC models for (4.2) are

$$\begin{cases} \max\limits_{x\in R^{n_1}} [E[F_1(x,y,\xi)], E[F_2(x,y,\xi)], \cdots, E[F_{m_1}(x,y,\xi)]] \\ \text{s.t.} \begin{cases} E[G_i(x,y,\xi)] \le 0, i = 1,2,\cdots,p_1 \\ \text{where } y \text{ solves:} \\ \max\limits_{y\in R^{n_2}} \{\bar{f}_1, \bar{f}_2, \cdots, \bar{f}_{m_2}\} \\ \text{s.t.} \begin{cases} Ch\{f_i(x,y,\xi) \ge \bar{f}_i\}(\alpha_i) \ge \beta_i, i = 1,2,\cdots,m_2 \\ Ch\{g_j(x,y,\xi) \le 0\}(\gamma_i) \ge \delta_i, i = 1,2,\cdots,p_2. \end{cases} \end{cases} \end{cases} \tag{4.5}$$

Remark 4.1. Since the there are different chances measures for Ra-Fu events, such as mean chance, equilibrium chance. So we can developed variants of EECC models according to different chance measures.

In the following, some solution concepts of (4.5) are defined as follows.

Definition 4.1. Let x^* be given by the upper-level decision maker. Then (x^*, y^*) is said to be a feasible solution to problem (4.5) if and only if (x^*, y^*) satisfies $E[G_i(x^*, y^*, \xi)] \le 0, i = 1,2,\cdots,p_1$ and is an efficient solution to the lower-level model at (α_i, β_i)-levels, $i = 1,2,\cdots,m_2$.

Remark 4.2. That (x^*, y^*) is an efficient solution to the lower-level model at (α_i, β_i)-levels ($i = 1,2,\cdots,m_2$) means (x, y) satisfies $Ch\{f_i(x^*, y^*, \xi) \ge \bar{f}_i\}(\alpha_i) \ge \beta_i, i = 1,2,\cdots,m_2$; $Ch\{g_j(x^*, y, \xi) \le 0\}(\gamma_i) \ge \delta_i, i = 1,2,\cdots,p_2$ and there is no other feasible solution (x, y) exists, such that $E[F_i(x, y)] \ge E[F_i(x^*, y^*)]$ and at least one $i = 1,2,\cdots,p_1$ is strict inequality.

Definition 4.2. Let (x^*, y^*) be a feasible solution to problem (4.5). Then (x^*, y^*) is the efficient solution to problem (4.5) if and only if there is no other feasible solution (x, y) exists, such that $E[F_i(x, y)] \ge E[F_i(x^*, y^*)]$ and at least one $i = 1,2,\cdots,m_1$ is strict inequality.

4.3.2 Constrained Variable Metric Method for Linear Model

If all the functions in (4.5) are linear form, then we have the Ra-Fu linear EECC (LEECC) model as follows.

$$\begin{cases} \max \left[E[\tilde{\bar{C}}_1^T x + \tilde{\bar{D}}_1^T y], E[\tilde{\bar{C}}_2^T x + \tilde{\bar{D}}_2^T y], \cdots, E[\tilde{\bar{C}}_{m_1}^T x + \tilde{\bar{D}}_{m_1}^T y]\right] \\ \text{s.t.} \begin{cases} E[\tilde{\bar{A}}_i^T x + \tilde{\bar{B}}_i^T y] \le E[\tilde{\bar{E}}_r], r = 1,2,\cdots,p_1 \\ \text{where } y \text{ solves:} \\ \max\limits_{y\in R^{n_2}} \{\bar{f}_1, \bar{f}_2, \cdots, \bar{f}_{m_2}\} \\ \text{s.t.} \begin{cases} Ch\{\tilde{\bar{c}}_i^T x + \tilde{\bar{d}}_i^T y \ge \bar{f}_i\}(\alpha_i) \ge \beta_i, i = 1,2,\cdots,m_2 \\ Ch\{\tilde{\bar{a}}_i^T x + \tilde{\bar{b}}_i^T y \le \tilde{\bar{e}}_i\}(\gamma_i) \ge \delta_i, i = 1,2,\cdots,p_2. \end{cases} \end{cases} \end{cases} \tag{4.6}$$

Crisp Equivalent Models

Due the complexity of (4.6) with bi-level structure, twofold uncertainties, multiple objectives, one way of solving problem (4.6) is to convert the objectives and constraints into their respective crisp equivalents and then reduce the solution complexity to some extent. In order to obtain the crisp equivalent models, the two theorem dealing with expected operators and chance operators are developed as follows:

Theorem 4.1. *Assume that random vector* $\tilde{\bar{\boldsymbol{C}}}_i = (\tilde{\bar{C}}_{i1}, \tilde{\bar{C}}_{i2}, \cdots, \tilde{\bar{C}}_{in})^T$ *is normally distributed with mean vector* $\bar{C}_i = (\bar{C}_{i1}, \bar{C}_{i2}, \cdots, \bar{C}_{in})^T$ *and positive definite covariance matrix* V_i^C *on the probability space* $(\Pi, \mathscr{A}, Pr)$*, written as* $\tilde{\bar{\boldsymbol{C}}}_i \sim \mathscr{N}(\bar{C}_i, V_i^C)(i = 1, 2, \cdots, m_1)$*. At the same time, assume* $\tilde{\bar{\boldsymbol{D}}}_i \sim \mathscr{N}(\bar{D}_i, V_i^D), \tilde{\bar{\boldsymbol{A}}}_r \sim \mathscr{N}(\bar{A}_r, V_r^A), \tilde{\bar{\boldsymbol{B}}}_r \sim \mathscr{N}(\bar{B}_r, V_r^B)$*. In addition, let* $\tilde{\bar{E}}_r \sim \mathscr{N}(\bar{E}_r, (\sigma_r^E)^2)(r = 1, 2, \cdots, p_1)$*.* $\bar{C}_{ij}, \bar{D}_{ij}, \bar{A}_{rj}, \bar{B}_{rj}, \bar{E}_r$ *are trapezoidal fuzzy variables characterized by*

$$\mu_{\bar{C}_{ij}} = [r_{1,C_{ij}}, r_{2,C_{ij}}, r_{3,C_{ij}}, r_{4,C_{ij}}],$$

$$\mu_{\bar{D}_{ij}} = [r_{1,D_{ij}}, r_{2,D_{ij}}, r_{3,D_{ij}}, r_{4,D_{ij}}],$$

$$\mu_{\bar{A}_{rj}} = [r_{1,A_{rj}}, r_{2,A_{rj}}, r_{3,A_{rj}}, r_{4,A_{rj}}],$$

$$\mu_{\bar{B}_{rj}} = [r_{1,B_{rj}}, r_{2,B_{rj}}, r_{3,B_{rj}}, r_{4,B_{rj}}], \text{ and}$$

$$\mu_{\bar{E}_r} = [r_{1,E_r}, r_{2,E_r}, r_{3,E_r}, r_{4,E_r}],$$

respectively. If, for all the i, j, $\tilde{\bar{C}}_{ij}(\theta), \tilde{\bar{D}}_{ij}(\theta)$ *are independently random variables for fixed* θ*. At the same time, for all the* r, j, $\tilde{\bar{A}}_{rj}(\theta), \tilde{\bar{B}}_{rj}(\theta), \tilde{\bar{E}}_r(\theta)$ *are independently random variables for fixed* θ *are independently random variables for fixed* θ*. Then the upper level model of (4.6) is equivalent to*

$$\begin{cases} \max[F_1, F_2, \cdots, F_{m_1}] \\ \text{s.t.} \begin{cases} \sum_{j=1}^{n}[(r_{1,A_{rj}} + r_{2,A_{rj}} + r_{3,A_{rj}} + r_{4,A_{rj}})x_j + (r_{1,B_{rj}} + r_{2,B_{rj}} + r_{3,B_{rj}} + r_{4,B_{rj}})y_j] \\ \leq r_{1,E_r} + r_{2,E_r} + r_{3,E_r} + r_{4,E_r}, r = 1, 2, \cdots, p_1, \end{cases} \end{cases} \tag{4.7}$$

where

$$F_i = \frac{1}{4}\left[\sum_{j}^{n}(r_{1,C_{ij}} + r_{2,C_{ij}} + r_{3,C_{ij}} + r_{4,C_{ij}})x_j + \sum_{j}^{n}(r_{1,D_{ij}} + r_{2,D_{ij}} + r_{3,D_{ij}} + r_{4,D_{ij}})y_j\right],$$

$i = 1, 2, \cdots, m_1.$

Proof. Since $\tilde{\bar{C}}_{ij}(\theta), \tilde{\bar{D}}_{ij}(\theta)$ normally distributed and are independent with each other, the linear combination of $\tilde{\bar{C}}_{ij}(\theta), \tilde{\bar{D}}_{ij}(\theta), \tilde{\bar{\boldsymbol{C}}}_i^T \boldsymbol{x} + \tilde{\bar{\boldsymbol{D}}}_i^T \boldsymbol{y}$ are also normally distributed characterized as

$$\tilde{\bar{\boldsymbol{C}}}_i^T \boldsymbol{x} + \tilde{\bar{\boldsymbol{D}}}_i^T \boldsymbol{y} \sim \mathcal{N}(\bar{C}_i^T x + \bar{D}_y^T y, x^T V_i^C x + y^T V_i^C y).$$

So the expected value of $\tilde{\bar{\boldsymbol{C}}}_i^T \boldsymbol{x} + \tilde{\bar{\boldsymbol{D}}}_i^T \boldsymbol{y}$ is the expected value of the fuzzy variable $\bar{C}_i^T x + \bar{D}_i^T y$. Note that $\bar{C}_i^T x + \bar{D}_i^T y$ is a trapezoidal fuzzy variables with form of:

$$\left(\sum_{j=1}^{n}(r_{1,C_{ij}}x_j + r_{1,D_{ij}}y_j), \sum_{j=1}^{n}(r_{2,C_{ij}}x_j + r_{2,D_{ij}}y_j),\right.$$
$$\left.\sum_{j=1}^{n}(r_{3,C_{ij}}x_j + r_{3,D_{ij}}y_j), \sum_{j=1}^{n}(r_{4,C_{ij}}x_j + r_{4,D_{ij}}y_j)\right).$$

In summary,

$$E[\tilde{\bar{\boldsymbol{C}}}_i^T \boldsymbol{x} + \tilde{\bar{\boldsymbol{D}}}_i^T \boldsymbol{y}] = E[\bar{C}_i^T x + \bar{D}_i^T y] = E\left(\sum_{j=1}^{n}(r_{1,C_{ij}}x_j + r_{1,D_{ij}}y_j), \sum_{j=1}^{n}(r_{2,C_{ij}}x_j + r_{2,D_{ij}}y_j),\right.$$
$$\left.\sum_{j=1}^{n}(r_{3,C_{ij}}x_j + r_{3,D_{ij}}y_j), \sum_{j=1}^{n}(r_{4,C_{ij}}x_j + r_{4,D_{ij}}y_j)\right)$$
$$= \frac{1}{4}\left(\sum_{j=1}^{n}(r_{1,C_{ij}}x_j + r_{1,D_{ij}}y_j) + \sum_{j=1}^{n}(r_{2,C_{ij}}x_j + r_{2,D_{ij}}y_j) + \sum_{j=1}^{n}(r_{3,C_{ij}}x_j + r_{3,D_{ij}}y_j)\right.$$
$$\left. + \sum_{j=1}^{n}(r_{4,C_{ij}}x_j + r_{4,D_{ij}}y_j)\right).$$

Similarly, for the constraints of the upper level model,

$$E[\tilde{\bar{\boldsymbol{A}}}_r^T \boldsymbol{x} + \tilde{\bar{\boldsymbol{B}}}_r^T \boldsymbol{y}] = \frac{1}{4}\left(\sum_{j=1}^{n}(r_{1,A_{rj}}x_j + r_{1,B_{rj}}y_j) + \sum_{j=1}^{n}(r_{2,A_{rj}}x_j + r_{2,B_{ij}}y_j)\right.$$
$$\left. + \sum_{j=1}^{n}(r_{3,A_{rj}}x_j + r_{3,B_{rj}}y_j) + \sum_{j=1}^{n}(r_{4,A_{rj}}x_j + r_{4,B_{rj}}y_j)\right)$$

and

$$E[\widetilde{\overline{\boldsymbol{E}}}_r^T] = \frac{1}{4}\left(\sum_{j=1}^{n}\left(r_{1,E_r} + \sum_{j=1}^{n}(r_{2,A_{rj}}x_j + r_{2,B_{ij}}y_j)\right.\right.$$
$$\left.\left.+ \sum_{j=1}^{n}(r_{3,A_{rj}}x_j + r_{3,B_{rj}}y_j) + \sum_{j=1}^{n}(r_{4,A_{rj}}x_j + r_{4,B_{rj}}y_j)\right)\right).$$

The theorem is proved. □

Theorem 4.2. *Assume that random vector* $\widetilde{\overline{\boldsymbol{C}}}_i = (\widetilde{\overline{\boldsymbol{C}}}_{i1}, \widetilde{\overline{\boldsymbol{C}}}_{i2}, \cdots, \widetilde{\overline{\boldsymbol{C}}}_{in})^T$ *is normally distributed with mean vector* $\overline{C}_i = (\overline{C}_{i1}, \overline{C}_{i2}, \cdots, \overline{C}_{in})^T$ *and positive definite covariance matrix* V_i^C *on the probability space* $(\Pi, \mathscr{A}, Pr)$*, written as* $\widetilde{\overline{\boldsymbol{C}}}_i \sim \mathscr{N}(\overline{C}_i, V_i^C)(i = 1, 2, \cdots, m_1)$*. At the same time, assume* $\widetilde{\overline{\boldsymbol{D}}}_i \sim \mathscr{N}(\overline{D}_i, V_i^D), \widetilde{\overline{\boldsymbol{A}}}_r \sim \mathscr{N}(\overline{A}_r, V_r^A), \widetilde{\overline{\boldsymbol{B}}}_r \sim \mathscr{N}(\overline{B}_r, V_r^B)$*. In addition, let* $\widetilde{\overline{E}}_r \sim \mathscr{N}(\overline{E}_r, (\sigma_r^E)^2)(r = 1, 2, \cdots, p_1)$*.* $\overline{C}_{ij}, \overline{D}_{ij}, \overline{A}_{rj}, \overline{B}_{rj}, \overline{E}_r$ *are trapezoidal fuzzy variables characterized by*

$$\mu_{\overline{C}_{ij}}(t) = \begin{cases} L\left(\dfrac{\mu_{ij}^C - t}{L_{ij}^C}\right) & t \le \mu_{ij}^C, L_{ij}^C > 0 \\ \\ R\left(\dfrac{t - \mu_{ij}^C}{R_{ij}^C}\right) & t \ge \mu_{ij}^C, R_{ij}^C > 0, \end{cases} \tag{4.8}$$

$$\mu_{\overline{D}_{ij}}(t) = \begin{cases} L\left(\dfrac{\mu_{ij}^D - t}{L_{ij}^C}\right) & t \le \mu_{ij}^D, L_{ij}^D > 0 \\ \\ R\left(\dfrac{t - \mu_{ij}^D}{R_{ij}^D}\right) & t \ge \mu_{ij}^D, R_{ij}^D > 0, \end{cases} \tag{4.9}$$

$$\mu_{\overline{A}_{rj}}(t) = \begin{cases} L\left(\dfrac{\mu_{rj}^A - t}{L_{rj}^C}\right) & t \le \mu_{rj}^A, L_{rj}^A > 0 \\ \\ R\left(\dfrac{t - \mu_{rj}^A}{R_{rj}^A}\right) & t \ge \mu_{rj}^A, R_{ij}^A > 0, \end{cases} \tag{4.10}$$

$$\mu_{\overline{B}_{rj}}(t) = \begin{cases} L\left(\dfrac{\mu_{rj}^B - t}{L_{rj}^B}\right) & t \le \mu_{rj}^B, L_{rj}^B > 0 \\ \\ R\left(\dfrac{t - \mu_{rj}^B}{R_{rj}^B}\right) & t \ge \mu_{rj}^B, R_{ij}^B > 0, \end{cases} \tag{4.11}$$

and

$$\mu_{\overline{E}_r}(t) = \begin{cases} L\left(\dfrac{\mu_r^E - t}{L_r^E}\right) & t \le \mu_r^E, L_r^E > 0 \\ \\ R\left(\dfrac{t - \mu_r^E}{R_j^E}\right) & t \ge \mu_{rj}^E, R_{ij}^E > 0. \end{cases} \tag{4.12}$$

Here, L_{ij}^C, R_{ij}^C are positive numbers expressing the left and right spreads of $\tilde{\mu}_{ij}^C$, respectively; L_{ij}^D, R_{ij}^D are positive numbers expressing the left and right spreads of $\tilde{\mu}_{ij}^D$, respectively; L_{rj}^A, R_{rj}^A are positive numbers expressing the left and right spreads of $\tilde{\mu}_{rj}^A$, respectively; L_{rj}^B, R_{rj}^B are positive numbers expressing the left and right spreads of $\tilde{\mu}_{rj}^A$, respectively; L_j^E, R_r^E are positive numbers expressing the left and right spreads of $\tilde{\mu}_r^E$, respectively. Reference functions $L, R : [0,1] \to [0,1]$ with $L(1) = R(1) = 0$ and $L(0) = R(0) = 1$ are non-increasing, continuous functions. Assume that for all the i,j, $\tilde{\overline{C}}_{ij}(\theta), \tilde{\overline{D}}_{ij}(\theta)$ are independently random variables for fixed θ. At the same time, for all the r,j, $\tilde{\overline{A}}_{ij}(\theta), \tilde{\overline{B}}_{rj}(\theta), \tilde{\overline{E}}_{rj}(\theta)$ are independently random variables for fixed θ. Then the upper level model of (4.6) is equivalent to

$$\begin{cases} \max[H_1(\boldsymbol{x},\boldsymbol{y}), H_2(\boldsymbol{x},\boldsymbol{y}), \cdots, H_{m_1}(\boldsymbol{x},\boldsymbol{y})] \\ s.t.\ \ K_r(\boldsymbol{x},\boldsymbol{y}) \le B_r, r = 1,2,\cdots,p_1, \end{cases} \tag{4.13}$$

where

$$\begin{aligned} H_i &= \frac{1}{2}(L_i^C\boldsymbol{x} + L_i^D\boldsymbol{y})[F(\mu_i^C\boldsymbol{x} + \mu_i^C\boldsymbol{y}) - F((\mu_i^C\boldsymbol{x} + \mu_i^c\boldsymbol{y}) - (L_i^C\boldsymbol{x} + L_i^D\boldsymbol{y}))] \\ &\quad + \frac{1}{2}(R_i^C\boldsymbol{x} + R_i^D\boldsymbol{y})[G((\mu_i^C\boldsymbol{x} + \mu_i^D\boldsymbol{y}) + (R_i^C\boldsymbol{x} + R_i^D\boldsymbol{y})) - G(\mu_i^C\boldsymbol{x} + \mu_i^D\boldsymbol{y})], \\ K_r(\boldsymbol{x},\boldsymbol{y}) &= \frac{1}{2}(L_r^A\boldsymbol{x} + L_r^B\boldsymbol{y})[F(\mu_r^A\boldsymbol{x} + \mu_r^B\boldsymbol{y}) - F((\mu_r^A\boldsymbol{x} + \mu_r^B\boldsymbol{y}) - (L_r^A\boldsymbol{x} + L_r^B\boldsymbol{y}))] \\ &\quad + \frac{1}{2}(R_r^A\boldsymbol{x} + R_r^B\boldsymbol{y})[G(\mu_r^A\boldsymbol{x} + \mu_r^B\boldsymbol{y}) + (R_r^A\boldsymbol{x} + R_r^B\boldsymbol{y}) - G(\mu_r^A\boldsymbol{x} + \mu_r^B\boldsymbol{y})], \\ B_r &= \frac{1}{2}(L_r^E)[F(\mu_r^E) - F(\mu_r^E - L_r^E)] + \frac{1}{2}R_r^E[G(\mu_r^E + R_r^b) - G(\mu_r^E)]. \end{aligned}$$

$F(x,y)$ and $G(x,y)$ are continuous about $\boldsymbol{x}$ on $[\mu_i^C\boldsymbol{x} - L_i^C\boldsymbol{x}, \mu_i^C\boldsymbol{x}]$ and $[\mu_i^C\boldsymbol{x}, \mu_i^C\boldsymbol{x} + R_i^C\boldsymbol{x}]$, respectively. Similarly, $F(x,y)$ and $G(x,y)$ are continuous about $\boldsymbol{y}$ on $[\mu_i^D\boldsymbol{y} - L_i^D\boldsymbol{y}, \mu_i^D\boldsymbol{y}]$ and $[\mu_i^D\boldsymbol{y}, \mu_i^D\boldsymbol{y} + R_i^D\boldsymbol{y}]$, respectively $F(x,y)$ and $G(x,y)$ satisfy:

$$\frac{\partial F}{\partial x} = L(\boldsymbol{x}),\ for\ x \in [\mu_i^C\boldsymbol{x} - L_i^C\boldsymbol{x}, \mu_i^C\boldsymbol{x}],$$
$$\frac{\partial F}{\partial y} = L(\boldsymbol{y}),\ for\ y \in [\mu_i^D\boldsymbol{y} - L_i^D\boldsymbol{y}, \mu_i^D\boldsymbol{y}],$$
$$\frac{\partial G}{\partial x} = R(\boldsymbol{x}),\ for\ x \in [\mu_i^C\boldsymbol{x}, \mu_i^C\boldsymbol{x} + R_i^C\boldsymbol{x}],$$
$$\frac{\partial G}{\partial y} = R(\boldsymbol{y}),\ for\ y \in [\mu_i^D\boldsymbol{y}, \mu_i^D\boldsymbol{y} + R_i^D\boldsymbol{y}].$$

Proof. This proof is derived from Theorems 4.1 and 4.6 in [43]. □

Consider the lower level of model (4.6). If the *Ch* measure is defined as $Pos - Pr$ measure, then the lower level of model (4.6) can be rewritten as

$$\begin{cases} \max\limits_{y \in R^{n_2}} \{\bar{f}_1, \bar{f}_2, \cdots, \bar{f}_{m_2}\} \\ \text{s.t.} \begin{cases} Pos\{\theta | Pr\{\tilde{\bar{c}}_i^T(\theta)\boldsymbol{x} + \tilde{\bar{d}}_i^T(\theta)\boldsymbol{y} \geq \bar{f}_i\} \geq \beta_i\} \geq \alpha_i, i = 1, 2, \cdots, m_2 \\ Pos\{\theta | Pr\{\tilde{\bar{a}}_r^T(\theta)\boldsymbol{x} + \tilde{\bar{b}}_r^T(\theta)\boldsymbol{y} \leq \tilde{\bar{e}}_r^T(\theta)\} \geq \delta_r\} \geq \gamma_r, r = 1, 2, \cdots, p_2. \end{cases} \end{cases} \tag{4.14}$$

Theorem 4.3. *If $\tilde{\bar{c}}_i^T$ and $\tilde{\bar{a}}_r^T$ are random vectors, $\tilde{\bar{d}}_i^T, \tilde{\bar{b}}_r^T$ are fuzzy vectors, $\tilde{\bar{e}}_r$ are fuzzy variable. Then (4.14) is equivalent to*

$$\begin{cases} \max\limits_{y \in R^{n_2}} \{\bar{f}_1, \bar{f}_2, \cdots, \bar{f}_{m_2}\} \\ s.t. \begin{cases} Pr\{Pos\{\tilde{\bar{c}}_i^T\boldsymbol{x} + \tilde{\bar{d}}_i^T\boldsymbol{y} \geq \bar{f}_i\} \geq \alpha_i\} \geq \beta_i, i = 1, 2, \cdots, m_2 \\ Pr\{Pos\{\tilde{\bar{a}}_r^T\boldsymbol{x} + \tilde{\bar{b}}_r^T\boldsymbol{y} \leq \tilde{\bar{e}}_r^T\} \geq \gamma_r\} \geq \delta_r, r = 1, 2, \cdots, p_2. \end{cases} \end{cases} \tag{4.15}$$

Proof. $Pos\{\theta | Pr\{\tilde{\bar{c}}_i^T(\theta)\boldsymbol{x} + \tilde{\bar{d}}_i^T(\theta)\boldsymbol{y} \geq \bar{f}_i\} \geq \beta_i\} \geq \alpha_i$ implies that there is a crisp vector d_i such that:

$$\mu(d_i) \geq \alpha_i, Pr\{\tilde{\bar{c}}_i^T\boldsymbol{x} + d_i^T\boldsymbol{y} \geq \bar{f}_i\} \geq \beta_i.$$

So $Pr\{Pos\{\tilde{\bar{c}}_i^T\boldsymbol{x} + \tilde{\bar{d}}_i^T\boldsymbol{y} \geq \bar{f}_i\} \geq \alpha_i\} \geq \beta_i, i = 1, 2, \cdots, m_2$.

Similarly $Pos\{Pr\{\tilde{\bar{a}}_r^T\boldsymbol{x} + \tilde{\bar{b}}_r^T\boldsymbol{y} \leq \tilde{\bar{e}}_r^T\} \geq \delta_r\} \geq \gamma_r$ is equivalent to $Pr\{Pos\{\tilde{\bar{a}}_r^T\boldsymbol{x} + \tilde{\bar{b}}_r^T\boldsymbol{y} \leq \tilde{\bar{e}}_r^T\} \geq \gamma_r\} \geq \delta_r$. Then the theorem is proved. □

In the follows, we consider a case that (4.14) can be converted into its crisp equivalent model.

Theorem 4.4. *Suppose that $\bar{\tilde{c}}_r = (\bar{\tilde{c}}_{i1}, \bar{\tilde{c}}_{r2}, \cdots, \bar{\tilde{c}}_{in})^T$ is a normally distributed Ra-Fu vector with the fuzzy mean vector $\tilde{c}_i) = (\tilde{c}_{i1}, \tilde{c}_{i2}, \cdots, \tilde{c}_{in})^T$ and covariance matrix V_i^c, written as $\bar{\tilde{c}}_i \sim \mathcal{N}(\tilde{c}_i, V_i^c)$. Similarly, $\bar{\tilde{d}}_i \sim \mathcal{N}(\tilde{d}_i, V_i^d)$. $\tilde{c}_i = (\tilde{c}_{i1}, \tilde{c}_2, \cdots, \tilde{c}_{in}), \tilde{d}_i = (\tilde{d}_{i1}, \tilde{d}_{i2}, \cdots, \tilde{d}_{in})$ are fuzzy vectors characterized as:*

$$\mu_{\tilde{c}_{ij}} = [r_{1,c_{ij}}, r_{2,c_{ij}}, r_{3,c_{ij}}, r_{4,c_{ij}}],$$

and

$$\mu_{\tilde{d}_{ij}} = [r_{1,d_{ij}}, r_{2,d_{ij}}, r_{3,d_{ij}}, r_{4,d_{ij}}].$$

Then $Pos\{\theta | Pr\{\bar{\tilde{\boldsymbol{c}}}_i^T(\theta)\boldsymbol{x} + \bar{\tilde{\boldsymbol{d}}}_i^T(\theta)\boldsymbol{y} \geq \bar{f}_i\} \geq \beta_i\} \geq \alpha_i$ is equivalent to

$$\bar{f}_i \leq \Phi^{-1}(1-\beta_i)\sqrt{x^T V_i^c x + y^T V_i^d y} + \alpha_i \sum_{j=1}^{n} (r_{3,c_{ij}} x_j + r_{3,d_{ij}} y_j)$$

$$+ (1-\alpha_i) \sum_{j=1}^{n} (r_{4,c_{ij}} x_j + r_{4,d_{ij}} y_j),$$

where Φ is standard normally distribution.

Proof. Since $\bar{\tilde{c}}_{ij}, \bar{\tilde{d}}_{ij}$ normally distributed and are independent with each other, the linear combination of $\bar{\tilde{c}}_{ij}, \bar{\tilde{d}}_{ij}, \bar{\tilde{c}}_i^T \boldsymbol{x} + \bar{\tilde{d}}_i^T \boldsymbol{y}$ are also normally distributed characterized as:

$$\bar{\tilde{c}}_i^T \boldsymbol{x} + \bar{\tilde{d}}_i^T \boldsymbol{y} \sim \mathcal{N}(\bar{c}_i^T x + \bar{d}_y^T y, x^T V_i^c x + y^T V_i^d y).$$

$$Pr\{\bar{\tilde{\boldsymbol{c}}}_i^T \boldsymbol{x} + \bar{\tilde{\boldsymbol{d}}}_i^T \boldsymbol{y} \geq \bar{f}_i\} \geq \beta_i$$

$$\Leftrightarrow Pr\left\{ \frac{\bar{\tilde{\boldsymbol{c}}}_i^T \boldsymbol{x} + \bar{\tilde{\boldsymbol{d}}}_i^T \boldsymbol{y} - (\bar{c}_i^T x + \bar{d}_y^T y)}{\sqrt{x^T V_i^c x + y^T V_i^d y}} \geq \frac{\bar{f}_i - (\bar{c}_i^T x + \bar{d}_y^T y)}{\sqrt{x^T V_i^c x + y^T V_i^d y}} \right\} \geq \beta_i$$

$$\Leftrightarrow Pr\left\{ \frac{\bar{\tilde{\boldsymbol{c}}}_i^T \boldsymbol{x} + \bar{\tilde{\boldsymbol{d}}}_i^T \boldsymbol{y} - (\bar{c}_i^T x + \bar{d}_y^T y)}{\sqrt{x^T V_i^c x + y^T V_i^d y}} \leq \frac{\bar{f}_i - (\bar{c}_i^T x + \bar{d}_y^T y)}{\sqrt{x^T V_i^c x + y^T V_i^d y}} \right\} \leq 1 - \beta_i$$

$$\Leftrightarrow \Phi\left(\frac{\bar{f}_i - (\bar{c}_i^T x + \bar{d}_y^T y)}{\sqrt{x^T V_i^c x + y^T V_i^d y}} \right) \leq 1 - \beta_i$$

$$\Leftrightarrow \frac{\bar{f}_i - (\bar{c}_i^T x + \bar{d}_y^T y)}{\sqrt{x^T V_i^c x + y^T V_i^d y}} \leq \Phi^{-1}(1-\beta_i)$$

$$\Leftrightarrow \bar{f}_i \leq \Phi^{-1}(1-\beta_i)\sqrt{x^T V_i^c x + y^T V_i^d y} + (\bar{c}_i^T x + \bar{d}_y^T y).$$

Note that $\Phi^{-1}(1-\beta_i)\sqrt{x^T V_i^c x + y^T V_i^d y}+(\bar{c}_i^T x+\bar{d}_y^T y)$ is a trapezoidal fuzzy variables with form of:

$$(\Phi^{-1}(1-\beta_i)\sqrt{x^T V_i^c x + y^T V_i^d y} + \sum_{j=1}^{n}(r_{1,c_{ij}}x_j + r_{1,d_{ij}}y_j) - \bar{f}_i,$$

$$\Phi^{-1}(1-\beta_i)\sqrt{x^T V_i^c x + y^T V_i^d y} + \sum_{j=1}^{n}(r_{2,c_{ij}}x_j + r_{2,d_{ij}}y_j) - \bar{f}_i,$$

$$\Phi^{-1}(1-\beta_i)\sqrt{x^T V_i^c x + y^T V_i^d y} + \sum_{j=1}^{n}(r_{3,c_{ij}}x_j + r_{3,d_{ij}}y_j) - \bar{f}_i,$$

$$\Phi^{-1}(1-\beta_i)\sqrt{x^T V_i^c x + y^T V_i^d y} + \sum_{j=1}^{n}(r_{4,c_{ij}}x_j + r_{4,d_{ij}}y_j) - \bar{f}_i),$$

are denoted by $(\hat{r}_{i,1}, \hat{r}_{i,2}, \hat{r}_{i,3}, \hat{r}_{i,4})$.

$$\begin{aligned}
&Pos\{\theta|Pr\{\tilde{\bar{\boldsymbol{c}}}_i^T(\theta)\boldsymbol{x} + \tilde{\bar{\boldsymbol{d}}}_i^T(\theta)\boldsymbol{y} \geq \bar{f}_i\} \geq \beta_i\} \geq \alpha_i \\
&\Leftrightarrow Pos\left\{\bar{f}_i \leq \Phi^{-1}(1-\beta_i)\sqrt{x^T V_i^c x + y^T V_i^d y} + (\bar{c}_i^T x + \bar{d}_y^T y)\right\} \geq \alpha_i \\
&\Leftrightarrow \alpha_i\hat{r}_{i,3} + (1-\alpha_i)\hat{r}_{i,3} \geq 0 \text{ (See Fig. 4.2)} \\
&\Leftrightarrow \alpha_i(\Phi^{-1}(1-\beta_i)\sqrt{x^T V_i^c x + y^T V_i^d y} + \sum_{j=1}^{n}(r_{3,c_{ij}}x_j + r_{3,d_{ij}}y_j) - \bar{f}_i) \\
&\quad +(1-\alpha_i)(\Phi^{-1}(1-\beta_i)\sqrt{x^T V_i^c x + y^T V_i^d y} + \sum_{j=1}^{n}(r_{4,c_{ij}}x_j + r_{4,d_{ij}}y_j) - \bar{f}_i) \geq 0 \\
&\Leftrightarrow \bar{f}_i \leq (\Phi^{-1}(1-\beta_i)\sqrt{x^T V_i^c x + y^T V_i^d y} + \alpha_i\sum_{j=1}^{n}(r_{3,c_{ij}}x_j + r_{3,d_{ij}}y_j) \\
&\quad +(1-\alpha_i)\sum_{j=1}^{n}(r_{4,c_{ij}}x_j + r_{4,d_{ij}}y_j).
\end{aligned}$$

The theorem is proved. □

Remark 4.3. It follows from Theorem 4.4 that to maximize $\bar{f}_i$ is equivalent to maximize

$$\Phi^{-1}(1-\beta_i)\sqrt{x^T V_i^c x + y^T V_i^d y} + \alpha_i\sum_{j=1}^{n}(r_{3,c_{ij}}x_j + r_{3,d_{ij}}y_j)$$

$$+ (1-\alpha_i)\sum_{j=1}^{n}(r_{4,c_{ij}}x_j + r_{4,d_{ij}}y_j).$$

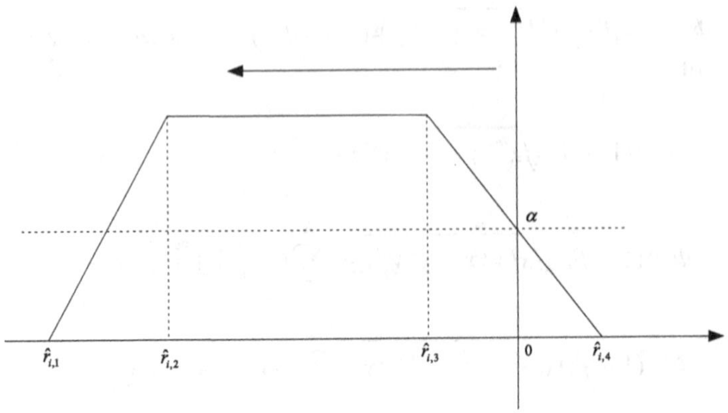

Fig. 4.2 Illustration of $(\hat{r}_{i,1}, \hat{r}_{i,2}, \hat{r}_{i,3}, \hat{r}_{i,4})$

Similarly, for the $Pos\{\theta|Pr\{\tilde{\bar{\boldsymbol{a}}}_r^T(\theta)\boldsymbol{x} + \tilde{\bar{\boldsymbol{b}}}_r^T(\theta)\boldsymbol{y} \leq \tilde{\bar{\boldsymbol{e}}}_r^T(\theta)\} \geq \delta_r\} \geq \gamma_r$, we have the following theorem.

Theorem 4.5. *Suppose that $\tilde{\bar{a}}_r = (\tilde{\bar{a}}_{r1}, \tilde{\bar{a}}_{r2}, \cdots, \tilde{\bar{a}}_{rn})^T$ is a normally distributed Ra-Fu vector with the fuzzy mean vector $\tilde{a}_r = (\tilde{a}_{r1}, \tilde{a}_{r1}, \cdots, \tilde{a}_{rn})^T$ and covariance matrix V_r^a, written as $\tilde{\bar{a}}_r \sim \mathcal{N}(\tilde{a}_r, V_r^a)$. Similarly, $\tilde{\bar{b}}_r \sim \mathcal{N}(\tilde{b}_r, V_r^b)$. $\tilde{\bar{e}}_r$ is normally distributed Ra-Fu variable defined by $\tilde{\bar{e}}_r \sim \mathcal{N}(\tilde{e}_r, (\sigma_r^e)^2)$. $\tilde{a}_{rj}, \tilde{b}_{rj}, \tilde{e}_r$ are fuzzy variables characterized as:*

$$\mu_{\tilde{a}_{rj}} = [r_{1,a_{rj}}, r_{2,a_{rj}}, r_{3,a_{rj}}, r_{4,a_{rj}}],$$
$$\mu_{\tilde{b}_{rj}} = [r_{1,b_{rj}}, r_{2,b_{rj}}, r_{3,b_{rj}}, r_{4,b_{rj}}], \text{ and}$$
$$\mu_{\tilde{e}_r} = [r_{1,e_r}, r_{2,e_r}, r_{3,e_r}, r_{4,e_r}],$$

Then $Pos\{\theta|Pr\{\tilde{\bar{\boldsymbol{a}}}_r^T(\theta)\boldsymbol{x} + \tilde{\bar{\boldsymbol{b}}}_r^T(\theta)\boldsymbol{y} \geq \tilde{\bar{\boldsymbol{e}}}_r^T(\theta)\} \geq \delta_r$ is equivalent to

$$\Phi^{-1}(\delta_r)\sqrt{x^T V_r^a x + y^T V_r^b y + (\sigma_r^e)^2} + (1-\gamma_r)\left(\sum_{j=1}^n (r_{1,a_{rj}}x_j + r_{1,b_{rj}}y_j) - r_{4,e_{rj}}\right)$$
$$+ \gamma_r\left(\sum_{j=1}^n (r_{2,a_{rj}}x_j + r_{2,b_{rj}}y_j) - r_{3,e_r}\right) \leq 0,$$

where Φ is standard normally distribution.

Proof. Since $\tilde{\bar{a}}_{rj}, \tilde{\bar{b}}_{rj}, \tilde{\bar{e}}_r$ are normally distributed and are independent with each other, the linear combination of $\tilde{\bar{a}}_{rj}, \tilde{\bar{b}}_{rj}, \tilde{\bar{e}}_r, \tilde{\bar{a}}_r^T\boldsymbol{x} + \tilde{\bar{d}}_i^T\boldsymbol{y} - \tilde{\bar{e}}_r$ are also normally distributed characterized as

$$\tilde{\bar{a}}_r^T \boldsymbol{x} + \tilde{\bar{b}}_r^T \boldsymbol{y} - \tilde{\bar{e}}_r \sim \mathcal{N}(\bar{a}_r^T x + \bar{b}_r^T y - \bar{e}_r, x^T V_i^c x + y^T V_i^d y + (\sigma_r^e)^2).$$

$$\begin{aligned}
& Pr\{\tilde{\bar{\boldsymbol{a}}}_r^T \boldsymbol{x} + \tilde{\bar{\boldsymbol{b}}}_r^T \boldsymbol{y} \le \tilde{\bar{\boldsymbol{e}}}_r\} \ge \delta_r \\
& \Leftrightarrow Pr\left\{ \frac{(\tilde{\bar{\boldsymbol{a}}}_r^T \boldsymbol{x} + \tilde{\bar{\boldsymbol{b}}}_r^T \boldsymbol{y} - \tilde{\bar{\boldsymbol{e}}}_r) - (\bar{a}_r^T x + \bar{b}_r^T y - \bar{e}_r)}{\sqrt{x^T V_i^c x + y^T V_i^d y + (\sigma_r^e)^2}} \le \frac{-(\bar{a}_r^T x + \bar{b}_r^T y - \bar{e}_r)}{\sqrt{x^T V_r^a x + y^T V_r^b y + (\sigma_r^e)^2}} \right\} \ge \delta_r \\
& \Leftrightarrow \Phi\left\{ \frac{-(\bar{a}_r^T x + \bar{b}_r^T y - \bar{e}_r)}{\sqrt{x^T V_r^a x + y^T V_r^b y + (\sigma_r^e)^2}} \right\} \ge \delta_r \\
& \Leftrightarrow \frac{-(\bar{a}_r^T x + \bar{b}_r^T y - \bar{e}_r)}{\sqrt{x^T V_r^a x + y^T V_r^b y + (\sigma_r^e)^2}} \ge \Phi^{-1}(\delta_r) \\
& \Leftrightarrow (\bar{a}_r^T x + \bar{b}_r^T y - \bar{e}_r) + \Phi^{-1}(\delta_r)\sqrt{x^T V_r^a x + y^T V_r^b y + (\sigma_r^e)^2} \le 0.
\end{aligned}$$

Note that $(\bar{a}_r^T x + \bar{b}_r^T y - \bar{e}_r) + \Phi^{-1}(\delta_r)\sqrt{x^T V_r^a x + y^T V_r^b y + (\sigma_r^e)^2}$ is a trapezoidal fuzzy variables with form of $(\hat{r}_{r,1}, \hat{r}_{r,2}, \hat{r}_{r,3}, \hat{r}_{r,4})$, where

$$\hat{r}_{r,k} = \Phi^{-1}(\delta_r)\sqrt{x^T V_r^a x + y^T V_r^b y + (\sigma_r^e)^2} + \sum_{j=1}^n (r_{k,a_{rj}} x_j + r_{k,b_{rj}} y_j) - r_{5-k,e_r},$$

$k = 1, 2, 3, 4.$

$$\begin{aligned}
& Pos\{\theta | Pr\{\tilde{\bar{\boldsymbol{a}}}_r^T(\theta)\boldsymbol{x} + \tilde{\bar{\boldsymbol{b}}}_r^T(\theta)\boldsymbol{y} \ge \tilde{\bar{\boldsymbol{e}}}_r^T(\theta)\} \ge \delta_r\} \ge \gamma_r \\
& \Leftrightarrow (1-\gamma_r)\hat{r}_{r,1} + \gamma_r \hat{r}_{r,2} \le 0 \\
& \Leftrightarrow (1-\gamma_r)(\Phi^{-1}(\delta_r)\sqrt{x^T V_r^a x + y^T V_r^b y + (\sigma_r^e)^2} + \sum_{j=1}^n (r_{1,a_{rj}} x_j + r_{1,b_{rj}} y_j) - r_{4,e_r}) \\
& \quad + \gamma_r(\Phi^{-1}(\delta_r)\sqrt{x^T V_r^a x + y^T V_r^b y + (\sigma_r^e)^2} + \sum_{j=1}^n (r_{2,a_{rj}} x_j + r_{2,b_{rj}} y_j) - r_{3,e_r}) \le 0 \\
& \Leftrightarrow \Phi^{-1}(\delta_r)\sqrt{x^T V_r^a x + y^T V_r^b y + (\sigma_r^e)^2} + (1-\gamma_r)\left(\sum_{j=1}^n (r_{1,a_{rj}} x_j + r_{1,b_{rj}} y_j) - r_{4,e_{rj}} \right) \\
& \quad + \gamma_r \left(\sum_{j=1}^n (r_{2,a_{rj}} x_j + r_{2,b_{rj}} y_j) - r_{3,e_r} \right) \le 0.
\end{aligned}$$

This theorem is proved. □

Constrained Variable Metric Method

It follows from Theorems 4.1∼4.4 that some Ra-Fu LEECC Model can be converted into crisp equivalent models, which are classical bi-level multiple decision making models. In this section, a interactive developed by Shi and Xia [32].

For convenience, the crisp equivalent model of (4.6) is rewritten as

$$\begin{cases} \max[F_1(\boldsymbol{x},\boldsymbol{y}), F_2(\boldsymbol{x},\boldsymbol{y})\cdots,F_{m_1}(\boldsymbol{x},\boldsymbol{y})] \\ \text{s.t. } (\boldsymbol{x},\boldsymbol{y}) \in \Omega_1 \\ \text{where } \boldsymbol{y} \text{ solves:} \\ \begin{cases} \max[f_1(\boldsymbol{x},\boldsymbol{y}), f_2(\boldsymbol{x},\boldsymbol{y})\cdots,f_{m_2}(\boldsymbol{x},\boldsymbol{y})] \\ \text{s.t.} (\boldsymbol{x},\boldsymbol{y}) \in \Omega_2, \end{cases} \end{cases} \tag{4.16}$$

where Ω_1 and Ω_2 are the constraint set of upper model and lower model, respectively.

The solution can be obtained by solving the upper and lower level decision-making problem separately. First, we consider the upper level multiobjective decision-making:

$$\begin{cases} \max[F_1(\boldsymbol{x},\boldsymbol{y}), F_2(\boldsymbol{x},\boldsymbol{y})\cdots,F_{m_1}(\boldsymbol{x},\boldsymbol{y})] \\ \text{s.t. } (\boldsymbol{x},\boldsymbol{y}) \in \Omega_1 \cap \Omega_2 = \Omega. \end{cases} \tag{4.17}$$

In order to achieve the preferred solution to the model, denote $(\boldsymbol{x},\boldsymbol{y})$ by $\tilde{\boldsymbol{x}}$, we change (4.17) into the following multi-objective decision-making problem:

$$\begin{cases} \max[F_1(\tilde{\boldsymbol{x}}), F_2(\tilde{\boldsymbol{x}})\cdots,F_{m_1}(\tilde{\boldsymbol{x}})] \\ \text{s.t.} \tilde{\boldsymbol{x}} \in \Omega_1 \cap \Omega_2. \end{cases} \tag{4.18}$$

We express quantitatively satisfactoriness and the preferred solution in view of singular-level multiobjective decision-making problem, and prove several theorems with the help of the quality of ε-constraint method [5].

Let $L_i = \min_{\tilde{\boldsymbol{x}}\in\Omega_1\cap\Omega_2} F_i(\tilde{\boldsymbol{x}})$, $U_i = \max_{\tilde{\boldsymbol{x}}\in\Omega_1\cap\Omega_2} F_i(\tilde{\boldsymbol{x}})$, on $u_i \in [L_i U_i]$ define $A_i \in F(u_i)$ whose membership function $\mu_{A_i}(F_i(\tilde{\boldsymbol{x}}))$ meets (i) and (ii) as follows:

(i) When the objective value $F_i(\tilde{\boldsymbol{x}})$ approaches or equals upper-level decision maker's ideal value, $\mu_{A_i}(F_i(\tilde{\boldsymbol{x}}))$ approaches or equals 1. Otherwise, 0.
(ii) If $F_i(\tilde{\boldsymbol{x}}_1) > F_i(\tilde{\boldsymbol{x}}_2)$, then $\mu_{A_i}(F_i(\tilde{\boldsymbol{x}}_1)) > mu_{A_i}(F_i(\tilde{\boldsymbol{x}}_2), i = 1,2,\cdots,m_1$.

Definition 4.3. If $\tilde{\boldsymbol{x}}^*$ is a non-inferior solution of (4.18), then $\mu_{A_i}(F_i(\tilde{\boldsymbol{x}}^*))$ is defined as the satisfactoriness of $\tilde{\boldsymbol{x}}^*$ for objective $F_i(\tilde{\boldsymbol{x}})$.

Definition 4.4. $\mu(\tilde{\boldsymbol{x}}^*) = \min_{1\le i\le m_1} \mu_{A_i}(F_i(\tilde{\boldsymbol{x}}^*))$ is defined as the satisfactoriness of non-inferior solution $\tilde{\boldsymbol{x}}^*$ to all objectives.

Definition 4.5. $\mu(\tilde{\boldsymbol{x}}^*) = \min_{1\le i\le m_1} \mu_{A_i}(F_i(\tilde{\boldsymbol{x}}^*))$ is defined as the satisfactoriness of non-inferior solution $\tilde{\boldsymbol{x}}^*$ to all objectives.

For s_0 given in advance by the upper-level decision maker, if non-inferior solution $\tilde{\boldsymbol{x}}^*$ satisfies $\mu_{\tilde{\boldsymbol{x}}^*} \ge s_0$, then $\tilde{\boldsymbol{x}}^*$ is the preferred solution corresponding to the satisfactoriness s_0.

We give the membership function $\mu_{A_i}(F_i(\tilde{x}))$ as below:

$$\mu_{A_i}(F_i(\tilde{x})) = 1 - \frac{U_i - F_i(\tilde{x})}{U_i - L_i}. \tag{4.19}$$

It is decided according to the decision maker's requirements. Obviously, (4.19) meets the requirements of (i) and (ii) for $\mu_{A_i}(F_i(\tilde{x}))$.

The ε-constraint method is effective for solving multi-objective decision-making problems. The formalization $P(\varepsilon - 1)$ is given as follows:

$$\begin{cases} \max F_1(\tilde{x}) \\ \text{s.t.} \begin{cases} F_i(\tilde{x}) \leq \varepsilon_i, i = 2, 3, \cdots, m_1 \\ \tilde{x} \in \Omega. \end{cases} \end{cases} \tag{4.20}$$

Assume: $\varepsilon_{-1} = (\varepsilon_2, \varepsilon_3, \cdots, \varepsilon_{m_1}), x'(\varepsilon_{-1}) = \{\tilde{x}, F_i(\tilde{x}) \leq \varepsilon_i, i = 2, 3, \cdots, m_1, \tilde{x} \in \Omega\}$ and $E_{-1} = \{\varepsilon_{-1} | x'(\varepsilon_{-1}) \neq \emptyset\}$.

Theorem 4.6 ([5]). *If* $\varepsilon_{-1} = (\varepsilon_2, \varepsilon_3, \cdots, \varepsilon_{m_1}) \in E_1$, *then the optimal solution to* $P(\varepsilon_{-1})$ *exits and includes the non-inferior solution of (4.20).*

Corollary, if x^1 is the only optimal solution to $P(\varepsilon_{-1})$, then x^1 is the non-inferior solution of (4.18).

Given satisfactoriness s, if $\mu_{A_i}(F_i(\tilde{x})) \geq s$, then by solving (4.19), we obtain:

$$_i(\tilde{x}) = (U_i - L_i))\mu_{A_i}(F_i(\tilde{x})) + L_i \geq (U_i - L_i)s + L_i.$$

Let:

$$\Delta_i = (U_i - L_i)s + L_i, i = 1, 2, \cdots, m_1,$$
$$\varepsilon_{-1}(s) = (\Delta_1, \Delta_2, \cdots, \Delta_{m_1}).$$

Therefore, we can obtain $P(\varepsilon_{-1}(s))$, the o-constraint problem including a satisfactoriness value:

$$\begin{cases} \max F_1(\tilde{x}) \\ \text{s.t.} \begin{cases} F_i(\tilde{x}) \geq \Delta_i, i = 2, 3, \cdots, m_1 \\ \tilde{x} \in \Omega. \end{cases} \end{cases} \tag{4.21}$$

Theorem 4.7. *If* $P(\varepsilon_{-1}(s))$ *has no solution or has the optimal solution* $\tilde{x}$ *and* $F_1(\tilde{x}) \leq \Delta_i$, *then no non-inferior solution* $\tilde{x}^*$ *exists, such that* $\mu(\tilde{x}^*) \geq s$.

Proof. If $\tilde{x}^*$ is a non-inferior solution of (4.18), such that $\mu(\tilde{x}^*) \geq s$, namely:

$$\mu_{A_i}(F_i(\tilde{x})) \geq s, i = 2, 3, \cdots, m_1.$$

Then $\tilde{\boldsymbol{x}}^*$ is a feasible solution of $P(\varepsilon_{-1}(s))$, and $F_1(\tilde{\boldsymbol{x}}) \geq \Delta_1$. Therefore, $P(\varepsilon_{-1}(s))$ has optimal solution $\bar{\boldsymbol{x}}$, such that $F_1(\bar{\boldsymbol{x}}) \geq F(\tilde{\boldsymbol{x}}^*) \geq \Delta_1$., which is in contradiction to the hypothesis.

Corollary, if $\tilde{\boldsymbol{x}}^*$ is a non-inferior solution of (4.18) such that $\mu(\tilde{\boldsymbol{x}}^*) \geq s$, then $\Delta_i \leq F(\tilde{\boldsymbol{x}}^*) \leq b_i, i = 1, 2, \cdots, m_1$.

Assume $s < s_1$, if there is no preferred solution to s, then so to s_1.

Theorem 4.8. *Assume $\bar{\boldsymbol{x}}$ is an optimal solution of $P(\varepsilon_{-1}(s))$ and $F_i(\bar{\boldsymbol{x}} \geq \Delta_i(i = 1, 2, \cdots, m_1)$, and $\varepsilon_{-1} = (\varepsilon_2, \varepsilon_3, \cdots, \varepsilon_{m_1})$, then $\bar{\boldsymbol{x}}$ is still an optimal solution of $P(\varepsilon_{-1})$.*

(a) If $\bar{\boldsymbol{x}}$ is the only optimal solution of $P(\varepsilon_{-1})$, then $\bar{\boldsymbol{x}}$ is non-inferior solution.
(b) If other optimal solution x' of $P(\varepsilon_{-1})$ exists, and $i_0 \in \{i = 1, 2, \cdots, m_1\}$ such that $F_{i_0}(x') \geq \varepsilon$, then $\bar{\boldsymbol{x}}$ is inferior solution.

Proof.

(a) $\varepsilon_i \geq \Delta_i(i = 1, 2, \cdots, m_1)$, namely:

$$\varepsilon_{-1} = (\varepsilon_2, \varepsilon_3, \cdots, \varepsilon_{m_1}) \geq (\Delta_2, \Delta_3, \cdots, \Delta_{m_1}), x'(\varepsilon_{-1}) \subset' (\varepsilon_{-1}(s)).$$

Let $\bar{x}$ to be an optimal solution of $P(\varepsilon_{-1}(s))$ and $\bar{x} \in x'(\varepsilon_{-1})$, then:

$$\max_{x \in x'(\varepsilon_{-1})(s)} F_1(\tilde{\boldsymbol{x}}) = \max_{x \in x'(\varepsilon_{-1})} F_1(\tilde{\boldsymbol{x}}).$$

Therefore, $\bar{x}$ is an optimal solution of $P(\varepsilon_{-1})$ and (a) is proven by the corollary of Theorem 4.6.

(b) $F_1(\bar{x}) = F_1(x')$, and $F_i(x') = F_i(\bar{x}) = \varepsilon_i$, which '>' holds when $i = i_0$, therefore, $\bar{x}$ is an inferior solution.

According to the definitions and theorems above, the steps to solve (4.18) are given as follows.

Step 1. Set the satisfactoriness s. Let $s = s_0$ at the beginning, and let $s = s_1, s_2, \cdots$,, respectively.
Step 2. Set the o-constraint problem $P(\varepsilon_{-1}(s))$. If $P(\varepsilon_{-1}(s))$ has no solution or has an optimal solution making $F_1(\bar{x}) < \Delta_i$, then, go to Step 1, and to adjust $s = s_{k+1} < s_k$. Otherwise, go to Step 3.
Step 3. Assume that $\bar{x}$ is an optimal solution to $P(\varepsilon_{-1}(s))$, judge by Theorem 4.8 whether or not $\bar{x}$ is a non-inferior solution to (4.18). If $\bar{x}$ is, turn to Step 4. If it is inferior, then there must be a $\bar{\bar{x}}$, such that $F_i(\bar{\bar{x}}) \geq F_i(\bar{x})$, and at least one ' >'. Repeat Step 3 with $\bar{\bar{x}}$ (also an optimal solution to $P(\varepsilon_{-1}(s))$.
Step 4. If the decision maker is satisfied with $\bar{x}$, then $\bar{x}$ is a preferred solution. Otherwise, go to Step 5.
Step 5. Adjust the satisfactoriness. Let $s = s_{k+1} < s_k$ and do Step 2.

A note to the steps above: The upper-level decision maker may achieve a set of acceptable preferred solutions according to different satisfactoriness, and put the solutions in rank order referring to the satisfactoriness. Therefore, the rule of thumb is to select the smaller satisfactoriness s first, and then to increase s, respectively, until the preferred solution is achieved.

Change Bi-Level Model into Singular Level Model

Suppose that upper-level decision maker presents s_k as his or her satisfactoriness, and the lower-level decision maker presents s', as his or her satisfactoriness, the bi-level model is given as follows:

$$\begin{cases} \max[F_1(\boldsymbol{x},\boldsymbol{y}), F_2(\boldsymbol{x},\boldsymbol{y}) \cdots, F_{m_1}(\boldsymbol{x},\boldsymbol{y})] \\ \text{s.t.} \begin{cases} F_i(\boldsymbol{x},\boldsymbol{y}) \geq \Delta_i(s_k), i = 1, 2, \cdots, m_1 \\ (\boldsymbol{x},\boldsymbol{y}) \in \Omega_1 \\ \text{where } \boldsymbol{y} \text{ solves:} \\ \begin{cases} \max[f_1(\boldsymbol{x},\boldsymbol{y}), f_2(\boldsymbol{x},\boldsymbol{y}) \cdots, f_{m_2}(\boldsymbol{x},\boldsymbol{y})] \\ \text{s.t.} \begin{cases} f_i(\boldsymbol{x},\boldsymbol{y}) \geq \Delta_i(s'_k), i = 1, 2, \cdots, m_2 \\ (\boldsymbol{x},\boldsymbol{y}) \in \Omega_2, \end{cases} \end{cases} \end{cases} \end{cases} \tag{4.22}$$

where $\Delta_i(s_k) = (U_i - L_i)s_k + L_i, L_i = \min_{\bar{x}\in\Omega_1} F_i(\bar{x}), U_i = \min_{\bar{x}\in\Omega_1} F_i(\bar{x}), i = 1, 2, \cdots, m_1$. $\Delta_i(s'_k) = (u_i - l_i)s_k + l_i, l_i = \min_{\bar{x}\in\Omega_1} f_i(\bar{x}), u_i = \min_{\bar{x}\in\Omega_2} f_i(\bar{x}), i = 1, 2, \cdots, m_2$.

Using the KKT condition, then (4.20) becomes:

$$\begin{cases} \max[F_1(\boldsymbol{x},\boldsymbol{y}), F_2(\boldsymbol{x},\boldsymbol{y}) \cdots, F_{m_1}(\boldsymbol{x},\boldsymbol{y})] \\ \text{s.t.} \begin{cases} F_i(\boldsymbol{x},\boldsymbol{y}) \geq \Delta_i(s_k), i = 1, 2, \cdots, m_1 \\ (\boldsymbol{x},\boldsymbol{y}) \in \Omega_1 \\ \nabla_{y_1} f_1(\boldsymbol{x},\boldsymbol{y}) + \sum_{i=1}^{m_2} u^{(i)} \nabla_{y_1} h^{(i)}(\boldsymbol{x},\boldsymbol{y}) = 0 \\ u^{(i)} h^{(i)}(\boldsymbol{x},\boldsymbol{y}) = 0 \\ u^{(i)} \geq 0, i = 1, 2, \cdots, m_2. \end{cases} \end{cases} \tag{4.23}$$

Let $Z = (\boldsymbol{x},\boldsymbol{y},\boldsymbol{u}), \boldsymbol{u} = (u^{(1)}, u^{(2)}, \cdots, u^{(m_1)})$. Then (4.23) becomes:

$$\max_{Z\in\Omega} F_1(Z). \tag{4.24}$$

The constrained variable metric method can be used to solve the optimization problem (4.24).

The constrained variable metric method is one of the most effective existing methods to solve constraint non-linear optimization problems. Like many of the other methods, it consists of two steps: determine the direction P and the step length moving along P.

Solving the following programming problem (HQP) we may determine P:

$$\begin{cases} \max \nabla F_1(Z^{(k)})^T P + \frac{1}{2} P^T H^{(K)} P \\ \text{s.t.} \begin{cases} J_i(Z^{(k)} + \nabla J_i Z^{(k)})^T P \geq 0 \\ h_j(Z^{(k)} + \nabla h_j Z^{(k)})^T P \geq 0, \end{cases} \end{cases} \tag{4.25}$$

where $H^{(k)}$ is a metric matrix, the initial value is $H^{(0)} = I$:

$$H^{(k+1)} = H^{(k)} - \frac{H^{(k)} \partial \partial^T H^{(k)}}{\partial^T H^{(k)} \partial} + \frac{\omega \omega^T}{\partial^T \omega}, \tag{4.26}$$

$$\omega = \theta R + (1 - \theta) H^{(k)} \partial, \tag{4.27}$$

$$\theta = \begin{cases} 1, & \partial^T R \geq 0.2 \partial^T H^{(k)} \partial \\ \dfrac{0.8 \partial^T H^{(k)} \partial}{\partial^T H^{(k)} \partial - \partial^T R}, & \partial^T R < 0.2 \partial^T H^{(k)} \partial, \end{cases} \tag{4.28}$$

$$R = \nabla_Z L(Z^{(k+1)}, \lambda^{(k+1)}, y^{(k+1)}) - \nabla_Z L(Z^{(k)}, \lambda^{(k+1)}, y^{(k+1)}). \tag{4.29}$$

L is the generalized Lagrange function of (4.24).

$$\partial = Z^{(k+1)} - Z^{(k)}. \tag{4.30}$$

It may be proved that $Z^{(k)}$ is the Kuhn-Tucker point of (4.24), if $P^* = 0, Z^{(k)}$ is the optimal solution to HQP:

$$P(Z, M) = F_1(Z) + M \left\{ \sum_{i=1}^{l} \max\{0, g_i(Z)\} + \sum_{i=1}^{m} h_j(Z) \right\}. \tag{4.31}$$

To achieve α_k, so that:

$$P(Z(\alpha_k)) = \max_{0 \leq \alpha \leq \beta} P(Z^{(k)} + \alpha p^{(k)}, m), \tag{4.32}$$

where m and β are two positive numbers that are selected.

The algorithm for solving (4.24) is given below:

Step 1. Give the original point $Z^{(0)}$ and $H^{(0)}, k = 0$.
Step 2. Solve the direction sub-problem HQP, to get $P^{(k)}$.
Step 3. According to Eq. (4.30), get α_k.
Step 4. End criteria check.
Step 5. Work out $H^{(k+1)}$ with the help of (4.26), (4.27), (4.28), (4.29), and (4.30), let $k = k + 1$, and go to Step 2.

Algorithm

Thus, the algorithm is presented as follows:
Step 1. Lower-level decision maker presents his smallest acceptable satisfactoriness, s', set $k = 1$.
Step 2. Solve the upper level decision-making problem to obtain a set of preferred solutions that are acceptable to the upper-level decision maker.
Step 3. Put $(x(k), y(k))$ into (4.23) to determine whether it satisfies. Otherwise, go to Step 5.
Step 4. Judge by Theorem 4.8 whether or not $(x(k), y(k)))$ is the non-inferior solution of the lower-level decision maker . If it is, then $(x(k), y(k)))$ is the final preferred solution to the bi-level problem. Otherwise, turn to Step 6.
Step 5. Let $k = 1$, and go to Step 3.
Step 6. The upper-level decision maker selects a satisfactoriness value, $s^{(k)^*} = s^{(k+1)} + \nabla d^{(k)}$. Then, put the satisfactoriness into (4.23), take $((x(k), y(k)))$ as the original point, and follow the Constrained variable metric method to work out the solution corresponding to $s^{(k)''}$, i.e. $((x(k'), y(k')))$.
Step 7. Judge by Theorem 4 whether or not $((x(k'), y(k')))$ is a non-inferior solution for both the upper-level decision maker and lower-level decision maker. If it is, then $((x(k'), y(k')))$ is the final preferred solution to the bi-level problem. Otherwise, turn to Step 5.

Notes to the algorithm above:

(1) In Step 1, the BMDMP problem is a leaderC follower Stackelberg game. The upper-level decision maker has authoritative power to require the lower-level decision makers to try their best solutions to meet upper-level decision maker's objectives. We assume that the lower-level decision makers meet the upper-level decision maker's objectives by presenting their smallest acceptable satisfactoriness..
(2) In Step 2, the upper-level decision maker determines a preferred solution with a satisfactoriness value, and arranges the preferred solution in rank order referring to the satisfactoriness.
(3) In Step 6, the lower-level Kuhn-Tucker condition gives the Kuhn-Tucker multiple $u^{(k)}$. It is difficult to get the global optimal solution to (4.24), a non-convex optimization problem, the existing methods to get global optimization solution are mostly directed to a certain particular form. Trying to avoid this difficulty, we take $(x^{(k)}, y^{(k)}, u^{(k)})$. as the original point, select $s^{(k+1)}$, adopt the constrained variable metric method to achieve the local optimal solution near $(x^{(k)}, y^{(k)}, u^{(k)})$, and then apply judgement on the solution by Step 7.

Numerical Examples

In this section, a numerical example is given to illustrate the effectiveness of the application of the models and algorithms proposed above.

Example 4.1. Consider the following problem

$$\begin{cases} \max_{x_1,x_2} F_1 = \tilde{\bar{\xi}}_1 x_1 - \tilde{\bar{\xi}}_2 x_2 + \tilde{\bar{\xi}}_3 y_1 + \tilde{\bar{\xi}}_4 y_2 \\ \max_{x_1,x_2} F_2 = \tilde{\bar{\xi}}_5 x_1 + \tilde{\bar{\xi}}_6 x_2 - \tilde{\bar{\xi}}_7 y_1 + \tilde{\bar{\xi}}_8 y_2 \\ \text{where } (y_1, y_2) \text{ solves} \\ \begin{cases} \max_{y_1,y_2} f_1 = 7\tilde{\bar{\xi}}_1 x_1 + 5\tilde{\bar{\xi}}_2 x_2 + 4\tilde{\bar{\xi}}_3 y_1 + \tilde{\bar{\xi}}_4 y_2 \\ \max_{y_1,y_2} f_2 = 3\tilde{\bar{\xi}}_5 x_1 + 4\tilde{\bar{\xi}}_6 x_2 + 5\tilde{\bar{\xi}}_7 y_1 + \tilde{\bar{\xi}}_8 y_2 \\ \text{s.t.} \begin{cases} x_1 + x_2 + y_1 + y_2 \geq 120 \\ x_1 + x_2 + y_1 + y_2 \leq 200 \\ x_1, x_2, y_1, y_2 \geq 0, \end{cases} \end{cases} \end{cases} \tag{4.33}$$

where $\xi_j, j = 1, 2, \cdots, 8$ are Ra-Fu variables characterized as

$$\begin{array}{ll} \tilde{\bar{\xi}}_1 \sim \mathcal{N}(\tilde{u}_1, 1), \text{ with } \tilde{u}_1 \sim (3, 6, 8, 11), & \tilde{\bar{\xi}}_2 \sim \mathcal{N}(\tilde{u}_2, 1), \text{ with } \tilde{u}_2 \sim (7, 10, 11, 16), \\ \tilde{\bar{\xi}}_3 \sim \mathcal{N}(\tilde{u}_3, 1), \text{ with } \tilde{u}_3 \sim (12, 16, 28, 40), & \tilde{\bar{\xi}}_4 \sim \mathcal{N}(\tilde{u}_4, 1), \text{ with } \tilde{u}_4 \sim (5, 20, 40, 63), \\ \tilde{\bar{\xi}}_5 \sim \mathcal{N}(\tilde{u}_5, 1), \text{ with } \tilde{u}_5 \sim (12, 15, 23, 38), & \tilde{\bar{\xi}}_6 \sim \mathcal{N}(\tilde{u}_6, 1), \text{ with } \tilde{u}_6 \sim (8, 16, 20, 36), \\ \tilde{\bar{\xi}}_7 \sim \mathcal{N}(\tilde{u}_7, 1), \text{ with } \tilde{u}_7 \sim (4, 12, 16, 20), & \tilde{\bar{\xi}}_8 \sim \mathcal{N}(\tilde{u}_8, 1), \text{ with } \tilde{u}_8 \sim (5, 25, 50, 100), \end{array}$$

Assume that all the confidence levels are 0.8, then he EECC model for (4.33) is

$$\begin{cases} \max_{x_1,x_2} E[F_1] = E[\tilde{\bar{\xi}}_1 x_1 - \tilde{\bar{\xi}}_2 x_2 + \tilde{\bar{\xi}}_3 y_1 + \tilde{\bar{\xi}}_4 y_2] \\ \max_{x_1,x_2} E[F_2] = E[\tilde{\bar{\xi}}_5 x_1 + \tilde{\bar{\xi}}_6 x_2 - \tilde{\bar{\xi}}_7 y_1 + \tilde{\bar{\xi}}_8 y_2] \\ \text{where } (y_1, y_2) \text{ solves} \\ \begin{cases} \max_{y_1,y_2} [\bar{f}_1, \bar{f}_2] \\ \text{s.t.} \begin{cases} Pos\{Pr\{7\tilde{\bar{\xi}}_1 x_1 + 5\tilde{\bar{\xi}}_2 x_2 + 4\tilde{\bar{\xi}}_3 y_1 + \tilde{\bar{\xi}}_4 y_2 \geq \bar{f}_1\} \geq 0.8\} \geq 0.8 \\ Pos\{Pr\{3\tilde{\bar{\xi}}_5 x_1 + 4\tilde{\bar{\xi}}_6 x_2 + 5\tilde{\bar{\xi}}_7 y_1 + \tilde{\bar{\xi}}_8 y_2 \geq \bar{f}_2\} \geq 0.8\} \geq 0.8 \\ x_1 + x_2 + y_1 + y_2 \geq 120 \\ x_1 + x_2 + y_1 + y_2 \leq 200 \\ x_1, x_2, y_1, y_2 \geq 0. \end{cases} \end{cases} \end{cases} \tag{4.34}$$

Since $\Phi^{(-1)}(1-0.8) = -0.842$, then it follows from Theorems 4.1 to 4.4 that (4.34) is equivalent to

$$
\begin{cases}
\max\limits_{x_1,x_2} 7x_1 - 11x_2 + 24y_1 + 32y_2 \\
\max\limits_{x_1,x_2} 22x_1 + 80x_2 + 13y_1 + 45y_2 \\
\text{where } (y_1, y_2, y_3) \text{ solves} \\
\begin{cases}
\max\limits_{y_1,y_2} -0.842\sqrt{49x_1^2 + 25x_2^2 + 16y_1^2 + y_2^2} + 48.2x_1 + 60x_2 + 121.6y_1 + 44.6y_2 \\
\max\limits_{y_1,y_2} -0.842\sqrt{9x_1^2 + 16x_2^2 + 25y_1^2 + y_2^2} + 78x_1 + 92.8x_2 + 84y_1 + 60y_2 \\
\text{s.t.} \begin{cases} x_1 + x_2 + y_1 + y_2 \geq 120 \\ x_1 + x_2 + y_1 + y_2 \leq 200 \\ x_1, x_2, y_1, y_2 \geq 0. \end{cases}
\end{cases}
\end{cases}
\tag{4.35}
$$

According to the algorithm in the section "Constrained Variable Metric Method", to solve this problem, when upper-level decision maker's satisfactoriness is 0.82, lower-level decision maker's satisfactoriness is 0.45 the preferred solution of the example is $(x_1^*, x_2^*, y_1^*, y_2^*,) = (26.9, 65.5, 55.2, 52.4)$. Correspondingly, the values of objectives are $(F_1^*, F_2^*) = (2469.4, 8907.4)$, $(f_1^*, f_2^*) = (13904.88, 15626.89)$.

4.3.3 Ra-Fu Simulation-Based PGSA for Nonlinear Model

If (4.2) is a nonlinear model, i.e., some functions in model (4.2) is nonlinear, the Ra-Fu features can not be converted into its crisp equivalent model. To solve the nonlinear (4.2), three issues are needed to solve: the trade-off of the multiple objectives, the Ra-Fu parameters and the bi-level structure. If the decision variables are integer, the bi-level Plant Growth Simulation Algorithm (PGSA) can be applied to solve the model (4.2). In this section, we design a hybrid method consisting of linear weighted-sum approach, Ra-Fu simulations for expected operator and chance operator, and PGSA.

Linear Weighted-Sum Approach

Without loss of generality, both of multiple objective models for upper model and lower model of (4.2) are written as

$$
\max_{x \in X}[f_1(\boldsymbol{x}), f_2(\boldsymbol{x}), \cdots, f_m(\boldsymbol{x})]. \tag{4.36}
$$

Assign a group of nonnegative coefficients $(w_1, w_2, \cdots, w_m)$ to each objective's function, where w_i can be interpreted as the relative emphasis or worth of that objective compared to other objectives. In other words, the weight can be interpreted as representing our preference over objectives. The bigger w_i is, the more important the objectives f_i is; The smaller w_i is, the less important the objectives f_i is. Then combine them into a single objective function $\sum_{i=1}^m w_i f_i$. We called these nonnegative coefficient *weight*. Sometimes, we require that $\sum_{i=1}^{m} w_i = 1$ to normalize it. The vector comprised by weight coefficients $w = (w_1, \cdots, w_m)$ is called it weigh vector. This approach is called linear weighted-sum approach for its characteristic. The procedure of linear weighted-sum approach is summarized as follows:

Step 1. Assign weight coefficients. Assign weight coefficients $w_1, \cdots, w_m$ to objectives according the degree of importance of the objective in problem. It is required that $w_i \geq 0$ and $\sum_{i=1}^{m} w_i = 1$.

Step 2. Solve the single objective programming problem, in which the objective is the weighted-sum objective function. Report the optimal solution as the result. The Linear weighted-sum approach can be represented as

$$\max_{x \in X} \sum_{i=1}^{m} w_i f_i(x). \tag{4.37}$$

The following theorem guarantees that the optimal solution of problem (4.37) is an efficient solution to the problem (4.36).

Theorem 4.9. *Assume that* $w_i > 0, i = 1, 2, \cdots, m$. *If* $\boldsymbol{x}^* \in X$ *is an optimal solution (4.37), then it is an efficient solution of problem (4.36).*

Proof. Suppose that $\boldsymbol{x}^*$ is not an efficient solution to problem (4.36). Then it follows from the definition of an efficient solution that there exists $\mathbf{x} \in X$ such that $f_i(\boldsymbol{x}) \geq f_i(\boldsymbol{x}^*)$ for $i = 1, 2, \cdots, n$ with strict inequality holding for at least one i. Observe that $w_i > 0$, then

$$w_i f_i(\boldsymbol{x}) \geq w_i f_i(\boldsymbol{x}^*),$$

for $i = 1, 2, \cdots, n$ with strict inequality holding for at least one i. Sum the above inequalities, and we have

$$\sum_{i=1}^{m} w_i f_i(\boldsymbol{x}) > \sum_{i=1}^{m} w_i f_i(\boldsymbol{x}^*),$$

which contradicts that $\boldsymbol{x}^* \in X$ is a optimal solution to (4.37). Thus, $\boldsymbol{x}^*$ is an efficient solution to problem (4.36). This proof is complete.

By changing $\mathbf{w}$, we can obtain a set composed of the efficient solutions of the problem (4.36) by solving the problem (4.37).

Ra-Fu Simulation for Expected Operator and Chance Operator

The Ra-Fu simulation is used to deal with those which cannot be converted into crisp ones. Next, let's introduce the process of the Ra-Fu simulation dealing with the expected value models.

Assume that $\boldsymbol{\xi}$ is an n-dimensional Ra-Fu vector defined on the possibility space $(\Theta, \mathscr{P}(\Theta), Pos)$, and $f : \mathbf{R}^n \to \mathbf{R}$ is a measurable function. One problem is to calculate the expected value $E[f(\boldsymbol{x}, \boldsymbol{\xi})]$ for given $\boldsymbol{x}$. Then $f(\boldsymbol{x}, \boldsymbol{\xi})$ is a Ra-Fu variable whose expected value $E[f(\boldsymbol{x}, \boldsymbol{\xi})]$ is

$$\int_0^{+\infty} Cr\{\theta \in \Theta | E[f(\boldsymbol{x}, \boldsymbol{\xi}(\theta))] \geq r\}dr - \int_{-\infty}^{0} Cr\{\theta \in \Theta | E[f(\boldsymbol{x}, \boldsymbol{\xi}(\theta))] \leq r\}dr.$$

A Ra-Fu simulation will be introduced to compute the expected value $E[f(\boldsymbol{x}, \boldsymbol{\xi})]$. We randomly sample θ_k from Θ such that $Pos\{\theta_k\} \geq \varepsilon$, and denote $v_k = Pos\{\theta_k\}$ for $k = 1, 2, \cdots, N$, where ε is a sufficiently small number. Then for any number $r \geq 0$, the credibility $Cr\{\theta \in \Theta | E[f(\boldsymbol{x}, \boldsymbol{\xi}(\theta))] \geq r\}$ can be estimated by

$$\frac{1}{2}\left(\max_{1\leq k\leq N}\{v_k | E[f(\boldsymbol{x}, \boldsymbol{\xi}(\theta_k))] \geq r\} + \min_{1\leq k\leq N}\{1 - v_k | E[f(\boldsymbol{x}, \boldsymbol{\xi}(\theta_k))] < r\}\right)$$

and for any number $r < 0$, the credibility $Cr\{\theta \in \Theta | E[f(\boldsymbol{x}, \boldsymbol{\xi}(\theta))] \leq r\}$ can be estimated by

$$\frac{1}{2}\left(\max_{1\leq k\leq N}\{v_k | E[f(\boldsymbol{x}, \boldsymbol{\xi}(\theta_k))] \leq r\} + \min_{1\leq k\leq N}\{1 - v_k | E[f(\boldsymbol{x}, \boldsymbol{\xi}(\theta_k))] > r\}\right)$$

provided that N is sufficiently large, where $E[f(\boldsymbol{x}, \boldsymbol{\xi}(\theta_k))], k = 1, 2, \cdots, N$ may be estimated by the stochastic simulation.

Then the procedure simulating the expected value of the function $f(\boldsymbol{x}, \boldsymbol{\xi})$ can be summarized as follows:

Step 1. Set $e = 0$
Step 2. Randomly sample θ_k from Θ such that $Pos\{\theta_k\} \geq \varepsilon$ for $k = 1, 2, \cdots, N$, where ε is a sufficiently small number
Step 3. Let $a = \min_{1\leq k\leq N} E[f(\boldsymbol{x}, \boldsymbol{\xi}(\theta_k))]$ and $b = \max_{1\leq k\leq N} E[f(\boldsymbol{x}, \boldsymbol{\xi}(\theta_k))]$
Step 4. Randomly generate r from $[a, b]$
Step 5. If $r \geq 0$, then $e \leftarrow e + Cr\{\theta \in \Theta | E[f(\boldsymbol{x}, \boldsymbol{\xi}(\theta_k))] \geq r\}$
Step 6. If $r < 0$, then $e \leftarrow e - Cr\{\theta \in \Theta | E[f(\boldsymbol{x}, \boldsymbol{\xi}(\theta_k))] \leq r\}$
Step 7. Repeat the fourth to sixth steps for N times
Step 8. $E[f(\boldsymbol{x}, \boldsymbol{\xi})] = a \vee 0 + b \wedge 0 + e \cdot (b - a)/N$.

Example 4.2. We employ the Ra-Fu simulation to calculate the expected value of $\sqrt{\tilde{\bar{\xi}}_1^2 + \tilde{\bar{\xi}}_2^2 + \tilde{\bar{\xi}}_3^2}$, where $\tilde{\bar{\xi}}_1$, $\tilde{\bar{\xi}}_2$ and $\tilde{\bar{\xi}}_3$ are random fuzzy variables defined as

$$\tilde{\bar{\xi}}_1 \sim \mathcal{N}(\tilde{\rho}_1, 1), \text{ with } \tilde{\rho}_1 = (1, 2, 3),$$
$$\tilde{\bar{\xi}}_2 \sim \mathcal{U}(\tilde{\rho}_2, 2), \text{ with } \tilde{\rho}_2 = (2, 5, 6),$$
$$\tilde{\bar{\xi}}_3 \sim \mathcal{U}(\tilde{\rho}_3, 1), \text{ with } \tilde{\rho}_3 = (2, 3, 4),$$

where $\tilde{\rho}_i$ are all triangular fuzzy numbers. We perform the Ra-Fu simulation with 1000 cycles and obtain that

$$E\left[\sqrt{\tilde{\bar{\xi}}_1^2 + \tilde{\bar{\xi}}_2^2 + \tilde{\bar{\xi}}_3^2}\right] = 6.0246.$$

Next, the Ra-Fu simulation technique is used to handle chance operator. Assume that $\boldsymbol{\xi}$ is an n-dimensional Ra-Fu vector defined on the possibility space $(\Theta, \mathcal{P}(\Theta), Pos)$, and $f : \mathbf{R}^n \to \mathbf{R}$ is a measurable function. For any given confidence level α and β, we need to design the Ra-Fu simulation to find the maximal value $\bar{f}$ such that

$$Ch\{f(\boldsymbol{x}, \boldsymbol{\xi}) \geq \bar{f}\}(\alpha) \geq \beta$$

holds. That is, we must find the maximal value $\bar{f}$ such that

$$Pos\{\theta | Pr\{f(\boldsymbol{x}, \boldsymbol{\xi}(\theta)) \geq \bar{f}\} \geq \beta\} \geq \alpha.$$

We randomly generate θ_k from Θ such that $Pos\{\theta_k\} \geq \varepsilon$, and write $v_k = Pos\{\theta_k\}, k = 1, 2, \cdots, N$, respectively, where ε is a sufficiently small number. For any number θ_k, we search for the maximal value $\bar{f}(\theta_k)$ such that $Pr\{f(\boldsymbol{x}, \boldsymbol{\xi}(\theta_k)) \geq \bar{f}(\theta_k)\} \geq \beta$ by stochastic simulation. For any number r, we have

$$H(r) = \frac{1}{2}\left(\max_{1 \leq k \leq N}\{v_k | \bar{f}(\theta_k) \geq r\} + \min_{1 \leq k \leq N}\{1 - v_k | \bar{f}(\theta_k) < r\}\right).$$

If follows from monotonicity that we may employ bisection search to find the maximal value r such that $H(r) \geq \alpha$. This value is an estimation of $\bar{f}$. Then the procedure simulating the critical value of $Pos\{\theta | Pr\{f(\boldsymbol{x}, \boldsymbol{\xi}(\theta)) \geq \bar{f}\} \geq \beta\} \geq \alpha$ can be summarized as follows:

Step 1. Generate θ_k from Θ such that $Pos\{\theta_k\} \geq \varepsilon$ for $k = 1, 2, \cdots, N$, where ε is a sufficiently small number
Step 2. Find the maximal value r such that $H(r) \geq \alpha$ holds;
Step 3. Return r.

Example 4.3. In order to find the maximal value $\bar{f}$ such that

$$Ch\left\{\sqrt{\tilde{\bar{\xi}}_1^2 + \tilde{\bar{\xi}}_2^2 + \tilde{\bar{\xi}}_3^2} \geq \bar{f}\right\}(0.8) \geq 0.8,$$

where $\tilde{\bar{\xi}}_1$, $\tilde{\bar{\xi}}_2$ and $\tilde{\bar{\xi}}_3$ are Ra-Fu variables defined as

$$\tilde{\bar{\xi}}_1 \sim \mathcal{N}(\tilde{\rho}_1, 1), \text{ with } \tilde{\rho}_1 = (1,2,3),$$
$$\tilde{\bar{\xi}}_2 \sim \mathcal{N}(\tilde{\rho}_2, 2), \text{ with } \tilde{\rho}_2 = (2,3,4),$$
$$\tilde{\bar{\xi}}_3 \sim \mathcal{N}(\tilde{\rho}_3, 1), \text{ with } \tilde{\rho}_3 = (3,4,5),$$

where $\tilde{\rho}_i (i = 1,2,3)$ are triangular fuzzy numbers. We perform the Ra-Fu simulation with 1000 cycles and obtain that $\bar{f} = 2.5604$.

PGSA

PGSA [26] is based on the plant growth process, where a plant grows a trunk from its root; some branches will grow from the nodes on the trunk; and then some new branches will grow from the nodes on the branches. Such process is repeated, until a plant is formed. Based on an analogy with the plant growth process, an algorithm can be specified where the system to be optimized first "grows" beginning at the root of a plant and then "grows" branches continually until the optimal solution is found [10, 29].

PGSA establishes a probability model by simulating the growth process of plant phototropism [17]. In PGSA, a function $f(Y)$ is introduced to describe the environment of the node Y on a plant. The smaller the value of $f(Y)$ is, the better the environment for the node in growing a new branch will be. The outline of the model is as follows: a plant grows a trunk M from its root B_o. Assuming there are k nodes $B_{M1}, B_{M2}, \cdots, B_{Mk}$ that have a better environment than the root B_o on trunk M, which means the function $f(Y)$ satisfies $f(B_{Mi}) < f(B_o)$. Then morphactin concentrations $P_{M1}, \cdots, P_{Mk}$ of nodes $B_{M1}, \cdots, B_{Mk}$ can be calculated using:

$$P_{Mi} = \frac{f(B_o) - f(B_{Mi})}{\sum_{i=1}^{k} (f(B_o) - f(B_{Mi}))}, \quad i = 1,2,\cdots,k. \tag{4.38}$$

The significance of Eq. (4.38) is that the morphactin concentration in a node not only depends on its environment but also on the environment of the other nodes on the plant.

Obviously, $\sum_{i=1}^{k} P_{Mi} = 1$, so the morphactin concentrations $P_{M1}, P_{M2}, \cdots, P_{Mk}$ of nodes $B_{M1}, B_{M2}, \cdots, B_{Mk}$ form a state space as shown in Fig. 4.3. A random number β is selected in the interval [0, 1]. β is like a ball thrown into the interval

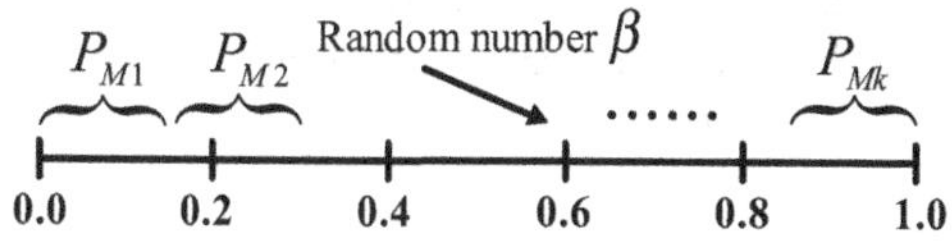

Fig. 4.3 Morphactin concentration state space

[0, 1] and drops into one of $P_{M1}, P_{M2}, \cdots, P_{Mk}$ in Fig. 4.3. The corresponding node called the preferential growth node takes priority in growing a new branch in the next step. For example, B_{Mh} takes priority in growing a new branch if the selected β satisfies:

$$\begin{cases} 0 \le \beta \le \sum_{i=1}^{h} P_{Mi}, & \text{when } h = 1, \\ \sum_{i=1}^{h-1} P_{Mi} < \beta \le \sum_{i=1}^{h} P_{Mi}, & \text{when } h = 2, 3, \cdots, k. \end{cases} \tag{4.39}$$

For instance, if $\sum_{i=1}^{2} P_{Mi} < \beta \le \sum_{i=1}^{3} P_{Mi}$, namely the random number β drops into P_{M3}, node B_{M3} grows a new branch m. Assuming there are q nodes $B_{m1}, B_{m2}, \cdots, B_{mq}$ on branch m, which have a better environment than the root B_o and the corresponding morphactin concentrations are $P_{m1}, P_{m2}, \cdots, P_{mq}$. Therefore, not only the morphactin concentrations of the nodes on branch m need to be calculated, but also the morphactin concentrations of the nodes on trunk M need to be recalculated after the growth of branch m. This calculation is done using Eq. (4.40) which is derived from Eq. (4.38) by adding the nodes on branch m:

$$\begin{aligned} P_{Mi} &= \frac{f(B_o) - f(B_{Mi})}{\sum_{i=1}^{k} (f(B_o) - f(B_{Mi})) + \sum_{j=1}^{q} (f(B_o) - f(B_{mj}))}, i = 1, 2, \cdots, k \text{ and} \\ P_{mj} &= \frac{f(B_o) - f(B_{mj})}{\sum_{i=1}^{k} (f(B_o) - f(B_{Mi})) + \sum_{j=1}^{q} (f(B_o) - f(B_{mj}))}, \quad j = 1, 2, \cdots, q. \end{aligned} \tag{4.40}$$

Obviously, $\sum_{i=1}^{k} P_{Mi} + \sum_{j=1}^{q} P_{mj} = 1$. Then, the morphactin concentrations of the nodes on trunk M and branch m form a new state space. A new preferential growth node, on which a new branch will grow in the next step, can be gained in a similar way. This process will be repeated until there is no new branch to be grown and the plant is formed eventually. The nodes on the plant represent the possible solutions; $f(Y)$ represents the objective function; the length of the trunk and the branch represents the search domain of possible solutions; the root represents the initial solution; the preferential growth node corresponds to the basic point of the next searching process. PGSA has already been successfully applied in transmission network optimal planning problems [41] and radial distribution system [26].

The advantages with the PGSA is that it treats the objective function and constraints separately, which averts the trouble to determine the barrier factors and makes the increase/decrease of constraints convenient, and that it does not need any external parameters such as crossover rate, mutation rate, etc. It adopts a guiding search direction that changes dynamically as the change of the objective function [29].

In the proposed solving method, solving Ra-Fu multi-objective bi-level model (4.2) is transformed to solve crisp single objective bi-level model by linear weighted-sum approach and Ra-Fu simulation. To handle the bi-level structure, we solve the upper-level and lower-level programming problem interactively by PGSA while determining respectively the decision variables of the upper level ($\boldsymbol{x}$) or the lower-level ($\boldsymbol{y}$). Note that the decision variables have to be integer, otherwise PGSA will not be appropriate to use.

Such that, the decision variables $(\boldsymbol{x}, \boldsymbol{y})$ are integer, and in the range of $[0, Range]$, the overall procedure for solving Ra-Fu multi-objective bi-level model (4.2) is as follows:

Step 1. Input all coefficients' values for model (4.2).
Set initial iteration $\tau = 0$. Initialize upper-level basis point $X^0 = \{x_j^0, j = 1, 2, \cdots, N\}$ in the range of $[0, Range]$.

Step 2. Solve the lower-level programming using the PGSA procedure with upper-level initialized basis point X^0.

Step 2.1. Initialize the lower-level basis point Y^0. Set the appropriate step length λ depending on the size of search region. λ shall be integer.

Step 2.2 Check the feasibility. If the feasibility criterion is met by the lower-level basis point, calculate the corresponding objective value $f'(Y^0)$, in which, $f' = w_1 f_1 + w_2 f_2 + \cdots, w_m f_m$. Let $Y_{\min} = Y^0, f'_{\min} = f'(S^0)$. Otherwise, return to Step 2.1.

Step 2.3. Find the trunk M and each preferential growth node $Y^0_{a1,b1}$ on the trunk in the lower-level programming, $a1 = 1, 2, \cdots, N, b1 = 1, 2, \cdots, m1$. $m1$ is the maximum number of growth nodes on the $a1$th trunk. Then eliminate infeasible growth nodes.

Step 2.4. Calculate the $f'(Y^0_{a1,b1})$ and if $f'(Y^0_{a1,b1}) < f'_{\min}$, replace $f'_{\min} = f'(S^0_{a1,b1})$.

Step 2.5. Calculate the morphactin concentrations

$$\begin{cases} P_{a1,b1} = \dfrac{f'(Y^0) - f'(Y^0_{a1,b1})}{\sum\limits_{a1=1}^{N} \sum\limits_{b1=1}^{m1} \left(f'(Y^0) - f'(Y^0_{a1,b1}) \right)}, & \text{if } f'(Y^0) > f'(Y^0_{a1,b1}) \quad (4.41a) \\ P_{a1,b1} = 0, & \text{otherwise.} \quad (4.41b) \end{cases}$$

Note that $a1, b1$ in Eq. (4.41a) do not contain the growth node when $P_{a1,b1} = 0$.

Step 2.6. Find the new basic node. Select a random number β_0 in the interval [0, 1], and β_0 satisfies

$$\sum_{a1=1}^{N} \sum_{b1=1}^{h1-1} P_{a1,b1} < \beta_0 \le \sum_{a1=1}^{N} \sum_{b1=1}^{h1} P_{a1,b1}, \tag{4.42}$$

then set $Y^0_{a1,h1}$ as the new lower-level basis point.

Step 2.7. Start from the new basis point, find new growth node and calculate the morphactin concentrations by Eq. (4.40). Replace $f'_{\min}$ constantly.

Step 2.8. When $f'_{\min}$ repeats appearing many times (e.g., 50 times), take $f^* = f_{\min}$ as the optimal solution. Otherwise, go back to Step 2.3.

Step 3. Solve the upper-level programming using the same PGSA procedure.

Step 4. When $f'_{\min}$, $F_{\min}$ repeats appearing many times (e.g., 50 times), take $F^* = F_{\min}$, $X^* = X_{\min}$ as the optimal solution. Otherwise, use the $X_{\min}$ as the parameter for lower-level programming and go back to Step 2.

The framework of PGSA is shown in Fig. 4.4.

Numerical Example

In this section, a numerical example is given to illustrate the effectiveness of the application of the models and algorithms proposed earlier.

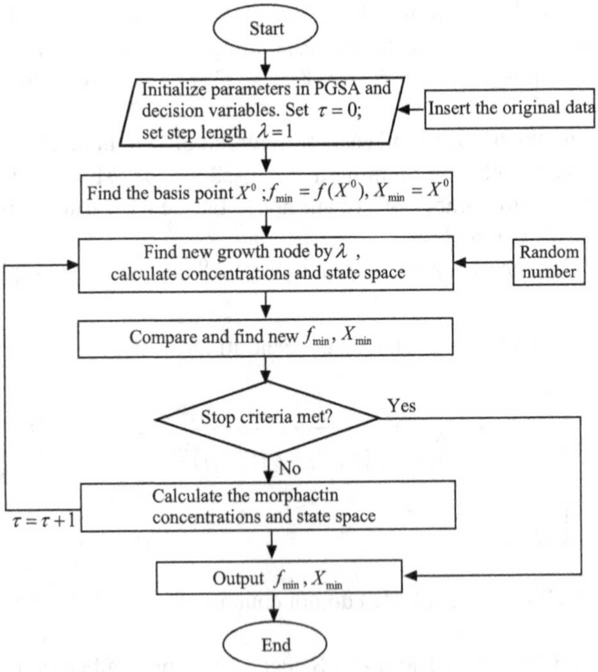

Fig. 4.4 The framework of PGSA

Example 4.4. Consider the problem

$$\begin{cases} \max\limits_{x_1,x_2,x_3} F_1 = 2\xi_1^2 x_1 + \sqrt{\xi_2 x_2 + \xi_3 x_3} + \xi_4 y_1 + \sqrt{\xi_5} y_2 + \sqrt{\xi_6} y_3 \\ \max\limits_{x_1,x_2,x_3} F_2 = \xi_1^2 x_1 + \xi_2^2 x_2 + x_3 + y_1 + y_2 + y_3 \\ \text{where } (y_1, y_2, y_3) \text{ solves} \\ \begin{cases} \max\limits_{y_1,y_2,y_3} f_1 = \sqrt{7\xi_1 x_1 + 3\xi_2 x_2 + 8\xi_3 x_3} + (12\xi_4 y_1 + 11\xi_5 y_2 + 5\xi_6 y_3)^2 \\ \max\limits_{y_1,y_2,y_3} f_2 = x_1 + 5x_2 + 7x_3 + y_1 + 11y_2 - 8\xi_6^3 y_3 \\ \text{s.t.} \begin{cases} \sqrt{x_1\xi_1 + x_2\xi_2} + x_3^2\xi_3 + \xi_4 y_2 + 17y_3\xi_6 \le 1540 \\ 3x_1 + 6x_2 + 12x_3 + 14y_1 + 19y_2 \le 2400 \\ x_1 + x_2 + x_3 + y_1 + y_2 - y_3 \ge 150 \\ x_1 + x_2 + x_3 + y_1 + y_2 + y_3 \le 200 \\ x_1, x_2, x_3, y_1, y_2, y_3 \ge 0, \text{ integer,} \end{cases} \end{cases} \end{cases} \tag{4.43}$$

where $\xi_j, j = 1, 2, \cdots, 6$ are Ra-Fu variables characterized as

$$\begin{aligned} &\tilde{\bar{\xi}}_1 \sim \mathcal{N}(\tilde{u}_1, 1), \text{ with } \tilde{u}_1 \sim (3, 6, 8, 11), && \tilde{\bar{\xi}}_2 \sim \mathcal{N}(\tilde{u}_2, 1), \text{ with } \tilde{u}_2 \sim (7, 10, 11, 16), \\ &\tilde{\bar{\xi}}_3 \sim \mathcal{N}(\tilde{u}_3, 1), \text{ with } \tilde{u}_3 \sim (12, 16, 28, 40), && \tilde{\bar{\xi}}_4 \sim \mathcal{N}(\tilde{u}_4, 1), \text{ with } \tilde{u}_4 \sim (5, 20, 40, 63), \\ &\tilde{\bar{\xi}}_5 \sim \mathcal{N}(\tilde{u}_5, 1), \text{ with } \tilde{u}_5 \sim (12, 15, 23, 38), && \tilde{\bar{\xi}}_6 \sim \mathcal{N}(\tilde{u}_6, 1), \text{ with } \tilde{u}_6 \sim (8, 16, 20, 36). \end{aligned}$$

The EECC model for (4.43) is

$$\begin{cases} \max\limits_{x_1,x_2,x_3} E[2\xi_1^2 x_1 + \sqrt{\xi_2 x_2 + \xi_3 x_3} + \xi_4 y_1 + \sqrt{\xi_5} y_2 + \sqrt{\xi_6} y_3] \\ \max\limits_{x_1,x_2,x_3} F_2 = E[\xi_1^2 x_1 + \xi_2^2 x_2 + x_3 + y_1 + y_2 + y_3] \\ \text{where } (y_1, y_2, y_3) \text{ solves} \\ \begin{cases} \max\limits_{y_1,y_2,y_3} [\bar{f}_1, \bar{f}_2] \\ \max\limits_{y_1,y_2,y_3} f_2 = x_1 + 5x_2 + 7x_3 + y_1 + 11y_2 - 8\xi_6^3 y_3 \\ \text{s.t.} \begin{cases} Pos\{Pr\{\sqrt{7\xi_1 x_1 + 3\xi_2 x_2 + 8\xi_3 x_3} + (12\xi_4 y_1 + 11\xi_5 y_2 + 5\xi_6 y_3)^2 \ge \bar{f}_1\} \\ \ge 0.8\} \ge 0.8 \\ Pos\{Pr\{\sqrt{x_1 + 5x_2 + 7x_3 + y_1 + 11y_2 - 8\xi_6^3 y_3} \ge \bar{f}_2\} \ge 0.8\} \ge 0.8 \\ Pos\{Pr\{x_1 + 5x_2 + 7x_3 + y_1 + 11y_2 - 8\xi_6^3 y_3\} \ge 0.8\} \ge 0.8 \\ 3x_1 + 6x_2 + 12x_3 + 14y_1 + 19y_2 \le 2400 \\ x_1 + x_2 + x_3 + y_1 + y_2 - y_3 \ge 150 \\ x_1 + x_2 + x_3 + y_1 + y_2 + y_3 \le 200 \\ x_1, x_2, x_3, y_1, y_2, y_3 \ge 0, \text{ integer.} \end{cases} \end{cases} \end{cases} \tag{4.44}$$

As a result, a run of the proposed algorithm (5000 cycles in 500 generations in GA, the probability of crossover $P_c = 0.3$, the probability of mutation $P_m = 0.2$) shows that the optimal solution is $(x_1^*, x_2^*, x_3^*, y_1^*, y_2^*, y_3^*) = (10, 27, 12, 21, 11, 0)$.

4.4 Ra-Fu CCDD Model

In this section, we use chance constraints technique and dependent chance to deal with leader's model and follower's model respectively. This kind of model is denoted by CCDD model.

4.4.1 General Form of Ra-Fu CCDD Model

The CCDD model (4.1) is formulated as follows:

$$\begin{cases} \max\limits_{x \in R^{n_1}} [\overline{F}(x,y)] \\ \text{s.t.} \begin{cases} Ch\{F(x,y,\xi) \geq \overline{F}\}(\alpha^U) \geq \beta^U \\ Ch\{G_i(x,y,\xi) \leq 0\}(\delta_i^U) \geq \gamma_i^U, i = 1,2,\cdots,p_1 \\ \text{where } y \text{ solves:} \\ \begin{cases} \max\limits_{y \in R^{n_2}} \alpha^L \\ \text{s.t.} \begin{cases} Ch\{f(x,y,\xi) \geq \bar{f}\}(\alpha^L) \geq \beta^L \\ Ch\{g_j(x,y,\xi) \leq 0\}(\delta_i^L) \geq \gamma_i^L, i = 1,2,\cdots,p_2, \end{cases} \end{cases} \end{cases} \end{cases} \tag{4.45}$$

where $\alpha^U, \beta^U, \delta_i^U, \gamma_i^U, \bar{f}, \beta^L, \delta_i^L$ and γ_i^L are predetermined parameters. Hence the follower need to obtain the (α^L, β^L) optimistic values to the return function $f(x,y,\xi)$.

Similarly, the CCDD models for (4.2) is

$$\begin{cases} \max\limits_{x \in R^{n_1}} [\overline{F}_1(x,y), \overline{F}_2(x,y), \cdots, \overline{F}_{m_1}(x,y)] \\ \text{s.t.} \begin{cases} Ch\{F_i(x,y,\xi) \geq \overline{F}_i\}(\alpha_i^U) \geq \beta_i^U, i = 1,2,\cdots,m_1 \\ Ch\{G_i(x,y,\xi) \leq 0\}(\delta_i^U) \geq \gamma_i^U, i = 1,2,\cdots,p_1 \\ \text{where } y \text{ solves:} \\ \begin{cases} \max\limits_{y \in R^{n_2}} [\alpha_1^L, \alpha_2^L, \cdots, \alpha_{m_2}^L] \\ \text{s.t.} \begin{cases} Ch\{f_i(x,y,\xi) \geq \bar{f}_i\}(\alpha_i^L) \geq \beta_i^L, i = 1,2,\cdots,m_2 \\ Ch\{g_j(x,y,\xi) \leq 0\}(\delta_i^L) \geq \gamma_i^L, i = 1,2,\cdots,p_2. \end{cases} \end{cases} \end{cases} \end{cases} \tag{4.46}$$

Since the there are different chances measures for Ra-Fu events, such as mean chance, equilibrium chance. So we can developed variants of CCDD models according to different chance measures.

This book focus on the case of multiple objectives, thus we present some solution concepts of (4.46) as follows:

Definition 4.6. Let x^* be given by the upper-level decision maker. Then (x^*, y^*) is said to be a feasible solution at (α_i^U, β_i^U)-levels to problem (4.46) if and only if

$(\boldsymbol{x}^*, \boldsymbol{y}^*)$ satisfies $Ch\{F_i(\boldsymbol{x}, \boldsymbol{y}, \boldsymbol{\xi}) \geq \overline{F}_i\}(\alpha_i^U) \geq \beta_i^U, Ch\{G_i(\boldsymbol{x}, \boldsymbol{y}, \boldsymbol{\xi}) \leq 0\}(\delta_i^U) \geq \gamma_i^U$, for each i, and is an efficient solution to the lower-level model.

Remark 4.4. That $(\boldsymbol{x}^*, \boldsymbol{y}^*)$ is an efficient solution to the lower-level model at (α_i, β_i)-levels $(i = 1, 2, \cdots, m_2)$ means $(\boldsymbol{x}^*, \boldsymbol{y}^*)$ satisfies $Ch\{f_i(\boldsymbol{x}, \boldsymbol{y}, \boldsymbol{\xi}) \geq \bar{f}_i\}(\alpha_i^L) \geq \beta_i^L$; $Ch\{g_j(\boldsymbol{x}, \boldsymbol{y}, \boldsymbol{\xi}) \leq 0\}(\delta_i^L) \geq \gamma_i^L$ for each i and there is no other feasible solution $(\boldsymbol{x}, \boldsymbol{y})$ exists, such that $\alpha_i^L(\boldsymbol{x}, \boldsymbol{y}) \geq \alpha_i^L(\boldsymbol{x}^*, \boldsymbol{y}^*)$ and at least one $i = 1, 2, \cdots, m_2$ is strict inequality. .

Definition 4.7. Let $(\boldsymbol{x}^*, \boldsymbol{y}^*)$ be a feasible solution to problem (4.5). Then $(\boldsymbol{x}^*, \boldsymbol{y}^*)$ is the efficient solution to problem (4.5) if and only if there is no other feasible solution $(\boldsymbol{x}, \boldsymbol{y})$ exists, such that $\overline{F}_i(\boldsymbol{x}, \boldsymbol{y}) \geq \overline{F}_i(\boldsymbol{x}^*, \boldsymbol{y}^*)$ and at least one $i = 1, 2, \cdots, m_1$ is strict inequality.

4.4.2 *Fuzzy Decision Method for Linear Model*

If all the functions in (4.46) are linear form, then we have the Ra-Fu linear CCDD (LCCDD) model as follows:

$$\begin{cases} \max\limits_{x \in R^{n_1}} [\overline{F}_1, \overline{F}_2, \cdots, \overline{F}_{m_1}] \\ \text{s.t.} \begin{cases} Ch\{\tilde{\overline{\boldsymbol{C}}}_i^T \boldsymbol{x} + \tilde{\overline{\boldsymbol{D}}}_i^T \boldsymbol{y} \geq \overline{F}_i\}(\alpha_i^U) \geq \beta_i^U, i = 1, 2, \cdots, m_1 \\ Ch\{\tilde{\overline{\boldsymbol{A}}}_i^T \boldsymbol{x} + \tilde{\overline{\boldsymbol{B}}}_i^T \boldsymbol{y} \leq 0\}(\delta_i^U) \geq \gamma_i^U, i = 1, 2, \cdots, p_1 \\ \text{where } \boldsymbol{y} \text{ solves:} \\ \begin{cases} \max\limits_{y \in R^{n_2}} [\alpha_1^L, \alpha_2^L, \cdots, \alpha_{m_2}^L] \\ \text{s.t.} \begin{cases} Ch\{\tilde{\bar{\boldsymbol{c}}}_i^T \boldsymbol{x} + \tilde{\bar{\boldsymbol{d}}}_i^T \boldsymbol{y} \geq \bar{f}_i\}(\alpha_i^L) \geq \beta_i^L, i = 1, 2, \cdots, m_2 \\ Ch\{\tilde{\bar{\boldsymbol{a}}}_i^T \boldsymbol{x} + \tilde{\bar{\boldsymbol{b}}}_i^T \boldsymbol{y} \leq 0\}(\delta_i^L) \geq \gamma_i^L, i = 1, 2, \cdots, p_2. \end{cases} \end{cases} \end{cases} \end{cases} \tag{4.47}$$

If the Ch measure is defined as $Pos - Pr$ measure, then model (4.47) can be rewritten as:

$$\begin{cases} \max\limits_{x \in R^{n_1}} [\overline{F}_1, \overline{F}_2, \cdots, \overline{F}_{m_1}] \\ \text{s.t.} \begin{cases} Pr\{Pos\{\tilde{\overline{\boldsymbol{C}}}_i^T \boldsymbol{x} + \tilde{\overline{\boldsymbol{D}}}_i^T \boldsymbol{y} \geq \overline{F}_i\} \geq \alpha_i^U\} \geq \beta_i^U, i = 1, 2, \cdots, m_1 \\ Pr\{Pos\{\tilde{\overline{\boldsymbol{A}}}_i^T \boldsymbol{x} + \tilde{\overline{\boldsymbol{B}}}_i^T \boldsymbol{y} \leq 0\} \geq \delta_i^U\} \geq \gamma_i^U, i = 1, 2, \cdots, p_1 \\ \text{where } \boldsymbol{y} \text{ solves:} \\ \begin{cases} \max\limits_{y \in R^{n_2}} [\alpha_1^L, \alpha_2^L, \cdots, \alpha_{m_2}^L] \\ \text{s.t.} \begin{cases} Pr\{Pos\{\tilde{\bar{\boldsymbol{c}}}_i^T \boldsymbol{x} + \tilde{\bar{\boldsymbol{d}}}_i^T \boldsymbol{y} \geq \bar{f}_i\} \geq \alpha_i^L\} \geq \beta_i^L, i = 1, 2, \cdots, m_2 \\ Pr\{Pos\{\tilde{\bar{\boldsymbol{a}}}_i^T \boldsymbol{x} + \tilde{\bar{\boldsymbol{b}}}_i^T \boldsymbol{y} \leq 0\} \geq \delta_i^L\} \geq \gamma_i^L, i = 1, 2, \cdots, p_2. \end{cases} \end{cases} \end{cases} \end{cases} \tag{4.48}$$

Crisp Equivalent Models

Due the complexity of (4.48) with bi-level structure, twofold uncertainties, multiple objectives, one way of solving problem (4.48) is to convert the objectives and constraints into their respective crisp equivalents and then reduce the solution complexity to some extent. In order to obtain the crisp equivalent models, some useful result are developed as follows.

Lemma 4.1. *Suppose that* $\overline{\widetilde{C}}_r = (\overline{\widetilde{C}}_{i1}, \overline{\widetilde{C}}_{r2}, \cdots, \overline{\widetilde{C}}_{in})^T$ *is a normally distributed Ra-Fu vector with the fuzzy mean vector* $\widetilde{C}_r = (\widetilde{C}_{i1}, \widetilde{C}_{r2}, \cdots, \widetilde{C}_{in})^T$ *and covariance matrix* V_i^C, *written as* $\overline{\widetilde{C}}_i \sim \mathcal{N}(\widetilde{C}_i, V_i^C)$. *Similarly,* $\overline{\widetilde{D}}_i \sim \mathcal{N}(\widetilde{D}_i, V_i^D), \overline{\widetilde{A}}_i \sim \mathcal{N}(\widetilde{A}_i, V_i^A), \overline{\widetilde{B}}_i \sim \mathcal{N}(\widetilde{B}_i, V_i^B)$. $\widetilde{C}_i = (\widetilde{C}_{i1}, \widetilde{C}_2, \cdots, \widetilde{C}_{in}), \widetilde{D}_i = (\widetilde{D}_{i1}, \widetilde{D}_{i2}, \cdots, \widetilde{D}_{in}), \widetilde{A}_i = (\widetilde{A}_{i1}, \widetilde{A}_2, \cdots, \widetilde{A}_{in})$, $\widetilde{B}_i = (\widetilde{B}_{i1}, \widetilde{B}_{i2}, \cdots, \widetilde{B}_{in})$, *are fuzzy vectors characterized as*

$$\mu_{\widetilde{C}_{ij}} = [r_{1,C_{ij}}, r_{2,C_{ij}}, r_{3,C_{ij}}, r_{4,C_{ij}}],$$

$$\mu_{\widetilde{D}_{ij}} = [r_{1,D_{ij}}, r_{2,D_{ij}}, r_{3,D_{ij}}, r_{4,D_{ij}}],$$

$$\mu_{\widetilde{A}_{ij}} = [r_{1,A_{ij}}, r_{2,A_{ij}}, r_{3,A_{ij}}, r_{4,A_{ij}}],$$

$$\mu_{\widetilde{B}_{ij}} = [r_{1,B_{ij}}, r_{2,B_{ij}}, r_{3,B_{ij}}, r_{4,B_{ij}}].$$

Then the upper model of model (4.48) is equivalent to

$$\begin{cases} \max\limits_{x \in R^{n_1}} [\overline{F}'_1, \overline{F}'_2, \cdots, \overline{F}'_{m_1}] \\ s.t.\ \Phi^{-1}(\delta_i^U)\sqrt{x^T V_i^A x + y^T V_i^B y} + (1-\gamma_i^U)\left(\sum\limits_{j=1}^{n}(r_{1,A_{ij}}x_j + r_{1,B_{ij}}y_j)\right) \\ +\gamma_i^U\left(\sum\limits_{j=1}^{n}(r_{2,A_{ij}}x_j + r_{2,B_{ij}}y_j)\right) \leq 0, i = 1, 2, \cdots, p_1, \end{cases} \tag{4.49}$$

where

$$\overline{F}'_i = \Phi^{-1}(1-\beta_i^U)\sqrt{x^T V_i^C x + y^T V_i^D y} + \alpha_i^U \sum_{j=1}^{n}(r_{3,C_{ij}}x_j + r_{3,D_{ij}}y_j)$$

$$+ (1-\alpha_i^U)\sum_{j=1}^{n}(r_{4,C_{ij}}x_j + r_{4,D_{ij}}y_j),$$

Φ *is the standard normally distribution.*

Proof. It is the direct from Theorems 4.4 and 4.5. □

For the lower model of (4.48), the following theorem holds.

Theorem 4.10. *Suppose that* $\bar{\tilde{c}}_i = (\bar{\tilde{c}}_{i1}, \bar{\tilde{c}}_{r2}, \cdots, \bar{\tilde{c}}_{in})^T$ *is a normally distributed Ra-Fu vector with the fuzzy mean vector* $\tilde{c}_r = (\tilde{c}_{i1}, \tilde{c}_{i2}, \cdots, \tilde{c}_{in})^T$ *and covariance* *matrix* V_i^c, *written as* $\bar{\tilde{c}}_i \sim \mathcal{N}(\tilde{c}_i, V_i^c)$. *Similarly,* $\bar{\tilde{d}}_i \sim \mathcal{N}(\tilde{d}_i, V_i^d), \bar{\tilde{a}}_i \sim \mathcal{N}(\tilde{a}_i, V_i^a), \bar{\tilde{b}}_i \sim \mathcal{N}(\tilde{b}_i, V_i^b)$. $\tilde{c}_i = (\tilde{c}_{i1}, \tilde{c}_2, \cdots, \tilde{c}_{in}), \tilde{d}_i = (\tilde{d}_{i1}, \tilde{d}_{i2}, \cdots, \tilde{d}_{in}), \tilde{a}_i = (\tilde{a}_{i1}, \tilde{a}_2, \cdots, \tilde{a}_{in}), \tilde{B}_i = (\tilde{b}_{i1}, \tilde{b}_{i2}, \cdots, \widetilde{b}_{in})$, *are fuzzy vectors characterized as*

$$\mu_{\tilde{c}_{ij}} = [r_{1,c_{ij}}, r_{2,c_{ij}}, r_{3,c_{ij}}, r_{4,c_{ij}}],$$

$$\mu_{\tilde{d}_{ij}} = [r_{1,d_{ij}}, r_{2,d_{ij}}, r_{3,d_{ij}}, r_{4,d_{ij}}],$$

$$\mu_{\tilde{a}_{ij}} = [r_{1,a_{ij}}, r_{2,a_{ij}}, r_{3,a_{ij}}, r_{4,a_{ij}}],$$

$$\mu_{\tilde{b}_{ij}} = [r_{1,b_{ij}}, r_{2,b_{ij}}, r_{3,b_{ij}}, r_{4,b_{ij}}].$$

Then the lower model of model (4.48) is equivalent to

$$\begin{cases} \max\limits_{x \in R^{n_1}} [\bar{f}_1', \bar{f}_2', \cdots, \bar{f}_{m_2}'] \\ \text{s.t. } \Phi^{-1}(\delta_i^L)\sqrt{x^T V_i^a x + y^T V_i^b y} + (1-\gamma_i^L)\left(\sum\limits_{j=1}^n (r_{1,a_{ij}}x_j + r_{1,b_{ij}}y_j)\right) \\ +\gamma_i^U\left(\sum\limits_{j=1}^n (r_{2,a_{ij}}x_j + r_{2,b_{ij}}y_j)\right) \leq 0, i = 1, 2, \cdots, p_2, \end{cases} \tag{4.50}$$

where

$$\bar{f}_i' = \frac{\left(\Phi^{-1}(1-\beta_i^U)\sqrt{x^T V_i^C x + y^T V_i^D y} + \sum\limits_{j=1}^n (r_{4,C_{ij}}x_j + r_{4,D_{ij}}y_j) - \bar{f}_i\right)}{\left(\sum\limits_{j=1}^n (r_{4,C_{ij}}x_j + r_{4,D_{ij}}y_j) - \sum\limits_{j=1}^n (r_{3,C_{ij}}x_j + r_{3,D_{ij}}y_j)\right)},$$

Φ *is the standard normally distribution.*

Proof. It follows from Lemma 4.1 that $Pr\{Pos\{\bar{\tilde{\mathbf{c}}}_i^T\mathbf{x} + \bar{\tilde{\mathbf{d}}}_i^T\mathbf{y} \geq \bar{f}_i\} \geq \alpha_i^L\} \geq \beta_i^L$ is equivalent to $\bar{f}_i \leq (\Phi^{-1}(1-\beta_i^L)\sqrt{x^T V_i^c x + y^T V_i^d y} + \alpha_i^L \sum\limits_{j=1}^n (r_{3,c_{ij}}x_j + r_{3,d_{ij}}y_j) + (1-\alpha_i^L)\sum\limits_{j=1}^n (r_{4,c_{ij}}x_j + r_{4,d_{ij}}y_j)$. Move α_i^L to the left side of the inequality, then

$$\alpha_i^L < \frac{\left(\Phi^{-1}(1-\beta_i^U)\sqrt{x^T V_i^C x + y^T V_i^D y} + \sum\limits_{j=1}^n (r_{4,C_{ij}}x_j + r_{4,D_{ij}}y_j) - \bar{f}_i\right)}{\left(\sum\limits_{j=1}^n (r_{4,C_{ij}}x_j + r_{4,D_{ij}}y_j) - \sum\limits_{j=1}^n (r_{3,C_{ij}}x_j + r_{3,D_{ij}}y_j)\right)}.$$

Denote

$$\alpha_i^L < \frac{\left(\Phi^{-1}(1-\beta_i^U)\sqrt{x^T V_i^C x + y^T V_i^D y} + \sum_{j=1}^{n}(r_{4,C_{ij}} x_j + r_{4,D_{ij}} y_j) - \bar{f}_i\right)}{\left(\sum_{j=1}^{n}(r_{4,C_{ij}} x_j + r_{4,D_{ij}} y_j) - \sum_{j=1}^{n}(r_{3,C_{ij}} x_j + r_{3,D_{ij}} y_j)\right)}$$

by $\bar{f}_i'$. The proof is complete. □

Fuzzy Decision Method

By adopting, the two-planner Stakelberg (see [16, 33]), and the well-known fuzzy decision model of Sakawa [28], one first gets the satisfactory solution that is acceptable to upper-level decision maker, and then give the upper-level decision maker decision variables and goals with some leeway to the lower level decision maker, for him or her to seek the satisfactory solution, and to arrive at the solution that is closest to the satisfactory solution of the upper-level decision maker. This due to the lower-level decision maker, who should not only optimize his or her objective functions but also try to satisfy the upper-level decision maker goals and preferences as much as possible.

In this way, the solution method simplifies a BLDM by transforming it into separate multi-objective decision making problems at two levels, and by that means the difficulty associated with nonconvexity to arrive at an optimal solution is avoided.

First, the upper-level decision maker solves the following multi-objective decision making problem:

$$\max_{(\boldsymbol{x},\boldsymbol{y})\in D} [F_1(\boldsymbol{x},\boldsymbol{y}), F_2(\boldsymbol{x},\boldsymbol{y}), \cdots, F_{m_1}(\boldsymbol{x},\boldsymbol{y})], \tag{4.51}$$

where $D := \{(\boldsymbol{x},\boldsymbol{y})|G_r(\boldsymbol{x},\boldsymbol{y}) \le 0, r = 1, 2, \cdots, p_1, g_r(x,y) \le 0, r = 1, 2, \cdots, p_2\}$. To build membership functions, goals and tolerances should be determined first. However, they can hardly be determined without meaningful supporting data.

We should first find the individual best solutions F_i^+ and individual worst solutions F_i^- for each objective of (4.51), where

$$F_i^+ = \max_{x\in D} F_i(\boldsymbol{x},\boldsymbol{y}), F_i^- = \min_{x\in D} F_i(\boldsymbol{x},\boldsymbol{y}), i = 1, 2, \cdots, P. \tag{4.52}$$

Goals and tolerances can then be reasonably set for individual solutions and the differences of the best and worst solutions, respectively. This data can then be formulated as the following membership functions of fuzzy set theory:

$$\mu_{F_i}(F_i(\boldsymbol{x},\boldsymbol{y})) = \begin{cases} 1, & \text{if } F_i(\boldsymbol{x},\boldsymbol{y}) > F_i^+ \\ \dfrac{F_i(\boldsymbol{x},\boldsymbol{y}) - F_i^-}{F_i^+ - F_i^-}, & \text{if } F_i^- \leq F_i(\boldsymbol{x},\boldsymbol{y}) \leq F_i^+ \\ 0, & \text{if } F_i(x,y) < F_i^-. \end{cases} \tag{4.53}$$

Now we can get the solution to the upper-level decision maker problem by solving the following Tchebycheff problem (see [1, 31]):

$$\begin{cases} \max \lambda \\ \text{s.t.} \begin{cases} (\boldsymbol{x},\boldsymbol{y}) \in D \\ \mu_{F_i}(F_i(\boldsymbol{x},\boldsymbol{y})) \geq \lambda, i = 1, 2, \cdots, m_1 \\ \lambda \in [0, 1], \end{cases} \end{cases} \tag{4.54}$$

whose solution is assumed to be $(\boldsymbol{x}^U, \boldsymbol{y}^U, \lambda^U, F_i^U, i = 1, 2, \cdots, m_1)$.

Second, in the same way, the lower-level decision maker independently solves

$$\max_{(\boldsymbol{x},\boldsymbol{y})\in D} [f_1(\boldsymbol{x},\boldsymbol{y}), f_2(\boldsymbol{x},\boldsymbol{y}), \cdots, f_{m_2}(\boldsymbol{x},\boldsymbol{y})], \tag{4.55}$$

where $D := \{(\boldsymbol{x},\boldsymbol{y})|G_r(\boldsymbol{x},\boldsymbol{y}) \leq 0, r = 1, 2, \cdots, P, g_r(\boldsymbol{x},\boldsymbol{y}) \leq 0, r = 1, 2, \cdots, p\}$. To build membership functions, goals and tolerances should be determined first. However, they could hardly be determined without meaningful supporting data.

We should first find the individual best solutions f_i^+ and individual worst solutions f_i^- for each objective of (4.55), where

$$f_i^+ = \max_{(\boldsymbol{x},\boldsymbol{y})\in D} f_i(\boldsymbol{x},\boldsymbol{y}) \text{ and } f_i^- = \min_{(\boldsymbol{x},\boldsymbol{y})\in D} f_i(\boldsymbol{x},\boldsymbol{y}), i = 1, 2, \cdots, p. \tag{4.56}$$

Goals and tolerances can then be reasonably set for individual solutions and the differences of the best and worst solutions, respectively. This data can then be formulated as the following membership functions:

$$\mu_{f_i}(f_i(\boldsymbol{x},\boldsymbol{y})) = \begin{cases} 1, & \text{if } f_i(\boldsymbol{x},\boldsymbol{y}) > f_i^+ \\ \frac{f_i(\boldsymbol{x},\boldsymbol{y}) - f_i^-}{f_i^+ - f_i^-}, & \text{if } f_i^- \leq f_i(\boldsymbol{x},\boldsymbol{y}) \leq f_i^+ \\ 0, & \text{if } f_i(\boldsymbol{x},\boldsymbol{y}) < f_i^-. \end{cases} \tag{4.57}$$

Now, we can get the solution of the lower-level decision maker problem by solving the following Tchebycheff problem:

$$\begin{cases} \max \delta \\ \text{s.t.} \begin{cases} (\boldsymbol{x},\boldsymbol{y}) \in D \\ \mu_{f_i}(f_i(\boldsymbol{x},\boldsymbol{y})) \geq \lambda, i = 1, 2, \cdots, m \\ \lambda \in [0, 1], \end{cases} \end{cases} \tag{4.58}$$

whose solutions is assumed to be $(\boldsymbol{x}^L, \boldsymbol{y}^L, f_i^L, \lambda^L, i = 1, 2, \cdots, m)$.

Now the solution to the upper-level decision maker and lower-level decision maker are disclosed. However, the two solutions are usually different because of the difference in nature between two level objective functions. The upper-level decision maker knows that using the optimal decisions $\boldsymbol{x}^U$ as a control factors for the lower-level decision maker are not practical. It is more reasonable to have some tolerance that gives the lower-level decision maker an extent of feasible region to search for his or her optimal solution, and also to reduce searching time or interactions.

In this way, the range of decision variables $\boldsymbol{x}$ and $\boldsymbol{y}$ should be around $\boldsymbol{x}^U$ with maximum tolerance t and the following membership function specifies x_1^U as

$$\mu_{\boldsymbol{x}^U} = \begin{cases} \frac{\boldsymbol{x}-(\boldsymbol{x}^U-t)}{t} & \boldsymbol{x}^U - t \leq \boldsymbol{x} \leq \boldsymbol{x}^U \\ \frac{(\boldsymbol{x}^U+t)-\boldsymbol{x}}{t} & \boldsymbol{x}^U \leq \boldsymbol{x} \leq \boldsymbol{x}^U + t, \end{cases} \tag{4.59}$$

where $\boldsymbol{x}^U$ is the most preferred solution; $(\boldsymbol{x}^U - t)$ and $(\boldsymbol{x}^U + t)$ are the worst acceptable decisions; and that satisfaction is linearly increasing with the interval of $[\boldsymbol{x}^U - t, x]$ and linearly decreasing with $[\boldsymbol{x}_1, \boldsymbol{x}^U + t]$, and other decision are not acceptable. In order to supervise the lower-level decision maker to search for solutions in the right direction, we will take the following steps.

First, the upper-level decision maker goals may reasonably consider that all $F_i \geq F_i^U, k = 1, 2, \cdots, N_1$ are absolutely acceptable and all $F_i \geq F_i' = F_i(\boldsymbol{x}, \boldsymbol{y}), k = 1, 2, \cdots, N_1$ are absolutely unacceptable, and that the preference with $[F_i', F_i^U], i = 1, 2, \cdots, M$ is linearly increasing. This is due to the fact that the lower-level decision maker obtained the optimum at $(\boldsymbol{x}^L, \boldsymbol{y}^L)$, which in turn provides the upper-level decision maker the objective function values F_i', which makes any $F_i < F_i', i = 1, 2, \cdots, M$ unattractive in practice.

The following membership functions of the upper-level decision maker can be stated as

$$\mu'_{\boldsymbol{x}^U} = \begin{cases} 1, & \text{if } F_i(\boldsymbol{x}, \boldsymbol{y}) > F_i^U \\ \frac{F_i(\boldsymbol{x}, \boldsymbol{y}) - F_i'}{F_i^U - F_i'}, & \text{if } F_i' \leq F_i(\boldsymbol{x}, \boldsymbol{y}) \leq F_i^U \\ 0, & \text{if } F_i(\boldsymbol{x}, \boldsymbol{y}) \leq F_i'. \end{cases} \tag{4.60}$$

Second, the lower-level decision maker may be willing to build a membership function for his or her objective functions, so that he or she can rate the satisfaction of each potential solution. In this way, the lower-level decision maker has the following membership functions for his/her goals:

$$\mu'_{x^U} = \begin{cases} 1, & \text{if } f_i(x, y) > f_i^L \\ \frac{f_i(\boldsymbol{x}, \boldsymbol{y}) - f_i'}{f_i^L - f_i'}, & \text{if } f_i' \leq f_i(\boldsymbol{x}, \boldsymbol{y}) \leq f_i^L \\ 0, & \text{if } f_i(\boldsymbol{x}, \boldsymbol{y}) \leq f_i', \end{cases} \tag{4.61}$$

where $f_i' = f_i(x^U, y^U)$.

Finally, in order to generate the satisfactory solution, which is also a Pareto-optimal solution with overall satisfaction for all DMs, we can solve the following Tchebycheff problem:

$$\begin{cases} \max \delta \\ \text{s.t.} \begin{cases} \frac{[(x^U+t)-x]}{t} \geq \delta I \\ \frac{[x-(x^U-t)]}{t} \geq \delta I \\ \mu'_{F_i}[F_i(\boldsymbol{x},\boldsymbol{y})] \geq \delta, i = 1,2,\cdots,M \\ \mu'_{f_i}[f_i(\boldsymbol{x},\boldsymbol{y})] \geq \delta, i = 1,2,\cdots,m \\ (\boldsymbol{x},\boldsymbol{y}) \in D \\ t \geq 0 \\ \delta \in [0,1], \end{cases} \end{cases} \tag{4.62}$$

where δ is the overall satisfaction, and I is the column vector with all elements equal to 1s. Equation (4.62) is actually a fuzzy problem by Sakawa [28].

By solving problem (4.62), if the upper-level decision maker is satisfied with solution, then a satisfactory solution is reached. Otherwise, he or she should provide a new membership function for the control variables and objectives to the lower-level decision maker until a satisfactory solution is reached.

Numerical Examples

In this section, a numerical example is given to illustrate the effectiveness of the application of the model and algorithm proposed above.

Example 4.5. Consider the problem

$$\begin{cases} \max\limits_{x_1,x_2} F_1 = \xi_1 x_1 + 2\xi_2 x_2 + 3\xi_3 y_1 + 4\xi_4 y_2 \\ \max\limits_{x_1,x_2} F_2 = 5\xi_1 x_1 + 3\xi_2 x_2 + 2\xi_3 y_1 + \xi_4 y_2 \\ \text{where } (y_1,y_2,y_3) \text{ solves} \\ \begin{cases} \max\limits_{y_1,y_2} f_1 = 7\xi_1 x_1 + 5\xi_2 x_2 + 4\xi_3 y_1 + \xi_4 y_2 \\ \max\limits_{y_1,y_2} f_2 = 3\xi_1 x_1 + 4\xi_2 x_2 + 5\xi_3 y_1 + 6\xi_4 y_2 \\ \text{s.t.} \begin{cases} x_1 + x_2 + y_1 + y_2 \geq 120 \\ x_1 + x_2 + y_1 + y_2 \leq 200 \\ x_1, x_2, y_1, y_2 \geq 10, \end{cases} \end{cases} \end{cases} \tag{4.63}$$

where $\xi_j, j = 1,2,3,4$ are Ra-Fu variable characterized as

$\tilde{\bar{\xi}}_1 \sim \mathcal{N}(\tilde{u}_1, 1)$, with $\tilde{u}_1 \sim (3,6,8,11)$, $\tilde{\bar{\xi}}_2 \sim \mathcal{N}(\tilde{u}_2, 1)$, with $\tilde{u}_2 \sim (7,10,11,16)$,
$\tilde{\bar{\xi}}_3 \sim \mathcal{N}(\tilde{u}_3, 1)$, with $\tilde{u}_3 \sim (2,6,8,10)$, $\tilde{\bar{\xi}}_4 \sim \mathcal{N}(\tilde{u}_4, 1)$, with $\tilde{u}_4 \sim (5,10,20,30)$,

If the decision makers determine the values of confidence levels and reference levels , the CCDD model for (4.63) is

$$\begin{cases} \max\limits_{x_1,x_2}[\overline{F}_1, \overline{F}_2] \\ \text{s.t.} \begin{cases} Ch\{\xi_1 x_1 + 2\xi_2 x_2 + 3\xi_3 y_1 + 4\xi_4 y_2 \geq \overline{F}_1\}(0.6) \geq 0.6 \\ Ch\{5\xi_1 x_1 + 3\xi_2 x_2 + 2\xi_3 y_1 + \xi_4 y_2 \geq \overline{F}_2\}(0.8) \geq 0.8 \end{cases} \\ \text{where } (y_1, y_2) \text{ solves} \\ \quad \begin{cases} \max\limits_{y_1,y_2}[\alpha_1^L, \alpha_2^L] \\ \text{s.t.} \begin{cases} Ch\{7\xi_1 x_1 + 5\xi_2 x_2 + 4\xi_3 y_1 + \xi_4 y_2 \geq 80\}(\alpha_1^L) \geq 0.8 \\ Ch\{3\xi_1 x_1 + 4\xi_2 x_2 + 5\xi_3 y_1 + 6\xi_4 y_2 \geq 100\}(\alpha_2^L) \geq 0.6. \\ x_1 + x_2 + y_1 + y_2 \geq 120 \\ x_1 + x_2 + y_1 + y_2 \leq 200 \\ x_1, x_2, y_1, y_2 \geq 10, \end{cases} \end{cases} \end{cases} \tag{4.64}$$

From Lemma 4.1 and Theorem 4.10, we know that problem (4.64) is equivalent to the model

$$\begin{cases} \max[0.8x_1 + 16.8x_2 + 19.6y_1 + 7.6y_2, -8x_1 + 16.8x_2 + 7.2y_1 - 2.8y_2] \\ \text{where } (y_1, y_2) \text{ solves} \\ \max[-14.2x_1 + 28x_2 + 14.4y_1 - 2.8y_2, 2.4x_1 + 28.8x_2 + 26y_1 + 2.4y_2] \\ \text{s.t.} \begin{cases} x_1 + x_2 + y_1 + y_2 \geq 120 \\ x_1 + x_2 + y_1 + y_2 \leq 200 \\ x_1, x_2, y_1, y_2 \geq 10. \end{cases} \end{cases} \tag{4.65}$$

Without causing ambiguity, we still use F_1, F_2, f_1, f_2 to represent $0.8x_1 + 16.8x_2 + 19.6y_1 + 7.6y_2, -8x_1 + 16.8x_2 + 7.2y_1 - 2.8y_2, -14.2x_1 + 28x_2 + 14.4y_1 - 2.8y_2$, and $2.4x_1 + 28.8x_2 + 26y_1 + 2.4y_2$, respectively.

First, the upper-level decision maker solves his or her problem as follows:

(1) Finding individual optimal solutions, we get $F_1^+ = 3584, F_1^- = 512, F_2^+ = 2820, F_2^- = -1148$.
(2) By using (4.55), build the membership functions μ_{F_1}, μ_{F_2}, hen solve (4.56) as follows:

$$\begin{cases} \max \lambda \\ \text{s.t.} \begin{cases} 0.8x_1 + 16.8x_2 + 19.6y_1 + 7.6y_2 \geq 3072\lambda + 512 \\ -8x_1 + 16.8x_2 + 7.2y_1 - 2.8y_2 \geq 3968\lambda - 1148 \\ x_1 + x_2 + y_1 + y_2 \geq 120 \\ x_1 + x_2 + y_1 + y_2 \leq 200 \\ x_1, x_2, y_1, y_2 \geq 10 \\ \lambda \in [0, 1]. \end{cases} \end{cases}$$

whose solution is $(x_1^U, x_2^U, y_1^U, y_2^U) = (10, 126.2, 53.8, 10), (F_1^U, F_2^U) = (3258.6, 2399.7), \lambda^U = 0.894$.

Then, in the same way, the lower-level decision maker solves his/her problem as follows:

Finding individual optimal solutions, we get $f_1^+ = 4734, f_1^- = 2018, f_2^+ = 5204, f_2^- = 788$.

By solving

$$\begin{cases} \max \lambda \\ \text{s.t.} \begin{cases} -14.2x_1 + 28x_2 + 14.4y_1 - 2.8y_2 \geq 6752\lambda - 2018 \\ 2.4x_1 + 28.8x_2 + 26y_1 + 2.4y_2 \geq 4416\lambda + 788x_1 + x_2 + y_1 + y_2 \geq 120 \\ x_1 + x_2 + y_1 + y_2 \leq 200 \\ x_1, x_2, y_1, y_2 \geq 10 \\ \lambda \in [0, 1], \end{cases} \end{cases}$$

whose solution is $(x_1^L, x_2^L, y_1^L, y_2^L) = (10, 170, 10, 10), (f_1^L, f_2^L) = (4734, 5204)$, $\lambda^L = 1$.

Finally, we assume that the upper-level decision makers control decision (x_1^U, x_2^U) is around 0 with tolerance 1.

By using following problem

$$\begin{cases} \max \delta \\ \text{s.t.} \begin{cases} x_1 - 9 \geq \delta \\ 11 - x_1 \geq \delta \\ x_2 - 125.2 \geq \delta \\ 127.2 - x_2 \geq \delta \\ 0.8x_1 + 16.8x_2 + 19.6y_1 + 7.6y_2 \geq 122.6\delta + 3136 \\ -8x_1 + 16.8x_2 + 7.2y_1 - 2.8y_2 \leq -420.38\delta + 2820 \\ -14.2x_1 + 28x_2 + 14.4y_1 - 2.8y_2 \geq 592.68\delta + 4138.32 \\ 2.4x_1 + 28.8x_2 + 26y_1 + 2.4y_2 \geq 122.64\delta + 5081.36 \\ x_1 + x_2 + y_1 + y_2 \geq 120 \\ x_1 + x_2 + y_1 + y_2 \leq 200 \\ x_1, x_2, y_1, y_2 \geq 10 \\ \delta \in [0, 1], \end{cases} \end{cases}$$

whose compromise solution is $(x_1^*, x_2^*, y_1^*, y_2^*) = (10, 127.2, 52.8, 10)$, $(F_1^*, F_2^*) = (3255.9, 2408.9), (f_1^*, f_2^*) = (4151.6, 5084, 1), \delta^* = 0.962$.

If the upper-level decision maker is satisfied with the above solution, then a satisfactory solution is obtained. Otherwise, he or she should provide new membership functions for the control variable and objectives to the lower-level decision maker until a satisfactory solution is reached. It is easy to see that there is an inverse correlation between t and δ.

4.4.3 Ra-Fu Simulation-Based PGSA-GA for Nonlinear Model

Similar to the Sect. 4.3.3, as (4.46) is a nonlinear model, i.e., some functions in model (4.46) is nonlinear, the Ra-Fu features can not be converted into its crisp equivalent model. To solve the nonlinear (4.46), we need to handle the trade-off between multiple objectives, the Ra-Fu parameters and the bi-level structure. In Sect. 4.3.3, we introduced the hybrid solution method consisting of linear weighted-sum approach, Ra-Fu simulations for expected operator and chance operator, and basic PGSA. The method is suitable when all decision variables are integer. If decision variables are not integer either in the upper or lower programming, the method needs to be adjusted. In this section, we design a solution method consisting of Minimax Point Method, Ra-Fu simulations techniques and PGSA-GA hybird algorithm.

Minimax Point Method

For a general multiple objective model (4.36), to maximize the objectives, the minimax point method first constructs an evaluation function by seeking the minimal objective value after computing all objective functions, that is, $u(\mathbf{H}(\boldsymbol{x})) = \min_{1\leq i\leq m} H_i(x)$, where $\mathbf{H}(\boldsymbol{x}) = (\boldsymbol{H}_1(\boldsymbol{x}), \boldsymbol{H}_2(\boldsymbol{x}), \cdots, \boldsymbol{H}_m(\boldsymbol{x}))^T$. Then the objective function of problem (4.36) comes down to solving the maximization problem as follows:

$$\max_{x\in X'} u(\mathbf{H}(\boldsymbol{x})) = \max_{x\in X'} \min_{1\leq i\leq m} \boldsymbol{H}_i(\boldsymbol{x}). \tag{4.66}$$

Sometimes, decision makers need to consider the relative importance of various goals, and then the weight can be combined into the evaluation function as follows:

$$\max_{x\in X'} u(\mathbf{H}(\boldsymbol{x})) = \max_{x\in X'} \min_{1\leq i\leq m} \{\omega_i \boldsymbol{H}_i(\boldsymbol{x})\}, \tag{4.67}$$

where the weight $\sum_{i=1}^{m} \omega_i = 1(\omega_i > 0)$ and is predetermined by decision makers.

Theorem 4.11. *Let x^* be the optimal solution to the problem (4.67). Then , x^* also is the weak efficient solution to the problem (4.36).*

Proof. Assume that $\boldsymbol{x}^* \in X'$ is the optimal solution to problem (4.67). If there exists an $\boldsymbol{x}$ such that $\boldsymbol{H}_i(\boldsymbol{x}) \geq \boldsymbol{H}_i(\boldsymbol{x}^*)(i = 1, 2, \cdots, m)$, we have

$$\min_{1\leq i\leq m} \{\omega_i \boldsymbol{H}_i(\boldsymbol{x}^*)\} \leq \omega_i \boldsymbol{H}_i(\boldsymbol{x}^*) \leq \omega_i \boldsymbol{H}_i(\boldsymbol{x}),\ 0 < \omega_i < 1.$$

Denote $\delta = \min_{1\leq i\leq m}\{\omega_i \boldsymbol{H}_i(\boldsymbol{x})\}$, then $\delta \geq \min_{1\leq i\leq m}\{\omega_i \boldsymbol{H}_i(\boldsymbol{x}^*)\}$. This means that $\boldsymbol{x}^*$ is not the optimal solution to problem (4.67). This conflicts with the condition.

Thus, there does not exist $\boldsymbol{x} \in X'$ such that $\boldsymbol{H}_i(\boldsymbol{x}) \geq H_i(\boldsymbol{x}^*)$, namely, $\boldsymbol{x}^*$ is a weak efficient solution to problem (4.36).

By introducing an auxiliary variable, the minimax problem (4.67) can be converted into a single objective problem. Let

$$\lambda = \min_{1 \leq i \leq m} \{\omega_i \boldsymbol{H}_i(\boldsymbol{x})\}.$$

Then problem (4.67) is converted into

$$\begin{cases} \max \lambda \\ \text{s.t.} \begin{cases} \omega_i \boldsymbol{H}_i(\boldsymbol{x}) \geq \lambda, i = 1, 2, \cdots, m \\ \boldsymbol{x} \in X'. \end{cases} \end{cases} \tag{4.68}$$

Theorem 4.12. *Problem (4.67) is equivalent to problem (4.68).*

Proof. Assume that $\boldsymbol{x}^* \in X'$ is the optimal solution to problem (4.67), and let $\lambda^* = \min_{1 \leq i \leq m}\{\omega_i \boldsymbol{H}_i(\boldsymbol{x}^*)\}$. Then it is apparent that $\boldsymbol{H}_i(\boldsymbol{x}^*) \geq \lambda^*$. This means that $(\boldsymbol{x}^*, \lambda^*)$ is a feasible solution to problem (4.68). Assume that $(\boldsymbol{x}, \lambda)$ is any feasible solution to problem (4.68). Since $\boldsymbol{x}^*$ is the optimal solution to problem (4.67), we have

$$\lambda^* = \min_{1 \leq i \leq m} \{\omega_i \boldsymbol{H}_i(\boldsymbol{x}^*)\} \geq \min_{1 \leq i \leq m} \{\omega_i \boldsymbol{H}_i(\boldsymbol{x})\} \geq \lambda,$$

namely, $(\boldsymbol{x}^*, \lambda^*)$ is the optimal solution to problem (4.68).

Conversely, assume that $(\boldsymbol{x}^*, \lambda^*)$ is an optimal solution to problem (4.68). Then, $\omega_i H_i(\boldsymbol{x}^*) \geq \lambda^*$ holds for any i, which means $\min_{1 \leq i \leq m}\{\omega_i H_i(\boldsymbol{x}^*)\} \geq \lambda^*$. It follows that, for any feasible $\boldsymbol{x} \in X'$,

$$\min_{1 \leq i \leq m} \{\omega_i \boldsymbol{H}_i(\boldsymbol{x})\} = \lambda \leq \lambda^* \leq \min_{1 \leq i \leq m} \{\omega_i \boldsymbol{H}_i(\boldsymbol{x}^*)\}$$

holds, namely, $\boldsymbol{x}^*$ is the optimal solution to problem (4.67).

The minimax point method can be summarized as follows:

Step 1. Compute the weight for each objective function by solving the two problems, $\max_{\boldsymbol{x} \in X'} H_i(\boldsymbol{x})$ and $\omega_i = \boldsymbol{H}_i(\boldsymbol{x}^*)/\sum_{i=1}^m \boldsymbol{H}_i(\boldsymbol{x}^*)$.
Step 2. Construct the auxiliary problem as follows:

$$\begin{cases} \max \lambda \\ \text{s.t.} \begin{cases} \omega_i \boldsymbol{H}_i(\boldsymbol{x}) \geq \lambda, i = 1, 2, \cdots, m \\ \boldsymbol{x} \in X'. \end{cases} \end{cases}$$

Step 3. Solve the above problem to obtain the optimal solution.

Ra-Fu Simulation for Dependent Chance

Assume that ξ is an n-dimensional Ra-Fu vector defined on the possibility space $(\Theta, \mathscr{P}(\Theta), Pos)$, and $f : \mathbf{R}^n \to \mathbf{R}$ is a measurable function. For any confidence level α, we design a Ra-Fu simulation to compute the α-chance $Ch\{f(\boldsymbol{x}, \boldsymbol{\xi}) \geq \bar{f}\}(\alpha)$. Equivalently, we should find the supremum $\bar{\beta}$ such that

$$Pos\{\theta | Pr\{f(\boldsymbol{x}, \boldsymbol{\xi}(\theta)) \geq \bar{f}\} \geq \bar{\beta}\} \geq \alpha.$$

We randomly generate θ_k from Θ such that $Pos\{\theta_k\} \geq \varepsilon$, and write $v_k = Pos\{\theta_k\}, k = 1, 2, \cdots, N$, respectively, where ε is a sufficiently small number. For any number θ_k, by using stochastic simulation, we can estimate the probability $g(\theta_k) = Pr\{f(\boldsymbol{x}, \boldsymbol{\xi}(\theta_k)) \geq \bar{f}\}$. For any number r, we set

$$L(r) = \frac{1}{2}\left(\max_{1 \leq k \leq N}\{v_k | g(\theta_k) \geq r\} + \min_{1 \leq k \leq N}\{1 - v_k | g(\theta_k) < r\}\right).$$

If follows from monotonicity that we may employ bisection search to find the maximal value r such that $L(r) \geq \alpha$. This value is an estimation of L. We summarize this process as follows.

Then the procedure simulating the α-chance $Ch\{f(\boldsymbol{x}, \boldsymbol{\xi}) \geq \bar{f}\}(\alpha)$ can be summarized as follows:

Input: The decision vector $\boldsymbol{x}$
Output: The α-chance $Ch\{f(\boldsymbol{x}, \boldsymbol{\xi}) \geq \bar{f}\}(\alpha)$
Step 1. Randomly sample θ_k from Θ such that
$Pos\{\theta_k\} \geq \varepsilon$ for $k = 1, 2, \cdots, N$, where
ε is a sufficiently small number
Step 2. Find the maximal value r such that $L(r) \geq \alpha$ holds
Step 3. Return r.

Example 4.6. We employ the Ra-Fu simulation to calculate the chance measure

$$Ch\left\{\sqrt{\tilde{\bar{\xi}}_1^2 + \tilde{\bar{\xi}}_2^2 + \tilde{\bar{\xi}}_3^2} \geq 2\right\}(0.9),$$

where $\tilde{\bar{\xi}}_1$, $\tilde{\bar{\xi}}_2$ and $\tilde{\bar{\xi}}_3$ are Ra-Fu variables defined as

$$\tilde{\bar{\xi}}_1 \sim \exp(\tilde{\rho}_1), \text{ with } \tilde{\rho}_1 = (1, 2, 3),$$
$$\tilde{\bar{\xi}}_2 \sim \exp(\tilde{\rho}_2), \text{ with } \tilde{\rho}_2 = (2, 3, 4),$$
$$\tilde{\bar{\xi}}_3 \sim \exp(\tilde{\rho}_3), \text{ with } \tilde{\rho}_3 = (3, 4, 5).$$

A run of Ra-Fu simulation with 1000 cycles shows that

$$Ch\left\{\sqrt{\tilde{\bar{\xi}}_1^2+\tilde{\bar{\xi}}_2^2+\tilde{\bar{\xi}}_3^2}\geq 2\right\}(0.9)=0.5801.$$

Hybrid PGSA-GA

In the proposed solving method, solving Ra-Fu multi-objective bi-level model (4.2) is transformed to solve crisp single objective bi-level model by Minimax Point Method and Ra-Fu simulation. To handle the bi-level structure, we solve the upper-level and lower-level programming problem interactively by hybrid PGSA-GA while determining respectively the decision variables of the upper level ($\boldsymbol{x}$) or the lower-level ($\boldsymbol{y}$). If the decision variable in the upper-level programming ($\boldsymbol{x}$) is non-integer, we can use the GA in the upper-level and PGSA in the lower-level. If the decision variable in the lower-level programming ($\boldsymbol{y}$) is non-integer, we can use the PGSA in the upper-level and GA in the lower-level.

Such that, the upper-level decision variable ($\boldsymbol{x}$) is integer and the lower-level decision variable is non-integer, the hybrid PGSA-GA overall procedure for solving Ra-Fu multi-objective bi-level model (4.2) is as follows:

Step 1. Input all coefficients' values for model (4.2). Set initial iteration $\tau = 0$. Initialize upper-level basis point $X^0 = \{x_j^0, j = 1, 2, \cdots, N\}$ in the range of $[0, Range]$. Let $X_{\min} = X^0$, $F_{\min} = F(X^0)$

Step 2. Solve the lower-level programming using the GA procedure with upper-level initialized basis point X^0.

Step 2.1. Initialization and representation. Choose an appropriate representation for the decision variables in the lower-level programming ($\boldsymbol{y}$), for example real number strings. Randomly initialize a population of $\boldsymbol{y}$ within the feasible region of lower-level programming.

Step 2.2. Fitness Evaluation. The fitness function in the lower level $f(\boldsymbol{x}, \boldsymbol{y})$ measures the quality of individuals. If there are multiple objectives in the lower level, we can apply the Minimax Point Method above to transfer the multiple objectives to a single objective. Then use it as the fitness function to evaluate the quality of individuals.

Step 2.3. Selection. Given a generation of feasible solutions, parents of the next generation are obtained by randomly selecting solutions of the current generation with probabilities proportional to their corresponding fitness. In this section, the selection procedure is based on simply spinning the roulette wheel.

Step 2.4. Crossover. Based on a pre-specified crossover rate, crossover is applied to couples of parent solutions to produce two solutions of the next generation.

Multi-point crossover operator is adopted. For two parents' chromosomes, exchange their right parts from the first cut-point to the last cut-point respectively.

Step 2.5. To avoid the loss of potentially useful genetic material, individuals are randomly mutated according to a pre-specified mutation rate. Multi-point mutation strategy is used. For a parents chromosome, invert the right part from the first cut-point to the last cut-point respectively.

Step 2.6. Feasibility. The solutions resulting from crossover and mutation operators may be infeasible and specifically tailored procedures are used to regain feasibility. In order to enforce feasibility of the upper-level constraints, infeasible individuals are modified by flipping a randomly selected gene according to the infeasibility level. This process is repeated until feasibility is attained. If the infeasibility was caused by the mutation operator, feasibility procedures are run without undoing what the mutation procedure did.

Step 2.7. The genetic algorithm is stopped after a pre-specified number of generations.

Step 3. Input the optimal solution found in the lower-level $bmy*$ to the upper-level, and solve the upper-level programming using the PGSA procedure.

Step 3.1. Set the appropriate step length λ depending on the size of search region. λ shall be integer.

Step 3.2. Calculate the objective function value $F(x, y)$ in the upper-level by the upper-level basis point X^0 and the optimal solution found in the lower-level $y*$. If there are multiple objectives in the upper-level, transform the multiple objectives to a single objective by the Minimax Point Method presented in section "Minimax Point Method". Set $X_{\min} = X^0, F_{\min} = F(X^0, y)$.

Step 3.3. Find the trunk M and each preferential growth node $X^0_{a1,b1}$, on the trunk in the upper-level programming, $a1 = 1, 2, \cdots, N, b1 = 1, 2, \cdots, m1$, $m1$ is the maximum number of growth nodes on the a1th trunk. Then eliminate infeasible growth nodes.

Step 3.4. Calculate the objective function value of each preferential growth node $X^0_{a1,b1}$. If any $F(X^0_{a1,b1}, y) < F_{\min}$, replace $F_{\min} = F(X^0_{a1,b1}, y)$.

Step 3.5. Calculate the morphactin concentrations by Eq. (4.40). Generate a random number β_0 in the interval [0, 1] and find a new basic node.

Step 3.6. Start from the new basis point, find new growth node and calculate the morphactin concentrations. Replace $F_{\min}$ constantly.

Step 4. When $f'_{\min}$, $F_{\min}$ repeats appearing many times (e.g., 50 times), take $F^* = F_{\min}, X^* = X_{\min}$ as the optimal solution. Otherwise, use the $X_{\min}$ as the parameter for lower-level programming and go back to Step 2.

The framework of the hybrid PGSA-GA is shown in Fig. 4.5.

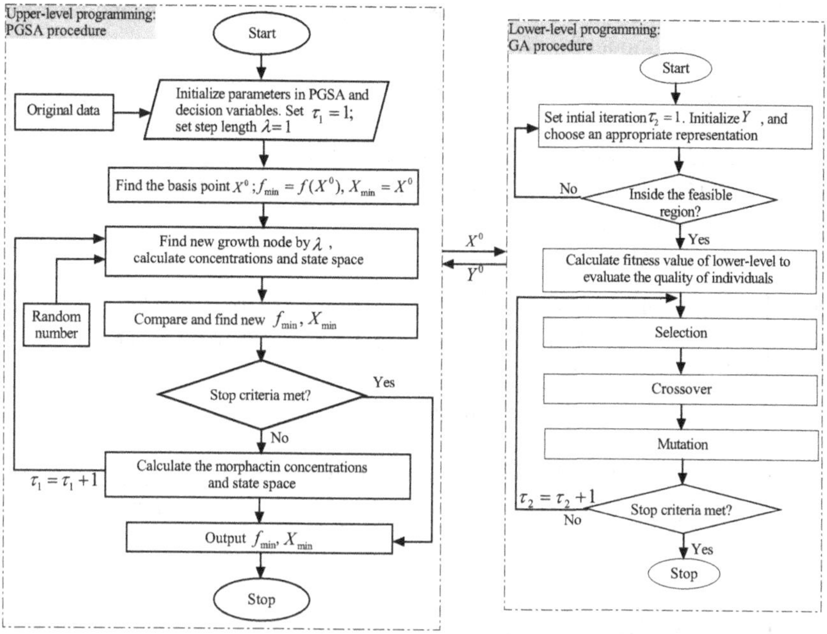

Fig. 4.5 The framework of the hybrid PGSA-GA

Numerical Examples

In this section, a numerical example is given to illustrate the effectiveness of the application of the model and algorithm proposed above.

Example 4.7. Consider the problem

$$\begin{cases} \max\limits_{x_1,x_2} F_1 = 2\xi_1^2 x_1 + 2\xi_2 x_2^2 + \sqrt{3\xi_3 y_1} + 4\xi_4 y_2 \\ \max\limits_{x_1,x_2} F_2 = 5\xi_1 x_1^2 + 3\xi_2^2 x_2 + \sqrt{2\xi_3 y_1} + \xi_4 y_2 \\ \text{where } (y_1, y_2, y_3) \text{ solves} \\ \quad \begin{cases} \max\limits_{y_1,y_2} f_1 = 7\xi_1^2 x_1 + 5\xi_2 x_2 + 4\xi_3 y_1 + \xi_4 y_2 \\ \max\limits_{y_1,y_2} f_2 = 3\xi_1 x_1 + 4\xi_2 x_2^2 + 5\xi_3 y_1 + 6\xi_4 y_2 \\ 2x_1 - x_2 + 3y_1 + y_2 \le 55 \\ x_1, x_2, y_1, y_2 \ge 0, \text{ integer}, \end{cases} \end{cases} \tag{4.69}$$

where $\xi_j, j = 1, 2, 3, 4$ are Ra-Fu variable characterized as

$\tilde{\bar{\xi}}_1 \sim \mathscr{N}(\tilde{u}_1, 1)$, with $\tilde{u}_1 \sim (3, 6, 8, 11)$, $\tilde{\bar{\xi}}_2 \sim \mathscr{N}(\tilde{u}_2, 1)$, with $\tilde{u}_2 \sim (7, 10, 11, 16)$,
$\tilde{\bar{\xi}}_3 \sim \mathscr{N}(\tilde{u}_3, 1)$, with $\tilde{u}_3 \sim (2, 6, 8, 10)$, $\tilde{\bar{\xi}}_4 \sim \mathscr{N}(\tilde{u}_4, 1)$, with $\tilde{u}_4 \sim (5, 10, 20, 30)$,

If the decision makers determine the values of confidence levels and reference levels , the CCDD model for (4.69) is

$$\begin{cases} \max\limits_{x_1,x_2}[\overline{F}_1,\overline{F}_2] \\ \text{s.t.} \begin{cases} Ch\{2\xi_1^2x_1+2\xi_2x_2^2+\sqrt{3\xi_3y_1}+4\xi_4y_2\geq \overline{F}_1\}(0.6)\geq 0.6 \\ Ch\{5\xi_1x_1^2+3\xi_2^2x_2+\sqrt{2\xi_3y_1}+\xi_4y_2\geq \overline{F}_2\}(0.8)\geq 0.8 \end{cases} \\ \text{where } (y_1,y_2) \text{ solves} \\ \quad \begin{cases} \max\limits_{y_1,y_2}[\alpha_1^L,\alpha_2^L] \\ \text{s.t.} \begin{cases} Ch\{7\xi_1^2x_1+5\xi_2x_2+4\xi_3y_1+\xi_4y_2\geq 80\}(\alpha_1^L)\geq 0.8 \\ Ch\{3\xi_1x_1+4\xi_2x_2^2+5\xi_3y_1+6\xi_4y_2\geq 100\}(\alpha_2^L)\geq 0.6 \\ 2x_1-x_2+3y_1+y_2\leq 55 \\ x_1,x_2,y_1,y_2\geq 0, \text{ integer.} \end{cases} \end{cases} \end{cases} \tag{4.70}$$

After running Ra-Fu Simulation-Based PGSA, we obtain $(x_1^*,x_2^*,y_1^*,y_2^*) = (10, 26, 13, 22)$ and $(\overline{F}_1^*,\overline{F}_2^*) = (16556.87, 12165.69)$, $(\alpha_1^{L*},\alpha_2^{L*}) = (0.88, 0.89)$.

4.5 Ra-Fu DDEE Model

In this section, we use dependent chance operators and dependent chanceeoperators to deal with leader's model and follower's model respectively. This kind of model is denoted by DDEE model.

4.5.1 General Form of Ra-Fu DDEE Model

The DDEE model (4.1) is formulated as follows:

$$\begin{cases} \max\limits_{\boldsymbol{x}\in R^{n_1}}[\alpha^U(\boldsymbol{x},\boldsymbol{y})] \\ \text{s.t.} \begin{cases} Ch\{F(\boldsymbol{x},\boldsymbol{y},\boldsymbol{\xi})\geq \overline{F}\}(\alpha^U)\geq \beta^U \\ Ch\{G_i(\boldsymbol{x},\boldsymbol{y},\boldsymbol{\xi})\leq 0\}(\delta_i^U)\geq \gamma_i^U, i=1,2,\cdots,p_1 \\ \text{where } \boldsymbol{y} \text{ solves:} \\ \begin{cases} \max\limits_{\boldsymbol{y}\in R^{n_2}} E[f(\boldsymbol{x},\boldsymbol{y},\boldsymbol{\xi})] \\ \text{s.t. } E[g_j(\boldsymbol{x},\boldsymbol{y},\boldsymbol{\xi})], i=1,2,\cdots,p_2, \end{cases} \end{cases} \end{cases} \tag{4.71}$$

where $\beta^U, \delta_i^U, \gamma_i^U$ and $\overline{F}$ are predetermined parameters.

Similarly, the DDEE models for (4.2)

$$\begin{cases} \max\limits_{x \in R^{n_1}} [\alpha_1^U(x,y), \alpha_2^U(x,y), \cdots, \alpha_{m_1}^U(x,y)] \\ \text{s.t.} \begin{cases} Ch\{F_i(x,y,\xi) \geq \overline{F}_i\}(\alpha_i^U) \geq \beta_i^U, i = 1,2,\cdots,m_1 \\ Ch\{G_i(x,y,\xi) \leq 0\}(\delta_i^U) \geq \gamma_i^U, i = 1,2,\cdots,p_1 \\ \text{where } y \text{ solves:} \\ \begin{cases} \max\limits_{y \in R^{n_2}} [E[f_1(x,y,\xi)], E[f_2(x,y,\xi)], \cdots, E[f_{m_1}(x,y,\xi)]] \\ \text{s.t. } Ch\{g_j(x,y,\xi) \leq 0\}(\delta_i^L) \geq \gamma_i^L, i = 1,2,\cdots,p_2. \end{cases} \end{cases} \end{cases} \tag{4.72}$$

Since the there are different chances measures for Ra-Fu events, such as mean chance, equilibrium chance. So we can developed variants of DDEE models according to different chance measures.

This book focus on the case of multiple objectives, thus we present some solution concepts of (4.2) as follows.

Definition 4.8. Let x^* be given by the upper-level decision maker. Then (x^*, y^*) is said to be a feasible solution at (α_i^U, β_i^U)-levels to problem (4.46), if and only if (x^*, y^*) satisfies $Ch\{F_i(x,y,\xi) \geq \overline{F}_i\}(\alpha_i^U) \geq \beta_i^U, Ch\{G_i(x,y,\xi) \leq 0\}(\delta_i^U) \geq \gamma_i^U$, for each i, and is an efficient solution to the lower-level model.

Remark 4.5. That (x^*, y^*) is an efficient solution to the lower-level model at (α_i, β_i)-levels $(i = 1,2,\cdots,m_2)$ means (x^*, y^*) satisfies $Ch\{f_i(x,y,\xi) \geq \bar{f}_i\}(\alpha_i^L) \geq \beta_i^L$; $Ch\{g_j(x,y,\xi) \leq 0\}(\delta_i^L) \geq \gamma_i^L$ for each i and there is no other feasible solution (x,y) exists, such that $\alpha_i^L(x,y) \geq \alpha_i^L(x^*, y^*)$ and at least one $i = 1,2,\cdots,m_2$ is strict inequality.

Definition 4.9. Let (x^*, y^*) be a feasible solution to problem (4.72). Then (x^*, y^*) is the efficient solution to problem (4.72) if and only if there is no other feasible solution (x,y) exists, such that $\overline{F}_i(x,y) \geq \overline{F}_i(x^*, y^*)$ and at least one $i = 1,2,\cdots,m_1$ is strict inequality.

4.5.2 *Interactive Method for Linear Model*

If all the functions in (4.72) are linear, then we have the Ra-Fu linear DDEE (LDDEE) model as follows:

$$\begin{cases} \max\limits_{x \in R^{n_1}} [\alpha_1^U, \alpha_2^U, \cdots, \alpha_{m_1}^U] \\ \text{s.t.} \begin{cases} Ch\{\tilde{\overline{C}}_i^T x + \tilde{\overline{D}}_i^T y \geq \overline{F}_i\}(\alpha_i^U) \geq \beta_i^U, i = 1,2,\cdots,m_1 \\ Ch\{\tilde{\overline{A}}_i^T x + \tilde{\overline{B}}_i^T y \leq 0\}(\delta_i^U) \geq \gamma_i^U, i = 1,2,\cdots,p_1 \\ \text{where } y \text{ solves:} \\ \begin{cases} \max\limits_{y \in R^{n_2}} [E[\tilde{\bar{c}}_1^T x + \tilde{\bar{d}}_1^T y], E[\tilde{\bar{c}}_2^T x + \tilde{\bar{d}}_2^T y], \cdots, E[\tilde{\bar{c}}_{m_1}^T x + \tilde{\bar{d}}_{m_1}^T y]] \\ \text{s.t. } E[\tilde{\bar{a}}_i^T x + \tilde{\bar{b}}_i^T y] \leq 0, i = 1,2,\cdots,p_2. \end{cases} \end{cases} \end{cases} \tag{4.73}$$

If the Ch measure is defined as $Pos - Pr$ measure, then model (4.47) can be rewritten as

$$\begin{cases} \max_{x\in R^{n_1}} [\alpha_1^U, \alpha_2^U, \cdots, \alpha_{m_1}^U] \\ \text{s.t.} \begin{cases} Pr\{Pos\{\overline{\widetilde{C}}_i^T x + \overline{\widetilde{D}}_i^T y \geq \overline{F}_i\} \geq \alpha_i^U\} \geq \beta_i^U, i = 1, 2, \cdots, m_1 \\ Pr\{Pos\{\overline{\widetilde{A}}_i^T x + \overline{\widetilde{B}}_i^T y \leq 0\} \geq \delta_i^U\} \geq \gamma_i^U, i = 1, 2, \cdots, p_1 \\ \text{where } y \text{ solves:} \\ \begin{cases} \max_{y\in R^{n_2}} [E[\bar{\tilde{c}}_1^T x + \bar{\tilde{d}}_1^T y], E[\bar{\tilde{c}}_2^T x + \bar{\tilde{d}}_2^T y], \cdots, E[\bar{\tilde{c}}_{m_1}^T x + \bar{\tilde{d}}_{m_1}^T y]] \\ \text{s.t. } E[\bar{\tilde{a}}_i^T x + \bar{\tilde{b}}_i^T y] \leq 0, i = 1, 2, \cdots, p_2. \end{cases} \end{cases} \end{cases} \tag{4.74}$$

Crisp Equivalent Models

Due the complexity of (4.74) with bi-level structure, twofold uncertainties, multiple objectives, one way of solving problem (4.74) is to convert the objectives and constraints into their respective crisp equivalents and then reduce the solution complexity to some extent. In order to obtain the crisp equivalent models, some useful result are developed as follows.

Suppose that $\overline{\widetilde{C}}_r = (\overline{\widetilde{C}}_{i1}, \overline{\widetilde{C}}_{r2}, \cdots, \overline{\widetilde{C}}_{in})^T$ is a normally distributed Ra-Fu vector with the fuzzy mean vector $\widetilde{C}_r = (\widetilde{C}_{i1}, \widetilde{C}_{r2}, \cdots, \widetilde{C}_{in})^T$ and covariance matrix V_i^C, written as $\overline{\widetilde{C}}_i \sim \mathcal{N}(\widetilde{C}_i, V_i^C)$. Similarly, $\overline{\widetilde{D}}_i \sim \mathcal{N}(\widetilde{D}_i, V_i^D), \overline{\widetilde{A}}_i \sim \mathcal{N}(\widetilde{A}_i, V_i^A), \overline{\widetilde{B}}_i \sim \mathcal{N}(\widetilde{B}_i, V_i^B)$. $\widetilde{C}_i = (\widetilde{C}_{i1}, \widetilde{C}_2, \cdots, \widetilde{C}_{in}), \widetilde{D}_i = (\widetilde{D}_{i1}, \widetilde{D}_{i2}, \cdots, \widetilde{D}_{in}), \widetilde{A}_i = (\widetilde{A}_{i1}, \widetilde{A}_2, \cdots, \widetilde{A}_{in})$, $\widetilde{B}_i = (\widetilde{B}_{i1}, \widetilde{B}_{i2}, \cdots, \widetilde{B}_{in})$, are fuzzy vectors characterized as

$$\mu_{\overline{C}_{ij}} = [r_{1,C_{ij}}, r_{2,C_{ij}}, r_{3,C_{ij}}, r_{4,C_{ij}}],$$

$$\mu_{\overline{D}_{ij}} = [r_{1,D_{ij}}, r_{2,D_{ij}}, r_{3,D_{ij}}, r_{4,D_{ij}}],$$

$$\mu_{\overline{A}_{ij}} = [r_{1,A_{ij}}, r_{2,A_{ij}}, r_{3,A_{ij}}, r_{4,A_{ij}}],$$

$$\mu_{\overline{B}_{ij}} = [r_{1,B_{ij}}, r_{2,B_{ij}}, r_{3,B_{ij}}, r_{4,B_{ij}}].$$

It follows from Theorem 4.10. Then the upper model of model (4.48) is equivalent to

$$\begin{cases} \max_{x\in R^{n_1}} [\overline{F}_1', \overline{F}_2', \cdots, \overline{F}_{m_2}'] \\ \text{s.t.} \Phi^{-1}(\delta_i^L)\sqrt{x^T V_i^A x + y^T V_i^B y} + (1 - \gamma_i^L)\left(\sum_{j=1}^n (r_{1,A_{ij}} x_j + r_{1,B_{ij}} y_j)\right) \\ +\gamma_i^U \left(\sum_{j=1}^n (r_{2,A_{ij}} x_j + r_{2,B_{ij}} y_j)\right) \leq 0, i = 1, 2, \cdots, p_1, \end{cases} \tag{4.75}$$

where

$$\overline{F}'_i = \frac{\left(\Phi^{-1}(1-\beta_i^U)\sqrt{x^T V_i^C x + y^T V_i^D y} + \sum_{j=1}^{n}(r_{4,C_{ij}}x_j + r_{4,D_{ij}}y_j) - \bar{f}_i\right)}{\left(\sum_{j=1}^{n}(r_{4,C_{ij}}x_j + r_{4,D_{ij}}y_j) - \sum_{j=1}^{n}(r_{3,C_{ij}}x_j + r_{3,D_{ij}}y_j)\right)},$$

Φ denotes the standard normal distribution.

Interactive Method

In the section we introduce the interactive method for the crisp equivalent model. The process can be summarized by five steps as follows:

Step 1. Set $k = 0$, solve the upper-level decision-making problem to obtain a set of preferred solutions that are acceptable to the upper-level decision maker; the upper-level decision maker then puts the solutions in order in the format as follows:
Preferred solution

$$(x^{(k)}, y^{(k)}), (x^{(k+1)}, y^{(k+1)}), \cdots, (x^{(k+p)}, y^{(k+p)}),$$

Preferred ranking

$$(x^{(k)}, y^{(k)}) \succ (x^{(k+1)}, y^{(k+1)}) \succ \cdots \succ (x^{(k+p)}, y^{(k+p)}).$$

Step 2. Given $x^{(k)}$ to the lower level, solve the lower-level decision-making problem and obtain $\bar{y}^{(k)}$.
Step 3. If

$$\frac{||F_0(x^{(k)}, y^{(k)}) - F_0(x^{(k)}, \bar{y}^{(k)})||}{||F_0(x^{(k)}, y^{(k)})||} \leq \sigma,$$

(where σ is a fairly small positive number) go to Step 4. Otherwise, go to Step 5.
Step 4. If the upper-level decision maker is satisfied with $(x^{(k)}, \bar{y}^{(k)})$ and $F_0(x^{(k)}, \bar{y}^{(k)})$, then $(x^{(k)}, \bar{y}^{(k)})$ is the preferred solution to the bi-level decision-making problem. Otherwise, go to Step 5.
Step 5. Let $k = k + 1$, and go to Step 2.

Numerical Examples

In this section, a numerical example is given to illustrate the effectiveness of the application of the model and algorithm proposed above.

Example 4.8. Consider the problem

$$\begin{cases} \max_{x_1,x_2} F_1 = \xi_1^2 x_1 + 2\xi_2 x_2^2 + \sqrt{3\xi_3 y_1} + 4\sqrt{\xi_4} y_2 \\ \max_{x_1,x_2} F_2 = 5\sqrt{\xi_1 x_1} + 3\xi_2 x_2^3 + 2\xi_3 y_1 + \xi_4 y_2 \\ \text{where } (y_1, y_2, y_3) \text{ solves} \\ \quad \begin{cases} \max_{y_1,y_2} f_1 = 7\xi_1 x_1 + 5\sqrt{\xi_2 x_2} + 4\xi_3 y_1 + \xi_4 y_2 \\ \max_{y_1,y_2} f_2 = 3\xi_1 x_1 + 4\xi_2 x_2 + 5\xi_3 y_1 + 6\xi_4^2 y_2 \\ \quad \text{s.t.} \begin{cases} x_1 + x_2 + y_1 + y_2 \geq 120 \\ x_1 + x_2 + y_1 + y_2 \leq 200 \\ x_1, x_2, x_3, y_1, y_2, y_3 \geq 10, \end{cases} \end{cases} \end{cases} \tag{4.76}$$

where $\xi_j, j = 1, 2, 3, 4$ are Ra-Fu variable characterized as

$\tilde{\bar{\xi}}_1 \sim \mathcal{N}(\tilde{u}_1, 1)$, with $\tilde{u}_1 \sim (3, 6, 8, 11)$, $\tilde{\bar{\xi}}_2 \sim \mathcal{N}(\tilde{u}_2, 1)$, with $\tilde{u}_2 \sim (7, 10, 11, 16)$,
$\tilde{\bar{\xi}}_3 \sim \mathcal{N}(\tilde{u}_3, 1)$, with $\tilde{u}_3 \sim (2, 6, 8, 10)$, $\tilde{\bar{\xi}}_4 \sim \mathcal{N}(\tilde{u}_4, 1)$, with $\tilde{u}_4 \sim (5, 10, 20, 30)$,

If the decision makers determine the values of confidence levels and reference levels , the DDEE model for (4.76) is

$$\begin{cases} \max_{x_1,x_2} [\alpha_1^U, \alpha_2^U] \\ \text{s.t.} \begin{cases} Ch\{\xi_1^2 x_1 + 2\xi_2 x_2^2 + \sqrt{3\xi_3 y_1} + 4\sqrt{\xi_4} y_2 \geq 100\}(\alpha_1^U) \geq 0.6 \\ Ch\{5\sqrt{\xi_1 x_1} + 3\xi_2 x_2^3 + 2\xi_3 y_1 + \xi_4 y_2 \geq 120\}(\alpha_2^U) \geq 0.8 \\ \text{where } (y_1, y_2) \text{ solves} \\ \quad \begin{cases} \max_{y_1,y_2} [E[7\xi_1 x_1 + 5\sqrt{\xi_2 x_2} + 4\xi_3 y_1 + \xi_4 y_2], \\ E[3\xi_1 x_1 + 4\xi_2 x_2 + 5\xi_3 y_1 + 6\xi_4^2 y_2]] \\ \quad \text{s.t.} \begin{cases} x_1 + x_2 + y_1 + y_2 \geq 120 \\ x_1 + x_2 + y_1 + y_2 \leq 200 \\ x_1, x_2, x_3, y_1, y_2, y_3 \geq 10. \end{cases} \end{cases} \end{cases} \end{cases} \tag{4.77}$$

We know that the problem (4.88) is equivalent to model

$$\begin{cases} \max\limits_{x_1,x_2} \left[\frac{8x_1+5x_2+12y_1+7y_2-1500}{6(2x_1+x_2+2y_1+y_2)}, \frac{2x_1+5x_2+6_1+14y_2-1500}{3(x_1+2x_2+2y_1+4y_2)}\right] \\ \text{where } (y_1, y_2, y_3) \text{ solves} \\ \begin{cases} \max\limits_{y_1,y_2} \left[\frac{4x_1+5x_2+12y_1+14y_2-1800}{6(x_1+2x_2+3y_1+2y_2)}, \frac{8x_1+5x_2+12y_1+14y_2-1800}{6(2x_1+x_2+2y_1+2y_2)}\right] \\ \text{s.t.} \begin{cases} x_1 + x_2 + y_1 + y_2 \geq 100 \\ x_1 + x_2 + y_1 + y_2 \leq 200 \\ x_1, x_2, y_1, y_2 \geq 10 \end{cases} \end{cases} \end{cases} \tag{4.78}$$

After running of interactive method, we obtain $(x_1^*, x_2^*, y_1^*, y_2^*) = (30.3, 32.4, 31.5, 83.8)$ and $(\alpha_1^{U*}, \alpha_2^{U*}) = (0.72, 0.93)$.

4.5.3 *Ra-Fu Simulation-Based PGSA-PSO for Nonlinear Model*

As (4.72) is a nonlinear model, i.e., some functions in model (4.72) is nonlinear, the Ra-Fu features can not be converted into its crisp equivalent model. In order solve nonlinear (4.72), three issues are needed to solve: the trade-off of the multiple objectives, the Ra-Fu parameters and the bi-level structure. In this section, we design a hybrid method consisting of Ideal Point Method, Ra-Fu simulations techniques and hybrid PGSA-PSO.

Ideal Point Method

In this section, we make use of the ideal point method proposed in [36] to resolve the multiple objective problem with crisp parameters.

$$\max_{x \in X} [f_1(\boldsymbol{x}), f_2(\boldsymbol{x}), \cdots, f_m(\boldsymbol{x})]. \tag{4.79}$$

If the decision maker can first propose an estimated value $\bar{F}_i$ for each objective function $\Psi_i^c \boldsymbol{x}$ such that

$$\bar{F}_i \geq \max_{x \in X'} \Psi_1^c \boldsymbol{x} i = 1, 2, \cdots, m, \tag{4.80}$$

where $X' = \{\boldsymbol{x} \in X | \Psi_r^e \boldsymbol{x} \leq \Psi_r^b, r = 1, 2, \cdots, p, \boldsymbol{x} \geq 0\}$, then $\bar{\mathbf{F}}_i = (\bar{F}_1, \bar{F}_2, \cdots, \bar{F}_m)^T$ is called the ideal point. Specially, if $\bar{F}_i \geq \max_{x \in X'} \Psi_1^c \boldsymbol{x}$ for all i, we call $\bar{\mathbf{F}}$ the most ideal point.

The basic theory of the ideal point method is to take an especial norm in the objective space $\mathbf{R}^m$ and obtain the feasible solution $\boldsymbol{x}$ that the objective value approaches the ideal point $\bar{\mathbf{F}} = (\bar{F}_1, \bar{F}_2, \cdots, \bar{F}_m)^T$ under the norm distance, that is, to seek the feasible solution $\boldsymbol{x}$ satisfying

$$\min_{x \in X'} u(\Psi^c(\boldsymbol{x})) = \min_{x \in X'} ||\Psi^c(\boldsymbol{x}) - \bar{F}||.$$

Usually, the following norm functions are used to describe the distance:

1. p-mode function

$$d_p(\Psi^c(\boldsymbol{x}), \bar{F}; \boldsymbol{\omega}) = \left[\sum_{i=1}^{m} \omega_i |\Psi_i^c \boldsymbol{x} - \bar{F}_i|^p \right]^{\frac{1}{p}}, \ 1 \le p < +\infty. \tag{4.81}$$

2. The maximal deviation function

$$d_{+\infty}(\Psi^c(\boldsymbol{x}), \bar{F}; \boldsymbol{\omega}) = \max_{1 \le i \le m} \omega_i |\Psi_i^c \boldsymbol{x} - \bar{F}_i|. \tag{4.82}$$

3. Geometric mean function

$$d(\Psi^c(\boldsymbol{x}), \bar{F}) = \left[\prod_{i=1}^{m} |\Psi_i^c \boldsymbol{x} - \bar{F}_i|^p \right]^{\frac{1}{m}}. \tag{4.83}$$

The weight parameter vector $\boldsymbol{\omega} = (\omega_1, \omega_2, \cdots, \omega_m)^T > 0$ needs to be predetermined.

Theorem 4.13 (F. Szidarovszky et al. [36]). *Assume that* $\bar{F}_i > \max_{x \in X'} \Psi_1^c \boldsymbol{x} (i = 1, 2, \cdots, m)$. *If* $\boldsymbol{x}^*$ *is the optimal solution to the following problem:*

$$\min_{x \in X'} d_p(\Psi^c(\boldsymbol{x}), \bar{F}; \boldsymbol{\omega}) = \left[\sum_{i=1}^{m} \omega_i |\Psi_i^c \boldsymbol{x} - \bar{F}_i|^p \right]^{\frac{1}{p}}, \tag{4.84}$$

then $\boldsymbol{x}^*$ *is an efficient solution to problem (4.79). On the contrary, if* $\boldsymbol{x}^*$ *is an efficient solution of problem (4.79), then there exists a weight vector* $\boldsymbol{\omega}$ *such that* $\boldsymbol{x}^*$ *is the optimal solution to problem (4.84).*

Next, we take the p-mode function to describe the procedure for solving the problem (4.84).

Step 1. Find the ideal point. If the decision maker can give the ideal objective value satisfying the condition (4.80), the value will be considered as the ideal point. However, decision makers themselves do not know how to give the objective value. Then we can get the ideal point by solving the following problem:

$$\begin{cases} \max \Psi_i^c \boldsymbol{x} \\ \text{s.t.} \begin{cases} \Psi_r^e x \le \Psi_r^b, r = 1, 2, \cdots, p \\ \boldsymbol{x} \in X. \end{cases} \end{cases} \tag{4.85}$$

Then the ideal point $\bar{\mathbf{F}} = (\bar{F}_1, \bar{F}_2, \cdots, \bar{F}_m)^T$ can be fixed by $\bar{F}_i = \Psi_i^c \boldsymbol{x}^*$, where $\boldsymbol{x}^*$ is the optimal solution to problem (4.85).

Step 2. Fix the weight. The method of selecting the weight can be referred to in the literature, and interested readers can consult them. We usually use the following function to fix the weight:

$$\omega_i = \frac{\bar{F}_i}{\sum_{i=1}^{m} \bar{F}_i}.$$

Step 3. Construct the minimal distance problem. Solve the following single objective programming problem to obtain the efficient solution to problem:

$$\begin{cases} \min \left[\sum_{i=1}^{m} \omega_i |\Psi_i^c \boldsymbol{x} - \bar{F}_i|^t \right]^{\frac{1}{t}} \\ \text{s.t.} \begin{cases} \Psi_r^e \boldsymbol{x} \le \Psi_r^b, r = 1, 2, \cdots, p \\ \boldsymbol{x} \in X. \end{cases} \end{cases} \tag{4.86}$$

Usually we take $t = 2$ to compute it.

Ra-Fu Simulation for Dependent Chance and Expected Operator

The Ra-Fu simulation is used to deal with those which cannot be converted into crisp ones. In this section, we need to simulate the dependent chance in the upper-level programming and the expected operator in the lower-level programming. The Ra-Fu simulation for the dependent chance can be seen in section "Ra-Fu Simulation for Dependent Chance" and the Ra-Fu simulation for the expected operator can be seen in section "Ra-Fu Simulation for Expected Operator and Chance Operator".

Hybrid PGSA-PSO

In the proposed solving method, solving Ra-Fu multi-objective bi-level model (4.2) is transformed to solve crisp single objective bi-level model by ideal Point Method and Ra-Fu simulation. To handle the bi-level structure, we solve the upper-level and lower-level programming problem interactively by hybrid PGSA-PSO while determining respectively the decision variables of the upper level ($\boldsymbol{x}$) or the lower-level ($\boldsymbol{y}$). If the decision variable in the upper-level programming ($\boldsymbol{x}$) is non-integer, we can use the PSO in the upper-level and PGSA in the lower-level. If the decision variable in the lower-level programming (y) is non-integer, we can use the PGSA in the upper-level and PSO in the lower-level.

Such that, the upper-level decision variable ($\boldsymbol{x}$) is integer and the lower-level decision variable is non-integer, the hybrid PGSA-PSO overall procedure for solving Ra-Fu multi-objective bi-level model (4.2) is as follows:

Step 1. Input all coefficients' values for model (4.2). Set initial iteration $\tau_1 = 1$. Initialize upper-level basis point $X^0 = \{x_j^0, j = 1, 2, \cdots, N\}$ in the range of $[0, Range]$. Let $X_{\min} = X^0$, $F_{\min} = F(X^0)$.

Step 2. Solve the lower-level programming using the PSO procedure with upper-level initialized basis point X^0.

Step 2.1. Set initial iteration $\tau_2 = 1$ and PSO parameters ω, c_1, c_2. Initialize l_2 particles to represent the lower-level programming $\boldsymbol{y}$, including positions $P^{l_2}(\tau_2)$ and $V^{l_2}(\tau_2)$.

Step 2.2. Calculate fitness value of lower-level. If there are multiple objectives in the lower-level, use the ideal point method to transfer the multiple objectives to a single objective. Find the personal best position $P^{l_2,best}(\tau_2)$ and the global best position $G^{best}(\tau_2)$.

Step 2.3. Update particles' positions and velocity by the updating equations of PSO:

$$
\begin{aligned}
V^{l_2}(\tau_2+1) &= \omega \times V^{l_2}(\tau_2) + c_1 \times rand() \times (P^{l_2,best}(\tau_2) \\
&\quad - P^{l_2}(\tau_2)) + c_2 \times rand() \times (G^{best}(\tau_2) - P^{l_2}(\tau_2)), \\
P^{l_2}(\tau_2+1) &= P^{l_2}(\tau_2) + V^{l_2}(\tau_2).
\end{aligned}
$$

Step 2.4. Check constraints. Eliminate or adjust infeasible $P^{l_2}(\tau_2)$. Evaluate updated particles by the fitness values and find the new $P^{l_2,best}(\tau_2)$ and $G^{best}(\tau_2)$.

Step 2.5. Check the stop criteria. If the stop criteria is met, output the $G^{best}(\tau_2)$ as the optimal solution of the lower level y. If not, $\tau_2 = \tau_2 + 1$ and go back to Step 2.2.

Step 3. Solve the upper-level programming using the PGSA procedure.

Step 3.1. Set the appropriate step length λ depending on the size of search region. λ shall be integer.

Step 3.2. Calculate the objective function value $F(x, y)$ in the upper-level by the upper-level basis point X^0 and the optimal solution found in the lower-level $y*$. If there are multiple objectives in the upper-level, transform the multiple objectives to a single objective by the Minimax Point Method presented in section "Minimax Point Method". Set $X_{\min} = X^0$, $F_{\min} = F(X^0, y)$.

Step 3.3. Find the trunk M and each preferential growth node $X^0_{a1,b1}$, on the trunk in the upper-level programming, $a1 = 1, 2, \cdots, N, b1 = 1, 2, \cdots, m1$, $m1$ is the maximum number of growth nodes on the $a1$th trunk. Then eliminate infeasible growth nodes.

Step 3.4. Calculate the objective function value of each preferential growth node $X^0_{a1,b1}$. If any $F(X^0_{a1,b1}, y) < F_{\min}$, replace $F_{\min} = F(X^0_{a1,b1}, y)$.

Step 3.5. Calculate the morphactin concentrations by Eq. (4.40). Generate a random number β_0 in the interval [0, 1] and find a new basic node.

Step 3.6. Start from the new basis point, find new growth node and calculate the morphactin concentrations. Replace $F_{\min}$ constantly.

Step 4. When $f'_{\min}$, $F_{\min}$ repeats appearing many times (e.g., 50 times), take $F^* = F_{\min}, X^* = X_{\min}$ as the optimal solution. Otherwise, use the $X_{\min}$ as the parameter for lower-level programming and go back to Step 2.

In the proposed hybrid PGSA-PSO, solving multi-objective bi-level programming is transformed to solve the upper-level and lower-level programming problem interactively while determining respectively the decision variables of the upper level or the lower-level. The solution information is exchanged between the two levels of programming. The output of one level is the input of another level, namely "**y**" — the output of the lower-level programming is the input of the upper-level programming and "**x**" — the outputs of the upper-level programming are the inputs of the lower-level programming. These form a hierarchical and sequential framework as shown in Fig. 4.6.

Numerical Examples

In this section, a numerical example is given to illustrate the effectiveness of the application of the model and algorithm proposed above.

Example 4.9. Consider the problem

$$\begin{cases} \max\limits_{x_1,x_2} F_1 = \xi_1 x_1 + 2\xi_2 x_2 + 3\xi_3 y_1 + 4\xi_4 y_2 \\ \max\limits_{x_1,x_2} F_2 = 5\xi_1 x_1 + 3\xi_2 x_2 + 2\xi_3 y_1 + \xi_4 y_2 \\ \text{where } (y_1, y_2, y_3) \text{ solves} \\ \quad \begin{cases} \max\limits_{y_1,y_2} f_1 = 7\xi_1 x_1 + 5\xi_2 x_2 + 4\xi_3 y_1 + \xi_4 y_2 \\ \max\limits_{y_1,y_2} f_2 = 3\xi_1 x_1 + 4\xi_2 x_2 + 5\xi_3 y_1 + 6\xi_4 y_2 \\ \text{s.t.} \begin{cases} x_1 + x_2 + y_1 + y_2 \geq 120 \\ x_1 + x_2 + y_1 + y_2 \leq 200 \\ x_1, x_2, y_1, y_2 \geq 10 \\ x_1, x_2 \text{ are integer,} \end{cases} \end{cases} \end{cases} \tag{4.87}$$

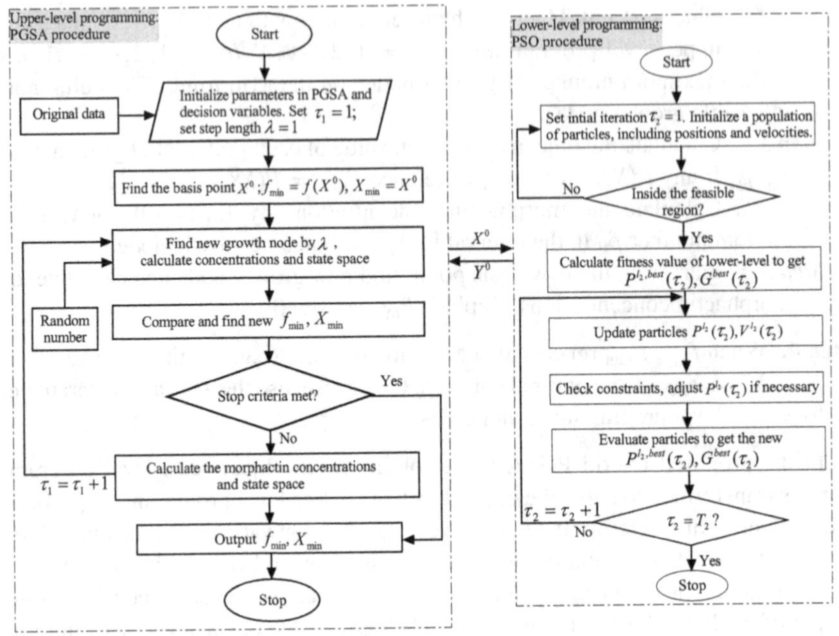

Fig. 4.6 The framework of the hybrid PGSA-PSO

where $\xi_j, j = 1, 2, 3, 4$ are Ra-Fu variable characterized as

$$\tilde{\tilde{\xi}}_1 \sim \mathcal{N}(\tilde{u}_1, 1), \text{ with } \tilde{u}_1 \sim (3, 6, 8, 11),\ \tilde{\tilde{\xi}}_2 \sim \mathcal{N}(\tilde{u}_2, 1), \text{ with } \tilde{u}_2 \sim (7, 10, 11, 16),$$
$$\tilde{\tilde{\xi}}_3 \sim \mathcal{N}(\tilde{u}_3, 1), \text{ with } \tilde{u}_3 \sim (2, 6, 8, 10),\ \tilde{\tilde{\xi}}_4 \sim \mathcal{N}(\tilde{u}_4, 1), \text{ with } \tilde{u}_4 \sim (5, 10, 20, 30),$$

If the decision makers determine the values of confidence levels and reference levels , the DDEE model for (4.87) is

$$\begin{cases} \max\limits_{x_1,x_2}[\alpha_1^U, \alpha_2^U] \\ \text{s.t.} \begin{cases} Ch\{\xi_1 x_1 + 2\xi_2 x_2 + 3\xi_3 y_1 + 4\xi_4 y_2 \geq \overline{F}_1\}(\alpha_1^U) \geq 0.6 \\ Ch\{5\xi_1 x_1 + 3\xi_2 x_2 + 2\xi_3 y_1 + \xi_4 y_2 \geq \overline{F}_2\}(\alpha_2^U) \geq 0.8 \\ \text{where } (y_1, y_2) \text{ solves} \\ \begin{cases} \max\limits_{y_1,y_2}[E[7\xi_1 x_1 + 5\xi_2 x_2 + 4\xi_3 y_1 + \xi_4 y_2], E[3\xi_1 x_1 + 4\xi_2 x_2 + 5\xi_3 y_1 + 6\xi_4 y_2]] \\ \text{s.t.} \begin{cases} x_1 + x_2 + y_1 + y_2 \geq 120 \\ x_1 + x_2 + y_1 + y_2 \leq 200 \\ x_1, x_2, y_1, y_2 \geq 10 \\ x_1, x_2 \text{ are integer.} \end{cases} \end{cases} \end{cases} \end{cases} \tag{4.88}$$

After running of above method, we obtain $(x_1^*, x_2^*, y_1^*, y_2^*) = (32, 36, 22.1, 82.4)$ and $(\alpha_1^{U*}, \alpha_2^{U*}) = (0.7, 0.9)$.

4.6 Security Planning of the LT Hydropower Construction Project

The case problem is from the LT hydropower construction project. In the following, the project description, case solution and discussions are presented.

The LT Hydropower Project [3] is located on Red River, Guangxi Autonomous Region. It is one of the top-ten key projects of the Great Western Development Plan and the strategic projects of "power transmission from west to east". The main functions of Longtan Hydropower Project are power generation incorporated with flood control, navigation, etc. It is designed as a grade-I project structure. The project has 6,300 MW of total installed capacity. Its layout contains: a roller compacted concrete gravity dam, a flood building with seven outlets, two bottom outlets on the river bed, a power stream system capacity with nine installations on the left bank, and navigation structures on the right, a 2-stage vertical ship lift which is used for navigation. During the construction process, tunnel diversion is used to divert the river with two diversion openings on the left and right banks respectively (see Fig. 4.7) [3]. There are 14 facilities involved in the principal part of the Longtan hydropower construction project (see Table 4.1 and Fig. 4.8).

4.6.1 *Modelling*

To model CSSP, the following assumptions are made.

(1) Offensive and protective resources are both limited so that at most R facilities can be attacked, and not all facilities can be secured.
(2) Attacker can only attack unsecured facilities.
(3) The goal of attackers is to inflict maximum effectiveness loss of the construction facilities system, while the goals of defender are to maximize the effectiveness of the construction facilities system and minimize the economic loss and security cost.
(4) The attack can be classified into five degrees, and the occurrence probabilities of different degrees are $(\{p_1, p_2, p_3, p_4, p_5\})$.
(5) The economic loss and efficiency loss are depicted by discrete Ra-Fu variables. The distribution and random parameters can be obtained by the defender.

Fig. 4.7 The principal part of the Longtan hydropower project

Table 4.1 Facilities involved in the principal part of the Longtan hydropower construction project

Index	Facility	M_j (CNY)
F_1	Reinforcing steel shop	18,000
F_2	Carpentry shop	21,000
F_3	Concrete precast shop	20,000
F_4	Drill tools repair shop	12,000
F_5	Equipment repairing workshop	23,000
F_6	Truck maintenance shop	21,000
F_7	Metal and electrical installing workshop	20,000
F_8	Oil depot	15,000
F_9	Explosive storage	15,000
F_{10}	Rebar storage	14,000
F_{11}	Steel storage	16,000
F_{12}	Integrated warehouse	18,000
F_{13}	Office and administrative area	23,000
F_{14}	Labor residence	23,000

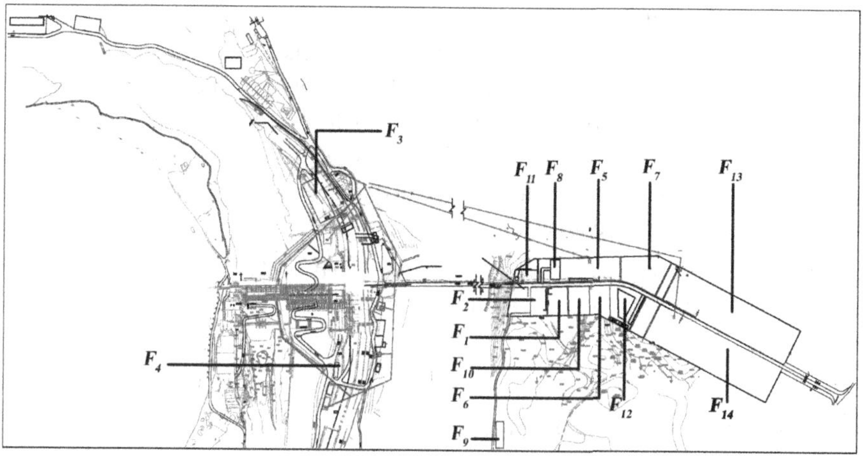

Fig. 4.8 The construction site layout of the Longtan hydropower construction project

The following symbols are used:

Indices:

i,j : facilities index, $i,j \in \Phi = \{1,2,\cdots,N\}$.

Variables:

$$z_j = \begin{cases} 1, \text{ if facility } j \text{ is secured} \\ 0, \text{ otherwise.} \end{cases}$$

$$s_j = \begin{cases} 1, \text{ if facility } j \text{ is attacked} \\ 0, \text{ otherwise.} \end{cases}$$

Certain parameters:

R : denotes that the attacker would be able to attack at most R facilities, $R \leq N$.

T : the limitation of funds for defender securing facilities.

M_j : the cost of securing facility j.

θ_{ij} : the weight of demand's importance, $0 \leq \theta_{ij} \leq 1$.

$$d_{ij} = \begin{cases} 1, \text{ if demand of facility } i \text{ is served by facility } j \\ 0, \text{ otherwise.} \end{cases}$$

C_0 : the acceptable maximum economic loss of defender.

D_0 : the acceptable minimum resources supply rate.

p_k : the occurrence probabilities of the kth degree attack, $k \in \{1,\cdots,5\}$.

Uncertain coefficients:

$\tilde{\bar{C}}_j$: the economic loss when facility j is attacked.
$\tilde{C}_{jk}$: the kth degree economic loss when facility j is attacked.
$\tilde{\bar{r}}_{ij}$: the fill rate of facility j to facility i when facility j is attacked. Obviously, when $d_{ij} = 0$, $\tilde{\bar{r}}_{ij} \equiv 0$, when $d_{ij} = 1$, $\tilde{\bar{r}}_{ij} \leq d_{ij}$.
$\tilde{r}_{ijk}$: the fill rate of facility j to facility i when facility j is attacked with the kth degree damage.

Objective functions and constraints are presented in this section. To make the model calculable, this section transforms uncertain formulations into calculable ones.

Upper-Level Model for the Bi-Level CSSP

1. Objective Functions

The decision maker on the upper-level is the defender. The defender has both effectiveness and economic objectives. In this chapter, facility efficiency is measured by the level of demand satisfaction. z_j denotes the security of facility j. $d_{ij} = 1$ denotes facility i is supplied by facility j. θ_{ij} denotes the weight of demand importance. Considering the level of different importance, the efficiency of secured facilities is $\sum_{j=1}^{N}\sum_{i=1,i\neq j}^{N}\theta_{ij}d_{ij}z_j$. When facilities are attacked, the validity or the efficiency is reduced. s_j denotes the attack, and $\tilde{\bar{r}}_{ij}$ denotes the fill rate of facility j to facility i when facility j is attacked. Considering the importance weight, the efficiency of attacked facilities is $\sum_{j=1}^{N}\sum_{i=1,i\neq j}^{N}\theta_{ij}d_{ij}\tilde{\bar{r}}_{ij}s_j$. The efficiency of unattacked and unsecured facilities is $\sum_{j=1}^{N}\sum_{i=1,i\neq j}^{N}\theta_{ij}d_{ij}(1-s_j)(1-z_j)$. Summing up the above, the efficiency objective (denoted as D) is to maximize the following:

$$\sum_{j=1}^{N}\sum_{i=1,i\neq j}^{N}\theta_{ij}d_{ij}z_j + \sum_{j=1}^{N}\sum_{i=1,i\neq j}^{N}\theta_{ij}d_{ij}\tilde{\bar{r}}_{ij}s_j + \sum_{j=1}^{N}\sum_{i=1,i\neq j}^{N}\theta_{ij}d_{ij}(1-s_j)(1-z_j). \tag{4.89}$$

Another objective in upper-level programming is to minimize the security cost and economic loss. M_j denotes the cost of securing facility j, so the total security cost is $\sum_{j=1}^{N}M_j z_j$. $\tilde{\bar{C}}_j$ denotes the economic loss when facility j is attacked. Therefore the economic loss in the potential attack is $\sum_{j=1}^{N}\tilde{\bar{C}}_j s_j$. The cost objective (denoted as C) is to minimize the following:

$$\sum_{j=1}^{N}\tilde{\bar{C}}_j s_j + \sum_{j=1}^{N}M_j z_j. \tag{4.90}$$

There are uncertain variables $\tilde{\tilde{r}}_{ij}$ and $\tilde{\tilde{C}}_j$ in the objective functions. The mathematical significance of objective functions are ambiguous unless they are transformed to equivalent crisp functions. This chapter applies the expected value method to transform the objective functions into crisp ones. The following will be used during the transformation.

Definition 4.10 ([43]). Let Ω be a nonempty set, $\mathscr{A}$ is a $\sigma-$algebra of subsets of Ω, Pr is a probability measure. Then the triplet $(\Omega, \mathscr{A}, Pr)$ is called a probability space.

Definition 4.11 ([43]). A Ra-Fu variable ξ is a mapping from a probability space $(\Omega, \mathscr{A}, Pr)$ to a collection $\mathscr{S}$ of random variables such that for any Borel subset B of the real line $\Re$, the induced function $\Pr\{\xi(\omega) \in B\}$ is a measurable function with respect to ω.

Theorem 4.14 ([46]). *(Expected value of the discrete Ra-Fu variable with finite events). Let p_i be the probability that a Ra-Fu event $\tilde{\tilde{\boldsymbol{\xi}}} = \xi_i, (i = 1, 2, \cdots, n)$ occurs. Where $\tilde{\tilde{\xi}}_i$ is a random variable defined on the probability space $(\Omega, \mathscr{A}, Pr)$, and the expected value of each $\tilde{\tilde{\xi}}_i$ is $E[\tilde{\tilde{\xi}}_i] = E_i$. Then the expected value of $\tilde{\tilde{\xi}}$ is $\sum_{i=1}^{n} p_i E_i$.*

Lemma 4.2 ([46]). *Assume that $\tilde{\tilde{\xi}}$ and $\tilde{\tilde{\zeta}}$ are Ra-Fu variables with finite expected values. Then for any real numbers a and b, we have*

$$E\left[a\tilde{\tilde{\xi}} + b\tilde{\tilde{\zeta}}\right] = aE\left[\tilde{\tilde{\xi}}\right] + bE\left[\tilde{\tilde{\zeta}}\right]. \tag{4.91}$$

From Theorem 3.1 and Lemma 3.1, the expected value of the two objectives on the upper level can be transformed as the following:

$$\begin{aligned} E[D] &= E\left\{\sum_{j=1}^{N}\sum_{i=1,i\neq j}^{N}\left(\theta_{ij}\tilde{\tilde{r}}_{ij}d_{ij}s_j + \theta_{ij}d_{ij}z_j + \theta_{ij}d_{ij}(1-s_j)(1-z_j)\right)\right\} \\ &= \sum_{j=1}^{N}\sum_{i=1,i\neq j}^{N}\left(\theta_{ij}E\left[\tilde{\tilde{r}}_{ij}\right]d_{ij}s_j + \theta_{ij}d_{ij}z_j + \theta_{ij}d_{ij}(1-s_j)(1-z_j)\right), \end{aligned} \tag{4.92}$$

$$E[C] = E\left\{\sum_{j=1}^{N}\tilde{\tilde{C}}_j s_j + \sum_{j=1}^{N}M_j z_j\right\} = \sum_{j=1}^{N}E\left[\tilde{\tilde{C}}_j\right]s_j + \sum_{j=1}^{N}M_j z_j. \tag{4.93}$$

Such that

$$\tilde{\tilde{C}}_j = \begin{cases} \tilde{C}_{j1} \sim \mathscr{N}(\mu_{j1}^c, \delta_{j1}^{c2}), & \text{with probability } p_1 \\ \vdots & \vdots \\ \tilde{C}_{j5} \sim \mathscr{N}(\mu_{j5}^c, \delta_{j5}^{c2}), & \text{with probability } p_5, \end{cases}$$

and

$$\tilde{\bar{r}}_{ij} = \begin{cases} \tilde{r}_{ij1} \sim \mathcal{N}(\mu^r_{ij1}, \delta^{r2}_{ij1}), & \text{with probability } p_1 \\ \vdots & \vdots \\ \tilde{r}_{ij5} \sim \mathcal{N}(\mu^r_{ij5}, \delta^{r2}_{ij5}), & \text{with probability } p_5, \end{cases}$$

therefore, $E[\tilde{\bar{C}}_j]$ in Eq. (4.93) and $E[\tilde{\bar{r}}_{ij}]$ in Eq. (4.92) can be further transformed into $\sum_{k=1}^5 p_k\mu^c_{jk}$ and $\sum_{k=1}^5 p_k\mu^r_{ijk}$, where μ^c_{jk} and μ^r_{ijk} are the mean values of $\tilde{\bar{C}}_j$ and $\tilde{\bar{r}}_{ij}$ under different degrees of damage respectively.

In conclusion, objective functions (4.89) and (4.90) in upper-level programming are transformed into:

$$\max D = \sum_{j=1}^{N}\sum_{i=1,i\neq j}^{N} \theta_{ij}d_{ij}\left(\sum_{k=1}^{n} p_k\mu^r_{ijk}s_j + \theta_{ij}d_{ij}z_j + \theta_{ij}d_{ij}(1-s_j)(1-z_j)\right), \tag{4.94}$$

$$\min C = \sum_{j=1}^{N}\sum_{k=1}^{n} p_k\mu^c_{jk}s_j + \sum_{j=1}^{N} M_jz_j. \tag{4.95}$$

2. Constraints

The constraints include funding constraint, acceptable economic loss constraint, acceptable resources supply rate constraint and logical constraints. It costs M_j to secure facility j, and the fund limitation of the defender is T. Therefore, the fund constraint is:

$$\sum_{j=1}^{N} M_jz_j \leq T, \forall j \in \Phi. \tag{4.96}$$

In addition, C_0 denotes the acceptable maximum economic losses of the defender, while D_0 denotes the acceptable minimum resources supply rate. Therefore, the economic loss of the defender $\sum_{j=1}^N \tilde{\bar{C}}_js_j$ should be no more than C_0, and the resources supply rate in the situation of attack $\sum_{i=1,i\neq j}^N \theta_{ij}\tilde{\bar{r}}_{ij}d_{ij}s_j$ should be no less than D_0. Technically, the decision maker cannot strictly ensure that the random event $\sum_{j=1}^N \tilde{\bar{C}}_js_j$ doesn't exceed C_0, similarly, the decision maker cannot strictly ensure that the random event $\sum_{i=1,i\neq j}^N \theta_{ij}\tilde{\bar{r}}_{ij}d_{ij}s_j$ isn't below D_0. In this case, the chance-constrained programming proposed by Charnes and Cooper [6] can serve as a useful tool. The constraints have to be satisfied to a certain extent, i.e., the probability of the corresponding random event shall be maximized under a given confidence level. Accordingly, the constraints are written in the following *Pr-Pr* format:

$$Pr\left\{\omega \left| Pr\left\{\sum_{j=1}^{N} \tilde{\bar{C}}_j(\omega)^T s_j \leq C_0\right\} \geq \theta_1\right.\right\} \geq \delta_1, \tag{4.97}$$

and

$$Pr\left\{\omega \left| Pr\left\{\sum_{i=1,i\neq j}^{N} \theta_{ij}\tilde{\bar{r}}_{ij}(\omega)^T d_{ij}s_j \geq D_0\right\} \geq \theta_2\right.\right\} \geq \delta_2, \tag{4.98}$$

where $\theta_1, \theta_2, \delta_1$ and δ_2 are predefined confidence levels. Hence, uncertain constraints are transformed into calculable ones.

Lower-Level Model for the Bi-Level CSSP

The lower-level decision maker is the attacker. The objective and constraints are as follows.

1. Objective Function

$\tilde{\bar{r}}_{ij}$ denotes the fill rate of facility j to i when facility j is attacked. Considering the importance weights of different demands, the effectiveness of facilities system is expressed as:

$$\sum_{j=1}^{N}\sum_{i=1,i\neq j}^{N} \theta_{ij}d_{ij}\tilde{\bar{r}}_{ij}s_j + \sum_{j=1}^{N}\sum_{i=1,i\neq j}^{N} \theta_{ij}d_{ij}(1-s_j). \tag{4.99}$$

The objective of attacker is to minimize the effectiveness. There is also a Ra-Fu variable $\tilde{\bar{r}}_{ij}$ in Eq. (4.99). Similar to the upper-level programming, Eq. (4.99) is transformed to the following:

$$\min_{s_j} D' = \sum_{j=1}^{N}\sum_{i=1,i\neq j}^{N}\sum_{k=1}^{n} \theta_{ij}p_k\mu^r_{ijk}d_{ij}s_j + \sum_{j=1}^{N}\sum_{i=1,i\neq j}^{N} \theta_{ij}d_{ij}(1-s_j). \tag{4.100}$$

1. Constraints

The attacker can only destroy a subset R of the facilities, therefore:

$$\sum_{j=1}^{N} s_j \leq R, \quad \forall j \in \Phi. \tag{4.101}$$

Not all of the unsecured facilities will be attacked. That is to say:

$$\sum_{j=1}^{N} s_j \leq N - \sum_{j=1}^{N} z_j, \quad \forall j \in \Phi. \tag{4.102}$$

Moreover, to obtain the feasible solutions, the logical constraints should be satisfied: $z_j, s_j \in \{0, 1\}$. The attacker can only attack unsecured facilities, i.e., $z_j + s_j \leq 1, \forall j \in \Phi$.

Global Model for the Bi-Level CSSP

Based on the objective functions and constraint formulas as well as the above method, the mathematical model for CSSP is stated as:

$$\begin{cases}
\min\limits_{z_j} C = \sum\limits_{j=1}^{N} \sum\limits_{k=1}^{5} p_k \mu_{jk}^{c} s_j + \sum\limits_{j=1}^{N} M_j z_j \\
\max\limits_{z_j} D = \sum\limits_{j=1}^{N} \sum\limits_{i=1, i\neq j}^{N} \sum\limits_{k=1}^{5} \theta_{ij} p_k \mu_{ijk}^{r} d_{ij} s_j + \sum\limits_{j=1}^{N} \sum\limits_{i=1, i\neq j}^{N} \theta_{ij} d_{ij} z_j \\
\quad + \sum\limits_{j=1}^{N} \sum\limits_{i=1, i\neq j}^{N} \theta_{ij} d_{ij} (1 - s_j)(1 - z_j) \\
\text{s.t.} \begin{cases}
\sum\limits_{j=1}^{N} M_j z_j \leq T \\
Pr\left\{\omega \,\middle|\, Pr\left\{\sum\limits_{j=1}^{N} \tilde{\bar{C}}_j(\omega)^T s_j \leq C_0\right\} \geq \theta_1\right\} \geq \delta_1 \\
Pr\left\{\omega \,\middle|\, Pr\left\{\sum\limits_{i=1, i\neq j}^{N} \theta_{ij} \tilde{\bar{r}}_{ij}(\omega)^T d_{ij} s_j \geq D_0\right\} \geq \theta_2\right\} \geq \delta_2 \\
z_j \in \{0, 1\}, \ \forall i, j \in \Phi \\
\min\limits_{s_j} D' = \sum\limits_{j=1}^{N} \sum\limits_{i=1, i\neq j}^{N} \sum\limits_{k=1}^{n} \theta_{ij} p_k \mu_{ijk}^{r} d_{ij} s_j + \sum\limits_{j=1}^{N} \sum\limits_{i=1, i\neq j}^{N} \theta_{ij} d_{ij} (1 - s_j) \\
\text{s.t.} \begin{cases}
\sum\limits_{j=1}^{N} s_j \leq R \\
\sum\limits_{j=1}^{N} s_j \leq N - \sum\limits_{j=1}^{N} z_j \\
s_j \in \{0, 1\} \\
z_j + s_j \leq 1, \ \forall i, j \in \Phi.
\end{cases}
\end{cases}
\end{cases} \tag{4.103}$$

4.6.2 Solving Procedure by PGSA Algorithm

Obviously, the model (4.103) is a kind of bi-level 0-1 integer programming. To solve it, this chapter applied PGSA presented above.

Multi-objective Procedure

There are two objective functions in the upper-level programming. The objective value which is used to evaluate the growth node in PGSA is the quotient of the two objective functions' weighted value. The weights (denoted by $\lambda_{\min C}$ and $\lambda_{\max D}$) reflect preferences of the decision maker and the importance of the two objectives [19]. Thus, the objective value O is:

$$O = \frac{\lambda_{\min C}\left(\sum\limits_{j=1}^{N}\sum\limits_{k=1}^{5} p_k\mu_{kj}^{c}s_j + \sum\limits_{j=1}^{N} M_j z_j\right)}{\lambda_{\max D}\sum\limits_{j=1}^{N}\sum\limits_{i=1,i\neq j}^{N}\left(\sum\limits_{k=1}^{5}\theta_{ij}p_k\mu_{ikj}^{r}d_{ij}s_j + \theta_{ij}d_{ij}z_j + \theta_{ij}d_{ij}(1-s_j)(1-z_j)\right)}.$$

Note: the realistic significance of O is the cost for a unit efficiency of the facility system.

Stochastic Simulation Based Constraints Checking Procedure

Although simulation is an imprecise technique which provides only statistical estimates rather than exact results, it is indeed a powerful tool when dealing with complex problems [43].

To check the feasibility of Eq. (4.97), H random vectors $\boldsymbol{\omega}^k = (\omega_1^k, \omega_2^k, \cdots, \omega_5^k)$, $(k = 1, 2, \cdots, H)$ are generated independently from Ω according to the probability measure Pr. For any given sample $\omega^k \in \Omega$, the technique of stochastic simulation can be applied to check the random constraints $\sum_{j=1}^{N} \tilde{\tilde{C}}_j(\omega^k)^T s_j \leq C_0$. First, this procedure generates $\sum_{j=1}^{N} C_j(\omega^k)$ from $\sum_{j=1}^{N} \tilde{\tilde{C}}_j(\omega^k)$ according to the probability measure Pr. If $\sum_{j=1}^{N} \tilde{\tilde{C}}_j(\omega^k)^T s_j \leq C_0$, the stochastic constraint is feasible. After a given number of cycles, if no feasible $\sum_{j=1}^{N} C_j(\omega^k)$ is generated, the stochastic constraint is infeasible. Let H' be the number of occasions on which

$$Pr\left\{\sum_{j=1}^{N} \tilde{\tilde{C}}_j(\omega)^T s_j \leq C_0\right\} \geq \theta_1. \tag{4.104}$$

By the probability measure definition,

$$Pr\left\{\omega \,\middle|\, Pr\left\{\sum_{j=1}^{N} \tilde{\tilde{C}}_j(\omega)^T s_j \leq C_0\right\} \geq \theta_1\right\}, \tag{4.105}$$

infeasible PGSA nodes can be estimated using H'/H provided that H is large enough. If $H'/H \geq \delta_1$, the constraint is feasible. Similarly, the feasibility of Eq. (4.98) can be checked.

Eliminating Infeasible Growth Nodes

The growth nodes on the trunk and branches continuously emerge. The growth nodes which do not meet the feasibility criterion and fail to pass the simulation procedure need to be eliminated. The elimination is executed by setting the morphactin concentration as 0 in the upper and lower programming.

Overall Procedure

The overall procedure for solving CSSP is as follows:

Step 1. Set initial iteration $\tau = 0$. Initialize upper-level basis point $Z^0 = \{z_1^0, z_2^0, \cdots, z_N^0\}$. $z_j^0 = \{0, 1\}, j = 1, 2, \cdots, N$.

Step 2. Solve the lower-level programming using the PGSA procedure with upper-level initialized basis point Z^0.

Step 2.1. Initialize the lower-level basis point $S^0 = \{s_1^0, s_2^0, \cdots, s_N^0\}$. $s_j^0 = \{0, 1\}, j = 1, 2, \cdots, N$. Set the step length $\lambda = 1$.

Step 2.2 Check the feasibility: for $i = 1, 2, \cdots, N$, if the feasibility criterion is met by the lower-level basis point, i.e. $\sum_{j=1}^{N} s_j \leq R, \sum_{j=1}^{N} s_j \leq N - \sum_{j=1}^{N} z_j, z_j + s_j \leq 1$, calculate the corresponding objective value $D'(S^0)$, let $S_{\min} = S^0, D'_{\min} = D'(S^0)$. Otherwise, return to Step 2.1.

Step 2.3. Find the trunk M and each preferential growth node $S_{a1,b1}^0$ on the trunk in the lower-level programming, $a1 = 1, 2, \cdots, N, b1 = 1, 2, \cdots, m1$. $m1$ is the maximum number of growth nodes on the $a1$th trunk. Then eliminate infeasible growth nodes.

Step 2.4. Calculate the $D'(S_{a1,b1}^0)$ and if $D'(S_{a1,b1}^0) < D'_{\min}$, replace $D'_{\min} = D'(S_{a1,b1}^0)$.

$$\begin{cases} \text{if } D'_{\min} = D'(S_{a1,b1}^0), \; S_{\min} = S_{a1,b1}^0 \\ \text{if } D'_{\min} = D'(S^0), \quad S_{\min} = S^0. \end{cases} \tag{4.106}$$

Step 2.5. Calculate the morphactin concentrations

$$P_{a1,b1} = \frac{D'(S^0) - D'(S_{a1,b1}^0)}{\sum\limits_{a1=1}^{N} \sum\limits_{b1=1}^{m1} \left(D'(S^0) - D'(S_{a1,b1}^0)\right)}, \quad \text{if } D'(S^0) > D'(S_{a1,b1}^0) \tag{4.107a}$$

$$P_{a1,b1} = 0, \quad \text{otherwise.} \tag{4.107b}$$

Note that $a1, b1$ in Eq. (4.107a) do not contain the growth node when $P_{a1,b1} = 0$.

Step 2.6. Find the new basic node. Select a random number β_0 in the interval [0, 1], and β_0 satisfies

$$\sum_{a1=1}^{N}\sum_{b1=1}^{h1-1} P_{a1,b1} < \beta_0 \le \sum_{a1=1}^{N}\sum_{b1=1}^{h1} P_{a1,b1}, \tag{4.108}$$

then set $S^0_{a1,h1}$ as the new lower-level basis point.

Step 2.7. Set $S^1 = S^0_{a1,h1}$ and find branches. Set step length $\lambda = 1$, search the potential growth node $S^1_{a2,b2}$ along the positive and negative directions, $a2 = 1, 2, \cdots, N, b2 = 1, 2, \cdots, m2$. $m2$ is the maximum number of growth nodes on the $a2$th branch. Then eliminate infeasible growth nodes.

Step 2.8. Calculate $D'(S^1_{a2,b2})$, and if $D'(S^1_{a2,b2}) < D'_{\min}$, replace $D'_{\min} = D'(S^1_{a2,b2})$.

$$\begin{cases} \text{if } D'_{\min} = D'(S^1_{a2,b2}),\ S_{\min} = S^1_{a2,b2}, \\ \text{if } D'_{\min} = D'(S^0_{a1,b1}),\ S_{\min} = S^0_{a1,b1}. \end{cases} \tag{4.109}$$

Step 2.9. Calculate the morphactin concentrations

(1) Calculate the morphactin concentrations on the trunk

$$\begin{cases} P_{a1,b1} = \dfrac{D'(S^0) - D'(S^0_{a1,b1})}{\sum\limits_{a1=1}^{N}\sum\limits_{b1=1}^{m1}\left(D'(S^0) - D'(S^0_{a1,b1})\right) + \sum\limits_{a2=1}^{N}\sum\limits_{b2=1}^{m2}\left(D'(S^0) - D'(S^1_{a2,b2})\right)}, \\ \quad \text{if } D'(S^0) > D'(S^0_{a1,b1}) & (4.110a) \\ P_{a1,b1} = 0, \text{ otherwise.} & (4.110b) \end{cases}$$

Note that, $a1, b1$ in Eq. (4.110a) do not include the situation in Eq. (4.110b), and $a2, b2$ do not include the growth node when $P_{a2,b2} = 0$.

(2) Calculate the morphactin concentrations on the branch

$$\begin{cases} P_{a2,b2} = \dfrac{D'(S^0) - D'(S^1_{a2,b2})}{\sum\limits_{a1=1}^{N}\sum\limits_{b1=1}^{m1}\left(D'(S^0) - D'(S^0_{a1,b1})\right) + \sum\limits_{a2=1}^{N}\sum\limits_{b2=1}^{m2}\left(D'(S^0) - D'(S^1_{a2,b2})\right)}, \\ \quad \text{if } D'(S^0) > D'(S^1_{a1,b1}) & (4.111a) \\ P_{a2,b2} = 0, \text{ otherwise.} & (4.111b) \end{cases}$$

Note that, $a1, b1$ in Eq. (4.111a) do not contain the situation in Eq. (4.111b), and $a2, b2$ do contain the growth node when $P_{a2,b2} = 0$.

Step 2.10. Find the new basic node. Select a random number β_1 in the interval $[0,1]$. If β_1 satisfies

$$\sum_{a1=1}^{a2}\sum_{b1=1}^{h2-1} P_{a1,b1} < \beta_1 \le \sum_{a1=1}^{a2}\sum_{b1=1}^{h2} P_{a1,b1}, \tag{4.112}$$

select the growth node $S^0_{a2,h2}$ on the trunk as the new basic node. If β_1 satisfies

$$\begin{aligned}&\left(\sum_{a1=1}^{N}\sum_{b1=1}^{m1} P_{a1,b1} + \sum_{a2=1}^{a2}\sum_{b2=1}^{h2-1} P_{a2,b2}\right)\\ &< \beta_1 \le \left(\sum_{a1=1}^{N}\sum_{b1=1}^{m1} P_{a1,b1} + \sum_{a2=1}^{a2}\sum_{b2=1}^{h2} P_{a2,b2}\right),\end{aligned} \tag{4.113}$$

select the growth node $S^1_{a2,h2}$ on the branch as the new basic node and $\tau = \tau + 1$.

Step 2.11. When $D'_{\min}$ repeats appearing 50 times, take $S^* = S_{\min}$ as the optimal solution. Otherwise, go back to Step 2.3.

Step 3. Solve the upper-level programming using the same PGSA procedure.

Step 4. When $D'_{\min}$, $O_{(\min C\&\max D)}$ repeats appearing 50 times, take $S^* = S_{\min}$, $Z^* = Z_{\min}$ as the optimal solution. Otherwise, use the $Z_{\min}$ as the parameter for lower-level programming and go back to Step 2.

The procedure of the PGSA framework is depicted by a flowchart in Fig. 4.9. Consider the following numerical example:

Example 4.10.

$$\left\{\begin{array}{l}\min\limits_{x} -\tilde{\bar{\xi}}_1 x_1^2 - 2\tilde{\bar{\xi}}_2 x_2^2 - \tilde{\bar{\xi}}_3 y_1^2 - 2\tilde{\bar{\xi}}_4 y_2^2\\ \text{s.t.}\left\{\begin{array}{l}-2\tilde{\bar{\xi}}_1 x_1 + 3\tilde{\bar{\xi}}_2 x_2 \le 9\\ 11x_1 + 7x_2 \le 77\\ x_1, x_2 \ge 0, \text{integer}\\ \min\limits_{y} -2\tilde{\bar{\xi}}_1 x_1^2 - \tilde{\bar{\xi}}_2 x_2^2 - 2\tilde{\bar{\xi}}_3 y_1^2 - \tilde{\bar{\xi}}_4 y_2^2\\ \text{s.t.}\left\{\begin{array}{l}-9\tilde{\bar{\xi}}_1 x_1 - 4\tilde{\bar{\xi}}_2 x_2 + 12y_1 + 7y_2 \le 100\\ -3x_1 - 2x_2 - 7\tilde{\bar{\xi}}_3 y_1 + 9\tilde{\bar{\xi}}_4 y_2 \le 70\\ x_1, x_2, y_1, y_2 \ge 0, \text{integer,}\end{array}\right.\end{array}\right.\end{array}\right. \tag{4.114}$$

where $\xi_j, j = 1, 2, \cdots, 4$ are normally distributed random variables characterized as

$$\begin{aligned}&\tilde{\bar{\xi}}_1 \sim \mathcal{N}(\mathcal{N}(1,2),1), \tilde{\bar{\xi}}_2 \sim \mathcal{N}(\mathcal{N}(1,1),2),\\ &\tilde{\bar{\xi}}_3 \sim \mathcal{N}(\mathcal{N}(1,3),1), \tilde{\bar{\xi}}_4 \sim \mathcal{N}(\mathcal{N}(1,2),2).\end{aligned}$$

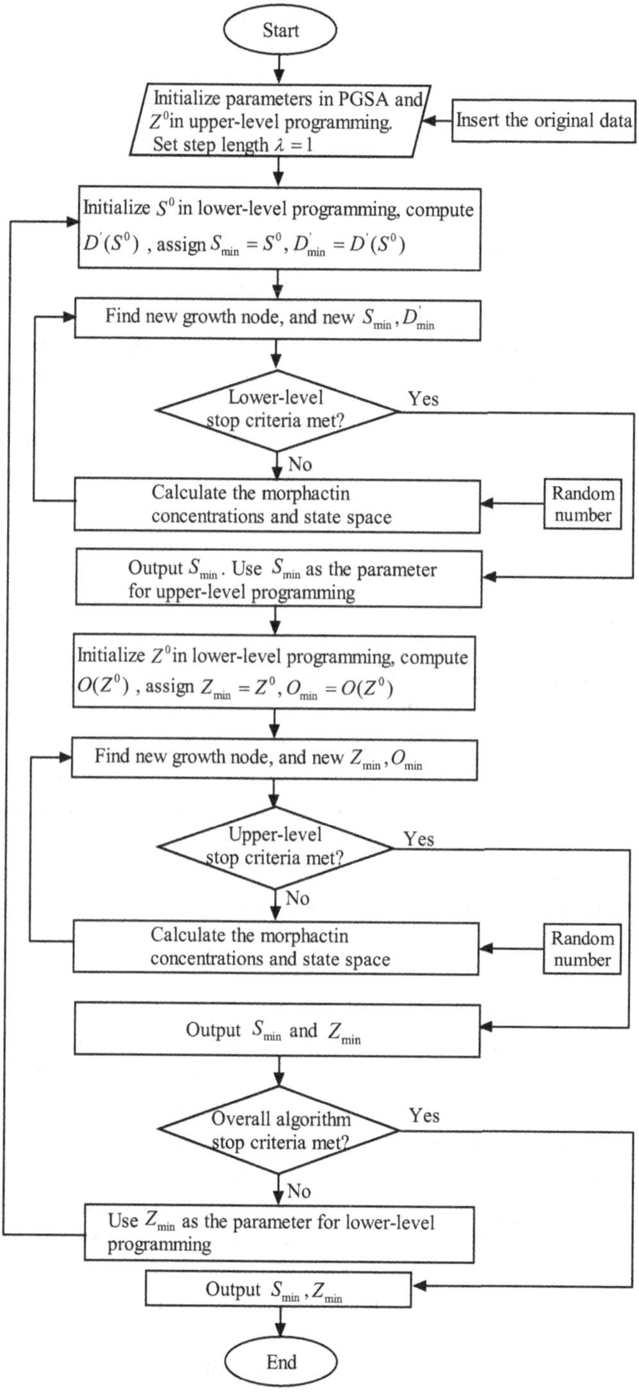

Fig. 4.9 Procedure of the PGSA framework

Apply the above PGSA algorithm, the following solutions can be obtained: $x_1^* = 7, x_2^* = 0, y_1^* = 13, y_2^* = 1$.

4.6.3 Data Collection

In this project, the attacker is assumed to be able to attack at most 4 facilities. The funds for defender securing facilities are 100 thousand CNY. The cost for securing each facility (M_j) is shown in Table 4.1. The Ra-Fu variables for economic loss when facilities are attacked are shown in Table 4.2.

The Ra-Fu variables for the fill rate of facility j to facility i when facility j is attacked are shown in Table 4.3. The importance weight of the different demands are as follows: $\theta_{1,3} = 0.1, \theta_{1,10} = 0.1, \theta_{2,3} = 0.2, \theta_{5,8} = 0.2, \theta_{6,8} = 0.1, \theta_{7,11} = 0.1, \theta_{7,12} = 0.2$. The confidence levels in Eqs. (4.97) and (4.98) are set as 96 %. In Longtan case, $C_0 = 1.0 \times 10^5$ CNY, $D_0 = 60\,\%$, $\lambda_{\min C} = 0.4$, $\lambda_{\max D} = 0.6$.

4.6.4 Solution and Discussion

The proposed PGSA algorithm was applied to CSSP of the Longtan hydropower construction project. By using MATLAB 7.0, the solution of the case and efficiency of the proposed algorithm were obtained.

The optimal solution was obtained after 50 PGSA iterations as shown in Table 4.4. The optimal solution is: $z_j = [1, 0, 0, 0, 0, 0, 1, 1, 1, 0, 0, 0, 0, 1]$, namely, the defender should secure the Reinforcing Steel Shop, the Metal and Electrical Installing Workshop, the Oil Depot, the Explosive Storage and the Labor Residence. The optimal solution for the lower-level decision maker is $s_j = [0, 1, 0, 0, 1, 1, 0, 0, 0, 0, 1, 0, 0, 0]$ which means the Carpentry Shop, the Equipment Repairing Workshop, the Truck Maintenance Shop and Steel Storage are estimated as the target of attack by the proposed model.

As shown above, the economic loss ($\tilde{\bar{C}}_j$) and the fill rate of facility j to facility i when facility j is attacked ($\tilde{\bar{r}}_{ij}$) are discrete random variables with the occurrence probabilities of five degrees denoted as $\{p_1, p_2, p_3, p_4, p_5\}$. Meanwhile, the value of $\tilde{\bar{C}}_j$ and $\tilde{\bar{r}}_{ij}$ under each degree of attack are also random variables. It is hard to describe this type of variables using simple random variables and crisp ones, especially when the attack degree reaches infinity, the probabilities are denoted as $p_i (i = 1, 2, \cdots, \infty), \sum_{i=1}^{\infty} p_i = 1$, it is even more difficult to describe this situation with simple random variables. Therefore this chapter employs the Ra-Fu variables to cope with the complex uncertainty in security problems.

Even when there have been 4 intentional attacks, the proposed model and method can guarantee the expected facilities efficiency be no less than 81.5 % and the expected security cost and economic loss in the potential attack be no more than 1.505 (1×10^5) CNY.

Table 4.2 The Ra-Fu variables for economic loss when facilities are attacked

$\tilde{\tilde{C}}_j$ (1×10^3 CNY)	$\tilde{\tilde{C}}_j$ (1×10^3 CNY)
$\tilde{\tilde{C}}_1 = \begin{cases} \tilde{C}_{11} \sim \mathcal{N}(1.9, 2), & p_1 = 25.3\% \\ \tilde{C}_{12} \sim \mathcal{N}(5.3, 4), & p_2 = 21.6\% \\ \tilde{C}_{13} \sim \mathcal{N}(15.6, 3), & p_3 = 20.8\% \\ \tilde{C}_{14} \sim \mathcal{N}(35.7, 4), & p_4 = 23.2\% \\ \tilde{C}_{15} \sim \mathcal{N}(45.1, 6), & p_5 = 9.1\% \end{cases}$	$\tilde{\tilde{C}}_8 = \begin{cases} \tilde{C}_{81} \sim \mathcal{N}(14.5, 6), & p_1 = 5.1\% \\ \tilde{C}_{82} \sim \mathcal{N}(48.8, 6), & p_2 = 15.8\% \\ \tilde{C}_{83} \sim \mathcal{N}(60.9, 8), & p_3 = 21.5\% \\ \tilde{C}_{84} \sim \mathcal{N}(87.3, 8), & p_4 = 24.4\% \\ \tilde{C}_{85} \sim \mathcal{N}(98.0, 10), & p_5 = 33.2\% \end{cases}$
$\tilde{\tilde{C}}_2 = \begin{cases} \tilde{C}_{21} \sim \mathcal{N}(3.2, 2), & p_1 = 20.3\% \\ \tilde{C}_{22} \sim \mathcal{N}(5.5, 2), & p_2 = 19.2\% \\ \tilde{C}_{23} \sim \mathcal{N}(8.3, 3), & p_3 = 31.2\% \\ \tilde{C}_{24} \sim \mathcal{N}(12.6, 4), & p_4 = 16.5\% \\ \tilde{C}_{25} \sim \mathcal{N}(15.5, 5), & p_5 = 12.8\% \end{cases}$	$\tilde{\tilde{C}}_9 = \begin{cases} \tilde{C}_{91} \sim \mathcal{N}(55.4, 3), & p_1 = 9.0\% \\ \tilde{C}_{92} \sim \mathcal{N}(68.5, 5), & p_2 = 14.2\% \\ \tilde{C}_{93} \sim \mathcal{N}(104.5, 6), & p_3 = 18.9\% \\ \tilde{C}_{94} \sim \mathcal{N}(120.6, 8), & p_4 = 28.6\% \\ \tilde{C}_{95} \sim \mathcal{N}(157.7, 10), & p_5 = 29.3\% \end{cases}$
$\tilde{\tilde{C}}_3 = \begin{cases} \tilde{C}_{31} \sim \mathcal{N}(5.1, 2), & p_1 = 23.1\% \\ \tilde{C}_{32} \sim \mathcal{N}(9.2, 2), & p_2 = 25.7\% \\ \tilde{C}_{33} \sim \mathcal{N}(15.4, 3), & p_3 = 20.4\% \\ \tilde{C}_{34} \sim \mathcal{N}(23.6, 3), & p_4 = 17.5\% \\ \tilde{C}_{35} \sim \mathcal{N}(46.9, 4), & p_5 = 13.3\% \end{cases}$	$\tilde{\tilde{C}}_{10} = \begin{cases} \tilde{C}_{10,1} \sim \mathcal{N}(4.5, 3), & p_1 = 24.3\% \\ \tilde{C}_{10,2} \sim \mathcal{N}(6.1, 3), & p_2 = 24.2\% \\ \tilde{C}_{10,3} \sim \mathcal{N}(9.3, 3), & p_3 = 20.5\% \\ \tilde{C}_{10,4} \sim \mathcal{N}(15.1, 4), & p_4 = 18.6\% \\ \tilde{C}_{10,5} \sim \mathcal{N}(25.2, 5), & p_5 = 12.4\% \end{cases}$
$\tilde{\tilde{C}}_4 = \begin{cases} \tilde{C}_{41} \sim \mathcal{N}(2.2, 1), & p_1 = 32.0\% \\ \tilde{C}_{42} \sim \mathcal{N}(6.2, 3), & p_2 = 26.4\% \\ \tilde{C}_{43} \sim \mathcal{N}(10.5, 3), & p_3 = 16.3\% \\ \tilde{C}_{44} \sim \mathcal{N}(15.6, 4), & p_4 = 16.4\% \\ \tilde{C}_{45} \sim \mathcal{N}(25.7, 5), & p_5 = 8.9\% \end{cases}$	$\tilde{\tilde{C}}_{11} = \begin{cases} \tilde{C}_{11,1} \sim \mathcal{N}(7.5, 5), & p_1 = 27.5\% \\ \tilde{C}_{11,2} \sim \mathcal{N}(15.7, 6), & p_2 = 26.3\% \\ \tilde{C}_{11,3} \sim \mathcal{N}(22.8, 6), & p_3 = 20.1\% \\ \tilde{C}_{11,4} \sim \mathcal{N}(35.1, 6), & p_4 = 17.5\% \\ \tilde{C}_{11,5} \sim \mathcal{N}(55.4, 5), & p_5 = 8.6\% \end{cases}$
$\tilde{\tilde{C}}_5 = \begin{cases} \tilde{C}_{51} \sim \mathcal{N}(4.4, 1), & p_1 = 32.5\% \\ \tilde{C}_{52} \sim \mathcal{N}(5.5, 2), & p_2 = 34.6\% \\ \tilde{C}_{53} \sim \mathcal{N}(6.2, 2), & p_3 = 20.8\% \\ \tilde{C}_{54} \sim \mathcal{N}(8.0, 1), & p_4 = 10.5\% \\ \tilde{C}_{55} \sim \mathcal{N}(10.0, 1), & p_5 = 1.6\% \end{cases}$	$\tilde{\tilde{C}}_{12} = \begin{cases} \tilde{C}_{12,1} \sim \mathcal{N}(2.4, 4), & p_1 = 28.4\% \\ \tilde{C}_{12,2} \sim \mathcal{N}(6.5, 3), & p_2 = 30.2\% \\ \tilde{C}_{12,3} \sim \mathcal{N}(12.8, 3), & p_3 = 24.3\% \\ \tilde{C}_{12,4} \sim \mathcal{N}(22.9, 5), & p_4 = 15.2\% \\ \tilde{C}_{12,5} \sim \mathcal{N}(44.0, 5), & p_5 = 1.9\% \end{cases}$
$\tilde{\tilde{C}}_6 = \begin{cases} \tilde{C}_{61} \sim \mathcal{N}(4.5, 3), & p_1 = 18.7\% \\ \tilde{C}_{62} \sim \mathcal{N}(7.0, 3), & p_2 = 21.4\% \\ \tilde{C}_{63} \sim \mathcal{N}(12.2, 4), & p_3 = 24.5\% \\ \tilde{C}_{64} \sim \mathcal{N}(21.4, 5), & p_4 = 25.3\% \\ \tilde{C}_{65} \sim \mathcal{N}(36.6, 5), & p_5 = 10.1\% \end{cases}$	$\tilde{\tilde{C}}_{13} = \begin{cases} \tilde{C}_{13,1} \sim \mathcal{N}(3.5, 5), & p_1 = 22.7\% \\ \tilde{C}_{13,2} \sim \mathcal{N}(7.3, 6), & p_2 = 21.5\% \\ \tilde{C}_{13,3} \sim \mathcal{N}(21.4, 6), & p_3 = 16.9\% \\ \tilde{C}_{13,4} \sim \mathcal{N}(58.3, 5), & p_4 = 19.9\% \\ \tilde{C}_{13,5} \sim \mathcal{N}(102.7, 8), & p_5 = 19.0\% \end{cases}$
$\tilde{\tilde{C}}_7 = \begin{cases} \tilde{C}_{71} \sim \mathcal{N}(5.7, 3), & p_1 = 17.6\% \\ \tilde{C}_{72} \sim \mathcal{N}(15.0, 5), & p_2 = 20.5\% \\ \tilde{C}_{73} \sim \mathcal{N}(21.6, 6), & p_3 = 24.4\% \\ \tilde{C}_{74} \sim \mathcal{N}(36.1, 8), & p_4 = 23.8\% \\ \tilde{C}_{75} \sim \mathcal{N}(75.4, 8), & p_5 = 13.7\% \end{cases}$	$\tilde{\tilde{C}}_{14} = \begin{cases} \tilde{C}_{14,1} \sim \mathcal{N}(32.2, 3), & p_1 = 18.7\% \\ \tilde{C}_{14,2} \sim \mathcal{N}(75.5, 3), & p_2 = 19.5\% \\ \tilde{C}_{14,3} \sim \mathcal{N}(145.4, 6), & p_3 = 22.4\% \\ \tilde{C}_{14,4} \sim \mathcal{N}(276.2, 8), & p_4 = 25.0\% \\ \tilde{C}_{14,5} \sim \mathcal{N}(365.7, 10), & p_5 = 14.4\% \end{cases}$

In this chapter, PGSA is applied to solve the bi-level CSSP. Compared with other traditional algorithms, PGSA has obvious advantages in solving CSSP for the following reasons: (1) PGSA deals with the objective function and constraints separately, which averts the trouble to determine the barrier factors and makes the increase/decrease of constraints convenient [29]. (2) PGSA does not need external parameters such as crossover rate and mutation rate in Genetic Algorithm

Table 4.3 The Ra-Fu variables for the fill rate of facility j to facility i when facility j is attacked

i, j	$\tilde{\tilde{r}}_{ij}(\%)$
$i = 1, j = 3$	$\tilde{\tilde{r}}_{1,3} = \begin{cases} \tilde{\tilde{r}}_{1,3}^1 \sim \mathcal{N}(89, 4), & p_1 = 34.4\% \\ \tilde{\tilde{r}}_{1,3}^2 \sim \mathcal{N}(68, 4), & p_2 = 29.5\% \\ \tilde{\tilde{r}}_{1,3}^3 \sim \mathcal{N}(45, 4), & p_3 = 15.8\% \\ \tilde{\tilde{r}}_{1,3}^4 \sim \mathcal{N}(33.5, 4), & p_4 = 12.4\% \\ \tilde{\tilde{r}}_{1,3}^5 \sim \mathcal{N}(12.4, 4), & p_5 = 7.9\% \end{cases}$
$i = 1, j = 10$	$\tilde{\tilde{r}}_{1,10} = \begin{cases} \tilde{\tilde{r}}_{1,10}^1 \sim \mathcal{N}(95, 4), & p_1 = 41.2\% \\ \tilde{\tilde{r}}_{1,10}^2 \sim \mathcal{N}(79, 4), & p_2 = 30.1\% \\ \tilde{\tilde{r}}_{1,10}^3 \sim \mathcal{N}(66, 4), & p_3 = 15.3\% \\ \tilde{\tilde{r}}_{1,10}^4 \sim \mathcal{N}(59.3, 4), & p_4 = 8.4\% \\ \tilde{\tilde{r}}_{1,10}^5 \sim \mathcal{N}(32, 4), & p_5 = 5\% \end{cases}$
$i = 2, j = 3$	$\tilde{\tilde{r}}_{2,3} = \begin{cases} \tilde{\tilde{r}}_{2,3}^1 \sim \mathcal{N}(92.5, 4), & p_1 = 34.8\% \\ \tilde{\tilde{r}}_{2,3}^2 \sim \mathcal{N}(78.2, 4), & p_2 = 30.2\% \\ \tilde{\tilde{r}}_{2,3}^3 \sim \mathcal{N}(62, 4), & p_3 = 15.6\% \\ \tilde{\tilde{r}}_{2,3}^4 \sim \mathcal{N}(53, 4), & p_4 = 13.5\% \\ \tilde{\tilde{r}}_{2,3}^5 \sim \mathcal{N}(29, 4), & p_5 = 5.9\% \end{cases}$
$i = 5, j = 8$	$\tilde{\tilde{r}}_{5,8} = \begin{cases} \tilde{\tilde{r}}_{5,8}^1 \sim \mathcal{N}(90.1, 4), & p_1 = 33.2\% \\ \tilde{\tilde{r}}_{5,8}^2 \sim \mathcal{N}(76, 4), & p_2 = 31.6\% \\ \tilde{\tilde{r}}_{5,8}^3 \sim \mathcal{N}(66.3, 4), & p_3 = 18.1\% \\ \tilde{\tilde{r}}_{5,8}^4 \sim \mathcal{N}(51.6, 4), & p_4 = 13.4\% \\ \tilde{\tilde{r}}_{5,8}^5 \sim \mathcal{N}(30, 4), & p_5 = 3.7\% \end{cases}$
$i = 6, j = 8$	$\tilde{\tilde{r}}_{6,8} = \begin{cases} \tilde{\tilde{r}}_{6,8}^1 \sim \mathcal{N}(88.4, 4), & p_1 = 26.2\% \\ \tilde{\tilde{r}}_{6,8}^2 \sim \mathcal{N}(73, 4), & p_2 = 30.6\% \\ \tilde{\tilde{r}}_{6,8}^3 \sim \mathcal{N}(59.5, 4), & p_3 = 24.3\% \\ \tilde{\tilde{r}}_{6,8}^4 \sim \mathcal{N}(50.1, 4), & p_4 = 14.5\% \\ \tilde{\tilde{r}}_{6,8}^5 \sim \mathcal{N}(42.5, 4), & p_5 = 4.4\% \end{cases}$
$i = 7, j = 11$	$\tilde{\tilde{r}}_{7,11} = \begin{cases} \tilde{\tilde{r}}_{7,11}^1 \sim \mathcal{N}(87.6, 4), & p_1 = 25.3\% \\ \tilde{\tilde{r}}_{7,11}^2 \sim \mathcal{N}(75.3, 4), & p_2 = 29.4\% \\ \tilde{\tilde{r}}_{7,11}^3 \sim \mathcal{N}(64.2, 4), & p_3 = 23.6\% \\ \tilde{\tilde{r}}_{7,11}^4 \sim \mathcal{N}(52, 4), & p_4 = 12.1\% \\ \tilde{\tilde{r}}_{7,11}^5 \sim \mathcal{N}(39.8, 4), & p_5 = 9.6\% \end{cases}$
$i = 7, j = 12$	$\tilde{\tilde{r}}_{7,12} = \begin{cases} \tilde{\tilde{r}}_{7,12}^1 \sim \mathcal{N}(86.4, 4), & p_1 = 26.4\% \\ \tilde{\tilde{r}}_{7,12}^2 \sim \mathcal{N}(78, 4), & p_2 = 28.4\% \\ \tilde{\tilde{r}}_{7,12}^3 \sim \mathcal{N}(69.4, 4), & p_3 = 19.6\% \\ \tilde{\tilde{r}}_{7,12}^4 \sim \mathcal{N}(58.5, 4), & p_4 = 18.5\% \\ \tilde{\tilde{r}}_{7,12}^5 \sim \mathcal{N}(43.2, 4), & p_5 = 7.1\% \end{cases}$

(GA), personal best position acceleration constant, global best position acceleration constant and inertia weight in the Particle Swarm Optimization algorithm (PSO). Therefore, PGSA can avoid the bias generated by parameters as in other traditional

Table 4.4 The iteration process of PGSA

Iteration	Growth nodes														Value of objectives			Morphactin concentration	State space	Random number
	1	2	3	4	5	6	7	8	9	10	11	12	13	14	C 10^5 CNY	D (%)	O (1×10^2)			
0	1	1	0	0	0	0	0	0	0	0	0	0	0	0	2.328	0.8	0.7275	1	1	
1	1	1	0	0	0	0	0	0	0	0	0	0	0	1	2.149	0.8	0.6716	1		
2	1	1	0	1	0	0	0	0	0	0	0	0	0	1	2.269	0.8	709.0625	0.096563	0.096563	
	1	1	0	0	0	0	0	1	0	0	0	0	0	1	2.299	0.8	718.4375	0.047463	0.144026	
	1	1	0	0	0	0	0	0	1	0	0	0	0	1	1.863	0.8	582.1875	0.761047	0.905074	0.9649
	1	1	0	0	0	0	0	0	0	1	0	0	0	1	2.289	0.8	715.3125	0.06383	0.968903	
	1	1	0	0	0	0	0	0	0	0	1	0	0	1	2.309	0.8	721.5625	0.031097	1	
3	1	1	0	1	0	0	0	0	0	0	0	0	0	1	2.269	0.8	709.0625	0.054833	0.054833	
	1	1	0	0	0	0	0	1	0	0	0	0	0	1	2.299	0.8	718.4375	0.026952	0.081784	0.1576
	1	1	0	0	0	0	0	0	1	0	0	0	0	1	1.863	0.8	582.1875	0.432156	0.513941	
	1	1	0	0	0	0	0	0	0	0	1	0	0	1	2.309	0.8	721.5625	0.01766	0.531599	
	1	1	0	0	0	0	0	0	1	1	0	0	0	1	2.003	0.8	625.9375	0.302045	0.833643	
	1	1	0	0	0	0	0	0	0	0	0	0	0	1	2.149	0.8	671.5625	0.166357	1	
4	1	1	0	1	0	0	0	0	0	0	0	0	0	1	2.269	0.8	709.0625	0.015902	0.015902	
	1	1	0	0	0	0	0	1	0	0	0	0	0	1	2.299	0.8	718.4375	0.007816	0.023718	
	1	1	0	0	0	0	0	0	0	0	1	0	0	1	2.309	0.8	721.5625	0.005121	0.028839	
	1	1	0	0	0	0	0	0	1	1	0	0	0	1	2.003	0.8	625.9375	0.087594	0.116433	
	1	1	0	0	0	0	0	0	0	0	0	0	0	1	2.149	0.8	671.5625	0.048244	0.164677	
	1	1	1	0	0	0	0	0	1	0	0	0	0	1	2.063	0.8	644.6875	0.071423	0.2361	
	1	1	0	1	0	0	0	0	1	0	0	0	0	1	1.983	0.8	619.6875	0.092985	0.329085	
	1	1	0	0	0	1	0	0	1	0	0	0	0	1	2.179	0.835	652.3952	0.064775	0.393861	

(continued)

Table 4.4 (continued)

Iteration	Growth nodes														Value of objectives			Morphactin concentration	State space	Random number
	1	2	3	4	5	6	7	8	9	10	11	12	13	14	C 10^5 CNY	D (%)	O (1×10^2)			
	1	1	1	0	0	0	0	0	1	0	0	0	0	1	2.063	0.8	644.6875	0.071423	0.2361	
	1	1	0	1	0	0	0	0	1	0	0	0	0	1	1.983	0.8	619.6875	0.092985	0.329085	
	1	1	0	0	0	1	0	0	1	0	0	0	0	1	2.179	0.835	652.3952	0.064775	0.393861	
	1	1	0	0	0	0	1	0	1	0	0	0	0	1	2.091	0.895	584.0782	0.123697	0.517557	
	1	1	0	0	0	0	0	1	1	0	0	0	0	1	1.632	0.8	510	0.187587	0.705144	
	1	1	0	0	0	0	0	0	0	0	0	0	0	1	2.149	0.8	671.5625	0.048244	0.753388	
	1	1	0	0	0	0	0	0	1	1	0	0	0	1	2.003	0.8	625.9375	0.087594	0.840983	
	1	1	0	0	0	0	0	0	1	0	1	0	0	1	2.023	0.8	632.1875	0.082204	0.923186	
	1	1	0	0	0	0	0	0	1	0	0	1	0	1	2.043	0.8	638.4375	0.076814	1	0.9706
⋮	⋮	⋮	⋮	⋮	⋮	⋮	⋮	⋮	⋮	⋮	⋮	⋮	⋮	⋮	⋮	⋮	⋮	⋮	⋮	⋮
50	1	1	0	1	0	0	0	0	0	0	0	0	0	1	2.269	0.8	709.0625	0.006092		
	1	1	0	0	0	0	0	1	0	0	0	0	0	1	2.299	0.8	718.4375	0.002994	0.009086	
	1	1	0	0	0	0	0	0	0	0	1	0	0	1	2.309	0.8	721.5625	0.001962	0.011048	
	1	1	0	0	0	1	0	0	1	0	0	0	0	1	2.179	0.835	652.3952	0.024814	0.035862	
	1	1	0	0	0	0	0	1	1	0	0	0	0	1	1.632	0.8	510	0.071861	0.107722	
	1	1	0	0	0	0	0	0	0	0	0	0	0	1	2.149	0.8	671.5625	0.018481	0.126204	
	1	1	0	0	0	0	0	0	1	0	0	1	0	1	2.043	0.8	638.4375	0.029426	0.155629	
	1	1	1	0	0	0	0	0	1	0	0	0	0	1	2.063	0.8	644.6875	0.027361	0.18299	
	1	1	0	1	0	0	0	0	1	0	0	0	0	1	1.983	0.8	619.6875	0.035621	0.218611	
	1	1	0	0	0	0	0	0	1	1	0	0	0	1	2.003	0.8	625.9375	0.033556	0.252166	

	1	1	0	0	0	0	0	0	1	0	1	0	0	1	2.023	0.8	632.1875	0.031491	0.283657	
	0	1	0	0	0	0	1	0	1	0	0	0	0	1	183,900	0.83	553.9157	0.057351	0.341008	
	1	0	0	0	0	0	1	0	1	0	0	0	0	1	173,600	0.815	532.5153	0.064422	0.40543	
	1	1	0	0	0	0	0	0	1	0	0	0	0	1	186,300	0.8	582.1875	0.04801	0.45344	
	1	1	0	0	0	0	1	0	1	0	0	0	0	1	2.091	0.895	584.0782	0.047386	0.500825	
	1	0	1	0	0	0	1	0	1	0	0	0	0	1	1.936	0.815	593.865	0.044152	0.544978	
	1	0	0	1	0	0	1	0	1	0	0	0	0	1	1.856	0.815	569.3252	0.05226	0.597237	
	1	0	0	0	1	0	1	0	1	0	0	0	0	1	2.032	0.885	574.0113	0.050712	0.647949	
	1	0	0	0	0	1	1	0	1	0	0	0	0	1	2.052	0.85	603.5294	0.040959	0.688908	
	1	**0**	**0**	**0**	**0**	**0**	**1**	**1**	**1**	**0**	**0**	**0**	**0**	**1**	**1.505**	**0.815**	**461.6564**	**0.087833**	**0.776741**	0.7577
	1	0	0	0	0	0	1	0	0	0	0	0	0	1	2.022	0.815	620.2454	0.035436	0.812177	
	1	0	0	0	0	0	1	0	1	1	0	0	0	1	1.876	0.815	575.4601	0.050233	0.86241	
	1	0	0	0	0	0	1	0	1	0	1	0	0	1	1.896	0.815	581.5951	0.048206	0.910616	
	1	0	0	0	0	0	1	0	1	0	0	1	0	1	1.916	0.815	587.7301	0.046179	0.956795	
	1	0	0	0	0	0	1	0	1	0	0	0	1	1	1.966	0.815	603.0675	0.041112	0.997907	
	1	0	0	0	0	0	1	0	1	0	0	0	0	0	2.351	0.815	721.1656	0.002093	1	

algorithms [26]. (3) PGSA adopts a guiding search direction that changes dynamically as the objective function changes [29]. (4) The representation of the growth nodes in PGSA corresponds to the model solutions. Therefore, there is no need to design the representation (like particles in PSO and genes in GA) in the beginning and transform the representation into solution in the end. (5) PGSA is efficient. The case problem is solved by the proposed algorithm with satisfactory solutions within 3 min 40 s on average, which is an acceptable time. The computer running environment is an inter core 2 Duo 2.00 GHz clock pulse with 2048 MB memory.

References

1. Abo-Sinna M (2001) A bi-level non-linear multi-objective decision making under fuzziness. Opsearch-New Delhi 38(5):484–495
2. Branch K, Baker K (2007) Security during the construction of critical infrastructure in the post 9/11 context in the u.s. In: 8th annual conference on human factors and power plants. Institute of Electrical and Electronics Engineers (IEEE), Monterey
3. Cai J (2009) Hydropower in China. Master's thesis, University of Gavle
4. Chadli O, Mahdioui H, Yao JC (2011) Bilevel mixed equilibrium problems in banach spaces: existence and algorithmic aspects. J Ind Manag Optim 1:549–561
5. Chankong V, Haimes Y (1983) Multiobjective decision making: theory and methodology. North-Holland, New York
6. Charnes A, Cooper WW (1959) Chance-constrained programming. Manag Sci 6(1):73–79
7. Church RL, Scaparra MP, Middleton RS (2004) Identifying critical infrastructure: the median and covering facility interdiction problems. Ann Assoc Am Geograph 94:491–502
8. Construction Industry Institute (2005) Implementing project security practices, implementation resource BMM-3, benchmarking and metrics, CII. University of Texas, Austin
9. FAA (2001) Recommended security guidelines for airport planning, design and construction. Report no DOT/FAA/AR-00/52, U.S. Deptartment of Transportation, Washington, DC
10. Fan Z (2016) Application of plant growth simulation algorithm. In: 4th international conference on machinery, materials and computing technology (ICMMCT 2016), Hangzhou. Atlantis Press, pp 1654–1657
11. Foreign Affairs Manual (1994) Construction security, construction materials, and transit security. Manual 12 FAM 350, Foreign Affairs Manual, Chapter 12 – Diplomatic Security, U.S. Department of State
12. Foreign Affairs Manual (1997) Industrial security program. Document 12 FAM 570, Foreign AffairsManual, Chapter 12 – Diplomatic Security, U.S. Department of State
13. Foreign Affairs Manual (2002) Construction security certification program. Manual 12 FAM 360, Foreign Affairs Manual, Chapter 12 – Diplomatic Security, U.S. Department of State
14. Hallowell MR, Gambatese JA (2009) Construction safety risk mitigation. J Constr Eng Manag 135:1316–1323
15. Khalafallah A, El-Rayes K (2008) Minimizing construction-related security risks during airport expansion projects. J Constr Eng Manag 134:40–48
16. Lee E, Shih H (2000) Fuzzy and multi-level decision making: and interactive computational approach. Springer, New York
17. Li T, Wang C, Wang W, Su W (2005) A global optimization bionics algorithm for solving integer programming-plant growth simulation algorithm. In: Proceedings of international conference of management science and engineering, Incheon, pp 13–15
18. Li Z, Shen W, Xu J et al (2015) Bilevel and multi-objective dynamic construction site layout and security planning. Autom Constr 57:1–16

19. Liang TF, Cheng HW (2011) Multi-objective aggregate production planning decisions using two-phase fuzzy goal programming method. J Ind Manag Optim 7:365–383
20. Liberatore F, Scaparra MP, Daskin MS (2011) Analysis of facility protection strategies against an uncertain number of attacks: the stochastic r-interdiction median problem with fortification. Comput Oper Res 38:357–366
21. Liu GS, Zhang J (2005) Decision making of transportation plan, a bilevel transportation problem approach. J Ind Manag Optim 1:305–314
22. Losada C, Scaparra MP, Church RL (2010) On a bilevel formulation to protect uncapacitated p-median systems with facility recovery time and frequent disruptions. Electron Notes Discret Math 36:591–598
23. Matthews B, Sylvie JR, Lee S, Thomas SR, Chapman RE, Gibson GE (2006) Addressing security in early stages of project life cycle. J Manag Eng 22:196–202
24. O'Hanley JR, Church RL (2011) Designing robust coverage networks to hedge against worst-case facility losses. Eur J Oper Res 209:23–36
25. (PEO), P.E.O.: National industrial security program. Fed Regist 58(5):1–6 (1993)
26. Rao RS, Narasimham SVL, Ramalingaraju M (2011) Optimal capacitor placement in a radial distribution system using Plant Growth Simulation Algorithm. Electr Power Energy Syst 33:1133–1139
27. Said H, El-Rayes K (2010) Optimizing the planning of construction site security for critical infrastructure projects. Autom Constr 19:221–234
28. Sakawa M (1993) Fuzzy sets and interactive multiobjective optimization. Plenum Press, New York
29. Sarma AK, Rafi KM (2011) Optimal selection of capacitors for radial distribution systems using Plant Growth Simulation Algorithm. Int J Adv Sci Technol 30:43–54
30. Scaparra MP, Church RL (2008) A bilevel mixed-integer program for critical infrastructure protection planning. Comput Oper Res 35(6):1905–1923
31. Shi X, Xia H (1997) Interactive bilevel multi-objective decision making. J Oper Res Soc 48(9):943–949
32. Shi X, Xia HS (2001) Model and interactive algorithm of bi-level multi-objective decision-making with multiple interconnected decision makers. J Multi-Criteria Decis Anal 10(1):27–34
33. Shih H, Lai Y, Stanley Lee E (1996) Fuzzy approach for multi-level programming problems. Comput Oper Res 23(1):73–91
34. Simaan M, Cruz JB (1973) On the Stackelberg strategy in nonzero-sum games. J Optim Theory Appl 11(5):533–555
35. Simpson S (2008) Airport security during construction. In: 4th annual international airfield operations area expo & conference. Airport Consultants Council (AAC), Milwaukee
36. Szidarovszky F, Gershon M, Duckstein L (1986) Techniques for multiobjective decision making in systems management. Elsevier Science, Amsterdam
37. Tarr CJ (1992) CLASP: a computerised aid to cost effective perimeter security. In: Security technology 28th international Carnahan conference. Institute of Electrical and Electronics Engineers (IEEE), Atlanta, pp 164–168
38. Toole TM (2002) Construction site safety roles. J Constr Eng Manag 128:203–210
39. Uno T, Katagiri H (2008) Single-and multi-objective defensive location problems on a network. Eur J Oper Res 188:76–84
40. Walker GH (2008) Securing the construction site. In: IRMI construction risk conference, Las Vegas
41. Wang C, Cheng H (2009) Transmission network optimal planning based on plant growth simulation algorithm. Eur Trans Electr Power 19:291–301
42. Xu JP, Wei P (2012) Production-distribution planning of construction supply chain management under fuzzy random environment for large-scale construction project. J Ind Manag Optim 9(1):31–56
43. Xu JP, Yao LM (2011) Random-like multiple objective decision making. Springer, Berlin/Heidelberg

44. Zhang L, Wu SY (2010) Robust solutions to Euclidean facility location problems with uncertain data. J Ind Manag Optim 6(4):751–760
45. Zhang G, Lu J, Gao Y (2015) Bi-Level programming models and algorithms. In: Multi-level decision making. Springer, Berlin/Heidelberg, pp 47–62
46. Zhou X, Xu J (2009) A class of integrated logistics network model under random fuzzy environment and its application to Chinese beer company. Int J Uncertain Fuzziness Knowl-Based Syst 17(6):807–831

Chapter 5
Methodology From an Equilibria Viewpoints

Abstract Since the Stackelberg game model (von Stackelberg, The theory of the market economy. Oxford University Press, New York, 1952) was proposed by German economist Heinrich Freiherr von Stackelberg in 1934, it has been widely applied (Peng and Lu, Appl Math Comput 271(C):259–268, 2015; Simaan and Cruz, J Optim Theory Appl 11(5):533–555, 1973; Shi et al., Ind Manag Data Syst 116(3):350–368, 2016; von Stengel, Games Econ Behav 69(2):512–516, 2010). The main idea of the Stackelberg game is equilibrium, which is the achievement of a state whereby the conflicting parties are in balance and do not seek any further changes. In real-world problems, decision makers often have several concurrent objectives, which are often in conflict and cannot be easily measured. Therefore, decision makers must consider trade-offs across the objectives. In addition to the hierarchical structure and the diverse objectives, imprecise data, inaccurate human decision making judgments, the decision maker's bounded rationality, and the decision making uncertainties all need to be taken into account. In this book, we focus on random-like phenomena, including random phenomena, the Ra-Ra phenomenon, and the Ra-Fu phenomenon, the latter two of which are considered to be two-fold uncertain phenomena. In the conclusion, Chap. 5 summarizes the methodologies adopted in this monograph from an equilibria viewpoint, including equilibrium motivation, equilibrium in real-world problems, model system in equilibrium, equilibrium algorithms, and perspectives for the models, theories, algorithms and applications.

Keywords Motivation for equilibrium • Equilibria in real-world problems • Model system in equilibria • Equilibria algorithms • RBMODM perspectives

5.1 Motivation for Equilibrium

In game theory terms, the two decision makers in the model are known as the leader and the follower. In a Stackelberg game, the leader makes the first move and the follower then seeks to maximize their move under consideration of the leader's move. The leader has the advantage as they can control the game by seeking to maximize their own gain in their move while knowing that the follower also seeks to always maximize their own gain. The main idea of the Stackelberg

J. Xu et al., *Random-Like Bi-level Decision Making*, Lecture Notes in Economics and Mathematical Systems 688, DOI 10.1007/978-981-10-1768-1_5

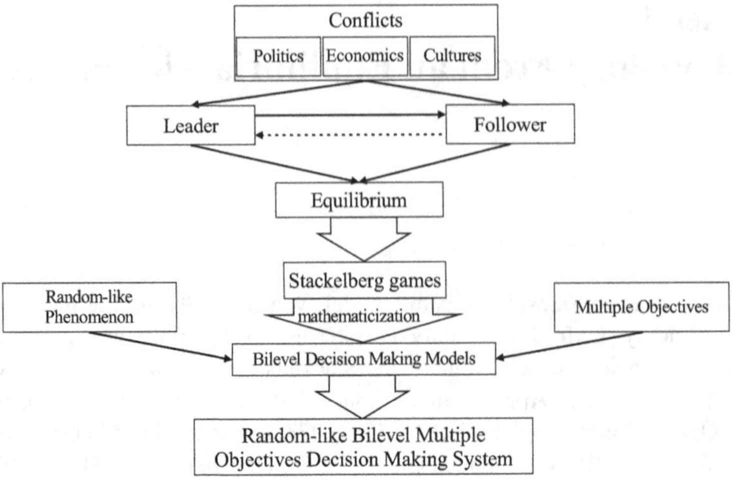

Fig. 5.1 Formulation of RBMODM systems

game is equilibrium, which is the achievement of a state whereby the conflicting parties are in balance and do not seek any further changes. Conflict refers to the friction, disagreements, or discord that may arise in a group when the beliefs or actions of one or more members of the group are either resisted or unacceptable to one or more members of another group, such as in politics, economics and across cultures. To resolve decision making conflicts in hierarchical organizations, multi-level mathematical programming has often been employed. In Stackelberg models, both leaders and followers are assumed to make rational economic judgments. As bi-level mathematical programming problems are interpreted as static Stackelberg games, bi-level programming can be used for noncooperative game theory, which is when there is no communication between the decision makers, or no binding agreements are made even if there is communication. In real-world problems, decision makers often have several concurrent objectives, which are often in conflict and cannot be easily measured. Therefore, decision makers must consider trade-offs across the objectives. In such situations, mathematical models which cater for multiple objectives can be used.

The integration of a hierarchical structure, diverse objectives and random-like phenomena can be developed into models such as the random-like bi-level multiple objective decision making (RBMODM) system. Figure 5.1 illustrates the elements and formulation process for the RBMODM system.

The concept of equilibrium (or plural form equilibria) can be found in many disciplines such as biology, physics, chemistry and economics, each of which views equilibria differently.

In biology, *genetic equilibrium* describes the condition of an allele or genotype in a gene pool (such as a population), in which the frequency remains unchanged from generation to generation. Genetic equilibrium, such as the Hardy-Weinberg equilibrium, is a theoretical state used to identify the reasons for and the types of population deviations, the models for which have the following assumptions:

- No gene mutations occurring at that locus or the loci associated with the trait
- A large population size
- Limited-to-no immigration, emigration, or migration (genetic flow)
- No natural selection on that locus or trait
- Random mating (panmixis)

These can also describe other types of equilibrium, especially in modeling contexts. In particular, many models use a variation of the Hardy-Weinberg principle as their basis. Instead of all the Hardy-Weinberg characteristics being present, these instead assume a balance between the diversifying effects of genetic drift and the homogenizing effects of migration between populations [7]. Population disequilibrium suggests that one of the assumptions in the model has been violated.

In physics, there are several equilibria measures such as equilibrium mode distribution, hydrostatic equilibrium, mechanical equilibrium, radiative equilibrium, and secular equilibrium [5, 13]. In classical mechanics, a particle is in mechanical equilibrium if the net force on that particle is zero. By extension, a physical system made up of many parts is in mechanical equilibrium if the net force on each individual part is zero. In addition to defining mechanical equilibrium in terms of force, there are many alternative definitions for mechanical equilibrium which are all mathematically equivalent. For momentum, a system is in equilibrium if the momentum of all parts is constant. For velocity, a system is in equilibrium if the velocity is constant. In rotational mechanical equilibrium, the angular momentum of the object is conserved and the net torque is zero. More generally in conservative systems, equilibrium is established at a point in the configuration space where the gradient of the generalized potential energy coordinates is zero. If a particle in equilibrium has zero velocity, that particle is considered to be in static equilibrium. Since all particles in equilibrium have constant velocity, it is always possible to find an inertial reference frame in which the particle is stationary with respect to the frame.

Chemistry has thermodynamic equilibrium, chemical equilibrium, diffusive equilibrium, thermal equilibrium, the Donnan equilibrium, dynamic equilibrium, equilibrium constants, partition equilibrium, quasistatic equilibrium, the Schlenk equilibrium, the solubility equilibrium, and vapor-liquid equilibrium, amongst others [3, 11, 17, 21]. In a chemical reaction, a chemical equilibrium is when both reactants and products are present in concentrations which have no further tendency to change overtime. Usually, this state occurs when the forward reaction proceeds at the same rate as the reverse reaction. The reaction rates of the forward and backward reactions are generally not equal, rather than zero. Therefore, there are no net changes in the concentrations of either the reactant(s) or the product(s); a state known as dynamic equilibrium.

In addition to the natural world, equilibria exist everywhere in human society. Economic equilibrium, for example, is a state in which the economic forces such as supply and demand are in balance and, in the absence of external influences, the (equilibrium) economic variable values do not change [22]. For example, in the standard text-book perfect competition model, equilibrium occurs when the quantities demanded and the quantities supplied are equal. Market equilibrium in this case refers to a condition in which a market price is established through competition so that the amount of goods or services sought by buyers is equal to the amount of goods or services produced by sellers. This price is often called the competitive price or the market clearing price which remains unchanged unless the demand or supply changes, and the quantity available is called a "competitive quantity" or a market clearing quantity.

Equilibria in the natural world or human society are states of harmony and stability. Once equilibrium is reached, no actions are taken to disturb the current state as disturbing the equilibrium often leads to extra costs or increased risk. For example, if the equilibrium of building components is disturbed due to earthquakes or man-made damage, there are dangers for the building's users. Economic equilibrium results instable economic growth, sufficient employment and a high happiness index. Conversely, a disturbance in economic equilibrium leads to unemployment, an increase in the gap between the rich and the poor and even social unrest. In this sense, equilibrium is optimum.

In decision-making theory, there is often more than one decision maker, who all tend to have different statuses and interests. A final decision, therefore, is made up of the decisions from all the involved decision makers. Because of the other stakeholders, a decision maker is unable to decide on an optimal decision without considering the other decision makers' responses. However, if the primary decision maker makes an optimal decision, the other decision makers also make their own optimal responses based on this initial decision, meaning that the initial decision is no longer optimal. For all decision makers, equilibrium is an acceptable state, because if one decision attempts to unilaterally break the equilibrium to find a more optimal solution, the equilibrium is disturbed and all decision makers are affected. Although various equilibria are discussed here, such as the Nash equilibrium and the Cournot equilibrium,this book focuses on the Stackelberg equilibrium, which has the following features:

- Two decision makers: a leader and a follower.
- The decision makers do not cooperate.
- The decision makers know each other's information.
- The decision makers make their respective decisions concurrently.

5.2 Equilibria in Real-World Problems

In the real-world, there are many decision problems that involve more than one decision maker. In these situations, it is assumed that both players know each other's objective functions and constraints. For bi-level programming, the decision maker on the upper level first specifies a strategy, and then the decision maker on the lower level specifies a strategy so as to optimize their objective with full knowledge of the actions of the decision maker on the upper level. Therefore, decision makers who have different benefits are in conflict, so to achieve the final compromise, both must find an *equilibrium* that can be fully accepted by all decision makers. This equilibrium is optimal for the decision makers if there is no rival; if there is only one decision maker, they can obtain a better solution. However, in a bi-level decision-making structure, all decision makers must accept the equilibrium.

In this book, three kinds of random-like bi-level decision making problems are discussed:

- A regional water resources allocation problem (RWRAP)
- A transport flow distribution problem (TFDP)
- A construction site security problem (CSSP)

In the following, the equilibria for these three problems are elaborated.

Water allocation is central to effective water resources management. Due to geographically and temporally unevenly distributed precipitation [2, 19], rapidly increasing water demands driven by growing populations and other stresses, and the degradation of the water environment, there is an increasing scarcity of water resources in many countries. Therefore, conflicts often arise when different water users(including the environment) compete for limited water supplies. The need to establish appropriate water allocation methodologies and associated management institutions and policies has been recognized by researchers, water planners and governments, and many studies have been conducted [23]. The RWRAP considers regional water planning design in sub-areas located beside the same river basin. In regional water planning under market mechanisms, because each sub-area is an independent decision maker, water resources allocation volume is no longer directly determined by the river basin regional authority. The regional authority allocates the water rights to the sub-areas and after obtaining the initial rights, each sub-area water manager makes water trading decisions under a market mechanism based on water use volume and water allocation to promote equitable cooperation in the river basin and to achieve efficient water use. Concurrently, water users pay for water withdrawal. Therefore, the regional authority influences the decision-making of each sub-area water manager by adjusting the initial water rights, while each sub-area water manager strives to meet their individual economic benefit goals by making rational water withdrawal decisions based on the decisions made by the regional authority. To ensure the sustainable development of the river basin, basin water resources are used mainly for ecological water, industrial water, municipal water and agricultural water, with ecological water taking priority. The regional authority has the higher authority, so the subarea water managers are the subordinates, even though they are relatively independent decision-makers. Therefore, the RWRAP can be abstracted as a bi-level programming problem, with the upper level programming being the main water distribution body from which the regional authority on the upper level (i.e., the leader) makes decisions about the initial water rights to n sub-areas to maximize the total social benefit. The following or lower level decision makers are the sub-area water managers (i.e., the followers),

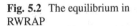

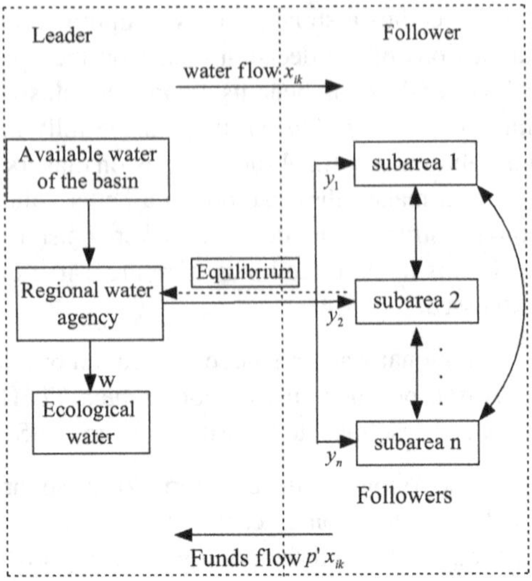

Fig. 5.2 The equilibrium in RWRAP

who decide on the water withdrawal volume to maximize their economic benefit. The RWRAP equilibrium is illustrated in Fig. 5.2.

The two participants in the TFDP are the construction contractor and the transportation manager. In a large-scale construction project, materials often have a supply (origin) and a receipt (destination) node, with the construction contractor generally assigning a specialized transportation company. The bi-level model concurrently considers the construction contractor and the transportation manager at the specialized transportation company, gives priority to contractor benefit and considers the influence of the contractor's decision-making on the transportation manager's carrier distribution flow. Costs and time control are also considered as these are vital to effective construction projects. The construction contractor assigns the material flow to each transportation path to minimize the direct costs and the transportation time costs, while the transportation manager aims to minimize transportation costs by making decisions about the material flows of each carrier on the transportation path based on the construction contractor's decision making, which in turn influences the contractor's decision-making through further adjustments to each carrier's material flow along the transportation path.

Therefore, the TFDP in this chapter can be abstracted as a bi-level programming problem. To model the problem conveniently, the transportation network is considered a bipartite network represented by a graph with sets of nodes and arcs. In the network, a node represents the facilities in the network; for instance, a station or a yard; and an arc represents a line between two adjacent facilities. The TFDP model structure is shown in Fig. 5.3.

The main purpose for developing the CSSP model is to deploy security counter measures in a way that collectively achieves the following four main security

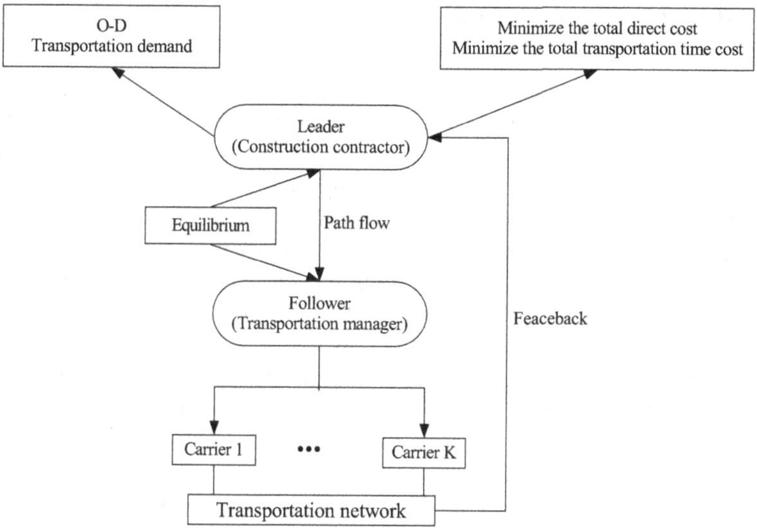

Fig. 5.3 The equilibrium in TFDP

Fig. 5.4 The equilibrium in a CSSP

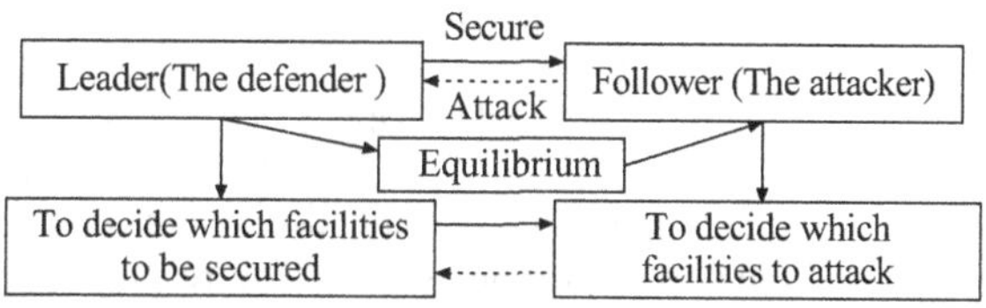

functions: deterrence, detection, delay, and detainment. Security countermeasures at a construction site can be grouped into three main layers:

- Site fences or the outer layer
- Site grounds or the intermediate layer
- Target fences or the inner layer

CSSP problems can be described as a leader-follower or a Stackelberg game. The leader is the project security officer (the defender), who must first decide which facilities should be secured with the limited funds, and which objectives are to maximize the effectiveness of the construction facilities system and minimize economic losses from potential attacks and security costs. The followers are the potential attackers, who are individuals or groups that aim to destroy onsite critical assets, inflict project efficiencies and destroy or attack a subset R of the facilities to inflict a maximum loss in construction facilities system effectiveness. Site targets can include a critical facility under construction, site office trailers that contain sensitive information, and/or storage areas for classified equipment or materials. The bi-level problem description is shown in Fig. 5.4.

5.3 Model System in Equilibria

When formulating mathematical programming problems, it is assumed that all decisions are made by a single person; however, game theory deals with the economic behavior of multiple decision makers who make fully rational judgments. Because two-level mathematical programming problems are interpreted as static Stackelberg games, multi-level mathematical programming is relevant to noncooperative game theory; in conventional multi-level mathematical programming models employing the Stackelberg equilibrium solution concept, it is assumed that there is no communication between the decision makers, or, they do not make any binding agreements even if such communication exists.

By considering the diverse evaluations and uncertainties, the general framework for the random-like bi-level multiple objective decision making model can be developed as:

$$\begin{cases} \min\limits_{x \in R^{n_1}} [F_1(\boldsymbol{x},\boldsymbol{y},\boldsymbol{\xi}), F_2(\boldsymbol{x},\boldsymbol{y},\boldsymbol{\xi}), \cdots, F_{m_1}(\boldsymbol{x},\boldsymbol{y},\boldsymbol{\xi})] \\ \text{s.t. } G_i(\boldsymbol{x},\boldsymbol{y},\boldsymbol{\xi}) \leq 0, i = 1,2,\cdots,p_1 \\ \text{where } \boldsymbol{y} \text{ solves:} \\ \begin{cases} \min\limits_{y \in R^{n_2}} [f_1(\boldsymbol{x},\boldsymbol{y},\boldsymbol{\xi}), f_2(\boldsymbol{x},\boldsymbol{y},\boldsymbol{\xi}), \cdots, f_{m_2}(\boldsymbol{x},\boldsymbol{y},\boldsymbol{\xi})] \\ \text{s.t. } g_j(\boldsymbol{x},\boldsymbol{y},\boldsymbol{\xi}) \leq 0, j = 1,2,\cdots,p_2, \end{cases} \end{cases} \tag{5.1}$$

where $\boldsymbol{x} \in R^{n_1}$ and $y \in R^{n_2}$ are decision vectors for the lower-level decision maker and the upper-level decision maker, respectively; $F(\boldsymbol{x},\boldsymbol{y},\boldsymbol{\xi})$ and $f(\boldsymbol{x},\boldsymbol{y},\boldsymbol{\xi})$ are objective function of the lower-level model and the upper-level model; $G_i(\boldsymbol{x},\boldsymbol{y},\boldsymbol{\xi})$ is the constraint of the upper-level programming and $g_j(\boldsymbol{x},\boldsymbol{y},\boldsymbol{\xi})$ is the constraint of the lower-level programming. $\boldsymbol{\xi}$ is a random-like (random, birandom, random fuzzy) vector.

It is necessary to understand that model (5.1) is a conceptual model rather than a mathematical model because it is not possible to optimize the objective functions with random-like parameters $\boldsymbol{\xi}$ nor judge whether a decision vector satisfies the random-like constraints, as there is no natural order to uncertain events. Therefore, it is not possible to compare alternatives directly with real numbers. Because they have uncertain variables, the above models are ambiguous, the meaning of the optimization $F_i(\boldsymbol{x},\boldsymbol{y},\boldsymbol{\xi}), f_i(\boldsymbol{x},\boldsymbol{y},\boldsymbol{\xi})$ is unclear, and the constraints $G_i(\boldsymbol{x},\boldsymbol{y},\boldsymbol{\xi}), g_i(\boldsymbol{x},\boldsymbol{y},\boldsymbol{\xi})$ do not define a deterministic feasible set. Therefore it is necessary to adopt methods to transform the above models into something mathematically meaningful.

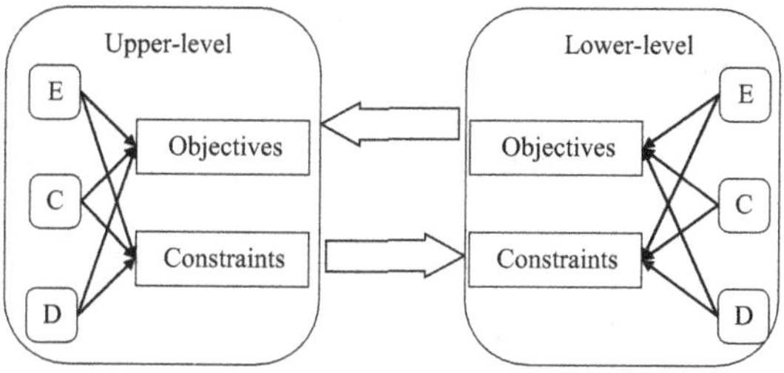

Fig. 5.5 Deterministic models on random-like operators

There are three random-like operators available to eliminate the uncertainties in the models, i.e., Expected Value Operator (E), Chance-Constrained Operators (C), and the Dependent Chance Operator(D). In theory, there are 36 combinations in all (see Fig. 5.5). There are some similarities when dealing with the 36 models, 9 of which have been selected to develop the relevant algorithms:

- Random EEEE Model
- Random CECC Model
- Random DEDC Model
- Ra-Ra EEDE Model
- Ra-Ra ECEC Model
- Ra-Ra DCCC Model
- Ra-Fu EECC Model
- Ra-Fu CCDD Model
- Ra-Fu DDEE Model

For example, a general Ra-Fu EECC model is formulated by

$$\begin{cases} \max\limits_{x \in R^{n_1}} [E[F_1(\boldsymbol{x},\boldsymbol{y},\boldsymbol{\xi})], E[F_2(\boldsymbol{x},\boldsymbol{y},\boldsymbol{\xi})], \cdots, E[F_{m_1}(\boldsymbol{x},\boldsymbol{y},\boldsymbol{\xi})]] \\ \text{s.t. } E[G_i(\boldsymbol{x},\boldsymbol{y},\boldsymbol{\xi})] \leq 0, i = 1, 2, \cdots, p_1 \\ \text{where } y \text{ solves:} \\ \begin{cases} \max\limits_{y \in R^{n_2}} \{\bar{f}_1, \bar{f}_2, \cdots, \bar{f}_{m_2}\} \\ \text{s.t. } \begin{cases} Ch\{f_i(\boldsymbol{x},\boldsymbol{y},\boldsymbol{\xi}) \geq \bar{f}_i\}(\alpha_i) \geq \beta_i, i = 1, 2, \cdots, m_2 \\ Ch\{g_j(\boldsymbol{x},\boldsymbol{y},\boldsymbol{\xi}) \leq 0\}(\gamma_i) \geq \delta_i, i = 1, 2, \cdots, p_2, \end{cases} \end{cases} \end{cases} \tag{5.2}$$

where α_i and β_i are predetermined confidence levels. Hence the follower need to obtain the (α_i, β_i)-optimistic values to the return function $f_i(\boldsymbol{x},\boldsymbol{y},\boldsymbol{\xi})$.

For the three kinds of problems listed in Sect. 5.2, the bi-level models are developed as

$$
(RWRAP)\left\{\begin{array}{l}
\max\limits_{x_i} E[V_0] = ew + \sum\limits_{i=1}^{n}\left[\sum\limits_{k=1}^{m}\left(b_{ik}y_{ik} - c_{ik}(E[\tilde{\omega}_{\tilde{d}_{ik}(r_i,\sigma_i)}] - y_{ik}) - p_w\lambda_{ik}y_{ik}\right)\right.\\
\qquad\qquad \left. + \left(x_i - \sum\limits_{k=1}^{m} y_{ik}\right)p(r)\right]\\
\text{s.t.}\left\{\begin{array}{l}
\sum\limits_{i=1}^{n} x_i + w = E[\tilde{Q}]\\
\sum\limits_{i=1}^{n}\sum\limits_{k=1}^{m} y_{ik} + w \le E[\tilde{Q}]\\
x_i \ge \theta_i\ (\theta_i > 0), \quad \forall i \in \Psi\\
w \ge \eta\ (\eta > 0)\\
\max\limits_{y_{ik}}\ E[V_i] = \sum\limits_{k=1}^{m}\left(b_{ik}y_{ik} - c_{ik}(E[\tilde{\omega}_{\tilde{d}_{ik}(r_i,\sigma_i)}] - y_{ik}) - p'_k y_{ik} - p_w\lambda_{ik}y_{ik}\right)\\
\qquad\qquad + \left(x_i - \sum\limits_{k=1}^{m} y_{ik}\right)p(r)\\
\text{s.t.}\left\{\begin{array}{l}
T_{ik\min} \le y_{ik} \le T_{ik\max}, \quad \forall i \in \Psi, k \in \Phi\\
y_{ik} \le E[\tilde{\omega}_{\tilde{d}_{ik}(r_i,\sigma_i)}], \quad \forall i \in \Psi,\ k \in \Phi\\
\sum\limits_{k=1}^{m} \lambda_{ik}y_{ik} \le q_i, \quad \forall i \in \Psi\\
p(r) = \kappa - \delta\left(\sum\limits_{i=1}^{n}\left(x_i - \sum\limits_{k=1}^{m} y_{ik}\right)\right) > 0,\ (\kappa > 0, \delta > 0)\\
y_{ik} \ge 0, \quad \forall i \in \Psi, k \in \Phi.
\end{array}\right.
\end{array}\right.
\end{array}\right.
\tag{5.3}
$$

$$
(TFDP)\left\{\begin{array}{l}
\min C(x_j, y_{kj}) = \sum\limits_{(o,d)\in E}\sum\limits_{j\in P_{od}} c_j x_j\\
\min\ T(x_j, y_{kj}) = E\left[\sum\limits_{k\in\Psi} w_k \sum\limits_{j\in\Omega}\sum\limits_{i\in A_j} \gamma_i^j \tilde{t}_{i0}^k\left[1 + \alpha\left(\frac{\lceil y_{kj}/v_k\rceil}{r_i}\right)^{\beta}\right]\right]\\
\text{s.t.}\left\{\begin{array}{l}
\sum\limits_{j\in P_{od}} x_j = Q_{od}, \quad (o,d) \in E\\
\sum\limits_{k\in\Psi} w_k = 1\\
x_j \ge 0, \quad j \in \Omega\\
\min\ C(y_{kj}) = E\left[\sum\limits_{k\in\Psi}\sum\limits_{j\in\Omega}\sum\limits_{i\in A_j} \gamma_i^j \tilde{e}_i^k \lceil y_{kj}/v_k\rceil\right]\\
\text{s.t.}\left\{\begin{array}{l}
Pr\left\{\omega\middle| Pr\left\{\sum\limits_{i\in A_j} \gamma_i^j \tilde{t}_{i0}^k(\omega)\left[1+\alpha\left(\frac{\lceil y_{kj}/v_k\rceil}{r_i}\right)^{\beta}\right] \le T_j\right\} \ge \theta\right\} \ge \delta\\
\sum\limits_{j\in\Omega}\sum\limits_{k\in\Psi} \gamma_i^j \lceil y_{kj}/v_k\rceil \le r_i, i \in \Phi, j \in \Omega, k \in \Psi\\
\gamma_i^j = \left\{\begin{array}{ll} 1, \text{ if } i \in A_j, & j \in \Omega\\ 0 \text{ otherwise} & \end{array}\right.\\
\sum\limits_{k\in\Psi} y_{kj} \ge x_j, \quad j \in \Omega\\
y_{kj} \ge 0, \quad j \in \Omega, k \in \Psi,
\end{array}\right.
\end{array}\right.
\end{array}\right.
\tag{5.4}
$$

and

$$
(CSSP)\left\{\begin{array}{l}
\min\limits_{z_j} C=\sum\limits_{j=1}^{N}\sum\limits_{k=1}^{5} p_k\mu_{jk}^{c}s_j+\sum\limits_{j=1}^{N}M_jz_j\\
\max\limits_{z_j} D=\sum\limits_{j=1}^{N}\sum\limits_{i=1,i\neq j}^{N}\sum\limits_{k=1}^{5}\theta_{ij}p_k\mu_{ijk}^{r}d_{ij}s_j+\sum\limits_{j=1}^{N}\sum\limits_{i=1,i\neq j}^{N}\theta_{ij}d_{ij}z_j\\
+\sum\limits_{j=1}^{N}\sum\limits_{i=1,i\neq j}^{N}\theta_{ij}d_{ij}(1-s_j)(1-z_j)\\
\text{s.t.}\left\{\begin{array}{l}
\sum\limits_{j=1}^{N}M_jz_j\leq T\\
Pr\left\{\omega\left|Pr\left\{\sum\limits_{j=1}^{N}\tilde{\bar{C}}_j(\omega)^Ts_j\leq C_0\right\}\geq\theta_1\right.\right\}\geq\delta_1\\
Pr\left\{\omega\left|Pr\left\{\sum\limits_{i=1,i\neq j}^{N}\theta_{ij}\tilde{\bar{r}}_{ij}(\omega)^Td_{ij}s_j\geq D_0\right\}\geq\theta_2\right.\right\}\geq\delta_2\\
z_j\in\{0,1\},\ \forall i,j\in\Phi\\
\min\limits_{s_j} D'=\sum\limits_{j=1}^{N}\sum\limits_{i=1,i\neq j}^{N}\sum\limits_{k=1}^{n}\theta_{ij}p_k\mu_{ijk}^{r}d_{ij}s_j+\sum\limits_{j=1}^{N}\sum\limits_{i=1,i\neq j}^{N}\theta_{ij}d_{ij}(1-s_j)\\
\text{s.t.}\left\{\begin{array}{l}
\sum\limits_{j=1}^{N}s_j\leq R\\
\sum\limits_{j=1}^{N}s_j\leq N-\sum\limits_{j=1}^{N}z_j\\
s_j\in\{0,1\}\\
z_j+s_j\leq 1,\ \forall i,j\in\Phi.
\end{array}\right.
\end{array}\right.
\end{array}\right.
\tag{5.5}
$$

They are the Random EEEE model, Ra-Ra EEEC model and Ra-Fu ECEC model, respectively.

In addition, there could be different reactions on the lower level towards each possible upper level action when multiple followers are involved in bi-level decision-making, as the different relationships between these followers could result in the need for many different processes when seeking to derive an optimal solution to the upper level decision-making. Therefore, the leader's decision can be affected by the reactions of the followers and also by the relationships between these followers. Basically, there are three different relationships between the followers, which are dependent on the share in the decision variables. The first relationship is an uncooperative situation, in which there is no sharing of decision variables between the followers. In this situation, there are no shared objectives or constraints. The second case is a cooperative situation, in which the followers all share the decision variables in their objectives and constraints. However, within this cooperative situation, there are several different sub-cases, which are determined by the relationships between the objectives (the second relationship factor) and the constraints (the third relationship factor) of the followers. Each follower may have an individual objective while sharing the constraints with the other followers. For

Table 5.1 A general framework for the bi-level decision problem

Relationship	Decision variables	Objectives	Constraints	Situations (S_i)
Uncooperative	Individual	Individual	Individual	S_1
Cooperative	Sharing	Sharing	Sharing	S_2
			Individual	S_3
		Individual	Sharing	S_4
			Individual	S_5
Partial cooperative	Partial individual and Partial sharing	Sharing	Sharing	S_6
			Individual	S_7
		Individual	Sharing	S_8
			Individual	S_9

example, one agricultural group may have an objective to maximize the agricultural profile, and another agricultural group objective may wish to maximize land sustainability based on government land-use policy. These two followers, therefore, share their decision variables, but have different objectives. Another pair of sub-cases is when followers have common objectives but have different constraints. For example, for any governmental agricultural policy, two agricultural groups may have a common objective to maximize the agricultural profile, but they may or may not share financial, environmental protection and cultural constraints when attempting to achieve an optimal solution. The last case is the partially cooperative situation, in which the followers partially share decision variables in their objectives or constraints or both. Similar to the second case, four sub-cases are involved within this case.

Based on these three cases and the respective sub-cases determined by the three relationship factors for the decision variables, objectives and constraints, there are nine potentially different situations between the followers; $S_1, S2, \cdots, S_9$. The framework for these situations is shown in Table 5.1. Regardless of whether followers share decision variables, the bi-level decision problem is dealt with as a variable sharing situation. Similarly, regardless of whether followers share objectives (or constraints), the bi-level decision problem is dealt with as an objective (or constraint) sharing situation.

5.4 Equilibria Algorithms

Due to the intrinsic model complexity, the development of the algorithms to solve the models has been one of the more difficult and challenging problems [20]. Therefore, an algorithmic framework (Fig. 5.6.) was developed to seek the models' equilibria.

More specifically, the models in this book have been divided into two categories: linear models and nonlinear models. The linear model algorithms are listed in Table 5.2.

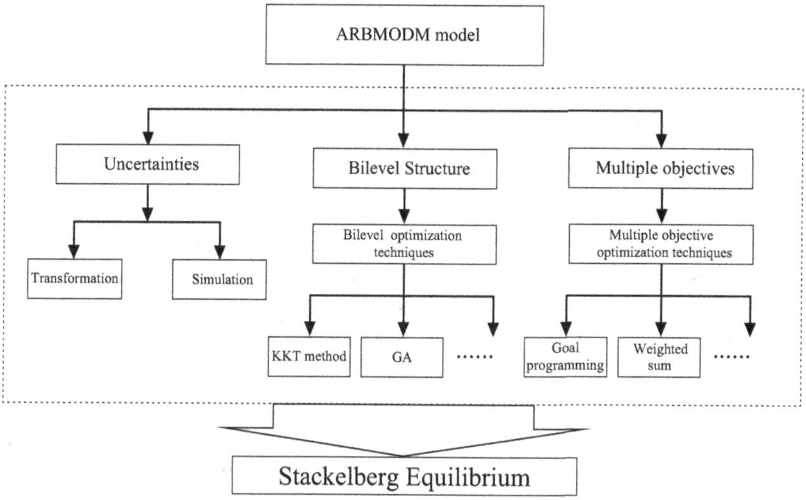

Fig. 5.6 The algorithm framework

Table 5.2 Algorithms for linear RBMODM models

Model	Algorithm
Random EEEE model	Reference point method
Random CECC model	KKT method
Random DEDC model	Interactive fuzzy goal programming method
Ra-Ra EEDE model	Steepest descent direction method
Ra-Ra ECEC model	Homotopy method
Ra-Ra DCCC model	Interactive balance space method
Ra-Fu EECC model	Constrained variable metric method
Ra-Fu CCDD Model	Fuzzy decision method
Ra-Fu DDEE model	Interactive method

In the nonlinear models, however, there are difficulties when calculating the expected values and the chance values for the random-like events. Therefore, simulation techniques are used to perform sampling experiments on the random-like system models. This technique is primarily based on the sampling of random variables from probability distributions, which is often referred to as the Monte Carlo simulation. Although simulation is an imprecise technique which provides only statistical estimates rather than exact results and is also a slow and costly way to study problems, it is a powerful tool for dealing with complex problems and does not require analytic techniques. In this book, random simulation, Ra-Ra simulation and Ra-Fu simulation are embedded into the algorithms for nonlinear RDMODM problems. These algorithms are listed in Table 5.3.

Apart from the metaheuristic algorithms mentioned above, there are also descent local searches (LS) [16], simulated annealing (SA) [14], tabu searches (TS) [10], evolutionary strategies (ES) [4], ant colony searches (AC) [6] and scatter searches

Table 5.3 Algorithm for nonlinear RBMODM models

Model	Simulation-based algorithm
Random EEEE model	Goal programming method + GA
Random CECC model	εconstraint method + EBS-based GA
Random DEDC model	Two-Stage method + STGA
Ra-Ra EEDE model	Lexicographic method + PSO
Ra-Ra ECEC model	STEM+CHK-PSO
Ra-Ra DCCC model	Satisfying trade-off method + CST-PSO
Ra-Fu EECC model	Linear weighted-sum method + PGSA
Ra-Fu CCDD Model	Minimax point method + PGSA-GA
Ra-Fu DDEE model	Ideal point method + PGSA-PSO

(SS) [9]. Metaheuristic algorithms are grouped into two categories; single-solution based metaheuristics (LS, SA, TS) and population-based metaheuristics (GA, ES, AC, PGSA, PSO). Single-solution based algorithms manipulate and transform a single solution during the search, while in population-based algorithms, the whole population of solutions is evolved. These two families have complementary characteristics: single-solution based metaheuristics are exploitation oriented and therefore, have the power to intensify searches in local regions. Population-based metaheuristics are exploration oriented, so allow for better diversification in the search space. Depending on the different approaches, these can be classified into the following strategy types [20]:

- Nested sequential approaches: in this class of metaheuristic strategies, the lower level optimization problem is solved in nested and sequential ways to evaluate the solutions generated on the upper-level of the bi-level optimization problem.
- Single-level transformation approaches: the main characteristic of this class of metaheuristics is the reformulation of the bi-level optimization problem into a single-level optimization problem. Then, any traditional metaheuristics can be used to solve the single-level problem.
- Multi-objective approaches: in this class of metaheuristics strategies, the bi-level optimization problem is transformed to a multi-objective optimization problem. Then, any multi-objective metaheuristics can be used to solve the generated problem.
- Co-evolutionary approaches: this is the most general methodology to solve the RBMODM, in which many metaheuristics coevolve in parallel and exchange information to solve different problem levels.

In the nested approaches, hierarchical optimization algorithms attempt to solve the two levels sequentially, gradually improving the solutions on each level to achieve a good overall solution on both levels. Such algorithms include [20]:

- Repairing approaches: this approach considers the lower-level problem as a constraint and solves it during the evaluation step. It is assumed that the lower level optimization problem has a given structure so the problem can be efficiently solved.

- Constructive approaches: this approach sequentially applies two improving algorithms on a population, one for each level, until the stopping criterion (e.g. given number of generations) is met.

As mentioned, the repairing approach considers the follower problem a constraint. In the first phase, a solution is generated on the upper-level (i.e., generation of $(\boldsymbol{x}, \boldsymbol{y})$), then this solution $(\boldsymbol{x}, \boldsymbol{y})$ is sent to the lower-level problem and is considered the given initial solution for the lower-level problem. Then, an optimization algorithm (e.g. any metaheuristic) can be used to finda "good" solution $\boldsymbol{y}$, according to the lower-level optimization problem. On the lower-level, a variable, x, is used as a parameter and is fixed. Then, the solution to the lower-level $(\boldsymbol{x}, \boldsymbol{y})$ is transmitted to the upper-level, after which the whole solution for the upper-level is replaced by $(\boldsymbol{x}, \boldsymbol{y})$ and the upper-level objective is evaluated. These three phases proceed iteratively and sequentially until the given stopping criteria is met.

In the constructive approach, the lower-level problem is solved to improve the population of solutions $(\boldsymbol{x}, \boldsymbol{y})$ generated on the upper-level using the objective function F. This approach is generally used for a population based metaheuristic (i.e. P-metaheuristic). The population is improved on the lower-level using the objective function f, in which the decision variables $\boldsymbol{x}$ are fixed. Finally, after a given stopping criteria, the improved population of solutions $(\boldsymbol{x}, \boldsymbol{y})$ constitutes the initial population on the upper-level. This processiterates until the given stopping criteria is met.

The main drawback of the nested approach is the computational complexity. The nested procedure has to solve an optimization problem (i.e. lower-level problem) for each solution to the problem generated on the upper-level. The efficiency of this class of strategies depends significantly on the difficulties encountered when solving the lower-level problem. For complex lower-level problems, more efficient metaheuristic strategies must be designed so that there is greater coordination between the two optimization levels.

The nested approach is characterized by [20]:

- An upper-level solution approach: the generation of solutions for the upper-level model needs to be considered carefully as it is the key to efficiently solving the complete bi-level optimization problem.
- Lower-level solution approach: as in the upper-level, depending on the difficulty of the inner problem, any traditional optimization strategy can be developed to solve the lower-level problem.

In the homogeneous nested approach, the same metaheuristic is used on both levels (upper-level and lower-level). For instance, a particle swarm optimization based metaheuristic has been designed and genetic algorithms developed to solve a continuous network design problem.

Over the years, the most popular metaheuristic approaches transform the bi-level optimization problem into a single-level optimization problem using approximate or exact methodologies to replace the lower-level optimization problem [20]. Several approaches, such as enumeration methods, penalty methods [1, 8], marginal functions [18], and method and trust-region methods, have been proposed for bi-level optimization problems, under the assumption that all functions are convex and twice differentiable [20].

For differentiable objectives and constraints in the lower-level problem, a popular approach has been to include the KKT conditions of the lower-level problem as the constraints for the upper-level optimization problem [12]. Additional variables for the upper-level problem are represented by the Lagrange multipliers from the lower-level problem. Other conditions must be satisfied to ensure that the KKT solutions are optimal solutions [20].

The KKT conditions for the lower-level optimization problem are generally used as constraints in the formulation of the KKT conditions for the upper-level optimization problem. This involves the use of the second derivatives of the objectives and constraints in the lower-level problem as the necessary conditions for the upper-level optimization problem. This methodology is difficult to apply in practical problems because of the many lower-level Lagrange multipliers and an abstract term which contains co-derivatives [20].

Once the RBMODM problem is transformed to a single-level optimization problem (e.g. using the KKT conditions), any traditional metaheuristic can be used to solve the single level problem: genetic algorithms, differential evolution, simulated annealing, evolutionary algorithms, local search algorithms, and hybrid metaheuristics (genetic algorithms with neural networks).

The basic idea of the penalty approach is the penalty function, which is used to generate an initial solution and improves the current solution using a Tabu search algorithm. The current solution always belongs to the admissible region. Marcotte et al. transformed the network design problem into a single-level equivalent differentiable optimization problem, in which the required constraints involved all extreme points of the closed convex polyhedron for the feasible acyclic multi-commodity flow patterns.

As this problem has two different objective functions, a natural approach for tackling bi-level optimization problems would be to use a Pareto-based multi-objective approach. However, bi-level optimization problems have a different structure. A good solution for a similar problem that approximates the Pareto frontier could be of poor quality when applied to a bi-level problem [20].

Therefore, any multi-objective optimization metaheuristic can be used to solve the problem. However, this approach is limited to differentiable problems as the derivatives of the objectives from the original bi-level optimization problem are used in the mathematical formulation of the MOP problem. These assumptions are clearly very restrictive and can seldom be satisfied [20].

In many cases, methodologies based on the nested, multi-objective or reformulation approaches may not be used or may be practically inefficient [20]. In fact, most of these traditional approaches were designed for specific bi-level optimization problems or were based on specific assumptions (e.g. upper-level or lower-level problem differentials, convex feasible regions, lower-level structural problems, or an upper-level reduced search space). Because of such deficiencies, these approaches cannot be used to solve real-life complex applications, such as RBMODM problems with non-differentiable objective functions and complex combinatorial RBMODM problems. Therefore, some co-evolutionary based metaheuristics approaches were developed to solve general RBMODM problems as bi-level programming problems without any transformations [20].

In co-evolutionary metaheuristics, the two levels proceed in parallel and on each level, an optimization strategy is applied. In general, the optimization strategy is a population-based metaheuristic, in which each level attempts to separately maintain and improve its own population. Therefore, the two populations evolve in parallel and these different populations evolve only a part of the decision variables, so complete solutions are built using a cooperative exchange of individuals from the populations, in which the two levels exchange information to maintain a global view of the RBMODM problems [20].

In designing a co-evolutionary model for any metaheuristic, the same design questions need to be answered [20]:

- The exchange decision criterion (When?): the exchange of information between the metaheuristics can be decided either in a blind (periodic or probabilistic) way or according to an "intelligent" adaptive criterion. Periodic exchanges occur in each algorithm after a fixed number of iterations, which indicates asynchronous communication. Probabilistic exchanges consist of performing a communication operation after each iteration using a given probability. Conversely, adaptive exchanges are guided by some run-time search characteristics. For instance, they may depend on an evolution in the quality of the solutions or the search memory. A classical criterion is related to improvements in the best found local solutions.
- The information exchanged (What?): this parameter specifies the information to be exchanged between the metaheuristics. In general, this may be composed of solutions and search memory.
- The integration policy (How?): symmetric with the information exchange policy, the integration policy deals with the use of the received information. In general, there is a local copy of the received information, where the local variables are updated using the received variable. For instance, the best found solution for a given solution $(\boldsymbol{x}, \boldsymbol{y})$ is recombined with the received solution $(\boldsymbol{x}, \boldsymbol{y}^*)$.

5.5 Perspectives

Since real-world decision making problems are becoming increasingly complex, RBMODM has become the subject of growing academic and application interest, so much so that new theories and research methods are constantly being developed. Therefore, future RBMODM research requires: (1) methods which surmount theory and method; and (2) connections with realistic problems. The importance of any further theories and methods can only be measured by their application to realistic problems. However, only a reliable theory can guarantee the correctness of any methods when dealing with realistic problems. Here, details about further research in this area are discussed.

5.5.1 Models

From a modelling aspect, the complexity of real-world problems results in models with significantly more complex structures, increased levels and a greater number of decision makers. In a real hierarchical organization, such as a large-scale enterprise, there are often more than two decision levels, so models which allow for three or more levels are necessary.

$$\left\{\begin{array}{l}\min\limits_{x_1\in R^{n_1}} [f_{11}(\boldsymbol{x},\boldsymbol{\xi}), f_{12}(\boldsymbol{x},\boldsymbol{\xi}),\cdots, f_{1m_1}(\boldsymbol{x},\boldsymbol{\xi})]\\ \text{s.t.}\left\{\begin{array}{l} g_{1i}(\boldsymbol{x},\boldsymbol{y},\boldsymbol{\xi})\le 0, i=1,2,\cdots,p_1\\ \text{where } \boldsymbol{x}_2,\boldsymbol{x}_3,\cdots,\boldsymbol{x}_l \text{ solves:}\\ \min\limits_{x_2\in R^{n_2}} [f_{21}(\boldsymbol{x},\boldsymbol{\xi}), f_{22}(\boldsymbol{x},\boldsymbol{\xi}),\cdots, f_{m_2}(\boldsymbol{x},\boldsymbol{\xi})]\\ \text{s.t.}\left\{\begin{array}{l} g_{2j}(\boldsymbol{x},\boldsymbol{\xi})\le 0, j=1,2,\cdots,p_2\\ \vdots\\ \text{where } \boldsymbol{x}_l \text{ solves:}\\ \min\limits_{x_l\in R^{n_2}} [f_{l1}(\boldsymbol{x},\boldsymbol{\xi}), f_{l2}(\boldsymbol{x},\boldsymbol{\xi}),\cdots, f_{m_l}(\boldsymbol{x},\boldsymbol{\xi})]\\ \text{s.t. } g_{lj}(\boldsymbol{x},\boldsymbol{\xi})\le 0, j=1,2,\cdots,p_l,\end{array}\right.\end{array}\right.\end{array}\right. \tag{5.6}$$

where $\boldsymbol{x}_i \in R^{n_i}$ is the decision vector for the i-th level decision maker, $i = 1,2,\cdots,l$, $\boldsymbol{x} = (\boldsymbol{x}_1,\boldsymbol{x}_2,\cdots,\boldsymbol{x}_l)$; $f_{ij}(\boldsymbol{x},\boldsymbol{\xi})$ are objective functions of the i-th-level model, $i = 1,2,\cdots,l, j = 1,2,\cdots,im_i$; $g_{ij}(\boldsymbol{x},\boldsymbol{\xi})$ is the constraint of the upper-level programming and $g_j(\boldsymbol{x},\boldsymbol{y},\boldsymbol{\xi})$ is the constraint of the i-th-level model,$i = 1,2,\cdots,l, j = 1,2,\cdots,im_i$; $\boldsymbol{\xi}$ is a random-like (random, birandom, random fuzzy) vector.

Although an n-level model can be regarded as a nested model with $n-1$ bi-level models, the computation complexity increases exponentially with the number of levels. Some efficient solution methods to bi-level situations may not work when dealing with 3 or more level models. Real-world problems promote the development of higher level models, including when there is more than one decision maker on the same level.

$$\left\{\begin{array}{l}\max\limits_{x}\ [F_1(\boldsymbol{x},\boldsymbol{y}^{(1)},\boldsymbol{y}^{(2)},\cdots,\boldsymbol{y}^J,\boldsymbol{\xi}), F_2(\boldsymbol{x},\boldsymbol{y}^{(1)},\boldsymbol{y}^{(2)},\cdots,\boldsymbol{y}^J,\boldsymbol{\xi}),\\ \cdots F_{m_1}(\boldsymbol{x},\boldsymbol{y}^{(1)},\boldsymbol{y}^{(2)},\cdots,\boldsymbol{y}^J,\boldsymbol{\xi})]\\ \text{s.t. } G_i(\boldsymbol{x},\boldsymbol{y},\boldsymbol{\xi})\le 0, i=1,2,\cdots,p_1\\ \text{where } \boldsymbol{y}^j (j=1,2,\cdots,J) \text{ solves}\\ \quad\left\{\begin{array}{l}\max\limits_{y^j} f_j(\boldsymbol{x},\boldsymbol{y}^{(1)},\boldsymbol{y}^{(}2),\cdots,\boldsymbol{y}^J,\boldsymbol{\xi})\\ \text{s.t. } g_{ji}(\boldsymbol{x},\boldsymbol{y},\boldsymbol{\xi})\le 0, i=1,2,\cdots,p_j.\end{array}\right.\end{array}\right. \tag{5.7}$$

Both (5.6) and (5.7) include uncertain vectors $\boldsymbol{\xi}$. There is a need to deal with these so that they are mathematically meaningful. The technologies EVM, CCM and DCM are effective. In addition, some other methods, such as variance minimization can be used to deal with uncertainty.

5.5.2 Theories

From a theoretical aspect, it is necessary that solution concepts be accurately defined. Compared with traditional decision-making models, increasingly complex structures mean that traditional optimal solutions are no longer suitable. Therefore, novel concepts need to be defined for these increasingly complex models (5.6) and (5.7). For example, the equilibrium between adjacent levels needs to be defined for model (5.6). In model (5.7), there is more than one follower, so as the leader determines their own decision variable, the followers must reach a consensus before reacting to the leader's decision. Since the followers are by nature competitive and make decisions concurrently, the consensus must be a Nash-equilibrium. Therefore, the leader and the followers together form a Stackelberg-Nash-equilibrium. Further, for model (5.7), because of the leader's multiple objectives, the term Pareto-Stackelberg-Nash-equilibrium is more suitable. To design effective and efficient solution algorithms, further analyses on the solutions are necessary, to cover the existence and/or the uniqueness of the solutions, the optimality conditions and regularity (see Fig. 5.7). Theoretical analyses build bridges between the models and the algorithms, and are the prerequisites for the design of a solution algorithm. Analyses of the existence of solutions and/or uniqueness assist the decision maker to estimate the number of solutions; analyses of optimality conditions provide hints for the solution algorithms.

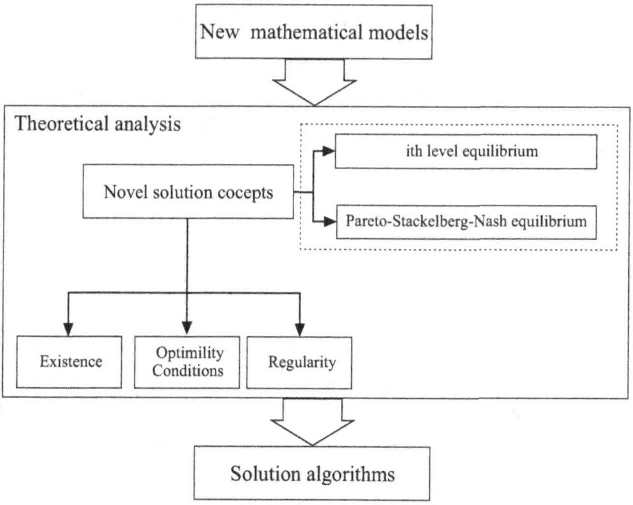

Fig. 5.7 Theoretical analysis aspects

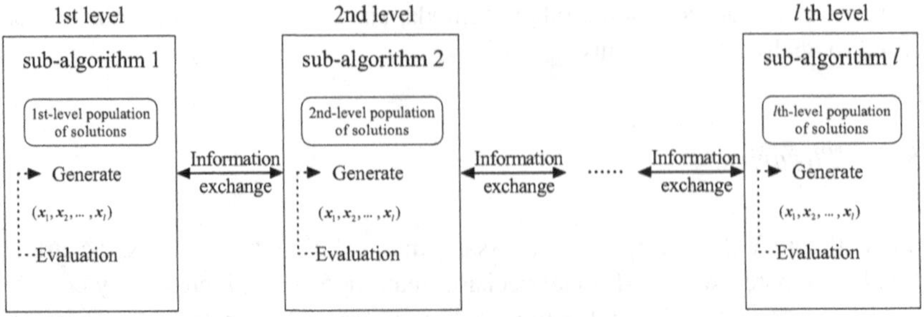

Fig. 5.8 Decentralized frame for multi-level models

5.5.3 *Algorithms*

Designing a solution algorithm may be the main bottleneck in applying these models to real-world problems. In this book, a series of solution algorithms for RBMODM problems are proposed. However, these algorithms may have no effect on novel models because of their more complex structures. While technologies for dealing with uncertain parameters (equivalent transformation and simulation) are available for the novel models, they occupy only a small part of the algorithmic framework. The main task when designing solution algorithms is deciding how to handle the hierarchal problems. In model (5.6), a decentralization idea is required for the construction of an effective framework, as under a decentralized framework, the whole model can be divided into several separate models (see Fig. 5.8).

This type of divide-and-conquer strategy decreases computation complexity. To obtain a solution to the whole model, it is necessary to build an effective information exchange mechanism between the adjacent level models, which is a significant challenge. For model (5.6), the idea of decentralization can still be used, but the Nash-equilibrium algorithms should be considered. As the solutions are obtained, some indicators are needed to assess the algorithms' performance, so error rates or rationality metrics [15] could be used. Convergence and convergence speed are also possible future research foci.

5.5.4 *Applications*

From an application aspect, there are three kinds of RBMODM problems; the regional water resources allocation problem, the transport flow distribution problem, and the construction site security planning problem, all of which are in engineering and management domains. However, the applications discussed in this book make up only a small part of the actual problems. In the management, finance, and manufacturing domains, there are several problems which require decision makers to deal with extremely complex situations such as complex leader-follower structures,

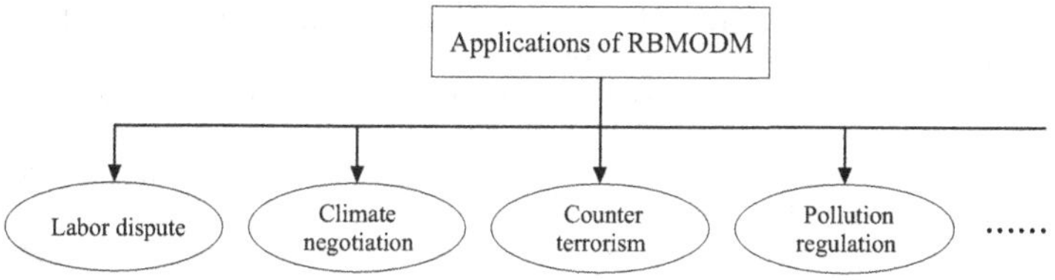

Fig. 5.9 More applications

multiple objectives and random-like phenomenon. At present, the RDMODM can be used to model these complexities, but the depiction of these objectives and associated information is worthy of deeper research.

The RBMODM could be scaled to address broader domains as conflict and equilibrium are the fundamental laws for all operations, and the RBMODM ably represents the decision making hierarchy, the diverse decision making objectives and the uncertain environment. For example, in a labor dispute, employers need to consider the employee wants and it is expected that the employees have a rational response to the employers. In addition to their salaries, the employees may require greater social security. Since economics is influenced by many uncertainties, the RBMODM would be able to assist employers and employees reach consensus. More problems, such as climate negotiations, counter terrorism and pollution regulations could also be effectively solved using the RBMODM methodology (see Fig. 5.9).

In conclusion, in this book, RBMODM systematic and scientific research is provided. However, as this book is only a small step in RBMODM research, additional research is required in the future to develop more effective decision methods.

References

1. Aiyoshi E, Shimizu K (1984) A solution method for the static constrained Stackelberg problem via penalty method. IEEE Trans Autom Control 29(12):1111–1114
2. Al-Radif A (1999) Integrated water resources management (IWRM): an approach to face the challenges of the next century and to avert future crises. Desalination 124(1):145–153
3. Atkins P, De Paula J (2013) Elements of physical chemistry. Oxford University Press, Oxford/New York
4. Beyer H (2013) The theory of evolution strategies. Springer Science & Business Media, Berlin/New York
5. De Pater I, Lissauer JJ (2014) Planetary sciences. Cambridge University Press, Cambridge/New York
6. Dorigo M, Blum C (2005) Ant colony optimization theory: a survey. Theor Comput Sci 344(2):243–278
7. Duvernell DD, Lindmeier JB, Faust KE, Whitehead A (2008) Relative influences of historical and contemporary forces shaping the distribution of genetic variation in the Atlantic killifish, Fundulus heteroclitus. Mol Ecol 17(5):1344–1360

8. Gendreau M, Marcotte P, Savard G (1996) A hybrid tabu-ascent algorithm for the linear bilevel programming problem. J Glob Optim 8(3):217–233
9. Glover F (1977) Heuristics for integer programming using surrogate constraints. Decis Sci 8(1):156–166
10. Glover F (1986) Future paths for integer programming and links to artificial intelligence. Comput Oper Res 13(5):533–549
11. Hefter GT, Tomkins R (2003) The experimental determination of solubilities. John Wiley, Chichester/Hoboken
12. Herskovits J, Leontiev A, Dias G, Santos G (2000) Contact shape optimization: a bilevel programming approach. Struct Multidiscip Optim 20:214–221
13. Johnston ER (2009) Vector mechanics for engineers: statics and dynamics. Tata McGraw-Hill Education, New York
14. Kirkpatrick S, Gelatt CD, Vecchi MP (1983) Optimization by simulated annealing. Science 220(4598):671–680
15. Legillon F, Liefooghe A, Talbi E (2012) Cobra: a cooperative coevolutionary algorithm for bi-level optimization. In: IEEE congress on evolutionary computation (CEC 2012), Brisbane, pp 1–8
16. Lenstra JK (2003) Local search in combinatorial optimization. Princeton University Press, Princeton
17. Lieb EH, Yngvason J (1999) The physics and mathematics of the second law of thermodynamics. Phys Rep 310(1):1–96
18. Meng Q, Yang H, Bell MGH (2000) An equivalent continuously differentiable model and a locally convergent algorithm for the continuous network design problem. Transp Res Part B 35(1):83–105
19. Nazemi A, Wheater HS: On inclusion of water resource management in Earth system models - Part 2: representation of water supply and allocation and opportunities for improved modeling. Hydrol Earth Syst Sci 19(1):63–90 (2015)
20. Talbi E (2014) Metaheuristics for bilevel optimization. Springer, Berlin/New York
21. Tschoegl NW (2000) Fundamentals of equilibrium and steady-state thermodynamics. Elsevier, Amsterdam/New York
22. Varian HR (1992) Microeconomic analysis. WW Norton, New York
23. Wang LZ, Fang L, Hipel KW (2003) Water resources allocation: a cooperative game theoretic approach. J Environ Inf 2(2):11–22

Appendix
MATLAB Codes

A.1 MATLAB® File for Example 2.3

```
N=10000;
f=0;
for i=1:N
    x1=exprnd(2);
    x2=normrnd(4,1);
    x3=unifrnd(4,8);
    f=f+sqrt(x1^2+x2^2+x3);
end
E=f/N
```

A.2 MATLAB® File for Example 2.7

```
N=10000;
alpha=0.8;
x1=exprnd(1,N,1);
x2=normrnd(3,1,N,1);
x3=unifrnd(0,1,N,1);
f=sqrt(x1.^2+x2.^2+x3.^2);
g=sort(f,1);
g(N*alpha)
```


J. Xu et al., *Random-Like Bi-level Decision Making*, Lecture Notes in Economics and Mathematical Systems 688, DOI 10.1007/978-981-10-1768-1

A.3 MATLAB® File for Example 2.10

```
N=10000;
N0=0;
x1=exprnd(6,N,1);
x2=normrnd(4,1,N,1);
x3=unifrnd(0,2,N,1);
f=sqrt(x1.^2+x2.^2+x3.^2);
for i=1:N
    if f(i)<=8
        N0=N0+1;
    end
end
Pr=N0/N
```

A.4 MATLAB® File for Example 3.2

```
M=10000;
N=10000;
E1=0;
rho1=normrnd(4,1,N,1);
rho2=unifrnd(3,5,N,1);
rho3=unifrnd(1,2,N,1);
for i=1:N
    x1=unifrnd(rho1(i), rho1(i)+2,N,1);
    x2=normrnd(rho2(i), 1, N,1);
    x3=exprnd(rho3(i), N,1);
    f=sqrt(x1.^2+x2.^2+x3.^2);
    E1=E1+sum(f)/N;
end
E=E1/M
```

A.5 MATLAB® File for Example 3.3

```
alpha=0.8;
N=10000;
N0=zeros(N,1);
N1=zeros(N,1);
rho1=normrnd(4,1,N,1);
rho2=unifrnd(3,5,N,1);
```

```
rho3=unifrnd(1,2,N,1);
for i=1:N
    x1=unifrnd(rho1(i),rho1(i)+2, N,1);
    x2=normrnd(rho2(i),1, N,1);
    x3=exprnd(rho3(i),N,1);
    for j=1:N
        if sqrt(x1(j)^2+x2(j)^2+x3(j)^2)>=6;
           N0(i)=N0(i)+1;
        end
    end
    N1(i)=N0(i)/N;
end
N2=sort(N1);
N2(N-alpha*N+1)
```

A.6 MATLAB® File for Example 3.6

```
alpha=0.9;
beta=0.9;
F=zeros(N,1);
x1=zeros(N,1);
x2=zeros(N,1);
N=10000;
rho1=normrnd(0,1,N,1);
rho2=unifrnd(3,5,N,1);
rho3=unifrnd(1,2,N,1);
for i=1:N
    x1=unifrnd(rho1(i),rho1(i)+2, N,1);
    x2=normrnd(rho2(i),1, N,1);
    x3=exprnd(rho3(i), N,1);
    f=sqrt(x1.^2+x2.^2+x3.^2);
    sort(f);
    F(i)=f(N-N*alpha+1);
end
G=sort(F);
G(N-N*beta+1)
```

A.7 MATLAB® File for Example 4.2

```
N=5000;
alpha=0.8;
```

```
e=0;
epsilon=0.0001;
U1=zeros(N,1);
U2=zeros(N,1);

pos1=zeros(N,1);
pos2=zeros(N,1);
pos3=zeros(N,1);
pos4=zeros(N,1);

theta1=unifrnd(1,3,N,1);
theta2=unifrnd(2,4,N,1);
theta3=unifrnd(3,5,N,1);
theta4=unifrnd(4,6,N,1);

theta01=zeros(N,1);
theta02=zeros(N,1);
theta03=zeros(N,1);
theta04=zeros(N,1);

v=zeros(N,1);

for i=1:N
    if (theta1(i)<=2)
        pos1(i)=theta1(i)-1;
    else
        pos1(i)=3-theta1(i);
    end

    if (theta2(i)<=3)
        pos2(i)=theta2(i)-2;
    else
        pos2(i)=4-theta2(i);
    end

    if (theta3(i)<=4)
        pos3(i)=theta3(i)-3;
    else
        pos3(i)=5-theta3(i);
    end

    if (theta4(i)<=5)
        pos4(i)=theta4(i)-4;
    else
        pos4(i)=6-theta4(i);
```

```
    end
    v(i)=min([pos1(i),pos2(i),pos3(i),pos4(i)]);
    if v(i)>epsilon
        theta01(i)=theta1(i);
        theta02(i)=theta2(i);
        theta03(i)=theta3(i);
        theta04(i)=theta4(i);
    else
        while 1
            theta1=unifrnd(1,3,N,1);
            theta2=unifrnd(2,4,N,1);
            theta3=unifrnd(3,5,N,1);
            theta4=unifrnd(4,6,N,1);
            if (theta1(i)<=2)
                pos1(i)=theta1(i)-1;
            else
                pos1(i)=3-theta1(i);
            end

            if (theta2(i)<=3)
                pos2(i)=theta2(i)-2;
            else
                pos2(i)=4-theta2(i);
            end

            if (theta3(i)<=4)
                pos3(i)=theta3(i)-3;
            else
                pos3(i)=5-theta3(i);
            end

            if (theta4(i)<=5)
                pos4(i)=theta4(i)-4;
            else
                pos4(i)=6-theta4(i);
            end
            v(i)=min([pos1(i),pos2(i),pos3(i),
            pos4(i)]);
            if v(i)>epsilon
                break
            end
            theta01(i)=theta1(i);
            theta02(i)=theta2(i);
            theta03(i)=theta3(i);
            theta04(i)=theta4(i);
```

```
        end
    end
end

f=zeros(N,1);
for i=1:N
    f(i)=sqrt(theta01(i)^2+theta02(i)^2+theta03(i)^2

    +theta04(i)^2);
end
a=min(f);
b=max(f);


for i=1:N
    r=unifrnd(a,b);
    for j=1:N
        if f(j)<=r
            U1(j)=v(j);
        else
            U2(j)=1-v(j);
        end
    end
    L1=max(U1);
    for k=1:N
        if U2(k)==0
           U2(k)=inf;
        end
    end
    L2=min(U2);
    L=(L1+L2)/2;
    e=e+L;
end

    E=a+e*(b-a)/N
```

A.8 MATLAB® File for Example 4.3

```
N=1000;
alpha=0.8;
beta=0.8;
e=0;
epsilon=0.001;
```

```
r=0;
L=0;
temp=zeros(N,1);
g=zeros(N,1);
U1=zeros(N,1);
U2=zeros(N,1);
v=zeros(N,1);
f=zeros(N,1);
xi1=zeros(N,1);
xi2=zeros(N,1);
xi3=zeros(N,1);
Pr=zeros(N,1);

pos1=zeros(N,1);
pos2=zeros(N,1);
pos3=zeros(N,1);

theta1=unifrnd(1,3,N,1);
theta2=unifrnd(2,4,N,1);
theta3=unifrnd(3,5,N,1);

theta01=zeros(N,1);
theta02=zeros(N,1);
theta03=zeros(N,1);
N0=0;

for i=1:N
    if (theta1(i)<=2)
        pos1(i)=theta1(i)-1;
    else
        pos1(i)=3-theta1(i);
    end
    if (theta2(i)<=3)
        pos2(i)=theta2(i)-2;
    else
        pos2(i)=4-theta2(i);
    end
    if (theta3(i)<=4)
        pos3(i)=theta3(i)-3;
    else
        pos3(i)=5-theta3(i);
    end
    v(i)=min([pos1(i),pos2(i),pos3(i)]);
    if v(i)>epsilon
        theta01(i)=theta1(i);
```

```
        theta02(i)=theta2(i);
        theta03(i)=theta3(i);
    else
        while 1
            theta1=unifrnd(1,3,N,1);
            theta2=unifrnd(2,4,N,1);
            theta3=unifrnd(3,5,N,1);
            if (theta1(i)<=2)
                pos1(i)=theta1(i)-1;
            else
                pos1(i)=3-theta1(i);
            end

            if (theta2(i)<=3)
                pos2(i)=theta2(i)-2;
            else
                pos2(i)=4-theta2(i);
            end

            if (theta3(i)<=4)
                pos3(i)=theta3(i)-3;
            else
                pos3(i)=5-theta3(i);
            end
            v(i)=min([pos1(i),pos2(i),pos3(i)]);
            if v(i)<epsilon
                break
            end
            theta01(i)=theta1(i);
            theta02(i)=theta2(i);
            theta03(i)=theta3(i);
        end
    end
end


for i=1:N
    while 1
        for j=1:N
            xi1(i)=exprnd(theta01(i));
            xi2(i)=exprnd(theta02(i));
            xi3(i)=exprnd(theta03(i));
            if (xi1(i)^2+xi2(i)^2+xi3(i)^2)>=f(i)
                N0=N0+1;
            end
```

```
            end
            Pr(i)=N0/N;
            if  Pr(i)<alpha
                break
            end
            f(i)=f(i)+4;
        end
    end


    while 1
        for i=1:N
            if f(i)>=r
                U1(i)=Pr(i);
            else
                U2(i)=1-Pr(i);
            end
        end
        L1=max(U1);
        for i=1:N
            if U2(i)==0
               U2(i)=inf;
            end
        end
        L2=min(U2);
        L=(L1+L2)/2;
        if L<alpha
            break
        end
        r=r+0.001;
    end
    r
```

A.9 MATLAB® File for Example 4.6

```
N=100;
alpha=0.9;
e=0;
epsilon=0.001;
r=0;
L=0;
temp=zeros(N,1);
```

```
g=zeros(N,1);
U1=zeros(N,1);
U2=zeros(N,1);
v=zeros(N,1);
xi1=zeros(N,1);
xi2=zeros(N,1);
xi3=zeros(N,1);
Pr=zeros(N,1);

pos1=zeros(N,1);
pos2=zeros(N,1);
pos3=zeros(N,1);
pos4=zeros(N,1);

theta1=unifrnd(1,3,N,1);
theta2=unifrnd(2,4,N,1);
theta3=unifrnd(3,5,N,1);
theta4=unifrnd(4,6,N,1);
theta01=zeros(N,1);
theta02=zeros(N,1);
theta03=zeros(N,1);

for i=1:N
    if (theta1(i)<=2)
        pos1(i)=theta1(i)-1;
    else
        pos1(i)=3-theta1(i);
    end
    if (theta2(i)<=3)
        pos2(i)=theta2(i)-2;
    else
        pos2(i)=4-theta2(i);
    end
    if (theta3(i)<=4)
        pos3(i)=theta3(i)-3;
    else
        pos3(i)=5-theta3(i);
    end
    v(i)=min([pos1(i),pos2(i),pos3(i)]);
    if v(i)>epsilon
        theta01(i)=theta1(i);
        theta02(i)=theta2(i);
        theta03(i)=theta3(i);
    else
        while 1
```

```
            theta1=unifrnd(1,3,N,1);
            theta2=unifrnd(2,4,N,1);
            theta3=unifrnd(3,5,N,1);
            if (theta1(i)<=2)
                pos1(i)=theta1(i)-1;
            else
                pos1(i)=3-theta1(i);
            end

            if (theta2(i)<=3)
                pos2(i)=theta2(i)-2;
            else
                pos2(i)=4-theta2(i);
            end

            if (theta3(i)<=4)
                pos3(i)=theta3(i)-3;
            else
                pos3(i)=5-theta3(i);
            end
            v(i)=min([pos1(i),pos2(i),pos3(i)]);
            if v(i)<epsilon
                break
            end
            theta01(i)=theta1(i);
            theta02(i)=theta2(i);
            theta03(i)=theta3(i);

        end
    end
end

for i=1:N
    for j=1:N
        xi1(i)=exprnd(theta01(i));
        xi2(i)=exprnd(theta02(i));
        xi3(i)=exprnd(theta03(i));
       if sqrt(xi1(i)^2+xi2(i)^2+xi2(i)^2)>=2
          temp(i)=temp(i)+1;
       end
    end
    Pr(i)=temp(i)/N;

end
```

```
while 1
    for i=1:N
        if Pr(i)>=r
            U1(i)=Pr(i);
        else
            U2(i)=1-Pr(i);
        end
    end
    L1=max(U1);
    for i=1:N
        if U2(i)==0
           U2(i)=inf;
        end
    end
    L2=min(U2);
    L=(L1+L2)/2;

    if L<=alpha
        break
    end
    r=r+0.0001;
end
r
```

Index

J. Xu et al., *Random-Like Bi-level Decision Making*, Lecture Notes in Economics and Mathematical Systems 688, DOI 10.1007/978-981-10-1768-1

Printed in the United States
By Bookmasters